T0214364

Lecture Notes in Computer Science 9196

Commenced Publication in 1973
Founding and Former Series Editors:
Gerhard Goos, Juris Hartmanis, and Jan van Leeuwen

More information about this series at http://www.springer.com/series/7409

Yu Wang · Hui Xiong
Shlomo Argamon · XiangYang Li
JianZhong Li (Eds.)

Big Data Computing and Communications

First International Conference, BigCom 2015
Taiyuan, China, August 1–3, 2015
Proceedings

 Springer

Editors

Yu Wang
University of North Carolina at Charlotte
Charlotte, NC
USA

Hui Xiong
Rutgers Business School
Newark, NJ
USA

Shlomo Argamon
Illinois Institute of Technology
Chicago, IL
USA

XiangYang Li
Illinois Institute of Technology
Chicago, IL
USA

JianZhong Li
Harbin Institute of Technology
Harbin
China

ISSN 0302-9743   ISSN 1611-3349   (electronic)
Lecture Notes in Computer Science
ISBN 978-3-319-22046-8   ISBN 978-3-319-22047-5   (eBook)
DOI 10.1007/978-3-319-22047-5

Library of Congress Control Number: 2015944451

LNCS Sublibrary: SL3 – Information Systems and Applications, incl. Internet/Web, and HCI

Springer International Publishing AG Switzerland is part of Springer Science+Business Media
(www.springer.com)

Preface

Welcome Message from the General Chairs

It is a great pleasure to welcome you to the proceedings of the inaugural BigCom 2015 conference held at Taiyuan, China. BigCom is an international symposium dedicated to addressing the challenges emerging from big date-related computing and networking architecture for supporting big data computing. This year, we were fortunate to receive many excellent papers covering a diverse set of research topics related to big data computing and communication. The conference brings together numerous delegates from around the globe to discuss the latest advances in this vibrant and constantly evolving field.

Undoubtedly, data technology and communication technology have fundamentally transformed our society in recent decades and the pace of change can only be described as disruptive. Progress in these technologies is rapid and new horizons are being explored.

We would like to express our sincere gratitude to everyone whose contributions made the inaugural BigCom 2015 conference a memorable and valuable event. The conference is the result of the hard work of many authors, reviewers, and conference committee members. We are very grateful for everyone's efforts in making this conference a success. Special thanks go to Program Co-chairs Sholomo Argamon, Hui Xiong, and Yu Wang, who put together a very strong technical program. We would also like to thank Yong Ge, Xufei Mao, and Shaojie Tang, who did an outstanding job in co-ordinating and helping the review process of three different tracks. Thanks also go to Fan Li for organizing the student research session. We would like to thank our local Organizing committee members Junjie Chen, Jiye Liang, and Dengao Li for their great job organizing the local arrangements and making the stay of every conference attendee a pleasant and memorable one. Special thanks are due to JianZhong Li and P.R. Kumar for their continuous support and guidance. Lan Yao, Xu Zhang, and Jianchao Zeng did a great job of attracting financial support for the conference. Dan Tao, Yuanfang Chen, Yanwei Wu, and Taeho Jung did a super job in promoting the conference and sending out the CFP regularly to attract more submissions to the conference. A special thanks to Jumin Zhao, Chunhong Zhang, Jinqiang Liu, and several other students for maintaining the conference website. Lan Zhang, Zenghua Zhao, and Xinghua Shi did an excellent job of coordinating the proceedings of the conference. Thanks are due to Gang Lu, Yonglie Yao, and Jumin Zhao for handling the conference registration. Our deepest thanks go to Dengao Li for his valuable work and advice on numerous occasions, often at short notice. We are very grateful to the special issues application by Yu Wang, Dan Tao, and Chunhong Zhang, which allow us to publish some selected papers in top-tier journals. Finally, we thank Taiyuan University of Technology for its support and for contributing student volunteers, and the Tsinghua

University Press for its grant in supporting the conference. Numerous other individuals were instrumental in the success of the conference. I may have missed your name here. Your contributions in making this a thought-provoking and memorable conference will be remembered.

It is with great pride that we hosted BigCom 2015 in Taiyuan. In addition to the stimulating program of the conference, Taiyuan, with its tourist attractions, the diversity and quality of its cuisine, is an unforgettable place to visit. It is our hope, therefore, that the participants got a chance to explore Taiyuan and its surrounding places, and enjoy the exotic and vibrant atmosphere of Taiyuan.

Welcome Message from the Technical Program Chairs

Welcome to the proceedings of the First International Conference on Big Data Computing and Communications (BigCom 2015). On behalf of the Technical Program Committee, we would like to thank all the authors for their high-quality papers that were accepted for the inaugural BigCom conference.

This year, BigCom received 74 paper submissions, out of which 41 were selected for publication as regular papers with an acceptance rate of 55.4%. All submissions received two or more peer reviews from our Technical Program Committee and external reviewers. We were only able to accept papers that received broad support from the reviewers. The final technical program included one keynote (by Prof. P.R. Kumar), nine technical sessions, and one student research session. We would like to thank our Program Committee members as well as the external reviewers, consisting of eminent researchers, whose dedication and hard work made the selection of papers for the proceedings possible.

We wish to thank all who contributed to the quality and success of BigCom 2015. We particularly appreciate the guidance and support from the general chairs, Prof. XiangYang Li and Prof. JianZhong Li. Special thanks also go to three Vice Program Co-chairs, Dr. Yong Ge, Dr. Xufei Mao, and Dr. Shaojie Tang, for their outstanding jobs in handling the review process of three research tracks, and Dr. Lan Zhang, Dr. Zenghua Zhao, and Dr. Xinghua Shi, for collecting the final versions and copyright forms of all accepted papers. We also thank other members of the Organizing Committee for their help and support.

June 2015

XiangYang Li
JianZhong Li
Shlomo Argamon
Hui Xiong
Yu Wang

Organization

Organizing Committee

Honorary General Chair

Ming Lv — Taiyuan University of Technology, China

General Co-chairs

JianZhong Li	Harbin Institute of Technology, China
XiangYang Li	Illinois Institute of Technology, USA

TPC Co-chairs

Shlomo Argamon	Illinois Institute of Technology, USA
Hui Xiong	Rutgers University, USA
Yu Wang	University of North Carolina at Charlotte, USA

TPC Track Chairs

Yong Ge	University of North Carolina at Charlotte, USA
Xufei Mao	Tsinghua University, China
Shaojie Tang	University of Texas at Dallas, USA

Local Arrangements Co-chairs

Junjie Chen	Taiyuan University of Technology, China
Jiye Liang	Taiyuan Normal University, China
Dengao Li	Taiyuan University of Technology, China

SRC Chair

Fan Li — Beijing Institute of Technology, China

Industry Liaison Co-chairs

Lan Yao	Northeastern University, China
Xu Zhang	Beijing University of Posts and Telecommunications, China
Jianchao Zeng	Taiyuan University of Technology, China

Publicity Co-chairs

Dan Tao	Beijing Jiaotong University, China
Yuanfang Chen	University of Pierre and Marie Curie, France

| Yanwei Wu | Western Oregon University, USA |
| Taeho Jung | Illinois Institute of Technology, USA |

Publication Co-chairs

Lan Zhang	Tsinghua University, China
Zenghua Zhao	Tianjin University, China
Xinghua Shi	University of North Carolina at Charlotte, USA

Submission Chair

| Lin Wang | Yanshan University, China |

Finance Chair

| Jumin Zhao | Taiyuan University of Technology, China |

Registration Co-chairs

| Gang Lu | Shaanxi Normal University, China |
| Yonglei Yao | Nanjing University of Information Science and Technology, China |

Web Co-chairs

| Jumin Zhao | Taiyuan University of Technology, China |
| Chunhong Zhang | Beijing University of Posts and Telecommunications, China |

Webmaster

| Jinqiang Liu | Taiyuan University of Technology, China |

Journal Special Issue Chairs

Yu Wang	University of North Carolina at Charlotte, USA
Chunhong Zhang	Beijing University of Posts and Telecommunications, China
Dan Tao	Beijing Jiaotong University, China
Lei Shu	Guangdong University of Petrochemical Technology, China

Steering Committee

JianZhong Li	Harbin Institute of Technology, China
XiangYang Li	Illinois Institute of Technology, USA
Yu Wang	University of North Carolina at Charlotte, USA

Program Committee

| Shlomo Argamon | Illinois Institute of Technology, USA |
| Jinsuk Baek | Winston-Salem State University, USA |

Wei Dong	Zhejiang University, China
Wei Gao	University of Tennessee, USA
Yong Ge	University of North Carolina at Charlotte, USA
Deke Guo	National University of Defense Technology, China
Xia Hu	Arizona State University, USA
Bo Ji	Temple University, USA
Donghyun Kim	North Carolina Central University, USA
Gene Moo Lee	University of Texas at Austin, USA
Fan Li	Beijing Institute of Technology, China
Qinghua Li	University of Arkansas, USA
Xiangyang Li	Illinois Institute of Technology, USA
Xin Li	Nanjing University of Aeronautics and Astronautics, China
Yanhua Li	University of Minnesota, USA
Zhanhuai Li	Northwestern Polytechnical University, China
Bo Liu	University of Massachusetts, Amherst, USA
Chuanren Liu	Rutgers Business School, USA
Hongbo Liu	Indiana University-Purdue University Indianapolis, USA
Kebin Liu	Tsinghua University, China
Liang Liu	Beijing University of Posts and Telecommunications, China
Li Lu	University of Electronic Science and Technology, China
Xufei Mao	Tsinghua University, China
Xin Miao	Tsinghua University, China
Xia Ning	Indiana University-Purdue University Indianapolis, USA
Diana Palsetia	Northwestern University, USA
Yang Panlong	PLA University of Science and Technology, China
Walid Saad	Virginia Tech, USA
Ganesh Ram Santhanam	Iowa State University, USA
Sumit Sarkar	University of Texas at Dallas, USA
Shaojie Tang	University of Texas at Dallas, USA
Dan Tao	Beijing Jiao Tong University, China
Guoren Wang	Northeastern University, China
Hanli Wang	Tong Ji University, China
Jiliang Wang	Tsinghua University, China
Lin Wang	Yanshan University, China
Yu Wang	University of North Carolina at Charlotte, USA
Ka-Chun Wong	University of Toronto, Canada
Fan Wu	Shanghai Jiao Tong University, China
Zhenyu Wu	The College of William and Mary, USA
Kai Xing	University of Science and Technology of China, China
Hui Xiong	Rutgers, the State University of New Jersey, USA
Jie Yang	Florida State University, USA

Qing Yang	Montana State University, USA
Lan Yao	Northeastern University, China
Ge Yu	Northeastern University, China
Chunhong Zhang	Beijing University of Posts and Telecommunications, China
Xu Zhang	Beijing University of Posts and Telecommunications, China
Jumin Zhao	Taiyuan University of Technology, China
Zenghua Zhao	Tianjing University, China
Zhibin Zhao	Northeastern University, China
Wenjun Zhou	University of Tennessee, USA
Hengshu Zhu	Baidu Research, China
Shiai Zhu	MCRLab, University of Ottawa, Canada
Tong Zhu	Tsinghua University, China

Keynote Speech
Big Dynamic Data and the Smart Grid

P.R. Kumar

Texas A&M University

Abstract. The smart grid is an important emerging area that is rich with big data. There are massive amounts of demand data from consumers and real-time price data. There are also data from potentially thousands of phasor measurement units measuring voltage, current, and frequency variables, several times each second, and generating hundreds of gigabytes of data every day. There is interest in preserving the privacy of data from consumers as novel arrangements are made between them and aggregators or load-serving entities. All these data are dynamic in nature, representing variables that are the outputs of dynamic systems, and hence changing with time. We explore several issues concerning such big dynamic data.

Contents

Wireless Communication and Networks

A Novel Markov Chain Model to Derive the Expected Contention Window
Size and Backoff Counter for IEEE 802.11 WLAN Nodes 3
  Yi-Hua Zhu, Chaoran Zhu, and Xianzhong Tian

RTDA: A Novel Reusable Truthful Double Auction Mechanism
for Wireless Spectrum Management 14
  Feng Tian, Di Li, Shuyu Li, Lei Wang, Naigao Jin, and Liang Sun

Dynamic Sparse Channel Estimation Using ℓ_0-constrained Kalman Filter
in OFDM Systems 28
  Nan Jing and Lin Wang

FOAM: Frequency-Offset Aware Multiple Client Selection for Cooperative
Packet Recovery System 43
  Ping Li, Panlong Yang, Yubo Yan, and Lei Shi

Database and Big Data

Research on Light-Weight Compression Schemes Based on Simulative
Column-Store 55
  Meng Huang, Xiaofeng Qiu, Shufang Li, and Daowei Liu

Study of Constructing Data Supply Chain Based on PROV 69
  Jiewei Lan, Xiyun Liu, Hong Luo, and Peng Li

Focused Deep Web Entrance Crawling by Form Feature Classification 79
  Lin Wang, Ammar Hawbani, and Xingfu Wang

A Framework for Optimization in Big Data: Privacy-Preserving Multi-agent
Greedy Algorithm 88
  Taeho Jung, Xiang-Yang Li, and Junze Han

Networking Big Data: Definition, Key Technologies and Challenging
Issues of Transmission 103
  Weigang Hou, Pengxing Guo, and Lei Guo

Smart Phone and Sensing Application

Gender Prediction Based on Data Streams of Smartphone Applications 115
  Yilei Wang, Yuanyang Tang, Jun Ma, and Zhen Qin

Anti-multipath Indoor Direction Finding Using Acoustic Signal
via Smartphones   126
   Xiaopu Wang, Yan Xiong, and Wenchao Huang

Crowdsourcing Based Event Reporting System Using Smartphones
with Accurate Localization and Photo Tamper Detection   141
   Tong Qin, Huadong Ma, Dong Zhao, Tianyuan Li, and Jianwei Chen

Parallel Accurate Localization from Cellular Network   152
   Chao Wu, Bin Xu, and Qi Li

A Vehicle Speed Estimation Algorithm Based on Wireless AMR Sensors . . .   167
   Zusheng Zhang, Tiezhu Zhao, and Huaqiang Yuan

Security and Privacy

Design and Evaluation of a Policy-Based Security Routing and Switching
System for Data Interception Attacks   179
   Yudong Zhao, Ke Xu, Rashid Mijumbi, and Meng Shen

Wireless Device Authentication Using Acoustic Hardware Fingerprints   193
   *Dajiang Chen, Xufei Mao, Zhen Qin, Weiyi Wang, Xiang-Yang Li,
   and Zhiguang Qin*

Strongly Secure and Cost-Effective Certificateless Proxy Re-encryption
Scheme for Data Sharing in Cloud Computing   205
   Zhiguang Qin, Shikun Wu, and Hu Xiong

Anomaly Detection of Single Sensors Using OCSVM_KNN   217
   Jing Su, Ying Long, Xiaofeng Qiu, Shufang Li, and Daowei Liu

An Efficient Method on Trajectory Privacy Preservation   231
   Zhiqiang Zhang, Yue Sun, Xiaoqin Xie, and Haiwei Pan

Architecture and Applications

R-Memcached: A Reliable In-Memory Cache System for Big Key-Value
Stores   243
   *Chengjian Liu, Kai Ouyang, Xiaowen Chu, Hai Liu,
   and Yiu-Wing Leung*

Performance Evaluation of NPB and SPEC CPU2006 on Various SIMD
Extensions   257
   *Bo Zhao, Wei Gao, Rongcai Zhao, Lin Han, Huihui Sun,
   and Yingying Li*

Prediction of High Resolution Spatial-Temporal Air Pollutant Map
from Big Data Sources   273
   *Yingyu Li, Yifang Zhu, Wotao Yin, Yang Liu, Guangming Shi,
   and Zhu Han*

A Graph Community Approach for Constructing microRNA Networks   283
   Benika Hall, Andrew Quitadamo, and Xinghua Shi

Sensor Networks and RFID

Distributed Multigrid Technique for Seismic Tomography
in Sensor Networks.   297
   Goutham Kamath, Lei Shi, Edmond Chow, and Wen-Zhan Song

Energy-Efficient and Smoothing-Sensitive Curve Recovery of Sensing
Physical World.   311
   Qian Ma, Yu Gu, Tiancheng Zhang, Fangfang Li, and Ge Yu

Feedback-Based Reduplicate Complex Event Processing in IoT   325
   Mingyue Cui, Chunhong Zhang, Yuewen Su, and Yang Ji

An Identification Algorithm in Grouping and Paralleling for Data-Intensive
RFID Systems   337
   Duan Litian, Wang John Zizhong, and Duan Fu

I Know When to Do the Replenishment.   347
   Jumin Zhao, Na Li, and Deng-ao Li

Social Networks and Recommendation

Tag-Based User Interest Discovery Though Keywords Extraction
in Social Network.   363
   Ping Yang, Yan Song, and Yang Ji

Implicit Feedback Mining for Recommendation   373
   Yan Song, Ping Yang, Chunhong Zhang, and Yang Ji

The Collaborative Filtering Algorithm with Time Weight Based
on MapReduce.   386
   Hongyi Su, Xianfei Lin, Bo Yan, and Hong Zheng

Recommendation Specially for Fanatic Fans in SNS   396
   Yuewen Su, Mingyue Cui, Yang Ji, and Yunxu Yuan

Signal Processing and Pattern Recognition

In-Line Monitoring of Belt Transport with Adaptive Bandwidth
Mean-Shift Hazard   409
   Tiezhu Qiao, Yanfei Duan, and Yusong Pang

A Method for Automated J Wave Detection and Characterisation
Based on Feature Extraction.   421
   Dengao Li, Yanfei Bai, and Jumin Zhao

Metadata Organization and Retrieval with Attribute Tree for Large-Scale
Traffic Surveillance Videos   434
   Yi Tang, Haitao Zhang, and Bin Xu

Development and Challenges of Crowdsourcing Quality of Experience
Evaluation for Multimedia   444
   Zhenji Wang, Dan Tao, and Pingping Liu

An Approach for J Wave Auto-Detection Based on Support Vector
Machine.   453
   Dengao Li, Xuebo Liu, and Jumin Zhao

Routing and Resource Management

Green and Fault-Tolerant Routing in Data Centers   465
   Liang Shi, Xintong Guo, Lailong Luo, and Yudong Qin

A Markov Chain Prediction Model for Routing in Delay
Tolerant Networks.   479
   Shuai Liu, Fan Li, Qian Zhang, and Meng Shen

RAM: Resource Allocation in Mobility for Device-to-Device
Communications.   491
   Weiyang Lin, Cuibo Yu, and Xu Zhang

Group Signature Based Trace Hiding in Web Query   503
   Jin Xu, Lan Yao, and Fuxiang Gao

Author Index   513

Wireless Communication and Networks

A Novel Markov Chain Model to Derive the Expected Contention Window Size and Backoff Counter for IEEE 802.11 WLAN Nodes

Yi-Hua Zhu$^{(\boxtimes)}$, Chaoran Zhu, and Xianzhong Tian

School of Computer Science and Technology, Zhejiang University of Technology,
Hangzhou 310023, Zhejiang, China
yhzhu@ieee.org, 150659925@qq.com, txz@zjut.edu.cn

Abstract. The Binary Exponential Backoff (BEB) algorithm in the distributed coordination function (DCF) introduced in the IEEE 802.11 Medium Access Control (MAC) layer is applied in controlling the nodes to access the channel. It is required to derive the Contention Window Size (CWS) and the Backoff Counter (BC) under the BEB since they are used in the nodes to attend for the channel. In this paper, a novel Markov chain model is presented to derive the joint probability distribution of the backoff stage and the BC that are picked by a node. Based on this model, the expected CWS, the expected BC, and the expected Number of Doubling Contention Window (NDCW) under the BEB are derived. Moreover, simulations are performed to validate the model. The derived expected CWS helps the node reset its CWS to a suitable value instead of the one presented in the IEEE 802.11 standard so that the oscillation in contention window is avoided.

Keywords: IEEE 802.11 standard · Wireless local area network · Binary exponential backoff algorithm · Markov chain

1 Introduction

In IEEE 802.11-based Wireless Local Area Network (WLAN), distributed coordinate function (DCF) is applied in the Medium Access Control (MAC) layer to support and control the nodes to access the channel [1]. To reliably deliver packets between the nodes, positive acknowledgement (ACK) and retransmission mechanism are adopted in the MAC layer. The sender of a frame considers retransmitting if no ACK from the receiver reaches the sender. To overcome the hidden node problem and reduce packet collision probability, Carrier Sense Multiple Access/Collision Avoidance (CSMA/CA) and Binary Exponential Backoff (BEB) are introduced in the DCF of the IEEE 802.11 standard.

In the BEB, a Contention Window (CW) is divided into time slots with equal length. The slots are numbered by $0, 1, \ldots, W$, where W is called CW parameter taking values in the interval $[CW_{\min}, CW_{\max}]$, in which $CW_{\min}$ and $CW_{\max}$ are

© Springer International Publishing Switzerland 2015
Y. Wang et al. (Eds.): BigCom 2015, LNCS 9196, pp. 3–13, 2015.
DOI: 10.1007/978-3-319-22047-5_1

relevant to the physical layer (PHY). We refer to the number of slots contained in the CW as Contention Window Size (CWS), which is equal to $W + 1$.

DCF requires that all nodes in the WLAN have to sense the channel for an idle period longer than DIFS (Distributed Inter-Frame Space) before transmitting. After an idle period of DIFS is sensed, the node performs the BEB by randomly picking a BC in the set $\{0, 1, \ldots, W\}$. If 0 is picked, the node is allowed to transmit immediately. Otherwise, it monitors the channel at the beginning of each time slot to see whether it is idle. If the channel is idle, the BC is reduced by 1; otherwise, the BC remains unchanged. The node transmits as soon as the BC goes to 0.

When the node transmits a frame, it sets a timer for the frame and then waits for the ACK frame from the receiver. If no ACK is received until the timer expires, the node considers retransmitting after the CWS of the node is doubled (the CWS remains unchanged if it reaches $CW_{\max} + 1$) and BEB is performed again. Whenever a frame is ACKed by the receiver, the CWS is reset to the initial one, i.e., $CW_{\min} + 1$, which is equivalent to setting CW parameter to $CW_{\min}$ [1]. Usually, for a given PHY, $CW_{\min}$ is fixed. For example, the values of $CW_{\min}$ in Direct Sequence Spread Spectrum (DSSS) and High Rate DSSS (HR/DSSS) are set to 31 while that in Orthogonal Frequency Division Multiplexing (OFDM) is set to 15. This means, with OFDM PHY, the parameter of CW in a node is always reset to 15 upon the node successfully transmits a frame.

Obviously, larger CWS causes the node to pick a larger BC with higher probability, i.e. the node may wait longer before transmitting. It is pointed out that, the standard BEB algorithm is far from satisfaction in improving the network performance in terms of packet delay and throughput [2], and it may cause unnecessary long delay and low channel utilization [3]. Moreover, resetting CW to $CW_{\min}$ after each successful transmission may cause an oscillation in CWS when the network has a large number of nodes, affecting the stability of the network performance. In this case, it is desirable not to reset CW to its initial value $CW_{\min}$ but to a suitable value after each successful transmission in order to stabilize the network performance [4]. In fact, performance fluctuation could be avoided if CW is reset to the value slightly larger than its expected value. The prerequisite to do so is that we have in hands the expected CWS under the BEB, which motivates us to derive the expected CWS.

The main contributions in this paper are as follows: 1) a novel Markov chain model is developed, which matches the BEB defined in the IEEE 802.11 standard very well; 2) the joint probability distribution of the backoff stage of the node and the BC picked by the node is derived; and 3) based on the derived probability distribution, the expected CWS, the expected BC, and the expected Number of Doubling Contention Window (NDCW) are derived, which help the node choose a suitable value rather than the fixed $CW_{\min}$ defined in the IEEE 802.11 standard when resetting its CW parameter such that the oscillation in CW is avoided.

The rest of the paper is organized as follows. In Section 2, related work are surveyed. Especially, Bianchi's model, a well-known model, is surveyed in detail.

The new Markov chain model is presented in Section 3, the statistics of the BEB under the proposed model are derived in Section 4, and simulations are conducted in Section 5 to validate our model and compare it with Bianchi's model. We conclude the paper in Section 6.

2   Related Work

As the BEB exerts much impact on the channel access and throughput in an IEEE 802.11 based WLAN, it is extensively investigated in research community. As well-known, Bianchi Giuseppe [5] presented a famous two-dimension Markov model (referred as Bianchi's model below) to analyze the performance of the BEB. This model has been cited many times by the researchers [6–8] around the world. A mathematical model for analyzing the performance of the BEB algorithm was presented in [9], which takes into account the packet loss probability of a wireless link. The CWSs that maximize the WLAN system throughput under both saturated and non-saturated conditions were derived in [10]. The semi-random backoff (SRB) method enabling resource reservation in contention-based WLANs was proposed in [11], which extends the traditional DCF in the IEEE 802.11 standard. The semi-distributed backoff (SDB) algorithm was presented in [12], and the SDB operates in two modes: the S-mode, in which sender-side backoff is performed, and the R-mode, in which receiver-side backoff is performed.

As mentioned in the previous section, CWS takes values in the interval $[CW_{\min} + 1, CW_{\max} + 1]$. Let L_i be the CWS of the node at backoff stage i i.e., L_i is the CWS resulted from the initial CWS being doubled i times. Thus, we have $L_0 = CW_{\min} + 1, L_i = 2^i L_0, i = 1, 2, \ldots, m$, where m is the maximum backoff stage, i.e., $L_m = CW_{\max} + 1$. Usually, $CW_{\min}$ and $CW_{\max}$ are medium dependent. For instance, $CW_{\min} = 15$ and $CW_{\max} = 1023$, i.e., $L_0 = 16$ and $L_m = 1024$, for the OFDM PHY [1]. Here, the maximum backoff stage $m = 6$. In addition, we use W_i to represent the maximum CW parameter at backoff stage i, i.e., $W_i = L_i - 1$.

At steady state, let $b_{i,j}$ be the joint probability of the node being at backoff stage i and its BC being j. This paper focuses on deriving $b_{i,j}$ so as to obtain the expected CWS and the expected BC.

First of all, we need to illustrate that $b_{i,j}$ could not be derived as the direct outcome of the Bianchi's model given in [5], which adopts the two-dimension Markov chain with state transitions shown in Fig. 1, where p is the packet collision probability, and the pair (i, j) represents the node's state that the node is at backoff stage i with BC j. Clearly, the state (i, j) takes values in the set $\Omega = \{(i, j) | i = 0, 1, \ldots, m; j \in \{0, 1, \ldots, L_i - 1\}\}$

It should be noted that, Fig. 1 and the figure given in [5] have a little difference. That is, we replace W_i in [5] with L_i to agree with the notations defined in this paper, because W_i in [5] stands for the CWS whereas in this paper it stands for the CW parameter equal to $L_i - 1, i = 0, 1, \ldots, m$.

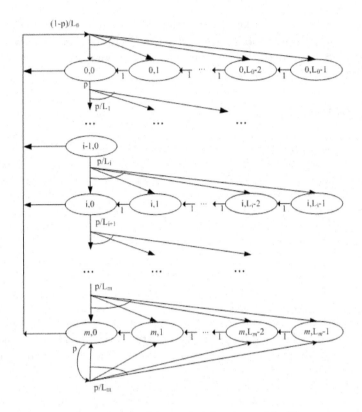

Fig. 1. Markov chain model presented in [5]

As given in [5], according to the flow balance regularity of Markov model, the first row in Fig. 1 leads to the following set of equations, where $b_{i,j}$ represents the probability of the state (i,j) and we define $a \equiv (1-p)/L_0 \sum_{j=0}^{m} b_{j,0}$.

$$
\begin{cases}
b_{0,L_0-1} = a \\
b_{0,L_0-2} = a + b_{0,L_0-1} = 2a \\
b_{0,L_0-3} = a + b_{0,L_0-2} = 3a \\
\vdots \\
b_{0,0} = a + b_{0,1} = L_0 a
\end{cases}
\tag{1}
$$

which leads to

$$
b_{0,0} > b_{0,1} > b_{0,2} > \cdots > b_{0,L_0-1}.
\tag{2}
$$

Equation (2) indicates that: at backoff stage 0, the probability that the node is with a smaller BC is higher than that with a larger BC. Moreover, from the other rows in Fig. 1, the same observation is captured when the node is at backoff stages $1, 2, \ldots, m$. The reason of these observations is as follows: Bianchi's model is based on time slot and already takes into account decreasing BC at each idle time slot. This is reflected in Fig. 1 in which the transition probability from state $(i, j + 1)$ to (i, j) is equal to 1 for $i = 0, 1, \ldots, m$ and $j = 0, 1, \ldots, L_i - 1$. As a result, the probability of the node being at state $(i, j + 1)$ includes the probability of the node being at state (i, j).

It is required in the IEEE 802.11 standard that, at any backoff stage, each BC in a contention window must be equally likely picked by the node, i.e., each BC shares the same probability. As a result, we could not expect to derive the expected CWS and the expected BC directly from Bianchi's model because (2) does not satisfy the above mentioned requirement. This is why we develop a new model.

3 The Novel Markov Chain Model

Now we move on to present our model. For a given backoff stage i, we combine all the states $(i, 0), (i, 1), \ldots, (i, W_i)$ into one state, called i-th backoff stage state (BSS) denoted by S_i. Meanwhile, the states $(i, 0), (i, 1), \ldots, (i, W_i)$ are referred as the sub-states of S_i. It should be stressed that all the sub-states share the same probability.

From the above definition, the node in state S_i means that the CWS of the node has been continuously doubled i times. We use Q_i to denote the probability that the node is at state S_i, i.e., the node is at backoff state i. Thus, we have the following equations.

$$Q_i \equiv Pr\{S_i\} = \sum_{j=0}^{W_i} b_{i,j} \tag{3}$$

$$b_{i,j} = Q_i/L_i, i = 0, 1, \ldots, m; j = 0, 1, \ldots, W_i \tag{4}$$

The transitions between the BSSes are depicted in Fig. 2, where the rectangle labeled with S_i stands for the i-th BSS ($i = 0, 1, \cdots, m$), and an ellipse represents a sub–state of the respective BSS.

Let $P\{S_j|S_i\}$ be the probability of one step transition from S_i to S_j. Then, the state transitions between BSSes in Fig. 2 can be summarized as the following.

$$\begin{cases} P\{S_0|S_k\} = 1 - p, k = 0, 1, \ldots, m \\ P\{S_{k+1}|S_k\} = p, k = 0, 1, \ldots, m - 1 \\ P\{S_m|S_m\} = p \end{cases} \tag{5}$$

where p is the packet collision probability (as in [5], we assume the node has the same packet collision probability at any backoff stage).

The first line in (5) means that the CWS is reset to its initial value L_0 when the node receives an ACK for a transmitted frame regardless of which BSS the node currently stays in; the second line indicates that the CWS is

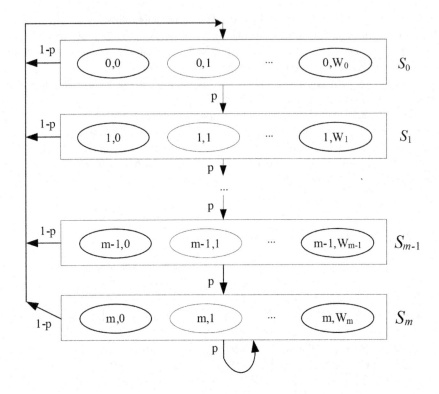

Fig. 2. State transmission diagram

doubled due to retransmission; and the last line reflects the fact that the CWS remains unchanged because the CWS takes the maximum value L_m. According to the regularity of Markov chain, the following equation can be easily derived from the flow balance equations.

$$Q_i = \begin{cases} (1-p)p^i, i = 0, 1, \ldots, m-1; \\ p^m, i = m. \end{cases} \tag{6}$$

Using (4) and (6) and noting $L_i = 2^i L_0$, we have

$$b_{i,j} = \begin{cases} \dfrac{1-p}{L_0}\left(\dfrac{p}{2}\right)^i, i = 0, 1, \ldots, m-1; j = 0, 1, \ldots, W_i \\ \dfrac{1}{L_0}\left(\dfrac{p}{2}\right)^m, i = m; j = 0, 1, \ldots, W_m. \end{cases} \tag{7}$$

4 Statistics of the BEB

Using (6), we obtain the expected CWS:

$$\bar{L} \equiv \sum_{i=0}^{m} L_i Q_i = L_0 \sum_{i=0}^{m} 2^i Q_i$$

$$= \begin{cases} L_0 \left(\dfrac{m}{2} + 1 \right), p = \dfrac{1}{2}; \\ \dfrac{L_0 \left[1 - p - p(2p)^m \right]}{1 - 2p}, p \neq \dfrac{1}{2}. \end{cases} \tag{8}$$

The relationship between the BC j and the allowable number of backoff stages i can be observed as follows: $i = 0, 1, 2, \ldots, m$ when $0 \leq j \leq W_0$; $i = 1, 2, \ldots, m$ when $W_0 + 1 \leq j \leq W_1$; $i = 2, 3, \ldots, m$ when $W_1 + 1 \leq j \leq W_2$; and so forth. This yields the probability that the node picks BC j, denoted by B_j, as follows:

$$B_j = \begin{cases} \sum_{i=0}^{m} b_{i,j}, W_{-1} + 1 \leq j \leq W_0; \\ \sum_{i=1}^{m} b_{i,j}, W_0 + 1 \leq j \leq W_1; \\ \vdots \\ \sum_{i=m}^{m} b_{i,j}, W_{m-1} + 1 \leq j \leq W_m \end{cases} \tag{9}$$

where we define $W_{-1} \equiv -1$. Thus, the expected BC picked by the node is:

$$\bar{B} \equiv \sum_{j=0}^{W_m} j B_j = \sum_{k=0}^{m} \sum_{j=W_{k-1}+1}^{W_k} j B_j \tag{10}$$

which can be determined by (7) and (9).

Let ξ be the number of doubling CWS. Noting CWS is doubled k times from its initial value if the node is in state $S_k (k = 0, 1, \ldots, m)$, we have the following probability:

$$P\{\xi = k\} = Q_k, k = 0, 1, \ldots, m. \tag{11}$$

Thus, from (6) and (11), we have the expected number of doubling CWS:

$$\bar{K} \equiv \sum_{k=0}^{m} k P\{\xi = k\} = \frac{p(1 - p^m)}{1 - p}. \tag{12}$$

5 Simulations

First, we validate the proposed model through simulation, and then we compare the proposed model with Bianchi's model to highlight the difference between

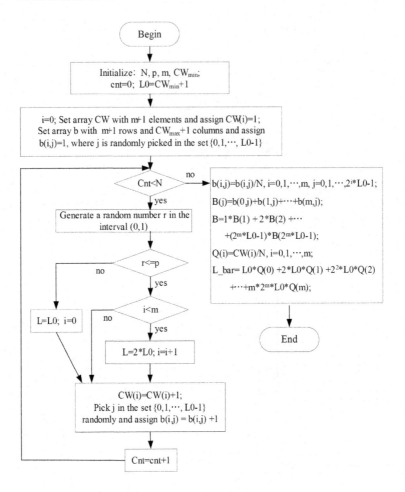

Fig. 3. The logic of simulation

them. The simulation logic is illustrated in Fig. 3, which mimics N times of frame transmissions. Besides, i and j are backoff stage and BC, respectively; L_0 is the minimum CWS and L keeps the current CWS; array $b(i,j)$ is used to hold the probability $b_{i,j}$; array $CW(i)$ keeps the total number of BSS S_i, into which the node enters ($i = 0, 1, \ldots, m$); array $B(j)$ holds the probability that the node picks j as a BC ($j = 0, 1, \ldots, W_m$), and B is the expected BC; array $Q(i)$ keeps the probability that the node enters BSS $S_i(i = 0, 1, \ldots, m)$, i.e., Q_i; and L_bar is the expected CWS, i.e., $\bar{L}$, given in (8). In addition, $r \leq p$ means a packet collision occurs, which triggers the CWS is doubled unless the node is at the maximum backoff stage.

From the perspective of the node, at a given backoff stage, picking a BC is irrespective of the underlying physical layer and the MAC. Hence, programming languages such as C++, MATLAB, and others are suitable for the simulation.

We use MATLAB. The simulation results for the case when $N = 10^{10}$ are shown in Figs. 4 - 6. In the figures, "Simulation" marks the simulation results. Moreover, "Bianchi" represent the results derived from Bianchi's model in [5] and "Ours" are the results calculated directly from the statistics presented in the previous section. In addition, "Our error" is defined as the difference of the simulation result minus our result, and "Bianchi error" is the difference of the simulation result minus Bianchi's result.

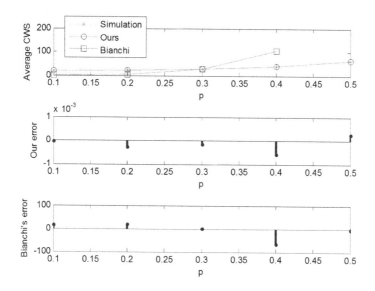

Fig. 4. Expected contention window size

From Fig. 4 and 5, we observe that our results be very close to the simulation results for the expected CWS and the expected BC whereas the results from Bianchi's model exhibit considerable errors. In addition, we found no CWS and BC is depicted when $p = 0.5$ for Bianchi's model in Figs. 4 and 5. The reason is that, equation (6) in [5], which calculates $b_{0,0}$, is mistaken when $p = 0.5$.

Next, Fig. 6 compares our model with Bianchi's model for probability $b_{i,j}$ when the node is at the first backoff stage and $p = 0.3$. It can be clearly seen from the figure that our results are also very close to the simulation results whereas Bianchi's results have much larger errors than ours. In addition, the error in Bianchi's model exhibits growing tendency as BC grows. Equivalently, probability $b_{i,j}$ decreases as BC grows, which agrees with (2).

The results of the simulations with other parameters also show that the above observations remain true.

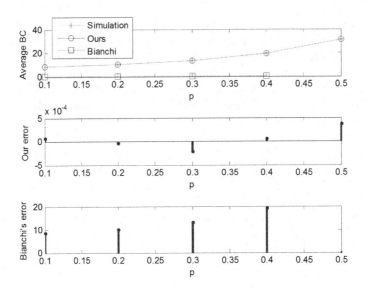

Fig. 5. Expected backoff counter

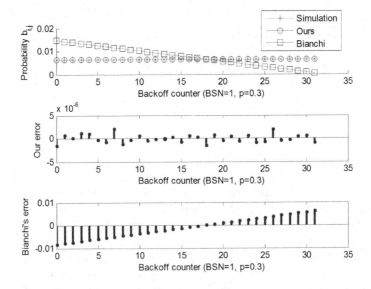

Fig. 6. Probability $b_{i,j}$

6   Conclusion

The proposed Markov chain model is able to match the requirement defined in the IEEE 802.11 standard that, at any backoff stage, each of the time slots in a contention window is picked by the node with equal probability. The derived expected CWS is applicable for the node to reset its contention window parameter to a suitable value rather than the one given in the IEEE 802.11 standard after a successful transmission so that oscillation in contention window size can be avoided.

Acknowledgments. This project is supported in part by National Natural Science Foundation of China under grant numbers 61379124 and 61432015, and in part by Zhejiang Provincial Natural Science Foundation of China under grant number LY15F020027.

References

1. LAN MAN Standards Committee of the IEEE Computer Society, IEEE 802.11 Standard -wireless LAN medium access control and physical layer specifications (2012)
2. Deng, D.-J., Ke, C.-H., Chen, H.-H., et al.: Contention Window Optimization for IEEE 802.11 DCF Access Control. IEEE Transactions on Wireless Communications. **7**(12), 5129–5135 (2008)
3. Pang, Q., Leung, V.C.M., Liew, S.C.: Improvement of WLAN contention resolution by loss differentiation. IEEE Transactions on Wireless Communications **5**(12), 3605–3615 (2006)
4. Ksentini, A., Nafaa, A., Gueroui, A., Naimi, M.: Deterministic contention window algorithm for IEEE 802.11. In: Proc. 2005 IEEE 16th International Symposium on Personal, Indoor and Mobile Radio Communications (PIMR 2005), Berlin, Germany, pp. 2712–2716 (September 2005)
5. Bianchi, G.: Performance analysis of the IEEE 802.11 distributed coordination function. IEEE Journal on Selected Areas in Communications **18**(3), 535–547 (2000)
6. Ye, S.-R., Tseng, Y.-C.: A multichain backoff mechanism for IEEE 802.11 WLANs. IEEE Transactions on Vehicular Technology 55(5), September 2006
7. Tickoo, O., Sikdar, B.: Modeling Queueing and Channel Access Delay in Unsaturated IEEE 802.11 Random Access MAC Based Wireless Networks. IEEE/ACM Transactions on Networking **16**(4), 878–891 (2008)
8. Jiang, L., Walrand, J.: A Distributed CSMA Algorithm for Throughput and Utility Maximization in Wireless Networks. IEEE/ACM Transactions on Networking **18**(3), 960–972 (2010)
9. Zhu, Y.-H., Tian, X.-Z., Zheng, J.: Performance Analysis of the Binary Exponential Backoff Algorithm for IEEE 802.11 Based Mobile Ad Hoc Networks. In: 2011 IEEE International Conference on Communications (ICC), pp. 1–6 (2011)
10. Hong, K., Lee, S.K., Kim, K., Kim, Y.H.: Channel Condition Based Contention Window Adaptation in IEEE 802.11 WLANs. IEEE Transactions on Communications **60**(2), 469–478 (2012)
11. He, Y., Sun, J., Ma, X., Vasilakos, A.V., Yuan, R., Gong, W.: Semi-Random Backoff: Towards Resource Reservation for Channel Access in Wireless LANs. IEEE/ACM Transactions on Networking **21**(1), 204–217 (2013)
12. Misra, S., Khatua, M.: Semi-Distributed Backoff: Collision-Aware Migration from Random to Deterministic Backoff. IEEE Transactions on Mobile Computing **14**(5), 1071–1084 (2015)

RTDA: A Novel Reusable Truthful Double Auction Mechanism for Wireless Spectrum Management

Feng Tian, Di Li, Shuyu Li, Lei Wang$^{(\boxtimes)}$, Naigao Jin, and Liang Sun

School of Software, Dalian University of Technology,
Dalian, People's Republic of China
{feng.den.tian,di.grace.li,lsytlsy,liang.sunry}@gmail.com,
lei.wang@ieee.org, ngjin@dlut.edu.cn

Abstract. In the secondary spectrum market, more and more primary users (PUs) release their idle spectrum to secondary users (SUs). While some of the existing auction mechanisms are truthful, few of them emphasize achieving a high usage rate. Even the SUs get the channel they require, the spectrum resource is still wasted in the spare time. In this paper, we propose a Reusable Truthful Double Auction (RTDA) mechanism for spectrum management, which considers temporal reuse and improve the usage rate significantly. Mathematical inference and game theory is used to prove that RTDA is economic-robust. The simulation results show that RTDA significantly improves the spectrum usage rate. In certain scenario, the usage rate can reach up to 100%.

Keywords: Double auction · Economic-robust · Temporal reuse · Usage rate

1 Introduction

With the development of new wireless technologies and applications, the demand for wireless spectrum is becoming larger than ever. Thus spectrum resources become scarce. Traditional Federal Communications Commission (FCC) spectrum auctions aim at long-term leases in a large area. It leads to 'white spectrum' where a large amount of spectrum is only used in a specific period in a small area. Furthermore, spectrum is a different commodity from the common ones[4], because two buyers can have the same spectrum as long as they don't interfere with each other.

As a result, we should pay more attention to the auctions in the secondary spectrum market where PUs sell their idle spectrum to SUs. Even though a buyer gets a channel, it may not occupy this channel all the time. At the same time, this buyer needs to pay for the whole channel with little usage.

In this paper, we propose an auction mechanism named Reusable Truthful Double Auction (RTDA) which is based on McAfee[10]. Different from TRUST[14], our auction mechanism achieves both spatial and temporal reuse.

© Springer International Publishing Switzerland 2015
Y. Wang et al. (Eds.): BigCom 2015, LNCS 9196, pp. 14–27, 2015.
DOI: 10.1007/978-3-319-22047-5_2

RTDA is also proved to be economic-robust[6], following the same three economic properties as TRUST[14]:

- Truthful. A double auction mechanism is truthful if the bids of all sellers and buyers are equal to their values.
- Individual Rationality. A double auction mechanism has individual rational if all winning buyers pay less than their bids and all winning sellers obtain more than their bids.
- Ex-post Budget Balance. A double auction mechanism is defined as ex-post budget balanced if the profit of an auctioneer is positive[14].

This paper makes the following key contributions:

- RTDA provides a temporally reusable mechanism for truthful double spectrum auctions.
- We prove RTDA by mathematical inference and game theory.
- We conduct simulations to confirm that RTDA can improve the usage rate of spectrum.

The structure of this paper is organized as follows. Section 2 introduces the auction model RTDA, which achieves temporal reuse. Section 3 presents a detailed description of our auction mechanism. Proof and analysis is provided in Section 5, and our auction mechanism is proved to be economic-robust. In Section 5, we show our simulation results by providing several figures. Section 6 shows some related work. Conclusions are in Section 7.

2   Model

RTDA is based on the secondary spectrum market in which reuse and fairness should be considered[3]. As shown in Fig. 1, We propose a model of M sellers and N buyers. An auctioneer who performs the auction is necessary.

We assume that each seller contributes one distinct channel while each buyer requiring one single channel. The channels are homogenous to everyone who takes apart in the auction. As sellers, PUs submit their bids to the auctioneer privately. Buyers submit the information of locations and the requirements of time slots besides bids privately to the auctioneer.

For a seller i, B_i^s is its bid of a channel, which is the minimum payment required to sell a channel. V_i^s is its true valuation of a channel. P_i^s is the exact payment received if it wins. The utility of a seller i is $U_i^s = P_i^s - V_i^s$ if it wins, and 0 otherwise.

As shown in Fig. 2, channel is divided by time. Every piece of spectrum is defined as a slot. Every buyer can ask for the slots it wants, and pay for them with the right price.

T is defined as the same total number of time slots for each channel. For a buyer j, b_j^b is its bid, the maximum price it is willing to pay for a time slot of a channel. V_j^b is its true valuation, and P_j^b is the price it pays if it wins. The utility

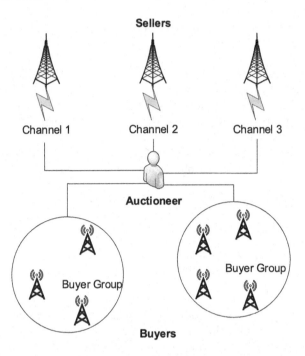

Fig. 1. System Description

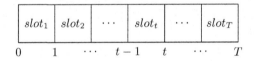

Fig. 2. The Time Slots of Channel

of buyer j is $U_j^b = V_j^b - P_j^b$ if it wins, and 0 otherwise. The buyer j's requirement vector of time slots is t_j, and j will set a TRUE if it wants a slot. We can count the elements of the vector to calculate the requirement of buyer j. We know that there is no buyer who can apply for less than one time slot or more than T time slots. The first norm $\|t_j\|_1$ is used to define the total requirement of buyer j as shown in 1.

$$1 \leq \|t_j\|_1 \leq T, \forall j \in [1, N]. \tag{1}$$

To easily count $\|t_j\|_1$, T and other properties of time slots, a time matrix $TIME$ is necessary which includes t_j as its rows as shown in Fig. 3.

In this auction mechanism, truthfulness and other economic properties are used to evaluate it. However, spectrum utilization should be paid more attention to. So usage rate will be the most important way to evaluate the performance of the auction. These should be discussed in the next sections.

	$slot_1$	$slot_2$	$slot_3$	$\cdots$	$\cdots$	$slot_T$
b_1	TRUE	FALSE	TRUE	$\cdots$	$\cdots$	TRUE
b_2	FALSE	TRUE	TRUE	$\cdots$	$\cdots$	FALSE
b_3	TRUE	TRUE	FALSE	$\cdots$	$\cdots$	TRUE
$\cdots$	$\cdots$	$\cdots$	$\cdots$	$\cdots$	$\cdots$	$\cdots$
b_j	TRUE	TRUE	TRUE	$\cdots$	$\cdots$	FALSE

Fig. 3. TIME: Matrix of Buyers' requirement

3 Design Details

In this section, we propose a truthful double spectrum auction mechanism named RTDA based on time reuse. Buyers are divided by different geographical locations. Channels are divided into several time slots. Temporal reuse is considered, while fairness and efficiency having been taken into consideration.

Three steps comprise RTDA which are buyer grouping, winner determination problem and pricing. First, buyers are supposed to be grouped by their location and time requirement information. Second, a time-dependent winner determination algorithm is redesigned to decide who are the winners. Finally, auctioneer gives out the right price to achieve fairness and efficiency in pricing step. The algorithm process is shown in Fig. 4.

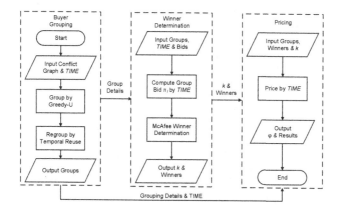

Fig. 4. Algorithm Process

3.1 Buyer Grouping

All channels are homogeneous to all buyers, so we can use one grouping algorithm for all buyers. The grouping algorithm is private to buyers and sellers. Before grouping, buyers submit their geographical information to the auctioneer, which is assumed to be truthful obviously. We use a conflict graph to describe conflicts among buyers. If two buyers are distant in geography, they don't conflict with each other. In another word, they won't share an edge. We use Greedy-U in [8] which recursively chooses a node with the minimum degree in the current conflict graph to group the buyers. By computing time slot vector t_j and $\|t_j\|_1$ of the buyer j, we will regroup buyers if they conflict with each other in geography but not in time slots. The conflict graph will be updated by considering the temporal requirement to achieve temporal reuse. The main algorithm of buyer grouping is shown in Algorithm 1, in which G_l is defined as the buyer number of group l. Function OPT is shown in Algorithm 2.

Algorithm 1. Allocation Algorithm

Require: *Conflict Graph, TIME, N*
Ensure: *Groups, L*
 1: **while** !*all* (|*Conflict Graph*| == *N*) **do**
 2:    *Call OPT(Conflict Graph, N) then Count the G_l and L.*
 3: **end while**
 4: *Regroup by TIME*
 5: **return** *Groups, L, T − N*

Algorithm 2. OPT

Require: *Conflict Graph, N*
Ensure: *Groups, L*
 1: **if** !*all* (|*Conflict Graph*| == *N*) **then**
 2:    *Find the node with minimum degree, store in result*
 3:    *Delete neighbors and node min.*
 4:    *recursive parameter ← list(Conflict Graph, N)*
 5:    *result ← c(result, OPT(recursive parameter))*
 6: **else**
 7:    **return** *result*
 8: **end if**

3.2 Winner Determination Problem

First, we define $G_1, G_2, \ldots, G_L$ which are the grouping results in the first step. For any group G_l, we define the bid of the group as π_l. The bid is aimed at a single time slot. By following McAfee[10], we will choose the minimum bid for a single time slot in a group G_l. Then we calculate the $TIME$ matrix to count the total requirement of G_l. We define the bid of group G_l as following.

$$\pi_l = min\{b_j^b | j \in G_l\} \cdot \sum_{n \in G_l} \|t_n\|_1. \tag{2}$$

In Winner Determination Problem (WDP), we follow the design of McAfee. The economic-robust properties are guaranteed which are also proved next. In details, we first sort buyer groups in descending order, while sorting seller groups in ascending order. We define the seller's bid of any channel as B_i^s.

$$B^b : \pi_1 \geq \pi_2 \geq \pi_3 \geq \cdots \geq \pi_L, \tag{3}$$

$$B^s : B_1^s \leq B_2^s \leq B_3^s \leq \cdots \leq B_M^s. \tag{4}$$

We are supposed to find a maximum k that makes auction revenue non-negative which is shown below.

$$k = argmax_{l \leq min\{L,M\}} \pi_l \geq B_l^s. \tag{5}$$

The $k - 1$ buyer groups and $k - 1$ sellers constitute the auction winners. Algorithm 3 shows how WDP works.

Algorithm 3. Winner Determination Problem-WDP

Require: t_N, G, L, b_N, B_M
Ensure: *Seller Winners, Buyer Winners, k*
1: **for** $l = 1$ to L **do**
2:    $\pi_l \leftarrow min\{b_j^b | j \in G_l\} \cdot \sum_{n \in G_l} \|t_n\|_1$
3: **end for**
4: $B1 \leftarrow sort(B_i), \forall i \in [1, M]$
5: $B2 \leftarrow sort(\pi, decreasing = TRUE)$
6: $k \leftarrow max(which(B2 - B1 \geq 0))$
7: **return** *Seller Winners, Buyer Winners, k*

3.3 Pricing

In order to make the auction truthful, we use B_k^s and π_k as the market clearing price. If a buyer wins a channel, the seller will be paid B_k^s. The buyer group will be charged π_k. The cost will be shared by all buyers in the buyer group as (6). For buyer j, the price is as followed.

$$P_j^b = \frac{\pi_k \cdot \|t_j\|_1}{\sum_{n \in G_l} \|t_n\|_1}, \forall j \in [1, N]. \tag{6}$$

Taking individual rationality into account, any buyer or any seller will not get a utility below 0. Thus, the buyer who does not win any channel or the seller who does not sell any channel, will not be charged anything. The revenue of auctioneer can be denoted as (7).

$$\varphi = (k - 1) \cdot (\pi_k - B_k^s). \tag{7}$$

4   Proof and Analysis

In this section, we will prove RTDA is economic-robust, especially truthful. First, we would define P_j^b as the clear price of a buyer j, and the price of a single time slot is p_j^b. They all satisfies the followed definition.

$$P_j^b = p_j^b \cdot \|t_j\|_1, \forall j \in [1, N] \tag{8}$$

4.1   Ex-post Budget Balanced

Theorem 1. *RTDA is budget-balanced, that is, $\varphi \geq 0$.*

Proof: In the second step Winner Determination Problem, we choose a k that satisfies $\pi_k \geq B_k^s$, and if the auction is successful which means there is at least one winner buyer, we have $k \geq 1$. According to the expression $\varphi = (k-1) \cdot (\pi_k - B_k^s)$, $\varphi \geq 0$. When $k = 0$, it is easy to show that $\varphi = 0$. So We can find that $\varphi \geq 0$ is always true. So RTDA is ex-post budget balanced. ∎

4.2   Individual Rational

Theorem 2. *RTDA is individual rational.*

Proof: In order to prove an auction mechanism is individual rational, that is, no participants will get a negative utility, we will prove that no buyer will pay more than its valuation, and no seller will be paid less than its valuation. Obviously, a buyer or seller who fails in the auction will not pay or be paid any fee, the utility of them is always equal to 0. So individual rationality is guaranteed in this situation.

First, we should prove the buyers are individual rational in RTDA. According to the design of WDP in the second step, buyers are sorted in a descending order of bids. For each winning buyer group G_l, the bid of buyer group G_l always satisfies $\pi_l \geq \pi_k$. For buyer j, the charged fee also always satisfies

$$P_j^b = \frac{\pi_k \cdot \|t_j\|_1}{\sum_{n \in G_l} \|t_n\|_1} \leq \frac{\pi_l \cdot \|t_j\|_1}{\sum_{n \in G_l} \|t_n\|_1}, \forall j \in [1, N]. \tag{9}$$

Then according to the definition of the bid of buyer group, we know that the bid of buyer j in group G_l also always satisfies

$$B_j^b = b_j^b \cdot \|t_j\|_1 \geq \frac{\pi_l \cdot \|t_j\|_1}{\sum_{n \in G_l} \|t_n\|_1} \geq p_j^b \cdot \|t_j\|_1 = P_j^b, \tag{10}$$

$B_j^b = \frac{\pi_l \cdot \|t_j\|_1}{\sum_{n \in G_l} \|t_n\|_1}$ when b_j^b is the minimum single slot bid of G_l. So $V_j^b = B_j^b - P_j^b \geq 0$, and we can prove that any buyer is individual rational.

Second, we will prove the sellers are individual rational. Different from the buyers, sellers are sorted in an ascending order of their bids. The clearing price P_i^s of any winning seller i is the bid of kth seller B_k^s, so $P_i^s = B_k^s \geq B_i^s$ is always true. So we can prove any seller is individual rational. ∎

4.3 Truthful

In order to prove the auction mechanism is truthful, we should prove that there is no buyer j or seller i can improve its utility by cheating when auction mechanism is strategy-proof[5].

We first verify two properties. One is that WDP is monotonous, the other is that pricing and bidding are uncorrelated. We first prove WDP is monotonous.

Lemma 3. *For a given bid collection*

$$b_1^b, \cdots, b_{j-1}^b, b_{j+1}^b, \cdots, b_N^b, \tag{11}$$

which excludes buyer j and the bid collection $\{B_i^s\}_{i=1}^M$ of sellers, if buyer j could win by the bid b_j^b, it will also win in the auction by bidding $b_j^{b'} > b_j^b$.

Proof: When the bid of buyer j is b_j^b, the bid of the buyer group is π_l. When its bid is $b_j^{b'}$, the bid of the buyer group is π_l'. We can easily find the limiting case is that only when the bid of buyer j is equal to the lowest bid of buyer group G_l, does $\pi_l' > \pi_l$ exists. So $\pi_l' > \pi_l$ is always true. In WDP, G_l is always the winning group, so Lemma 4.1 is true. ∎

Lemma 4. *For a given bid collection*

$$B_1^s, \cdots, B_{i-1}^s, B_{i+1}^s, \cdots, B_M^s, \tag{12}$$

which excludes seller i and the bid collection $\{b_j^b\}_{j=1}^N$ of buyers, if seller i could win by the bid B_i^s, it will also win the auction by bidding $B_i^{s'} > B_i^s$.

Proof: The same to Lemma 3. ∎

Then we will prove that pricing and biding are uncorrelated by proving Lemma 5 and Lemma 6 is true.

Lemma 5. *For a given bid collection*

$$b_1^b, \cdots, b_{j-1}^b, b_{j+1}^b, \cdots, b_N^b, \tag{13}$$

which excludes buyer j and the bid collection $\{B_i^s\}_{i=1}^M$ of sellers, if buyer j could win by the bid b_j^b and $b_j^{b'}$, the clearing price P_j^b and $P_j^{b'}$ is the same.

Proof: When $b_j^{b'} > b_j^b$, we have proven that P_j^b is the same under different bids. Thus we only prove the case that $b_j^{b'} < b_j^b$. As long as the clearing price P_j^b is the same, seller i will win the auction by different bids. If a buyer wins the auction in the case that $b_j^{b'} < b_j^b$, there exists an extreme condition that the bid of buyer j is equal to the lowest bid of buyer group G_l. The bid of buyer j can influence the bid of the buyer group G_l, but can not influence the size of the group.

At the same time, the clearing price of every winning group is always π_k, and the clearing price P_j^b of buyer j will not change with their bids. So Lemma 5 is true.   ∎

Lemma 6. *For a given bid collection*

$$B_1^s, \cdots, B_{i-1}^s, B_{i+1}^s, \cdots, B_M^s \tag{14}$$

which excludes seller i and the bid collection $\{b_j^b\}_{j=1}^N$ of buyers, if seller i could win by the bid B_i^s and $B_i^{s'}$, the clearing price P_i^s and $P_i^{s'}$ is the same.

Proof: The same to Lemma 5.   ∎

Theorem 7. *RTDA is truthful.*

Proof: First, We should prove that no buyer or seller can improve their utilities by bidding untruthfully. The dominant strategy to every one is bidding according to their valuation, that is $B_j^b = V_j^b$, $B_i^s = V_i^s$.

Table 1. Bidding Cases

Case	1	2	3	4
Untruthful	Fail	Fail	Win	Win
Truthful	Fail	Win	Fail	Win

We consider the four cases in Table 1. If RTDA is truthful in all the four cases, then we can prove the theorem 7.

- Case 1: Whether buyer j bids truthfully or untruthfully, it will not win the channels, and the utility is always equal to 0. No one can improve it's utility by cheating. So it is obviously truthful.
- Case 2: In this case, if buyer j bids untruthfully, it bids less than its valuation and will fail in the auction. Thus utility is 0, and the utility is positive if it wins by biding truthfully. Untruthful bid leads to less utility, which violates individual rationality. So it is also truthful in this situation.
- Case 3: This case is the main case. We define $b_j^{b'}$ as the untruthful bidding of buyer j. Only when buyer j bids untruthfully, that is, $B_j^{b'} = b_j^{b'} \cdot \|t_j\|_1 > V_j^b = b_j^b \cdot \|t_j\|_1$, will cause this case. Now we suppose when the bid of buyer j is b_j^b, the bid of the buyer group is π_l. When its bid is $b_j^{b'}$, the bid of the buyer group is π_l'. If we want to make the bid of buyer j the dominant factor of the bid of the group, we must make b_j^b the lowest bid of the group. That is, $\pi_l = b_j^b \cdot \sum_{n \in G_l} \|t_n\|_1$, if a buyer wins the auction by shilling, the clearing price P of the winning buyer group it is in satisfies that $\pi_l' \geq P \geq \pi_l$. So the utility is:

$$V_j^b - \frac{P \cdot \|t_j\|_1}{\sum_{n \in G_l} \|t_n\|_1} = B_j^b - \frac{P \cdot \|t_j\|_1}{\sum_{n \in G_l} \|t_n\|_1},$$

$$B_j^b = \frac{\pi_l \cdot \|t_j\|_1}{\sum_{n \in G_l} \|t_j\|_1},$$

$$\frac{\pi_l \cdot \|t_j\|_1}{\sum_{n \in G_l} \|t_n\|_1} - \frac{P \cdot \|t_j\|_1}{\sum_{n \in G_l} \|t_n\|_1} < 0,$$

$$\forall j \in [1, N].$$

Because the utility of its true bid is 0, and the utility is negative if it bids untruthfully. Thanks to individual rationality, RTDA is truthful in this case.

- Case 4: We have proved that whether it bids truthfully or untruthfully in the proof of lemma 5, the clearing price will not change when they win the channels. As a result of the uniqueness of its true value, the utility is also the same. So RTDA is also truthful in this case.

The proof of sellers is similar to the buyers. Then RTDA is a truthful double auction which has been proved.   ∎

5   Evaluations

First, we discuss the influence of the number of time slots of a single channel by the time slot experiment. We compare RTDA with TRUST to show the improvement of utilization in the usage rate experiment. Then we discuss the simulation results, and make a conclusion.

5.1   Setup of Simulation

First, we will declare three parameters in the simulation.

- Bids Distribution. We assume that the bid b_j^b of a buyer j is randomly distributed, where 0 is the minimum while 1 is the maximum value. To simplify the simulation experiment, we let $b_j^b = \frac{B_j^b}{\|t_j\|_1}$. For sellers, we assume that their bids B_i^s are randomly distributed over $(0, 2]$, just as the setup in TRUST[14].
- Time Slot. We assume that each channel is cut into 4 time slots by default. Each buyer will ask for at least one time slot of one channel by random distribution.
- Interference Condition. We assume that our experiments are under cluster network topology. We randomly place 50% buyers in the center of a given area to create a hot-spot and randomly deploy the rest 50% buyers in the whole space.

Second, we define two parameters to show the performance of RTDA.

- Average Usage Rate f. The usage rate of a channel can be defined as the number of the time slots which have been sold divided by the total number of the time slots of one channel. So we consider that the mean sold channels' usage rate as average usage rate.
- Per-channel Utilization c[14]. We consider that the mean number of users who share the same channel as the Per-channel Utilization, which shows the average utilization of the channels.

5.2 Simulation Result

There are two main simulation experiments which are time slot experiment and usage rate experiment. Our work performs better, especially in terms of usage rate.

(a) Average Usage Rate f (b) Per-channel Utilization c (c) Profit of Auctioneer φ

Fig. 5. Time Slot Experiment

Time Slot Experiment. In this experiment, we can find out what effect will be caused by the number of the time slots. It could help us choose the proper T in RTDA.

As shown in Fig. 5a, it is easily to find out that with the increasing of the number of the time slots T, average usage rate f increases quickly. When T becomes too large such as 9, f falls down fast. Because when T increases, there will be much more channels which can not be sold. Once when a channel has been sold, it will get almost 100% usage rate. The channels which are unsold can be sold again in another auction. So RTDA wastes less channels and performs better.

In Fig. 5b, c holds steadily around 40 with lower T. Higher than 20 slots, with the increasing of T, c decreases gradually. We infer that due to the increasing of time slots, the bids buyers offer go lower and lower. But the price of the channel stays in the distribution over $(0, 2]$, so there will be more and more users who join in the bigger group with lower π_l and who can not gain a channel.

As shown in Fig. 5c, with the increasing of T, φ decreases obviously. The less the buyer paid, the smaller φ will be when B_i^s keep unchanged.

We can find out that T can not be setup to a high value. RTDA will perform better when T is approximately from 4 to 10. Too many slots will make the performance of RTDA worse, which can be observed in Fig. 5.

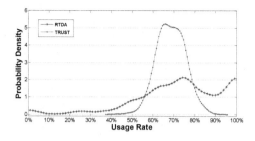

Fig. 6. The PDF of Usage Rate

Usage Rate Experiment. Usage rate should be an appropriate parameter to evaluate auction mechanism while considering reuse. About 1000 round experiments and statistic are taken in our paper. As shown in Fig. 6, we can find out that RTDA can product higher usage rates which are up to 100%. The usage rate of TRUST is mainly distributed around 65%. Meanwhile RTDA has higher usage rates distribution. TRUST do not reach 100% usage rate. It proves that RTDA can achieve high usage rate for each channel. It also means that RTDA can significantly improve the usage rate of the channel. As a result, RTDA is a better mechanism to achieve higher level reuse of the spectrum.

5.3 Discussions

After analyzing the result of the simulation experiments, we conclude that RTDA is a better auction mechanism for temporal reuse. By graphical analysis, RTDA is proved to improve the usage rate of the channels. However, temporal reuse can not be infinite, it should be set under threshold which depends on the realities. Until now, our work could be a guidance for setting an auction, especially in temporal reuse. The details under certain scenarios should be considered in the future.

6 Related Work

McAfee[10] and VCG[1] are among the most famous double auction mechanisms which provide an economic-robust method, which are the basic work for spectrum auction. Spectrum is different from the traditional commodities in traditional auctions. It can be used by different users at the same time. VERITAS[13] proposed by Zhou is the first truthful auction while considering reusability. Furthermore, Zhou and Zheng extended their work. They propose a general framework for truthful double spectrum auction TRUST[14], in which the both reusability of buyer and seller are considered.

There are also authors take heterogeneous channels into consideration such as TAHES[2]. These designs consider sorts of properties except temporal reuse, but it should be taken advantage of in modern circumstance. Although TASG[7] considers the reuse of spectrum while being economic-robust, the conflicts among SUs are ignored. Some mechanisms are temporal reusable, but incomprehensive. For example in TODA[9], Wang and others choose the begin point in the field of temporal reuse while RTDA selects the time slots arbitrarily. Some mechanisms such as TASC[12], even ignore temporal reuse. Sometimes requests from SUs fail to arrive simultaneously. The auction design in [11] proposed by Xu and others resolves this problem without reusing channels. Core-selecting auction[15] proposed by Li allows secondary users to bid for combinations of channels which improves seller revenue. Based on these problems and work, a spacial and temporal reusable mechanism such as RTDA is needed. It is also supposed to be efficient and truthful.

7   Conclusions

In this paper, we propose a truthful double auction mechanism named RTDA, which achieves temporal reuse. RTDA is proved to be truthful, individual rational and ex-post budget balanced. We use mathematical inference and game theory to validate our auction mechanism is economic-robust. RTDA makes an important contribution on maximizing usage rate. It can achieve 100% usage rate under some scenarios. To deploy RTDA in practice, several practical issues must also be addressed. Heterogeneous and complex demands of channels should be considered in the future.

Acknowledgment. This work is supported by Natural Science Foundation of China under Grants No. 61070181, No. 61272524 and No. 61202442 and the Fundamental Research Funds for the Central Universities No.DUT15QY05 and No.DUT15QY51. This work is also supported by the Guangdong University of Petrochemical Technology Internal Project No. 2012RC0106 and Open Fund of Guangdong Provincial Key Laboratory of Petrochemical Equipment Fault Diagnosis No.GDUPTKLAB 201323.

References

1. Babaioff, M., Nisan, N.: Concurrent auctions across the supply chain. In: Proceedings of the 3rd ACM Conference on Electronic Commerce, EC 2001, pp. 1–10. ACM, New York (2001)
2. Feng, X., Chen, Y., Zhang, J., Zhang, Q., Li, B.: TAHES: truthful double auction for heterogeneous spectrums. In: 2012 Proceedings IEEE INFOCOM, pp. 3076–3080, March 2012
3. Gopinathan, A., Li, Z.: Strategyproof auctions for balancing social welfare and fairness in secondary spectrum markets. In: 2011 Proceedings IEEE INFOCOM, pp. 3020–3028, April 2011
4. Huang, J., Berry, R., Honig, M.: Auction mechanisms for distributed spectrum sharing. In: Proceedings of 42nd Allerton Conference (2004)

5. Huang, Q., Tao, Y., Wu, F.: Spring: a strategy-proof and privacy preserving spectrum auction mechanism. In: 2013 Proceedings IEEE INFOCOM, pp. 827–835, April 2013
6. Klemperer, P.: What really matters in auction design. The Journal of Economic Perspectives **16**(1), 169–189 (2002)
7. Lin, P., Feng, X., Zhang, Q., Hamdi, M.: Groupon in the air: a three-stage auction framework for spectrum group-buying. In: 2013 Proceedings IEEE, pp. 2013–2021, April 2013
8. Ramanathan, S.: A unified framework and algorithm for $(t/f/c)$dma channel assignment in wireless networks. In: Proceedings IEEE Sixteenth Annual Joint Conference of the IEEE Computer and Communications Societies. Driving the Information Revolution, INFOCOM 1997, vol. 2, pp. 900–907, April 1997
9. Wang, S., Xu, P., Xu, X., Tang, S., Li, X., Liu, X.: TODA: truthful online double auction for spectrum allocation in wireless networks. In: 2010 IEEE Symposium on New Frontiers in Dynamic Spectrum, pp. 1–10, April 2010
10. Wurman, P.R., Wellman, M.P., Walsh, W.E.: A parametrization of the auction design space. Games and Economic Behavior **35**(1C2), 304–338 (2001)
11. Xu, P., Xu, X., Tang, S., Li, X.Y.: Truthful online spectrum allocation and auction in multi-channel wireless networks. In: 2011 Proceedings IEEE INFOCOM, pp. 26–30, April 2011
12. Yang, D., Fang, X., Xue, G.: Truthful auction for cooperative communications. In: Proceedings of the Twelfth ACM International Symposium on Mobile Ad Hoc Networking and Computing, MobiHoc 2011, pp. 9:1–9:10. ACM, New York (2011)
13. Zhou, X., Gandhi, S., Suri, S., Zheng, H.: eBay in the sky: strategy-proof wireless spectrum auctions. In: Proceedings of the 14th ACM International Conference on Mobile Computing and Networking, MobiCom 2008, pp. 2–13. ACM, New York (2008)
14. Zhou, X., Zheng, H.: TRUST: a general framework for truthful double spectrum auctions. In: IEEE INFOCOM 2009, pp. 999–1007, April 2009
15. Zhu, Y., Li, B., Li, Z.: Core-selecting combinatorial auction design for secondary spectrum markets. In: 2013 Proceedings IEEE INFOCOM, pp. 1986–1994, April 2013

Dynamic Sparse Channel Estimation Using ℓ_0-constrained Kalman Filter in OFDM Systems

Nan Jing[✉] and Lin Wang

School of Information Science and Engineering, Yanshan University,
Qinhuangdao 066004, Hebei, People's Republic of China
{jingnan,wlin}@ysu.edu.cn

Abstract. The conventional Compressed Channel Sensing (CCS) methods are concerned with the static sparse channel which is modeled as time invariable path number and path delays, but they fail to work in a dynamic scenario which allows the channel parameters to vary over time. To solve this problem, we introduce a simple Dynamic Sparse Channel Estimation (DSCE) algorithm for Orthogonal Frequency-Division Multiplexing (OFDM) systems. Exact reconstruction is provided by interchanging the constrained part and the optimization part of minimum ℓ_0-norm recovery, and then minimum ℓ_0-norm recovery is transformed into a ℓ_0-norm constrained Kalman Filter (KF) problem. The key idea of the proposed algorithm is the linearization of the non-linear state constraint. In this case, four types of fictitious observations are developed to substitute for the ℓ_0-norm constraint based on four families of continuous functions which are used to approximate to the discontinuous ℓ_0-norm. Therefore, DSCE is performed by employing KF in a stand alone manner. Numerical evaluations are presented to demonstrate the effectiveness of the proposed algorithm in practice.

1   Introduction

The Orthogonal Frequency-Division Multiplexing (OFDM) technique has been widely employed in the current and future wireless communication systems, in which the Channel State Information (CSI) obtained by applying pilot tones is used for coherent detection. The channel estimation technique can be classified into two categories in OFDM systems: training-based methods and blind methods. Blind methods attempt to estimate the channel based on second order statistics of the unknown data only [10]. Obviously, blind channel estimation saves bandwidth by avoiding the use of pilots when the channel remains constant over a large number of OFDM symbols, however, their implementation on a mobile system is fairly difficult in practice because the underlying channel changes from one block to the next block. Training-based methods estimate the CSI using training symbols known at both the transmitter and the receiver. The training symbols may be inserted at different subcarriers of different OFDM blocks, which are more often called pilots. The CSI corresponding to the pilot subcarriers is first estimated, and that corresponding to the data subcarriers is obtained

© Springer International Publishing Switzerland 2015
Y. Wang et al. (Eds.): BigCom 2015, LNCS 9196, pp. 28–42, 2015.
DOI: 10.1007/978-3-319-22047-5_3

by interpolation. This is called Pilot-Assisted Channel Estimation (PACE) for which a tremendous research effort is yet being made [9].

Numerous experimental studies undertaken by researchers in the recent past have shown that wireless channels are often characterized by multipath propagation and tend to exhibit sparse structures with a few significant paths in many wideband OFDM systems, such as underwater acoustic systems, digital television systems and residential ultra-wideband systems [1]. Conventional linear PACE methods mentioned above, based on Least Square (LS), or Minimum Mean Square Error(MMSE) etc. criterion, fail to capitalize on the sparse structure of the multipath channel in wideband OFDM systems. More recently, the Compressed Sensing (CS) techniques have been applied to Sparse Channel Estimation (SCE) [1,15,18]. However, these methods perform in an unchanging sparsity pattern with fixed delays and number of significant paths, in fact, both of them may change along with the moving transmitter or the receiver. The sparse channel which allows the path number and the path delays to vary over time is called dynamic sparse channel and often modeled as a Gauss-Markov random process [4]. Therefore, the methods addressing the Static SCE (SSCE) are no longer suitable for the scenario of Dynamic Sparse Channel Estimation (DSCE).

In light of pioneering work on[20], some practical approaches for treating the time-varying sparse channel estimate have been developed over the last five years [5–8,11,13,16]. The minimum ℓ_1 norm is converted into an Maximum Likelihood(ML) problem augmented by an ℓ_1-penalty, and four adaptive algorithms are proposed to carry out the likelihood function by incorporating Expectation-Maximization(EM) with Kalman Filter(KF), Least Mean Squares(LMS), Recursive Least squares (RLS) or Fast RLS(FRLS)[13]. Variations of the LMS and Least Mean Forth (LMF) algorithms with ℓ_1-penalty, ℓ_p-penalty or ℓ_0-penalty has been developed in [5–7]. A simple low-complexity DSCE method is introduced utilizing the greedy search algorithms, in which adaptive path delay grids are constructed to represent the dynamic sparse channel, then two reduced-order Orthogonal Matching Pursuit (OMP) algorithms are utilized to estimate and track the path delays according to the changes in the path number and the path delays between two consecutive OFDM blocks [8]. A ℓ_1-regularized least-absolutes algorithm (ℓ_1-LA) is investigated based on the Linear Programming to recover the dynamic sparse channel in non-Gaussian impulsive noise scenarios[11]. Moreover, a geometric mixed norm approach is proposed for shallow water acoustic channel estimation and tracking, which reformulates the well-known ℓ_1-ℓ_2 sparse optimization metric as an equivalent mixture of the ℓ_2 norm of the complex squared root of the channel coefficients and a fourth-order complex function based on the estimation error, and employs a gradient descent algorithm to estimate and track sparse coefficients[16].

However, the methods mentioned in [5–8,11,13,16] are hybrid which refers to that two or more techniques are combined to deal with DSCE, contrary to the hybrid methods, we present a simple self-reliant DSCE algorithm based on the work outlined in [3] in which KF endowed with a Pseudo-Measurement (PM) equation can be straightforwardly implemented to recover the dynamic sparse signal in a stand-alone manner. The proposed algorithm performs in a batch form without any additional increments in computational complexity unlike the description in [20]. In the proposed method, the convex constrained minimization for SCE is converted into a CS-embedded KF with ℓ_0 norm nonlinear equality constraint problem, defined as ℓ_0 constrained CS-embedded KF (ℓ_0-CSKF). Therefore, the DSCE is divided into two steps: First, an unconstrained solution is obtained by using of the standard KF. Next, a constrained solution is achieved by employing KF again to a PM equation which approximates to the ℓ_0 norm constraint. Searching the minimum ℓ_0 norm is an intractable problem as the dimension increases, and it is too sensitive to noise, thus, four types of continuous functions are designed to approximate to the ℓ_0 norm, based on which, four PM observations are constructed for ℓ_0 norm constraint. We analyze the computational complexity of the proposed algorithms, and demonstrate their practical feasibility. Simulation results show that channel estimation Mean Square Error (MSE) and Bit Error Ratio (BER) of the proposed algorithms are always significantly smaller than that achievable using the classical ℓ_1 optimization program, i.e., Dantzig Selector (DS)[1], Basis Pursuit (BP)[18]. Stated differently, the algorithms presented here are slightly superior to the ones proposed in [3] in the approximate static scenario (without any prior knowledge of the channel) and considerable better in the dynamic scenario (a part of path number and path delays is available according to the previous channel estimation).

The paper is organized as follows. In Section 2, we introduce the system model and the problem definition for DSCE of OFDM systems. Some basic facts about CS are reviewed in Section 3. The ℓ_0-CSKF DSCE algorithms are investigated in Section 4. Performance comparisons and numerical results of our proposed algorithms are given in Section 5. Finally, the conclusions are drawn in the last Section.

Notation: $(\cdot)^H$ will represent conjugate transpose operator. diag$\{x\}$ stands for the diagonal matrix whose diagonal is the entry of the vector x. I_K is the $K \times K$ identity matrix. $F_{K \times K}$ is the $K \times K$ unitary Discrete Fourier Transform (DFT) matrix. For vectors $x \in \mathbb{C}^N$, the function $\|x\|_2$ represents the ℓ_2 norm, given by $\|x\|_2 = \sqrt{\sum_{i=0}^{N-1} x^2(i)}$. In the sequel, we will also make use of the ℓ_1 norm and ℓ_p norm, which is given by $\|x\|_1 = \sum_{i=0}^{N-1} |x(i)|$ and $\|x\|_p = \sum_{i=0}^{N-1} (|x(i)|^p)^{1/p}$ respectively, and we will use the ℓ_0 norm, $\|x\|_0$, to denote the number of nonzero entries of the vector x. The set operations $\cup$, $\cap$, and $\setminus$ have the usual meanings. Let T denote a set, then $T^c = [1 : N] \setminus T$ is the complement of T. For a matrix F, $(F)_{T_1, T_2}$ denotes the submatrix of F containing rows and columns corresponding to the entries in T_1 and T_2 respectively.

2   Channel and System Model

2.1   Sparse Approximation for Channel

$h(t, \tau)$ is the continuous-time Channel Impulse Response(CIR), thus $h(t, \tau) \approx h(0, \tau) = h(\tau)$ if it remains constant over the time duration of interest.

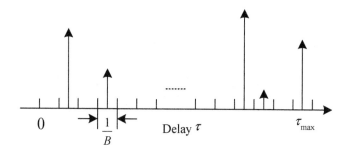

Fig. 1. A schematic illustrating the h_k by uniformly-spaced sampling within the maximum delay spread τ_{max} at a resolution B

The channel generates multiple delayed and attenuated copies of the transmitted waveform, for such multipath channel, $h(\tau)$ can be described as [1]

$$h(\tau) = \sum_{i=0}^{S-1} \alpha(i)\delta(\tau - \tau(i)) \tag{1}$$

where $\delta(\cdot)$ is the Dirac delta, S is the number of multipaths, $\alpha(i)$ and $\tau(i)$ are the ith path's complex gain and path delay, respectively. Each complex gain $\alpha(i)$, $i = 0, \ldots, S-1$ is modeled as a wide-sense stationary, zero-mean complex Gaussian random variable. As mentioned in Section 1, S and $\tau(i)$ are not always fixed because they may vary with the movement of the mobile station, therefore, the dynamic sparse channel is modeled as

$$h_k(\tau) = \sum_{i=0}^{S_k-1} \alpha_k(i)\delta(\tau - \tau_k(i)) \tag{2}$$

where k is the kth OFDM block duration, $k = 1, \ldots, \infty$. Based on the formulation in [8], the proposed method is carried out block by block. Each block consists of two or more consecutive OFDM symbols. We assume that S_k and $\{\tau_k(i), i = 0, \ldots, S_k\}$ do not vary within an OFDM block but they change from block to block.

The discrete path model (2) is realistic, however, difficult to analyze and identify due to nonlinear dependence on the delay parameters $\{\tau_k(i)\}$. A discrete multipath channel vector $h_k = \{h_k(\tau - l/B)\}_{l=0}^{L-1}$ can be obtained by uniformly sampling $h_k(\tau)$ within the maximum delay spread τ_{max} at a resolution B as illustrated in Fig.1, where B is the communication bandwidth, $L = \tau_{\mathrm{max}}B + 1$ is the number of path delay samplings. $h_k \in \mathbb{C}^L$ is approximately sparse if $S_k \ll L$, hence, increasing B can improve the sparsity of h_k.

2.2   System Model

The channel state h_k is usually modeled as a first-order Gauss-Markov process [4]. In order to characterize the time-varying sparse channel, a spatially i.i.d. Gaussian random walk model is given by [20]

$$h_k = \xi h_{k-1} + u_k \tag{3}$$

where the process noise $u_k \overset{i.i.d}{\sim} N(0, Q_k)$ is the zero-mean white Gaussian driving noise with covariance matrix $Q_k = \sigma_u^2 I$. Parameter $\xi \in [0,1]$ is the fading coefficient characterizing the degree of time variation[4]. Based on the models presented in [20], a time-varying sparse channel state h_k can be obtained with the following setting: Q_k is divided into two subsections, one is to produce the time-invariant part of h_k by using of $(Q_k)_{\{\tau_k\}\cap\{\tau_{k-1}\},\{\tau_k\}\cap\{\tau_{k-1}\}}$ $= \sigma_{sys}^2 I$, and the other is to generate the time-varying part of h_k utilizing $(Q_k)_{\{\tau_k\}\setminus\{\tau_{k-1}\},\{\tau_k\}\setminus\{\tau_{k-1}\}}$ $= \sigma_{init}^2 I$ at each time step. Moreover $(Q_k)_{\{\tau_k\}^c,\{\tau_k\}^c} = 0$, thus we have the modified model, let $h_0 = 0$

$$
\begin{aligned}
h_k &= \xi h_{k-1} + u_k, u_k \overset{i.i.d}{\sim} N(0, Q_k) \\
(Q_k)_{\{\tau_k\}\cap\{\tau_{k-1}\},\{\tau_k\}\cap\{\tau_{k-1}\}} &= \sigma_{sys}^2 I \\
(Q_k)_{\{\tau_k\}\setminus\{\tau_{k-1}\},\{\tau_k\}\setminus\{\tau_{k-1}\}} &= \sigma_{init}^2 I \\
(Q_k)_{\{\tau_k\}^c,\{\tau_k\}^c} &= 0
\end{aligned}
\tag{4}
$$

We assume that the data $s_k = [s_k(0), \ldots, s_k(N-1)]^T \in \mathbb{C}^N$ are transmitted over the N OFDM subcarriers. To avoid Inter-Block Interference (IBI), we also assume that each OFDM symbol is preceded by a cyclic prefix (CP) whose minimum length is $L-1$. After removing the CP and performing FFT, the received data vector can be given by

$$y_k = H_k s_k + w_k, k = 1, 2 \ldots \tag{5}$$

where $y_k \in \mathbb{C}^N$ is the observation sequence, the observation noise $w_k \overset{i.i.d}{\sim} N(0, R)$ is a AWGN vector with covariance $R \in \mathbb{C}^{N\times N}$. $H_k \in \mathbb{C}^{N\times N}$ is a diagonal Channel Frequency Response (CFR) matrix comprising of the N OFDM channel coefficients that are related to h_k as $H_k(n) = \sum_{l=0}^{L-1} h_k(l) e^{-j(2\pi/N)nl}$, $n = 0, 1 \ldots, N-1$. First, in order to perform PACE, we select a set of indices $P \subset \{0, 1, \ldots, N-1\}$ with cardinality $|P| = N_p$. Next, we construct a training data vector $s_k^{tr} \in \mathbb{C}^{N_p}$ having energy $\|s_k^{tr}\|_2^2 = \varepsilon^{tr}$ and transmit this vector using the pilot subcarriers specified by P. Finally, the received training data vector $y_k^{tr} \in \mathbb{C}^{N_p}$ can be written as

$$y_k^{tr} = diag(s_k^{tr}) F_{N_p \times L} h_k + w_k^{tr}, k = 1, 2 \ldots \tag{6}$$

where $y_k^{tr} = \{y_k^{tr}(p)\}_{p\in P}$, $s_k^{tr} = \{s_k^{tr}(p)\}_{p\in P}$ and $w_k^{tr} = \{w_k^{tr}(p)\}_{p\in P} \in \mathbb{C}^{N_p}$ respectively. $F_{N_p \times L}$ denotes the $N_p \times L$ submatrix of the N-point Discrete Fourier Transform (DFT) matrix with the entries $F(p,l) = e^{-j(2\pi/N)lp}$, $p \in P$, $0 \le l \le L-1$. Let $A_k = diag(s_k^{tr}) F_{N_p \times L}$, $A_k \in \mathbb{C}^{N_p \times L}$ denotes the observation (or measurement) matrix, then (6) can be rewritten as

$$y_k^{tr} = A_k h_k + w_k^{tr} \tag{7}$$

Our goal is to get the "best" causal estimate of h_k at the kth time step based on the state equation (3) and the measurement equation (7).

3 ℓ_0 Constrained CS-Embedded-KF Algorithm

The underlying assumption is that h_k is sparse or at least approximately so. This is usually based on the model that h_k only consists of $S \ll L$ nonzero significant path coefficients. If the set P is generated randomly, then A_k indexed by the elements of P, obeys RIP, and we can recover h_k accurately by solving the combinatorial optimization problem

$$\hat{h}_k \overset{\Delta}{=} \underset{h_k \in H_\varepsilon}{\arg\min} \|h_k\|_0, k = 1, \ldots, \infty \tag{8}$$

where $\hat{h}_k$ is the sparest solution to (18), H_ε is the set of all $h_k \in \mathbb{C}^L$ satisfying $\sum_{i=1}^{k} \|y_i^{tr} - A_i h_i\|_2 \leq \varepsilon$. Since the measurement errors in (7) are stochastic [2], (8) can be given by

$$\hat{h}_k \overset{\Delta}{=} \underset{h_k \in H_\varepsilon}{\arg\min} \|h_k\|_0 s.t. E_{h_k|y_k} \left[\left\| h_k - \hat{h}_k \right\|_2^2 \right] \leq \varepsilon \tag{9}$$

We interchange the optimization part and the constrained part in (9) leveraging the dual problem [19]

$$\underset{\hat{h}_k}{\min} E_{h_k|y_k} \left[\left\| h_k - \hat{h}_k \right\|_2^2 \right] s.t. \left\| \hat{h}_k \right\|_0 \leq \varepsilon' \tag{10}$$

The constrained optimization problem (10) can be solved in the framework of KF[17]. Corresponding to (3), (7) and (10), DSCE can be transformed into a KF problem with nonlinear state constraint: ℓ_0 norm constraint. Therefore, the estimation is divided into two steps: one is unconstrained estimation stage, the other is constrained estimation stage. To begin with, the standard KF is employed in the unconstrained stage, and a filtered result is obtained. Next, a PM equation is constructed to substitute for the constrained function. Finally, the first filtered result is adopted as the initial value of the constrained stage, and the standard KF is applied to the PM equation again in order to attain a constrained estimate which is the sparsest solution to (10).

3.1   The Unconstrained Estimation Stage

The standard KF is a Minimum Mean-Squared Error (MMSE) estimator, then the unconstrained estimate $\hat{h}_{k|k}$ is obtained by solving the ℓ_2 minimization problem: $\underset{\hat{h}_{k|k}}{\min} E_{h_k|y_k} \left[\left\| h_k - \hat{h}_{k|k} \right\|_2^2 \right]$ with the standard KF based on the following five equations:

$$\hat{h}_{k|k-1} = \xi \hat{h}_{k-1|k-1}$$

$$\Sigma_{k|k-1} = \xi^2 \Sigma_{k-1|k-1} + \sigma_u^2 I$$

$$K_k = \Sigma_{k|k-1} A_k^H \left(A_k \Sigma_{k|k-1} A_k^H + \sigma_w^2 I \right)^{-1} \tag{11}$$

$$\hat{h}_{k|k} = \hat{h}_{k|k-1} + K_k \left(y_k^{tr} - A_k \hat{h}_{k|k-1} \right)$$

$$\Sigma_{k|k} = \left(I - K_k A_k \right) \Sigma_{k|k-1}$$

where $\hat{h}_{k|k-1}$ denotes the priori estimate of h_k at the kth time step. $\Sigma_{k|k-1}$ and $\Sigma_{k|k}$ represent the priori covariance of $\hat{h}_{k|k-1}$ and posteriori covariance of $\hat{h}_{k|k}$ respectively. $\hat{h}_{k|k}$ is the unconstrained estimate. $K_k \in \mathbb{C}^{L \times N_p}$ is the KF gain.

3.2 The Constrained Estimation Stage

To begin with, PM equations are developed to substitute for $\left\| \hat{h}_k \right\|_0 \le \varepsilon'$ in (10). The inequality-constrained $\left\| \hat{h}_k \right\|_0 \le \varepsilon'$ can be substituted with equality-constrained $\left\| \hat{h}_k \right\|_0 = \varepsilon'$ [17], then a PM equation can be written as

$$\left\| \hat{h}_k \right\|_0 = \tilde{H} \hat{h}_k = \varepsilon' \tag{12}$$

where $\tilde{H}$ is a fictitious matrix, called as the PM matrix. ε' is regarded as a constrained measurement noise. The covariance R_ε of ε' regulates the tightness of the constraint. Here, let $R_\varepsilon = \gamma^2 I_{N_p \times N_p}$ with $\gamma \ge 100$ [3]. Unfortunately, the solution to (9) is both numerically unstable and a NP-hard problem, nevertheless, there is no accurate function to represent the $\left\| \hat{h}_k \right\|_0$ because it is discontinuous, thus we can not find a PM matrix directly to establish a PM equation for $\left\| \hat{h}_k \right\|_0$. Instead, we propose four families of continuous functions to approximate to the ℓ_0 norm. The four families of functions are given below:
Quadratic function

$$f_{2order} \left(\hat{h}_k \right) = \frac{\sigma^2}{\hat{h}_k^2 + \sigma^2} \tag{13}$$

Arc Tangent function

$$f_{tan} \left(\hat{h}_k \right) = \frac{2}{\pi} \arctan \left(\frac{\hat{h}_k^2}{2\sigma^2} \right) \tag{14}$$

Hyperbolic Tangent function

$$f_{tanh} \left(\hat{h}_k \right) = \frac{e^{\frac{\hat{h}_k^2}{2\sigma^2}} - e^{-\frac{\hat{h}_k^2}{2\sigma^2}}}{e^{\frac{\hat{h}_k^2}{2\sigma^2}} + e^{-\frac{\hat{h}_k^2}{2\sigma^2}}} \tag{15}$$

Gaussian function

$$f_{gaussian}\left(\hat{h}_k\right) = \exp(-\frac{\hat{h}_k^2}{2\sigma^2})$$

(16)

It should be noted that $\lim_{\sigma \to 0} f_{2order}\left(\hat{h}_k\right) = 1 - v\left(\hat{h}_k\right)$, $\lim_{\sigma \to 0} f_{gaussian}$ $\left(\hat{h}_k\right) = 1 - v\left(\hat{h}_k\right)$, however, $\lim_{\sigma \to 0} f_{\tan}\left(\hat{h}_k\right) = v\left(\hat{h}_k\right)$ and $\lim_{\sigma \to 0} f_{\tanh}\left(\hat{h}_k\right) = v\left(\hat{h}_k\right)$. Based on the functions, we establish four PM matrix which are given by

$$\tilde{H}_{2order}(l) = \left[1 - \frac{\sigma^2}{\hat{h}_k^2(l) + \sigma^2}\right] \bigg/ \hat{h}_k(l)$$

(17)

$$\tilde{H}_{\tan}(l) = \frac{2}{\pi} \arctan\left(\frac{\hat{h}_k^2(l)}{2\sigma^2}\right) \bigg/ \hat{h}_k(l)$$

(18)

$$\tilde{H}_{\tanh}(l) = \tanh\left(\frac{\hat{h}_k^2(l)}{2\sigma^2}\right) \bigg/ \hat{h}_k(l)$$

(19)

$$\tilde{H}_{gaussian}(l) = \left[1 - \exp\left(-\frac{\hat{h}_k^2(l)}{2\sigma^2}\right)\right] \bigg/ \hat{h}_k(l)$$

(20)

where $l = 0, \ldots, L - 1$.

Next, the standard KF is initialized by the unconstrained estimate $\hat{h}_{k|k}$ and its covariance $\Sigma_{k|k}$, and then applied to the PM equations again.

Finally, the second filtered results are the final sparse channel estimates. The constrained estimation stage can be completed iteratively to better enforce the constraint. The number of iterations is strongly dependent on the trade-off between the accuracy and the complexity. The entire ℓ_0-CSKF is summarized as Algorithm 1.

3.3 Convergence Analysis

The proposed algorithms ℓ_0-CSKF convert the minimum ℓ_0 norm problem into a KF with nonlinear equality constraints problem, therefore, we investigate the statistical steady properties of the KF to present the convergence of the proposed ℓ_0-CSKF. The statistical steady properties of the KF is defined by Defination 4.1 [12]:

Statistical Steady Properties. Suppose there is a positive integer Z for some ε_e, we have $\left\| h_k - \hat{h}_{k|k} \right\|_2^2 < \varepsilon_e$ when $k \geq Z$, then, the KF is a statistical steady procedure.

The channel state equation is a dynamic linear system driven by stochastic process in essence, therefore, we will use notations $\bar{h}_{k|k} = \mathrm{E}\left[\hat{h}_{k|k}\right]$ and $\Sigma_{k|k} = \mathrm{E}\left[\left(\hat{h}_{k|k} - \bar{h}_{k|k}\right)\left(\hat{h}_{k|k} - \bar{h}_{k|k}\right)^H\right]$ to represent the mean of

---

Algorithm 1.. ℓ_0-CSKF

1: **Initialization**

  Set $\hat{h}_0 = 0$ and $\Sigma_0 = 100$

2: **Prediction**

  $\hat{h}_{k|k-1} = \xi \hat{h}_{k-1|k-1}$

  $\Sigma_{k|k-1} = \xi^2 \Sigma_{k-1|k-1} + \sigma_u^2 I$

3: **Measurement Update**

  $K_k = \Sigma_{k|k-1} A_k^H \left(A_k \Sigma_{k|k-1} A_k^H + \sigma_w^2 I \right)^{-1}$

  $\hat{h}_{k|k} = \hat{h}_{k|k-1} + K_k \left(y_k^{tr} - A_k \hat{h}_{k|k-1} \right)$

  $\Sigma_{k|k} = (I - K_k A_k) \Sigma_{k|k-1}$

4: ℓ_0 **Constrained estimate**

  Let $\tilde{h}_k^i$ and $\tilde{\Sigma}_k^i$ denote the ℓ_0 norm constrained estimate and its covariance, thus, initialized with $\tilde{h}_k^0 = \hat{h}_{k|k}$ and $\tilde{\Sigma}_k^0 = \Sigma_{k|k}$.

  Let $\tilde{H} = \tilde{H}_{2order}$, or $\tilde{H} = \tilde{H}_{tan}$, or $\tilde{H} = \tilde{H}_{tanh}$, or $\tilde{H} = \tilde{H}_{gaussian}$

   for $i = 1, 2, \ldots, Iter$ **do**

$$\tilde{K}_k^i = \tilde{\Sigma}_k^i \left(\tilde{H} \right)^H \left[\tilde{H} \tilde{\Sigma}_k^i \left(\tilde{H} \right)^H + R_\varepsilon \right]^{-1}$$

$$\tilde{h}_k^i = \left(I - \tilde{K}_k^i \tilde{H} \right) \tilde{h}_k^i$$

$$\tilde{\Sigma}_k^i = \left(I - \tilde{K}_k^i \tilde{H} \right) \tilde{\Sigma}_k^i$$

   end for

5: **Output** $\hat{h}_{k|k} = \tilde{h}_k^{Iter}$ and $\Sigma_{k|k} = \tilde{\Sigma}_k^{Iter}$. **Increment k and go to Prediction.**

$\hat{h}_{k|k}$ and its covariance respectively. It is obvious that $\left\| h_k - \hat{h}_{k|k} \right\|_2^2 = \mathrm{E}\left[\left(\hat{h}_{k|k} - \bar{h}_{k|k} \right) \left(\hat{h}_{k|k} - \bar{h}_{k|k} \right)^H \right]$, then, only if $\Sigma_{k|k}$ is bounded, $\hat{h}_{k|k}$ will converge.

Considering the time-varying Lyapunov iteration of the covariance:

$$\Sigma_{k|k} = C \Sigma_{k-1|k-1} C^H + \sigma_u^2 I \tag{21}$$

where C is a state transition matrix. The stability criterion on the time-varying Lyapunov equation is provided by means of Lemma 4.1[14]. The time-varying Lyapunov equation is given by

$$\Sigma_k = C \Sigma_{k-1} C^H + U_k \tag{22}$$

where C is stable if $\|C\|_2 \leq \beta$ when $k \geq 1$ for $\forall \beta > 0$. If U_k is bounded, then Σ_k is bounded, and if $U_k \to 0$, then $\Sigma_k \to 0$. It can be seen from Algorithm 1 that the state transform matrix are the same both in the unconstrained estimate stage and the constrained estimate stage although the measurement matrix are different when KF is employed twice respectively. As shown in (3) C is a constant $\xi \in [0, 1]$ and in (22) is bounded whether or not $\sigma_u^2 = \sigma_{sys}^2$ or $\sigma_u^2 = \sigma_{init}^2$, which satisfies Lemma 4.1, therefore, the proposed ℓ_0-CSKF is convergent.

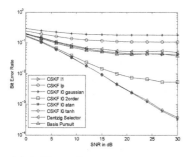

(a) BER performance comparison

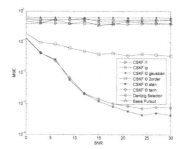

(b) MSE performance comparison

Fig. 2. Robust reception performance of evaluated schemes

4   Simulation Results

In this section, we present results from numerical experiments which illustrate and verify our results. We consider a 3GPP LTE standard alike scenario with OFDM system parameters:

- The channel bandwidth is $B = 5MHz$.
- FFT size $N = 512$.
- The active subcarriers 301.
- The subcarrier spacing is 15kHz.
- The sample interval is $T_s = 130ns$.
- There is a cyclic prefix of 26us.
- We consider that maximum delay spread $\tau_{\max}$ of the dynamic sparse channel is $25.4us$.
- The modulation scheme used is 4QAM.

Furthermore, the prior model on h_k is (3) with $\sigma_{sys}^2 = 1$ and $\sigma_{init}^2 = 9$, the channel length is $L = 128$ with a varying sparsity S_k and the fixed fading coefficient $\xi = 1$ at each time step. We assume that the pilot subcarriers are inserted in a random pattern, therefore, the observation matrix $A_k \in \mathbb{C}^{N_p \times L}$ in (7) consists of random subselections of Fourier matrices, which are known to have near-optimal RIP guarantees [2]. MSE and BER are used to evaluate the performance of the proposed ℓ_0-CSKF, which are averaged over 500 Monte Carlo trials. MSE is given by

$$MSE = 10 \log_{10} \left(\left\| h_k - \hat{h}_k \right\|_2^2 \Big/ \left\| h_k \right\|_2^2 \right) \tag{23}$$

In particular, DS, BP, examples of the classical ℓ_1 optimization program, which have been used for sparse channel estimation are considered in simulations for comparison [1, 15, 18]. Moreover, CSKF-ℓ_1 and CSKF-ℓ_p investigated in [3] are both evaluated for comparison.

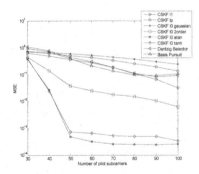
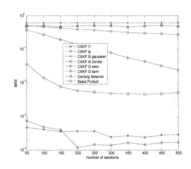

Fig. 3. the averaged MSE performance of evaluated schemes versus the number of pilots

Fig. 4. the averaged MSE performance of evaluated schemes versus the number of PM iterations

Let the observation noise covariance $R = \sigma_n^2 I_{N_p \times N_p}$, where $I_{N_p \times N_p}$ is an identity matrix, σ_n^2 is the noise variance. The constrained measurement noise covariance in (12) is considered as $R_{\varepsilon'} = 200^2[3]$, and each constrained estimate result is obtained after 50 iterations. Let $p = 0.7$ for CSKF-ℓ_p and $\sigma = 0.02$ for the proposed ℓ_0-CSKF.

Two dynamic scenarios are examined to evaluate the performance of the proposed DSCE algorithms which perform all in a batch form. 1) The approximate static case: DSCE is applied at each time step separately without any prior knowledge of the channel parameters. The initial value of the CIR and its covariance is fixed at each time step. This is extremely similar to the conventional SSCE although the path number and path delays vary over time. 2) The dynamic case: DSCE can be obtained utilizing a part of prior knowledge of the channel parameters available from the previous time instant. The previous estimate is the initial value of this time step, for instance, $\hat{h}_{k|k-1} = \xi \hat{h}_{k-1|k-1}$ and $\Sigma_{k|k-1} = \Sigma_{k-1|k-1} + \sigma_u^2 I$ presented in Algorithm 1.

4.1   Approximate Static Recover Without Channel Prior Knowledge

The approximate static performances of the proposed ℓ_0-CSKF and the other evaluated algorithms are examined in this section with a fixed initial value $\hat{h}_0 = 0$ and $\Sigma_0 = 100$ at each time step. Because the sparsity S_k varies over time, we assume that $S_k = 16$ at the kth time step corresponding to $N_p \geq cS_k \log(L/S_k)[2]$.

We investigate the computational complexity listed in Table 1 where $Iter$ is the number of PM iterations, L is the length of the channel, and N_p is the number of pilot subcarriers. The proposed methods have complexity $O(Iter \cdot L)$, the same as CSKF-ℓ_p and slightly more mathematical operations than CSKF-ℓ_1 if the number of PM iterations is small enough. The DS is a Linear Programming (LP) so it has the theoretical complexity of an LP, which is about $O\left((\min(N_p, L))^2\right)$ less than BP. In order to comply with RIP, N_p should be

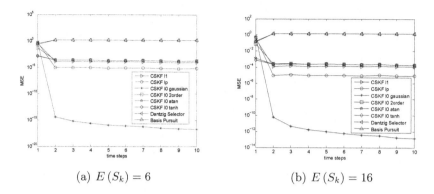

(a) $E\left(S_k\right)=6$ (b) $E\left(S_k\right)=16$

Fig. 5. Channel estimation MSE comparisons of different reconstruction algorithms with different variations in sparsity

Table 1. Computational complexity of different reconstruction algorithms

Algorithm	Complexity
ℓ_0-gaussian CSKF	$O\left(Iter \cdot L\right)$
ℓ_0-2order CSKF	$O\left(Iter \cdot L\right)$
ℓ_0-atan CSKF	$O\left(Iter \cdot L\right)$
ℓ_0-tanh CSKF	$O\left(Iter \cdot L\right)$
CSKF-ℓ_1	$O\left(Iter\right)$
CSKF-ℓ_p	$O\left(Iter \cdot L\right)$
DS	$O\left(\left(\min\left(N_p, L\right)\right)^2\right)$
BP	$O\left(L^3\right)$

more than 50, furthermore, $Iter = 50$ is considered to reduce the number of mathematical operations on condition that there is no obvious deterioration on performance. Based on these parameters, it can be declared that CSKF-ℓ_1 runs faster than the other methods and BP is slowest. The proposed ℓ_0-CSKF may run faster than DS and BP when $N_p \geq 80$.

Robust reception performance of the proposed schemes is demonstrated in Fig.2. Fig. 2(a) illustrates the BER versus Signal Noise Ratio (SNR) performance of the proposed ℓ_0-CSKF and the other methods, and Fig.2(b) demonstrates the MSE versus SNR performance of these schemes for $S_k = 16$ and $N_p = 50$ identically. It can be seen that ℓ_0-atan CSKF, ℓ_0-tanh CSKF and ℓ_0-2order CSKF outperform the other methods more than 3dB both in BER and MSE performance comparison, however, ℓ_0-Gaussian CSKF fails to work due to that there is a time uncorrelated initial value $\hat{h}_0 = 0$ and $\Sigma_0 = 100$ forced to be used at each time step without any prior knowledge of the channel sparsity.

Fig.3 illustrates in detail the averaged MSE performance of evaluated schemes versus the number of pilots for $S_k = 16$, $Iter = 50$, and $SNR = 30dB$. It is shown in Fig.3 that ℓ_0-atan CSKF, ℓ_0-tanh CSKF achieve the lowest MSE along with the increments of the number of pilots, ℓ_0-2order CSKF is worse than the two schemes but better than the others. Satisfied with RIP constraint $N_p \geq S_k \log_2 (L/S_k)$, more than 3dB MSE improvements are achieved by employing the proposed schemes when $N_p \geq 50$. After that, the rate of convergence of the proposed schemes becomes slower than before.

Fig.4 shows the effects of the number of PM iterations on the estimation performance of the evaluated schemes for $S_k = 16$, $N_p = 50$ and $SNR = 30dB$. It is depicted in Fig.4 that the MSE performance of the proposed schemes improves slightly when the number of PM iterations increases, based on which, we let $Iter = 50$ with the purpose of reducing the computational complexity when considering the simulation parameters. We can see that the proposed schemes give at most 3dB better performance than the other methods.

4.2 Dynamic Updating with Partially Known Channel Prior Knowledge

In the dynamic scenario, instead of a fixed initial value at each time step, the initial value of this time step is the previous estimate. We let the channel path number be a Poisson random variable [8] with mean $E(S_k) = \lambda$. We test two cases with $\lambda = 6$ illustrated in Fig.5(a) and $\lambda = 16$ illustrated in Fig.5(b). Consistent with the former experiments, $N_p = 50$,$Iter = 50$, and SNR=30. As shown in Fig.5, compared with that in the approximate static scenario, the performance of the proposed schemes can be improved significantly by utilizing a part of prior knowledge of the channel sparsity from the previous time instant at each time step. It is interesting that contrast to the worse performance in the approximate static scenario, ℓ_0-gaussian CSKF outperforms the other three proposed methods remarkably at least 10dB MSE improvements when $\lambda = 6$ and 6dB when $\lambda = 16$. It also can be seen that the other three proposed schemes have the same performance with CSKF-ℓ_1 and CSKF-ℓ_p , even so, they are about 1dB superior to that in the approximate static scenario shown in Fig.3 when $N_p = 50$. It is notable that both in the approximate static scenario and the dynamic scenario, DS and BP deteriorate thoroughly because the sparsity of the channel varies rapidly from block to block.

5 Conclusions

We propose a ℓ_0-CSKF algorithm for DSCE in OFDM systems. The newly derived scheme is based on ℓ_0-constrained KF employing a PM observation to substitute for the ℓ_0 norm constraint. For comparison, four fictitious equations are developed in terms of four families of continuous functions which are employed to approximate to the ℓ_0 norm. Simulations have demonstrated that the proposed ℓ_0-CSKF can provide remarkable performance improvements than the other evaluated methods although it has a litter more mathematical operations. Furthermore, numerical experiments also illustrate that the proposed ℓ_0-CSKF always performs well both in the approximate static case and in the dynamic case compared with DS, BP and other two methods CSKF-ℓ_1, CSKF-ℓ_p. It also can be seen that the performance of the proposed ℓ_0-CSKF is improved significantly by utilizing a part of prior knowledge of the channel sparsity from the previous time instant at each time step.

Acknowledgments. I would like to express my sincere gratitude to Prof. E. J. Candès for his encouragement and constructive feedback. This work is supported by the National Natural Science Foundation of China (Grant No.61201263, No.61102110, and No.61303233), the Natural Science Research Programs of Hebei Educational Committee for University Young Teachers (Grant No. QN20131058).

References

1. Bajwa, W.U., Haupt, J., Sayeed, A.M., Nowak, R.: Compressed channel sensing: A new approach to estimating sparse multipath channels. Proceedings of the IEEE **98**(6), 1058–1076 (2010)
2. Candès, E.J., et al.: Compressive sampling. In: Proceedings of the International Congress of Mathematicians, Madrid, Spain, vol. 3, pp. 1433–1452 (2006)
3. Carmi, A., Gurfil, P., Kanevsky, D.: Methods for sparse signal recovery using kalman filtering with embedded pseudo-measurement norms and quasi-norms. IEEE Transactions on Signal Processing **58**(4), 2405–2409 (2010)
4. Dong, M., Tong, L., Sadler, B.M.: Optimal pilot placement for time-varying channels. In: 4th IEEE Workshop on Signal Processing Advances in Wireless Communications, SPAWC 2003, pp. 219–223. IEEE (2003)
5. Gui, G., Adachi, F.: Sparse least mean fourth algorithm for adaptive channel estimation in low signal-to-noise ratio region. International Journal of Communication Systems (2013)
6. Gui, G., Adachi, F.: Stable adaptive sparse filtering algorithms for estimating mimo channels. IET Communications **8**(7), 1032–1040 (2014)
7. Gui, G., Peng, W., Adachi, F.: Improved adaptive sparse channel estimation based on the least mean square algorithm. In: 2013 IEEE Wireless Communications and Networking Conference (WCNC), pp. 3105–3109. IEEE (2013)
8. Hu, D., Wang, X., He, L.: A new sparse channel estimation and tracking method for time-varying ofdm systems. IEEE Transactions on Vehicular Technology **62**(9), 4648–4653 (2013)
9. Hwang, T., Yang, C., Wu, G., Li, S., Li, G.Y.: Ofdm and its wireless applications: a survey. IEEE Transactions on Vehicular Technology **58**(4), 1673–1694 (2009)
10. Jagannatham, A.K., Rao, B.D.: Cramer-rao bound based mean-squared error and throughput analysis of superimposed pilots for semi-blind multiple-input multiple-output wireless channel estimation. International Journal of Communication Systems (2012)
11. Jiang, X., Kirubarajan, T., Zeng, W.J.: Robust sparse channel estimation and equalization in impulsive noise using linear programming. Signal Processing **93**(5), 1095–1105 (2013)
12. Kalman, R.E.: A new approach to linear filtering and prediction problems. Journal of basic Engineering **82**(1), 35–45 (1960)
13. Kalouptsidis, N., Mileounis, G., Babadi, B., Tarokh, V.: Adaptive algorithms for sparse system identification. Signal Processing **91**(8), 1910–1919 (2011)

14. Kamen, E.W., Su, J.K.: Introduction to optimal estimation. Springer (1999)
15. Li, W., Preisig, J.C.: Estimation of rapidly time-varying sparse channels. IEEE Journal of Oceanic Engineering **32**(4), 927–939 (2007)
16. Gupta, Sen: A., Preisig, J.: A geometric mixed norm approach to shallow water acoustic channel estimation and tracking. Physical Communication **5**(2), 119–128 (2012)
17. Simon, D.: Kalman filtering with state constraints: a survey of linear and nonlinear algorithms. IET Control Theory & Applications **4**(8), 1303–1318 (2010)
18. Taubock, G., Hlawatsch, F.: A compressed sensing technique for ofdm channel estimation in mobile environments: Exploiting channel sparsity for reducing pilots. In: IEEE International Conference on Acoustics, Speech and Signal Processing, ICASSP 2008, pp. 2885–2888. IEEE (2008)
19. Tibshirani, R.: Regression shrinkage and selection via the lasso: a retrospective. Journal of the Royal Statistical Society: Series B (Statistical Methodology) **73**(3), 273–282 (2011)
20. Vaswani, N.: Kalman filtered compressed sensing. In: 15th IEEE International Conference on Image Processing, ICIP 2008, pp. 893–896. IEEE (2008)

FOAM: Frequency-Offset Aware Multiple Client Selection for Cooperative Packet Recovery System

Ping Li$^{(\boxtimes)}$, Panlong Yang, Yubo Yan, and Lei Shi

PLA University of Science and Technology, Nanjing, China
{pingli0112,panlongyang,yanyub,shilei9018}@gmail.com

Abstract. Software radio systems are playing fundamentally important role for developing intelligent and advanced wireless network system. For cooperative network design, especially loosely coupled system, e.g., the multi-user MIMO system for clients, and the coordinations over ethernet for APs, frequency offset is one of the most important issues needs to be addressed.

In this study, we investigate the role of frequency offset in cooperative packet recover system, especially when client transmissions are overwhelming the APs. In our design, the frequency offset is compensated with the frequently and periodically transmitted preambles. In our design, the impacts of frequency offset on cooperative packet recovery are studied. Further, we make evaluations on the cooperation performance across clients with different frequency offset. We find that, there are significant differences of cooperative packet recovery among clients. We show the performance of PDR and SINR in various carrier frequency offset levels. Our conclusion is, when multiple clients' transmissions are involved for cooperative packet recovery, the frequency offset should be fully considered to enhance the network performance. Such feature could not be seen in regular transmission, where cooperative packet recovery is not applied.

Keywords: Software radio · Frequency offset · Cooperative packet recovery

1 Introduction

Recent years have witnessed the emergence of software radio network design and implementation. Especially for cooperative wireless network design, such as SYMPHONY[4], CRMA[5], FSA[6], JMB [7], BigStation [8], and MOZART [9], the cooperations heavily relying on the performance of software radio platform, such as Ettus USRP [1], WARP [2], and SORA [3].

Instead of improving the capacity and availability of a single AP (access point), these systems enable the cooperation among distributed stations to enhance the capability of cooperative wireless transmissions[10] [11]. Unfortunately, these cooperations need relatively strong coordinations, which will incur

© Springer International Publishing Switzerland 2015
Y. Wang et al. (Eds.): BigCom 2015, LNCS 9196, pp. 43–52, 2015.
DOI: 10.1007/978-3-319-22047-5_4

extremely large amount of network overhead, and inevitably offset the benefits gained through cooperative transmission. Moreover, and more importantly, rigid clock synchronization becomes the fundamental issue need to be investigated for efficient cooperations. According to the evaluations and design disciplines in previous studies [17][7][16][9], most of the overheads are used for accurate synchronization and frequency offset compensation.

In this paper, we investigate the role of frequency offset in cooperative packet recover system, especially when client transmissions are overwhelming the APs. In this case, simple cooperation is not enough, and conventional synchronization scheme is not applicable anymore, because there are no reference symbols for synchronization during packet reception.

In our preliminary experimental study, we propose a cooperative recovery scheme, which is built upon the accurate channel coefficient, and interested signal augmentation among interferences. The basic idea is, we leverage the channel coefficient values to enhance the received signal, where interested signals are multiplied by its invert form of the channel coefficient value, which means that, the signals are restored to the its original form. Meanwhile, the interfering signals, as opposed to the interested signals, are projected to relatively random directions. Moreover, leveraging many APs, these interested signal augmentation is beneficial. Thus, multiple copies could enhance the interested signal and suppress the others.

We make evaluations on the cooperation performance across APs with different frequency offset. We find that, there are significant differences of cooperative packet recovery among clients. We show the performance of PDR and SINR in various carrier frequency offset levels. The contribution of our work is two-folds: First, we propose to use signal lever cooperation among signals to enhance the signal reception for enterprise network. Second, we are the first to explore the frequency offset effects of cooperative packet recover scheme, especially when the APs are overwhelmed by interfering signals.

Our conclusion is, when multiple clients' transmissions are involved for cooperative packet recovery, the frequency offset should be fully considered to enhance the network performance.

The remainder of the paper is organized as follows: In Sec. 2, we introduce the basic rationale of inter-carrier interference and the basic knowledge about constellation diagram. Sec. 3 presents our design discipline for cooperative packet recovery, which is followed by extensive evaluation results in Sec. 4. Finally, we conclude our paper in Sec. 5.

2   Preliminary

2.1   Inter-Carrier Interference

Inter-carrier Interference (ICI) is a phenomenon well known to deteriorate performance of communication systems, especially in Orthogonal Frequency Division Multiplexing (OFDM) transmissions. ICI has a negative impact on the data throughput, as it breaks the orthogonality of subcarriers and contributes additional noise to the received signal, thus effectively decreases the Signal-To-Inter-carrier-Interference-Plus-Noise Ratio (SINR) [13].

Both the carrier frequency offset (CFO) and the sampling frequency offset (SFO) can introduce ICI. All the roles which SFO, CFO and other factors such as transmitter movement play in the ICI, have been well investigated. Both theoretical and experimental results show that the power of ICI caused by SFO is usually can be ignored since it is several orders of magnitude smaller than the power of the received signal [12]. CFO, which caused by the difference on Local Oscillator (LO) between transmitter and receiver, on the other hand, can introduce ICI that greatly distorts the received signal, and could thus introduce a significant bit error ratio (BER) degenerate [14].

2.2   Equalization

The purpose of equalization is to eliminate inter-symbol interference so the receivers can recover the transmitted symbols accurately. Consider R_k and T_k is the received and transmitted symbols of subcarrier k (k=1,2,...K) in frequency domain, where K is the number of subcarriers in OFDM modulation. Let H_k is the channel impulse response of subcarrier k. Then, we have the received signal model:

$$R_k = H_k T_k + N_k \tag{1}$$

where N_k is the additive noise at receiver. The channel impulse response estimation $\hat{H}_k$ is:

$$\hat{H}_k = R_k/T_k \tag{2}$$

The equalization process is to find the coefficient C_k, which can meet the condition:

$$\min C_k R_k - T_k \tag{3}$$

To use the minimum mean square error (MMSE) solution to this problem, we have:

$$C_k = T_k/R_k = 1/\hat{H}_k = \hat{H}_k^{-1} \tag{4}$$

where $(\cdot)^{-1}$ and $(\cdot)^*$ is the inverse and conjugate operation, respectively. Then, the recovered transmitted signal $\hat{T}_k$ is:

$$\hat{T}_k = \hat{H}_k^{-1} R_k = \frac{1}{|\hat{H}_k|^2} \hat{H}_k^* R_k \tag{5}$$

2.3   Constellation Basic

Constellation diagram is a good representation of a signal modulated by a digital modulation scheme such as quadrature amplitude modulation or phase-shift keying. It displays the signal as a two-dimensional scatter diagram in the complex plane at symbol sampling instants. Measured constellation diagrams can be used to recognize the type of interference and distortion in a signal.

As the symbols are represented as complex numbers, they can be visualized as points on the complex plane. The real and imaginary axes are often called 'I-axis' and 'Q-axis' for In-phase and quadrature, respectively. For example, a BPSK modulation could map the signals to the 'I-Q' space with tuples, *i.e.*, (-1,0) and (1,0). Plotting several symbols in a scatter diagram produces the constellation diagram. The points on the constellation diagram are called constellation points.

As the signals may have been corrupted by the channel or the receiver (*e.g.*, additive white Gaussian noise, distortion, phase noise or interference), the receiver performs maximum likelihood detection to decide what was actually transmitted. It selects that point on the constellation diagram, which is closest (in euclidian distance) to the predefined constellation area. If the interference is large enough to move it closer to another constellation area than the one transmitted, it will demodulate incorrectly.

For the purpose of analyzing received signal quality, some types of corruption are very evident in the constellation diagram. For example: Gaussian noise shows as fuzzy constellation points, while Non-coherent single frequency interference shows as circular constellation points. Especially in our concern, the phase noise shows rotationally spreading constellation points, and the attenuation causes the corner points to move towards the center.

3 System Design

3.1 Synchronization Scheme

The customized preamble design aims at dealing with the frequency offset between clients and APs. In our design, We periodically add the training symbols during data transmission, which could effectively ensure the synchronization process could work frequently. In the client side design, we consider typical and practical system, where transmissions are grouped into frames. Before data transmission begins, the preambles of each client should be transmitted sequentially without interference. No collision is allowed during preamble transmission. For this design, the AP station needs to coordinate the clients for sequential preamble transmission. The sequence of preamble transmission could be predefined or being sorted in a random way [8]. Note that, such design also need no

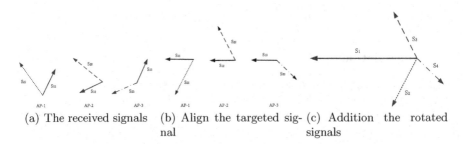

(a) The received signals   (b) Align the targeted signal   (c) Addition the rotated signals

Fig. 1. Illustration of cooperative packet recovery

scheduling. Because the sequential transmission could be achieved through the priority among clients, or the MAC address, ID, *etc.*.

3.2 Cooperative Packet Recovery

In our cooperative decoding with signal restoration scheme, we intentionally rotate the received mixed signals to ensure the targeted signal is aligned. Thus simple addition of these rotated signals from different APs will ensure the consistency of signals. The alignment could be achieved by multiplying the received signals with channel coefficient. As shown in Fig. 1(a), there are two vectors that indicate the two received signals at each AP. Assume that the vectors s_{1i} which labeled in solid line are the interested signals. Cause of the spatial diversity, all these signals have a different direction. Then we rotate the two signals to align the interested signal for each AP, as shown in Fig. 1(b). After all, the rotated signals are added together. As shown in Fig. 1(c), the interested signal is enhanced, while others are not. As the channel coefficients of clients are independent, the signals except interested signal are projected to random directions. This phenomenon could improves the SINR of interested signal, since the interested signals could be improved, while the non-interested signals are counteracting among themselves. Comparing to conventional schemes, ours could intentionally improve the SINR of interested signal in the received interfered signals by the cooperation among APs.

4 Experimental Study

4.1 Experiment Setups

To study the impact of carrier frequency offset on our cooperative packet recovery scheme, we use the GNURadio/USRP N210 software radio platforms with SBX daughter board, since such experiments need total control of wireless physical layer. To study the impact of different CFOs, we use an external reference clock to synchronize USRP N210. Then, various CFOs are set to different clients by adjusting the center frequency. In our experiment, three USRP N210 devices are involved, which are marked with label D_1, D_2, and D_3. Device D_1 and D_2 are serving as two clients, and D_3 act as an AP. To form an Enterprise WLAN with large number of APs, we do our experiments in a large 66 feet × 49 feet office as shown in Fig. 2, the office has multiple tables and chairs. We place two clients in fixed locations which labeled as a regular and an inverted triangle respectively in the figure. 16 locations for D_3 are randomly selected to guarantee the spatial diversity among APs as shown by the solid points in the figure. We let these clients transmit signal repeatedly while the AP receives these signals at different locations, for each location, we execute the experiment with 10 runs.

We set the central frequency at 2.55 GHz, and the bandwidth at 20 MHz. We implement the physical layer of IEEE 802.11g according to the IEEE standard 802.11 [15]. In our experiment, we use BPSK modulation combined with 1/2

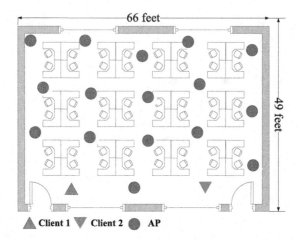

Fig. 2. Office Testbed: The solid points are APs location, regular and inverted triangles are the two clients

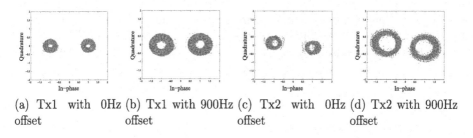

(a) Tx1 with 0Hz offset (b) Tx1 with 900Hz offset (c) Tx2 with 0Hz offset (d) Tx2 with 900Hz offset

Fig. 3. Constellation when CFO offset is within 1KHz

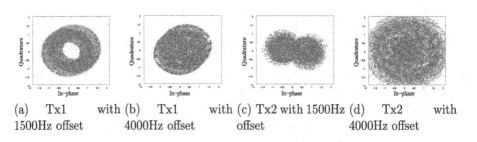

(a) Tx1 with 1500Hz offset (b) Tx1 with 4000Hz offset (c) Tx2 with 1500Hz offset (d) Tx2 with 4000Hz offset

Fig. 4. Constellation when CFO is greater than 1KHz

convolution encoding. To investigate the impact of carrier frequency offset on the cooperative packet recovery scheme under the typical traffics, we set the length of packet to 1500 bytes, and set the interval to 1 ms. For each run, 100 packets are sent in the network for measurable data collection. Among these experiments, we intentionally set one of the clients with carrier frequency offset at various levels, and the other one as the control group without carrier frequency offset. This method is equivalent to measuring the performance impacts among different levels of carrier frequency offset.

4.2 Experiment Details and Results

The experiments were carried out according to the following steps. First, we set D_1's central frequency to 2.55GHz, and set D_2's central frequency with various

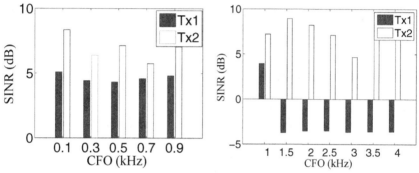

(a) SINR evaluations when tx1 with small CFO

(b) SINR evaluations when tx1 with large CFO

Fig. 5. Evaluations of SINR with different CFO

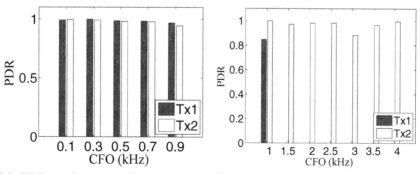

(a) PDR evaluations when tx1 with small CFO

(b) PDR evaluations when tx1 with large CFO

Fig. 6. Evaluations of PDR with different CFO

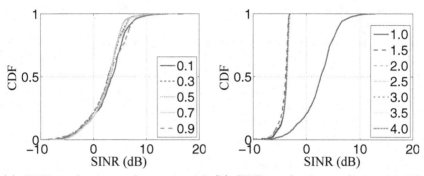

(a) SINR evaluations when tx1 with small CFO

(b) SINR evaluations when tx1 with large CFO

Fig. 7. CDF of SINR with different CFO

offset. Then we let these two devices transmit signals simultaneously. Thus, the received signal in D_3 is a mixed signal of D_1 and D_2. Then, we exchange the carrier frequency setting in D_1 and D_2, and repeat the above experiments again. When the mixed signal was received, we use our proposed scheme to recover the transmitted signals. After all, we measure the SINR of these received signals, mapping the constellation diagrams and compute the packet delivery ratio (PDR) of both D_1 and D_2.

The corresponding results with low level frequency offsets are shown in Fig. 5(a), Fig. 6(a), Fig. 7(a) and Fig. 3, respectively. When the CFO's level become higher, the constellation points in the constellation diagram become more spread and more fuzzy. This trend shows that the orthogonality between the two subcarriers has being destroyed. Even through the orthogonality between subcarriers is broken, the offset signals in the constellation diagram of both D_1 and D_2 are still separate enough, and there is no significant change on the system performance. As it shown in Fig. 5(a), the SINR of both signals are above 5 dB, and the PDR of both signals are almost 100% according to Fig. 6(a). This is because even the orthogonality between subcarriers are destroyed, the interference caused by CFO still under the level that can fully destroy the orthogonality. As a result, the AP still can decodes these interfered signals. But what will happen if we set a higher level of CFO?

In order to get a deep sense of what would happen of our system in a higher level of CFO, we do some extended experiment studies. In these extended experiments, we set the carrier frequency offset in a higher level. The CFOs are set in the range from 1 kHz to 4 kHz with interval of 0.5 kHz, then we repeat the above experiments. The corresponding results are shown in Fig. 5(b), Fig. 6(b), Fig. 7(b) and Fig. 4. As Fig. 4 illustrates, when the carrier frequency offset is above 1.5 kHz, the constellation points in the constellation diagrams of both D_1 and D_2 are rotational spread too terribly to distinguish them at all, when the CFO achieved 4.0 kHz, there is only some fuzzy blobs in the figure. This phenomenon shows that the high level of CFO destroy the orthogonality totally that beyond the level which our system can tolerate. No doubtful, the corresponding SINR and PDR performance are also degrade dramatically. As the Fig. 5(b) and Fig. 6(b) illustrated, when the frequency offset is above 1.5 kHz, the SINR of received signals are decreasing dramatically, which are about 12 dB, and the PDR are also dropping to zero. Hence, when the frequency offset is greater than 1 kHz, there should be a frequency offset aware scheme for client selection, because client with large carrier frequency offset will deteriorate the network performance significantly.

5   Conclusion

In this work, we evaluate the impact of frequency offset, when cooperative packet recovery is concerned.

We find that, when the CFO is below 1 kHz, although the received signal are distorted, there are still at least 85 % PDR and 4 dB SINR, but when the CFO is above 1.5 kHz, the orthogonality between subcarriers are totally destroyed, thus the PDR is drop to 0 % and the SINR is below 0 dB.

The highly coupled signals could be treated in a cooperative way, where many copies across different APs are processed for aligning the interested signals. Such improvement could be additive, but effective. In our experimental study, we find the important factor of the frequency offset, and propose to notice the frequency offset among clients, since it will affect the decoding performance significantly.

References

1. http://home.ettus.com/
2. http://warp.rice.edu/trac/wiki/about
3. http://research.microsoft.com/en-us/projects/sora/
4. Bansal, T., Chen, B., Sinha, P., Srinivasan, K.: Symphony: cooperative packet recovery over the wired backbone in enterprise WLANS. In: Proceedings of the Annual International Conference on Mobile Computing and Networking MobiCom 2013 (September 2013)
5. Li, T., Han, M.K., Bhartia, A., Qiu, L., Rozner, E., Zhang, Y., Zarikoff, B.: CRMA: Collision-resistant Multiple Access. In: Proceedings of The Annual International Conference on Mobile Computing and Networking MobiCom 2011 (September 2011)
6. Yun, S., Kim, D., Qiu, L.: Fine-grained spectrum adaptation in wifi networks. In: Proceedings of the Annual International Conference on Mobile Computing and Networking MobiCom 2013 (September 2013)
7. Rahul, H.S., Kumar, S., Katabi, D.: JMB: scaling wireless capacity with user demands. In: Proceedings of Special Interest Group on Data Communication (ACM SigComm 2012) (August 2012)
8. Yang, Q., Li, X., Yao, H., Fang, J., Tan, K., Hu, W., Zhang, J., Zhang, Y.: Bigstation: enabling scalable real-time signal processing in large MU-MIMO systems. In: Proceedings of Special Interest Group on Data Communication (ACM SigComm 2013) (August 2013)

9. Bansal, T., Chen, B., Sinha, P., Srinivasan, K.: Mozart: orchestrating collisions in wireless networks. In: Proceedings of the IEEE International Conference on Computer Communications (INFOCOM 2013) (April 2013)

10. Miu, A., Balakrishnan, H., Koksal, C.E.: Improving loss resilience with multi-radio diversity in wireless networks. In: Proceedings of the Annual International Conference on Mobile Computing and Networking MobiCom 2005 (September 2005)

11. Gowda, M., Sen, S., Choudhury, R., Lee, S.-J.: Cooperative packet recovery in enterprise WLANs. In: Proceedings of the IEEE International Conference on Computer Communications, INFOCOM 2013 (April 2013)

12. Garcia, M., Oberli, C.: Intercarrier Interference in OFDM - A General Model for Transmissions in Mobile Environments with Imperfect Synchronization. J. EURASIP (2009)

13. Stamoulis, A., Diggavi, S.N., Al-Dhahir, N.: Inter-Carrier interference in MIMO OFDM. In: Proceedings of the IEEE International Conference on Communications, ICC 2002 (June 2002)

14. Gjengset, J., Xiong, J., Mcphillips, G., Jamieson, K.: Phaser: enabling phased array signal processing on commodity WiFi access points. In: Proceedings of The Annual International Conference on Mobile Computing and Networking MobiCom 2014 (September 2014)

15. IEEE Standard, Wireless Medium Access Control (MAC) and Physical Layer (PHY) Specifications for Further Higher Data Rate Extension in the 2.4 GHz Band (2006)

16. Murty, R., Padhye, J., Chandra, R., Wolman, A., Zill, B.: Designing high performance enterprise WiFi networks. In: Proceedings on USENIX Symposium on Networked Systems Design and Implementation, NSDI 2008 (January 2008)

17. Sundaresan, K., Arslan, M.Y., Singh, S., Rangarajan, S., Krishnamurthy, S.V.: Fluidnet: A flexible cloud-based radio access network for small cells. In: Proceedings of the Annual International Conference on Mobile Computing and Networking MobiCom 2013 (September 2013)

Database and Big Data

Research on Light-Weight Compression Schemes Based on Simulative Column-Store

Meng Huang[1]([⊠]), Xiaofeng Qiu[1], Shufang Li[1], and Daowei Liu[2]

[1] Beijing University of Posts and Telecommunications, Beijing 100876, China
{hmtttank,qiuxiaofeng,lisf}@bupt.edu.cn
[2] China Electric Power Research Institute, Beijing 100192, China
liudaowei@epri.sgcc.com.cn

Abstract. Column-stores have achieved significant improvement over row-oriented databases and various light-weight compression schemes make this architecture more efficient. However these performance benefits cannot be utilized directly by most relational DBMSs though they still occupy the largest share of database market. In this paper we explore the potential of row-stores by simulating column-store within a traditional engine and additionally introduce light-weight compression algorithms that are commonly used in column-stores. Besides we propose a novel light-weight algorithm in the context of simulative column-store. Experimental analysis demonstrates that these optimization techniques bring encouraging performance gains in certain OLAP query scenarios.

Keywords: Column-stores · Simulative column-store · Light-weight compression · Dictionary encoding

1 Introduction

In recent years a series of database applications have been produced to cope with the challenge of big data processing. Though traditional row-oriented database systems (*row-store* for short) are widely used in OLTP tasks that are characterized by update-intensive workloads, they become clumsy when it comes to read-intensive analytical processing workloads such as those tasks in OLAP and data-warehouses. To address this performance trouble, novel methods have been proposed in database industry such as column-oriented database (*column-store* for short)[3][11].

Researches on column-oriented database implements such as Sybase IQ [15], MonetDB [6] and C-Store [16] have achieved significant performance. Whereas, traditional row-store database (e.g., Oracle, IBM DB2 and SQL Server) still plays an important role in a large number of companies and organizations, most of whom cannot share the beneficial features of column-stores. And deploying two kinds of databases at the same time seems too expensive to the majority of these companies [17]. Therefore whether these performance gains in column-stores can be accessed by row-store databases using a column-oriented design is a practical issue.

© Springer International Publishing Switzerland 2015
Y. Wang et al. (Eds.): BigCom 2015, LNCS 9196, pp. 55–68, 2015.
DOI: 10.1007/978-3-319-22047-5_5

Simulating column-oriented database within row-store system provides an approach to resolving this issue. The most straightforward method is vertically partitioning relations, which is first proposed in 1987 [12]. Since then it developed several techniques for column-store simulation designs such as index-only plan and materialized view. Unfortunately, these explorations do not bring much improvement [2]. In fact, the results of previous researches indicate that column-store simulation even lead to worse performance than conventional row-stores: among these approaches materialized view performs best. It brings approximate 6.2x speedups on average compared with traditional row-stores. But vertical partitioning and index-only plan lower average query execution speed by 3.1x and 8.6x respectively. On the other hand, C-Store is on average 164x faster than a commercial row-store database [16].

There are two important optimizations in column-store that cause its efficiency: wise compression schemes and later materialization [4]. As the most essential factor, compression schemes can reduce the amount of disk space. Therefore they shorten the seek distance and seek time as well as bring improvement to I/O performance. Particularly, compression schemes implemented in column-stores often focus on light-weight schemes [9] (e.g., run-length encoding) that directly operate on compressed data rather than decompressing all data before operating queries as heavy-weight compression schemes do (e.g., Huffman encoding, Lempel-Ziv encoding). Besides, light-weight schemes are more inclined to lower CPU overhead so that they lead to encouraging performance advantages. And novel light-weight algorithms have been proposed within column-stores. These optimizations leave an open question: *Will the light-weight compression schemes help facilitate a faster simulative column-store system?*

Actually, these sequence-oriented compression schemes often do not work well in row-stores mainly because an attribute within row-stores is stored as part of an entire tuple, so combining one attribute from each tuple together into one value requires special techniques to reconstruct tuples [1], meanwhile logical/physical design and the query executor of DBMS need to be redesigned. Consequently the authors of [4] believed that a successful column-oriented simulation required important system improvements.

In this paper we demonstrate the potential of simulative column-store design with flexible light-weight compression schemes in a row-store system to explore the above open question. Experimental results prove that simulative column-store with compression schemes will significantly improve the performance of row-stores for read-intensive analytical workloads. The contributions of this paper are:

1. This paper innovatively introduces compression algorithms into simulative column-store system and compares their performance of enhancing row-store databases.
2. This paper proposes a novel light-weight compression algorithm and implements it within simulative column-store. This algorithm achieves more outstanding capability on execution time and disk space required.

The rest of this paper is organized as follows: Section 2 introduces prior efforts on simulative column-store and light-weight compression schemes. Details of simulative column-store will be given in Section 3. In Section 4 we outline several light-weight schemes and describe how to introduce them into simulative column-store. In Section 5 a novel light-weight compression algorithm is proposed, as well the multi-column process will be discussed. An experimental evaluation of the performance of the methods is given in Section 6. And Section 7 concludes this paper.

2   Related Work

In this section we briefly present some important prior work on simulative column-store and column-oriented light-weight compression schemes.

In [1] a series of light-weight compression algorithms were overviewed and introduced into the column-oriented system C-Store. The work demonstrated the performance of operating directly on compressed data in column-store using light-weight schemes was much outstanding than querying from row-oriented system (an average 10.3x speedup). But since all schemes were implemented within C-Store, simulative column-store could not directly learn these methods.

In [5] innovative compression algorithms based on dictionary encoding were proposed to improve the performance of column-stores (speedup 7.0x at most). And the column-store system PowerDrill [10] of Google used a double dictionary encoding to compress its web data, which gave a performance boost of 10-100x compared to traditional column-store databases.

The work in [4] first systematically researched simulating column-oriented physical designs in a commercial row-store DBMS. This work summarized three column-oriented simulation designs: vertical partitioning, index-only plan and materialized views. However these techniques did not result in much performance improvement than row-stores. And this work concluded that flexible compression was one of most significant elements that make difference to column-store.

A column-oriented query processing mechanism on the basis of index-only plans was put forward to improve the speed of row-stores in [8]. This method avoided the cost of reading the entire table rows when only several columns were referenced in the query. And experiments suggested these operators improve the performance of row-store DBMSs by 5.8x, which was a promising result on simulative column-store research.

In [17], the authors proposed optimized simulative models based on clustering relative columns. This model analyzed the relevance of each column by clustering the query workloads, then combining the columns with high degree of relevance, which worked well in OLAP applications.

The work in [7] explored simulative column-store by introducing compression algorithm, a specialized logical design called C-Table was implemented upon row-oriented database to extend the vertical partitioning. And experimental evidence showed that the performance with C-Tables is just on average 2.7x slower than the lower-bound of column-store. That is, compression schemes indeed reveal the possibility of speeding up row-stores databases.

3   Simulative Column Store

In this section, we introduce our method of simulating column-store as well as some important definitions for simulative column-store.

Vertical partitioning is the most fundamental way of simulating column-store, which means fully vertically partitioning every physical table into a set of column-tables. In other words, this approach generates one table for each column. To maintain the order (or *position*) of data, a column that records the position information is added into the column-tables (having the same order for all columns is essential thus the original tuples can be correctly reconstructed by operating join to related column-tables). Therefore, the scheme of the ith column is physically stored as: *Column_i(position_id, value)*. Moreover, creating clustered indices for *position_id* column of every column-table so that rows are reconstructed by index joins can improve the performance for this method.

In our implementation of simulative column-store, only vertical partitioning is taken into consideration for the reason that other methods such as materialized views need particular function supports of DBMS, so they are not common in different row-store database systems.

Before vertically partitioning original tables, some kinds of row-reordering operations should be performed firstly so that light-weight schemes can be brought into column-tables. For a standard relation such as *EMP(name, age, salary, dept)*, we indicate one of the projections as *EMP1(name, age)*. And we present the sorted projection on certain sort key(s) by appending the sort key to the projection (e.g. *EMP1(name, age | age)*). Any column(s) of the projection can be regarded as sort key(s). After reordering the tables, we can partition them into column-tables. Note that we use simple dictionary order to sort tables which is the most intuitive reordering choice. Although there are other more effective reordering rules (e.g. see [14][13]) to get better-sorted table, we will not consider them right now.

We also define *sorted-runs(X)* to reflect the level of reordering and repetition in a column-table (X is a variable). For example, to a schema *EMP(age, salary, dept | age, salary)*, the projection is primarily sorted by attribute *age* and secondly sorted by *salary*. If this relation has 90 million tuples (rows), the cardinality of column *age* is 90 (that is, the number of distinct values within a column is 90) and the cardinality of column *salary* is 1000 then column *dept* will have an average sorted-runs of size 90000000/ (90*1000) = 1000 (donated as sorted-runs (1000)). It means if the cardinality of column *dept* is 50 then each value will repeatedly appears 20 times (calculated by 1000/50) within every sorted-run (here we suppose the possibility of every distinct value in one column is same). Hence different size of sorted-runs and cardinality can result in different compression ratio. And this estimate will be discussed in Section 5.

4   Light Weight Schemes

In this section, we describe the light-weight compression algorithms included by simulative column-store. The emphasis of this section is how to design every scheme in simulative column-store. The implementations can differ from their conventional fulfilments, which is mostly determined by the architecture of simulative column-store introduced in Section 3.

4.1 Run-Length Encoding

Run-length encoding (RLE) compresses data by storing runs (that is, data of same value that occurs continuously in a sequence) as a more compact form. These runs can be represented as *(value, run_length)*, where *run_length* means the number of repeated values. For example, consider a sequence of values: {A, A, A, B, B, B, B, B, B, C, C, C, C}, we encode the sequence as {(A, 3), (B, 6), (C, 4)}.

To implement run-length encoding in simulative column-store, position information should be reserved in compressed sequence to reconstruct tuples. Thus the above sequence can be encoded as: (1, A, 3), (4, B, 6), (10, C, 4) in which each triple stands for *(start_id, value, run_length)*. With this data structure we compress an original table as follow: consider a sorted column-table *Column_i(position_id, value)*, we represent it with aggregating every sequence of the same value that agrees with the sorted column-table. Then we create a table with the schema of *(start_id, value, run_length)* for each group of aggregation, where *start_id* is the minimum id in the original column-table for one element in the group and run_length is the number size of every group (see details in Fig.1).

position_id	value
1	A
2	A
3	A
4	B
5	B
6	B
7	B
8	B
9	C
10	C
11	C
12	C

(a)

value	run_length
A	3
B	5
C	4

(b)

start_id	value	run_length
1	A	3
4	B	5
9	C	4

(c)

Fig. 1. (a) Original column-table, (b) traditional RLE and (c) RLE with triple in simulative column-store

RLE can be applied to improve the performance of row-stores by compressing large string attributes. While in column-stores it appears more powerful because attributes that stored continuously result in more redundancy, and the length of runs can be long especially after data sorting (column-stores often lead to a high ratio of columns being sorted or secondarily sorted). However, when one column-table is not sorted well, the performance of RLE will worsen.

4.2 Dictionary Encoding

In column-store there are two dictionary schemes employed, Lempel-Ziv (a heavy-weight scheme) and simple dictionary scheme. The core of dictionary encoding is to replace large and frequent patterns with smaller ones. Some examples can be found in [1][9].

The dictionary encoding we apply in simulative column-store creates a compressed data table to store encoded data and a dictionary table to store every distinct value and the corresponding fixed-length of token (similar to the hashmap in traditional dictionary encoding in Fig.2(b)) for each column-table. It will check the dictionary table for each value to retrieve its corresponding token. If one value does not exist in that table, the token counter will increase by one, as well the value and its token will be added into the dictionary table. And the corresponding token of every data value will be written into the compressed-data table.

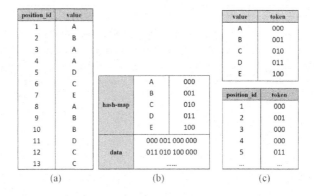

Fig. 2. (a) Column-table, (b) traditional dictionary encoding and (c) dictionary encoding in simulative-column-store)

To encode the tokens, cardinality of a column-table should be calculated first. And then the bit length of token will be known. If a column-table has X distinct values, $\lceil \log_2 X \rceil$ bits are needed for every token (e.g. {A, B, C, D, E} is encoded as {000, 001, 010, 011, 100}). An example for dictionary encoding we used inside simulative column-store is shown in Fig.2(c). Note that traditional simple dictionary encoding fit a certain number of encoded tokens to bytes. In the last instance, 2 values will be fit in 1 byte, 5 values in 2 bytes and so on. This consideration makes sense to make the dictionary suitable to the machine caches (1 or 2 bytes are used specifically in some architectures like C-store), which can raise the efficiency of reading data from disk immediately and obtain even modest CPU savings.

4.3   Bit-Vector Encoding

Bit-vector encoding associates each distinct value of one column with a sequence of bit-string, where the bit-value in corresponding position will be set to one if the original value there equals the value this sequence stands for or set to zero otherwise. Lower cardinality is suggested because every distinct value needs one bit-string to represent with. For example, a sequence of values: {A, A, A, B, C, C, B, A, B, B, A, C, C} will be encoded as:

(A): 1110000100100
(B): 0001001011000
(C): 0000110000011

Bit-vector encoding in simulative column-store is implemented as following: we consider a sorted column-store and take the size of average sorted-runs as one compression unit to avoid the bit-string too long. That is to say when we have a column-table of average sorted-runs (100), the length of every bit-string will be controlled to 100. Thus for $Column\_i(position\_id, value)$ we encode it and create a new table with the schema of $(value, bit\_string)$, where bit_string stores the sequence of bits described above.

5   RLE-Dictionary Encoding

This section describes the details of the novel light-weight compression optimization algorithm based on dictionary encoding. Then we discuss how to perform queries involving more than one column which is the most common within actual OLAP workloads.

5.1   Description of Algorithm

Above three light-weight algorithms discussed in Section 4 have been widely accepted in the field of database compression and especially deployed into the architecture of C-store. Nevertheless, when executing dictionary encoding into column-tables, it is noticed that there exists redundancy in the compressed data. For example in Fig.2, repeated data continuously appears in the data table all the same, which inspires more suitable light-weight schemes to be designed.

Thus a novel encoding algorithm is proposed in the context of simulative column-store by integrating run-length encoding into dictionary encoding, called RLE-dictionary encoding. Suppose that column-tables have been generated according to the method described in Section 3 therefore a set of tables $Column\_i(position\_id, value)$ are obtained. Firstly, each distinct value of one column-table is coded with a hash-map token of fixed-size and inserted into the dictionary table with the schema of $(value, token)$ (see Section 4.3 for more details). Then all original values in $Column\_i$ will be substituted by the hash-map tokens and written into the compressed-data table. Then we take the compressed-data table as the input of run-length encoding and compress the repeated values with the triple of $(start\_position, token, run\_length)$. We obtain a dictionary table

---

Algorithm 1. RLE-dictionary encoding

Input: Relation $C(position\_id, value)$ with n tuples where $C.value_i$ is the *value* when
$position\_id=i$, $1 \leq i \leq n$

Output: RLE-compressed relation R and dictionary relation D

1: Initialize $R(start\_id, token, run\_length)$, $D(value, token)$ and let $checkValue = null$
2: Calculate $sizeToken = \lceil \log_2(cardinality) \rceil$, where $cardinality = getCar(C.value)$
3: **for** $i = 1 \rightarrow n$ **do**
4:      **if** $chekValue \neq C.value_i$ **then**
5:          Check for the existence of $C.value_i$ in the list of $D.value$
6:          **if** no existence of $value_i$ **then**
7:              Calculate $token_i = calToken(C.value_i, sizeToken)$
8:              $D.write(C.value_i, token_i)$
9:              $R.write(i, token_i, 1)$
10:              $checkValue = C.value_i$
11:          **else**
12:              $R.write(i, D.token_{D.value=value_i}, 1), checkValue = C.value_i$
13:          **end if**
14:      **else**
15:          $R.run_length_m \leftarrow R.run_length_{m+1}$, where m is the latest inserted tuple
16:      **end if**
17: **end for**
18: **return** R,D

---

and a RLE-compressed table right now and finish the RLE-dictionary encoding.
The detail of this algorithm is shown in Algorithm 1.

Compared to dictionary encoding, this method eliminate the redundancy due
to data repetition and more importantly it borrows the benefits of RLE, which
has aggregated some helpful preparatory information. This avoids repeated calculation when performing relevant queries and results in speedups. On the other
hand, compared to run-length encoding, this scheme acquires the property of
low-overhead in dictionary encoding and further reduce size of whole data as well
as CPU costs. Consequently, RLE-dictionary encoding shares the advantages of
both run-length encoding and dictionary encoding, meanwhile operating directly
on compressed data makes this scheme free of the costs of decompression, which
is indeed one of the most significant features of light-weight schemes.

5.2   Multi-column Query Processing

Section 4 and Section 5.1 presents the compression methods aiming at single column-table. However, practical OLAP or data-warehousing applications
rarely involve single column but several ones. RLE-dictionary encoding covers
the advantage of run-length encoding and it contains the position information
within the compressed data. This feature facilitates multi-column queries in simulative column-store.

Consider the original TPC-H fact table *lineitem*, we vertically partition this table according to the schema *S1: (L_orderkey, L_suppkey, L_shipdate — L_shipdate, L_suppkey)*. So far we generate three column-tables that each one corresponds an attribute in the projection of schema S1. We encode these column-tables with RLE-dictionary encoding and obtain eight RLE-compressed tables (called *S1_L_orderkey, S1_L_suppkey*, etc.) with the schema of (start_id, token, run_length) and their dictionary tables (called *S1_dict_L_orderkey, S1_dict_L_suppkey*, etc.) with the schema of *(value, token)*. The main route of multi-column query is join-ing different RLE-compressed tables with -join and joining each RLE-compressed table with the corresponding dictionary table with equi-join. To the **Example Query** below, the execution process can be redesigned as **Redesigned Query Execution Process** presents. We use inner join to complete the equi-join (nested loops for joining by same values of token) and -join process (utilizing predicates *BETWEEN* and *AND* to determine the range of start_id) firstly. Then a hash-match operator is executed to perform the aggregation and grouping. Here we use *SUM (T1.run_length)* to reduce I/O overhead and the output of this execution plan is what is required.

Example Query (count of items shipped for each supplier before day D):

```
SELECT l_suppkey, COUNT (*)
FROM lineitem
WHERE l_shipdate < D
GROUP BY l_suppkey
```

Redesigned Query Execution Process:

```
SELECT D1.value, SUM (T1.run_length)
FROM S1_dict_l_suppkey D1, S1_dict_l_shipdate D0,
S1_l_suppkey T1, S1_l_shipdate T0
WHERE D0.value < D
AND T1.token = D1.token
AND T0.token = D0.token
AND T1. start_id BETWEEN T0.start_id
AND (T0.start_id + T0.run_length - 1)
GROUP BY D1.value
```

Optimized Query Execution Process:

```
SELECT D1.value, SUM (T1.run_length)
FROM (SELECT MIN (T0.start_id) AS tMIN,
MAX (T0.start_id + T0.run_length - 1) AS tMAX)
FROM S1_l_shipdate T0, S1_dict_l_shipdate D0
WHERE T0.token = D0.token
AND D0.value < D) T0Agg,
S1_dict_l_suppkey D1,
S1_l_suppkey T1
WHERE T1.token = D1.token
```

```
AND T1. start_id BETWEEN TOAgg.tMIN AND TOAgg.tMAX
GROUP BY D1.value
```

In terms of SQL query optimization, we can design an alternative execution plan to further enhance the performance of above strategies. Since the column-table S1_1_shipdate and its relevant tables do not appear in the query projection but only be used to constraint tuples, the faster query plan can be created as **Optimized Query Execution Process** shows. This process reduces context-switches as a result of smaller outer table for nested loop join (it suggests a smaller outer table in NL-join to drive a larger inner table to get more efficient query processing).

6   Experimental Evaluation

All experiments in this paper were conducted on a Windows Server 2008 system using CPU of 2.27GHz Intel i5 with 4GB of RAM, 1TB of hard drive and our implementation was designed inside Microsoft SQL Server 2012. Firstly we evaluate the performance of light-weight schemes discussed in Section 4 and RLE-dictionary encoding in Section 5. The purpose is to validate these schemes indeed bring improvement in the context of simulative column-store and detail their benefits to traditional row-stores. Therefore we generate column-tables so that we can vary cardinality and average size of sorted-runs to observe their influence to query performance. Then we consider the condition of multi-column queries and emphasize the potential of RLE-dictionary encoding. In this part we use TPC-H data and several query workloads provided by prior researches to measure our algorithm. Also we choose the performance of C-Store as our optimization target though we do not expect that our design can beat C-Store for the limitation of underlying DBMS.

6.1   Performance of Light-Weight Schemes

In this part a series of experiments are designed to measure the performance of the four light-weight schemes within simulative column-store introduced in prior sections. The queries we use are simple aggregation operations (AGG $(value)$) such as $AVG()$ and $SUM()$:

```
SELECT AGG(value)
FROM table
GROUP BY value
```

The data we use are generated column-tables (the schema is $(position\_id, value)$) so that we could adjust parameters such as cardinality and average size of sorted-runs. And the attribute of value has 20 million integer values of 32-bit. We store the data into tables with different light-weight encoding schemes and run the above query for each table. To explore the influence of data diversity to query performance, we vary column cardinality from 2 to 40 and average size of sorted-runs between 100 and 1000 (the definition can be seen in Section 3).

We compress each column-table with five schemes: no compression, RLE, bit-vector encoding, dictionary encoding and our RLE-dictionary encoding. Firstly we investigate the sizes of each compressed table, which reflects the compression ratio of every scheme. Fig.3 presents the trends of data sizes following the number of distinct column values (or cardinality). Fig.3(a) and Fig.3(b) respectively depict the conditions of average sorted-runs (100) and sorted-runs (1000). Among the three light-weight schemes, RLE brings the highest compression ratio when the column cardinality is low or the average size of sorted-runs is high, this is because that low-cardinality and high average size of sorted-runs both result in the occurring of large runs and RLE is suited when there are lots of appeared continuously values. Additionally, bit-vector encoding gives a similar trend plot to run-length encoding but it does not perform so well as RLE, because bit-vector encoding will not aggregate the information of repeated values in the processing of encoding. The compression ratios of both RLE and bit-vector are linear with the cardinalities of columns for we modify the original data according to a linear rule. However, the compression ratio of dictionary encoding brings a more flat tendency thus in certain conditions such as the cardinality is more than 30 and the average size of sorted-run is 100, dictionary encoding leads to the highest compression ratio of all schemes. With sorted-run (1000) these three light-weight schemes have almost entirely separately polylines of data sizes, which RLE perform better bit-vector encoding and better than dictionary encoding. Especially we observed the compression ratio of RLE-dictionary encoding, obviously it works better than all of other light-weight schemes except several conditions of high cardinality in Fig.3(a). As in most data-warehouse scenarios there are plenty of column-tables with low cardinality (for example, in the TPC-H fact table *lineitem*, attributes with no more than 40 distinct values are 23%), the potential of RLE-dictionary encoding is apparent in terms of compression ratio.

The focal point of our experiments is the query performance of light-weight schemes in simulative column-store that is measured by execution time. We run the aggregation query above in each compressed table and recorded specific durations. It is noted that we imitate the methods of Section 5.2 and operate directly on compressed data but not eagerly decompress them firstly to embody the power of light-weight schemes. We plot the detailed trends for condition of sorted-run (100) and sorted-run (1000) in Fig.3(c) and Fig.3(d). It is unsurprising that the query performance is indicated by compression ratio to a certain degree but not entirely consistent with it. In Fig.3(c) we discover that with an average size of sorted-run (100), RLE results in 3.8x performance improvement on average (compared with the execution time of no compression), bit-vector achieves 3.0x speedup, dictionary encoding is 1.8x. Despite these schemes all provide obvious benefits of light-weight schemes, RLE-dictionary encoding acquire 4.8x improvement, which is more efficient to other schemes.

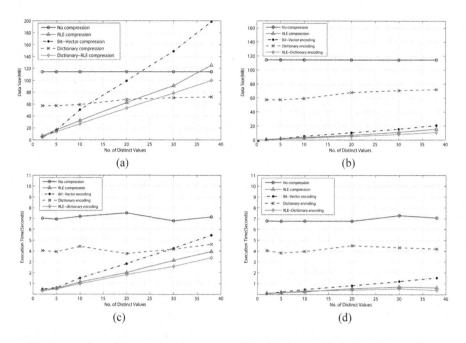

Fig. 3. Data size trends of sorted-runs (100)(a) and sorted-runs (1000)(b). Execution time trends of sorted-runs (100)(c) and sorted-runs (1000)(d).

6.2   RLE-Dictionary Encoding for TPC-H

To measure the performance of RLE-dictionary encoding for multi-column tasks, we use generated TPC-H data (similar to general real-world data) with scale factor of ten, in which the fact table *lineitem* has an approximately item number of 60,000,000. And we choose the seven queries used in [16][7] (we do not display them due to space limitation). Then we deploy simulative column-store and implement necessary schemas with RLE-dictionary encoding. As following experiments performed according to the optimized execution processes given in Section 5.2, we show the results in Table 1. In this table RS stands for the traditional row-store system, CS for the column-store system C-store and S-CS for our simulative column-store with the light-weight compression scheme of RLE-dictionary encoding.

In this workload C-store obtain an average 27.9x faster improvement to row-store system. And our simulative column-store with RLE-dictionary encoding is on average 13.1x speedups. Hence compared to C-store, simulative column-store is 2.1x slower than C-store. This is mainly because that C-store adopts over-lapping projections while row-store need to store tuple overheads. Considering the simulative column-store has rich index strategies, the results is believed a meaningful indicator to close the gap between row-store and column-store.

Table 1. Average execution time in different systems (in second)

Query	RS	CS	S-CS
1	137.56	0.67	1.44
2	107.29	4.64	4.53
3	187.17	7.84	7.21
4	237.83	9.58	18.01
5	198.51	0.97	25.24
6	287.70	15.73	32.44
7	205.64	10.26	15.11
AVG	194.52	6.95	14.84

7   Conclusions

We consider although traditional row-store systems become inflexible and slow when confronting read-intensive workloads, they cannot be completely abandoned for the robustness and extendibility. This encourages us to explore innovative simulative methods to share the advantages of column-store systems. Especially we are aware that compression schemes make column-store faster prominently, thus we focus on light-weight compression algorithms which can be introduced into the environment of simulative column-store. An RLE-dictionary encoding algorithm is proposed and implemented for further optimization in our work. We expect to narrow the gap between row-store and column-store via these simulative techniques and appropriate light-weight schemes, and it is shown that these indeed produce promising performance improvements. Hence we believe this work will help the design and development of wiser row-store DBMSs.

Acknowledgments. The work is supported by the Research Project of Power System Online Security Assessment Big Data based (No.XT71-15-056).

References

1. Abadi, D., Madden, S., Ferreira, M.: Integrating compression and execution in column-oriented database systems. In: Proceedings of the 2006 ACM SIGMOD International Conference on Management of Data, pp. 671–682. ACM (2006)
2. Abadi, D.J.: Query execution in column-oriented database systems. Ph.D. thesis, Massachusetts Institute of Technology (2008)
3. Abadi, D.J., Boncz, P.A., Harizopoulos, S.: Column-oriented database systems. Proceedings of the VLDB Endowment **2**(2), 1664–1665 (2009)

4. Abadi, D.J., Madden, S.R., Hachem, N.: Column-stores vs. row-stores: how different are they really? In: Proceedings of the 2008 ACM SIGMOD International Conference on Management of Data, pp. 967–980. ACM (2008)
5. Apte, T., Ingle, M., Goyal, A.: Dictionary symbol encoding: A column-store case study. In: 2014 World Congress on Computer Applications and Information Systems (WCCAIS), pp. 1–4. IEEE (2014)
6. Boncz, P.A., Kersten, M.L.: Monet: An impressionist sketch of an advanced database system. In: Proc. IEEE BIWIT Workshop. Citeseer (1994)
7. Bruno, N.: Teaching an old elephant new tricks. arXiv preprint (2009). 0909.1758
8. El-Helw, A., Ross, K.A., Bhattacharjee, B., Lang, C.A., Mihaila, G.A.: Column-oriented query processing for row stores. In: Proceedings of the ACM 14th International Workshop on Data Warehousing and OLAP, pp. 67–74. ACM (2011)
9. Ferreira, M.C.: Compression and query execution within column oriented databases. Ph.D. thesis, Massachusetts Institute of Technology (2005)
10. Hall, A., Bachmann, O., Büssow, R., Gănceanu, S., Nunkesser, M.: Processing a trillion cells per mouse click. Proceedings of the VLDB Endowment 5(11), 1436–1446 (2012)
11. Harizopoulos, S., Liang, V., Abadi, D.J., Madden, S.: Performance tradeoffs in read-optimized databases. In: Proceedings of the 32nd International Conference on Very Large Data Bases, pp. 487–498. VLDB Endowment (2006)
12. Jagodits, S.K.G.C.T., Valduriez, H.B.P.: A query processing strategy for the decomposed storage model. In: Proceedings, p. 636. Order from IEEE Computer Society (1987)
13. Lemire, D., Kaser, O.: Reordering columns for smaller indexes. Information Sciences 181(12), 2550–2570 (2011)
14. Lemire, D., Kaser, O., Gutarra, E.: Reordering rows for better compression: Beyond the lexicographic order. ACM Transactions on Database Systems (TODS) 37(3), 20 (2012)
15. MacNicol, R., French, B.: Sybase iq multiplex-designed for analytics. In: Proceedings of the Thirtieth International Conference on Very Large Data Bases, vol. 30, pp. 1227–1230. VLDB Endowment (2004)
16. Stonebraker, M., Abadi, D.J., Batkin, A., Chen, X., Cherniack, M., Ferreira, M., Lau, E., Lin, A., Madden, S., O'Neil, E., et al.: C-store: a column-oriented dbms. In: Proceedings of the 31st International Conference on Very Large Data Bases, pp. 553–564. VLDB Endowment (2005)
17. Yu, L., Zhang, Y., Wang, S., Zhang, Q.: Research on simulative column-storage model policy based on row-storage model. J. Comput. Res. Dev. 47, 78–885 (2010)

Study of Constructing Data Supply Chain Based on PROV

Jiewei Lan[✉], Xiyun Liu, Hong Luo, and Peng Li

Department of Computer Science, Beijing University of Posts
and Telecommunication, Beijing 100876, China
{lanvivian,spooons,luoh,lipeng1106}@bupt.edu.cn

Abstract. In the era of big data, the value of data can be better
explored during data flowing and processing. If a data supply chain from
the source to the destination is constructed across data platforms where
data flows through, then it will help users analyze and use these data
more safely and effectively. Due to the complexity and diversity of data
platforms, there is no uniform data supply chain model specification. To
solve the problem, we construct a distributed data supply chain model
based on PROV, a data provenance specification presented by W3C to
standardize information records of data activities in corresponding data
platforms. On this basis, we design Data Supply Chain Service Module
(DSCSM), so as to provide effective accessing methods for data traceability information on distributed platforms. Finally, we deploy the proposed
model to real data platforms we built to verify the effectiveness and feasibility of solution.

Keywords: PROV · Data supply chain · Distributed

1 Introduction

In the era of big data, decision-making in commercial, military or other fields
is increasingly dependent on big data analysis rather than experience and intuition, which puts forward higher requirements on the data management. To meet
needs from various applications, data stored in dispersed data platforms flows
frequently and is complicatedly processed. In order to ensure the security and
authentication of data, researches on data provenance are more extensive [1].
Cause records of data activities are stored in distributed data platforms, it is
difficult to get it effectively and safely. If we can sort out data relationships
among platforms using provenance information, it will not only bring benefits
to data analysis but also provide a safe and efficient method to access the data.

There is no uniform standard for data provenance record until now, which
is not conducive for data integration and analysis in distributed environment.
PROV [2] presented by W3C is a data provenance specfication. Because of its
strong analyticity and semantic feature, PROV is suitable to be used as data
provenance standard. Data provenance researches mostly focus on how to record

© Springer International Publishing Switzerland 2015
Y. Wang et al. (Eds.): BigCom 2015, LNCS 9196, pp. 69–78, 2015.
DOI: 10.1007/978-3-319-22047-5_6

provenance information and how to analyze it. The information are usually stored in a central server or follow the data flows. However, provenance information increases with the multiple data transfer. When circulation gets more frequent, provenance data may be more massive than the data itself. And security of centralized storage is poor. So this model does not apply to information environment in the era of big data. Thus, considering about the reliability and security, we store data provenance information in distributed platforms and construct a dynamic and efficient data supply chain to obtain information quickly and safely. It will also do good to dynamic data management and access control.

Since massive provenance information are stored in distributed data platforms, how to get these information becomes a big challenge. As provenance is analogous to object tracking information of supply chain and discovery service (DS) in EPCGlobal network is used to access tracking information, we learn from SCOR (Supply Chain Operations Reference model) and analyze discovery service. It turns out that distributed index model in DS works better for data acquisition. Therefore, we design data supply chain service model combined with distributed index model.

Based on the PROV specification, we propose a data provenance model to formally describe how the data is received, processed and provided, which standardizes records of data activities in corresponding data platforms. Further more, we construct data supply chain service model to obtain provenance information combined with distributed index model in EPCGlobal framework and new characteristics of data supply chain. It can organize scattered data stored in the heterogeneous platforms to one or more data supply chains, which will provide convenience to form data ecosystem and further explore data value.

The remainder of this paper is organized as follows. We discuss the related work in Section 2. Section 3 describes the prov data model. The proposed Data Supply Chain Service Module is detailed in Section 4. In Section 5, we give our experimental study and simulations. We conclude the paper in Section 6.

2   Related Work

Currently, researches about big data mainly aim at data mining [3], data credibility [4], privacy protection [5], access control [6] and etc. However, there are relatively fewer studies about data traceability information among distributed platforms. We intend to construct data supply chain based on data provenance. And here we present the research states of data provenance.

2.1   Data Provenance

In the era of information data mostly includes raw data and derived data. Due to the complexity and variety of derivation, data users usually have doubt on the reliability of data. In fact, many mistakes of derived data have nothing to do with the raw data. Therefore, it is necessary to record provenance information of data so that people can build data supply chain for scattered data.

The data supply chain construct model mainly includes the record and acquisition of data provenance. When it comes to the record of data provenance, how to get provenance information and which information should be recorded are taken into consideration. Early studies stored data and intermediate data together, which would cause mismanagement easily. With the expansion of data volume and penetrating of study, Buneman et al presented a provenance model [7] of Why and Where. Based on the model, Vansummeren proposed a model called "how provenance" [8]. Then Sudha et al brought a 7w model [9] which means who, when, where, how, which, what and why. 7w model records comprehensive details of data provenance but it will increase storage space.

Although researches about data provenance model are quite aplenty, it is still difficult to consolidate variety information from different platforms and build data supply chain cause lacking of uniform data provenance standard. The PROV specification presented by W3C gives a standard description of data provenance. It has strong analyticity and semantic feature cause it can be recorded in forms like XML, JSON, OWL2 and etc. What's more, PROV defines inference rule to consolidate provenance information, which makes it possible to construct data supply chain and achieve data provenance among distributed platforms. As a conclusion, we choose PROV as the standard of data supply chain model.

However, cause massive provenance information are stored on distributed data platforms, how to get these information becomes a big challenge. Since provenance is analogous to object tracking information of supply chain in EPC-Global network [10], it brings a good reference to this issue.

2.2 Supply Chain

Provenance information of data supply chain has some similarities with the object tracking information of supply chain. Discovery service technology in EPCGlobal framework is designed to achieve object tracking information dynamically. However, data relationships and data processes in data supply chain are more complex and diverse. So we should build flexible data supply chain to acquire provenance information based on supply chain research combined with characteristics of data supply chain. There are three kinds of discovery service technology in EPCGlobal framework.

Centralized Server Model. There is a central server in this model [11]. Information of supply chain will not only be put in local-storage but also sent to central server, which is suitable for logistic query. It is easy to accomplish this method but there is no guarantee on safety since central server can be overload easily.

Centralized Index Model. In Centralized index model, supply chain data is put in the local-storage and at the same time the index related will be sent to central server [12]. By using this model, storage pressure will be reduced and privacy of supply chain data can also be protected. But once the central server is broken down, information traceability will be influenced heavily.

Distributed Index Model. In this model, central server is replaced by distributed structure. Each local server is bound with a query engine. When users want to query for data of supply chain, only need to send the query to the related query engine. Then a process of distributed index model called "Process and Forward" [13] will send queries to related participants spontaneously and recursively. Distributed index model reduces storage and safety demand to central server and intercurrent pattern can also improve query efficiency. In conclusion, this model offers valuable references to data supply chain construction.

Combined with characteristics of data supply chain and distributed index model, we propose data supply chain service model to get provenance information stored in distributed system. It will bring benefits to build flexible and dynamic data supply chain so as to manage and use big data in better ways.

3 PROV Data Model

Provenance is information about entities, activities, and people involved in producing a piece of data or thing, which can be used to form assessments about its quality, reliability or trustworthiness. PROV is a standard data model defined by W3C. It allows users to standardly describe provenance that exist in different information systems by using widely available formats like XML, JSON, RDF, etc, which offers great convenience in operations. Meanwhile, the PROV specification also presents the standardized and readable description format-PROV Nation (PROV-N). The conceptions of the PROV data model are introduced in the following contents.

Entity. In PROV, an entity is a physical, digital or other kind of thing whose provenance information is to be recorded. The provenance of one entity might be related to other entities. For instance, an document entity includes a data table and the data is derived from a database. Therefore, the document is in connection with the table and the database.

Activity. Activities are how entities come into existence and how their attributes change to become new entities, often making use of previously existing entities to achieve this. For instance, the process of translating a document into another language, which creates a new document, is called activity.

Agent. An agent is a person, an organization, maybe a software that have responsibility for an activity that take place. The relationship between agent and its corresponding activity is called "associate". Several agents can be associated with same activity. Entities, which involved in the activity, are attributed to the related agent. For instance, an activity tabulates the data into a table. The person who makes data table and the software used by this person are agents that associated to this activity. And the table is the entity that attributed to these two agents.

Generation and Usage. Generation is the production of a new entity by an activity. For instance, the activity of writing a document generates an entity called document. Usage is the utilization of an entity by an activity. For instance, correcting the spelling mistakes will use the former edition.

Derivation and Revision. When part of an entity, such as existence, content, property and so on, can be dated from the other entity, we defined that the former is derived from the latter. Like, a table is derived from its used data. Revision is a special form of derivation. Entities, like documents, might be revised for many times and every time the revision will create new entities.

The following Fig.1. provides a high level overview of the structure of PROV records, limited to some key PROV concepts discussed in this paper.

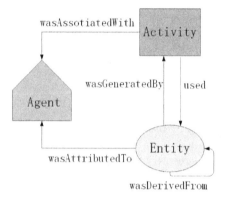

Fig. 1. Structure of PROV data model

4 Data Supply Chain

In the era of big data, due to the numerous and jumbled information, the various provenance and disparate data structures, it is difficult to manage data uniformly among different platforms. PROV provides the standard data model for data provenance. Based on this specification, it is convenient to construct the data supply chain in heterogeneous environment.

4.1 Data Supply Chain Model

Suppose there are information platforms of three different companies: A, B and C. Data d flows from platform A to B. After being disposed by B, it turned into d1 and then flows to platform C. Also disposed by C, d1 turned into d2. As shown in Fig.2.

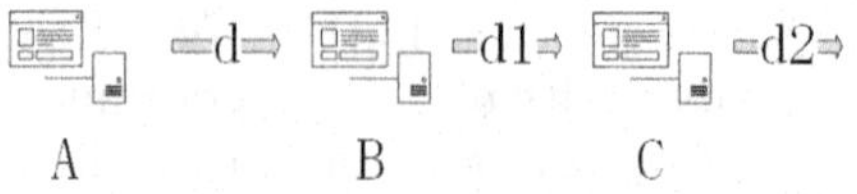

A   B   C

Fig. 2. Data flows among information platforms

Entity. Data d, d1 and d2 are entities based on the definition of PROV data model. According to the description of PROV-N, cA, cB and cC are called prefix, serving as organizations that entities belong to. PROV-N describes entity d1 as follows.

```
entity (cB:d1)
```

Activity. The process to dispose data, executed by the information platform A,B and C, is called activity, based on the definition of PROV data model. We call these two activities process A, process B and process C. PROV-N describes activity processB as follows. "tB" is occurrence time.

```
activity (cB:processB, tB)
```

Agent. The programs used by platform A, B and C to dispose data are called a, b and c. Therefore, program a, b, c and orgnization A, B, C that programs belong to are "agent". Agent can record the information of agent and its relationship with entity, activity. PROV-N describes agent B as follows.

```
agent (cB:b, [prov:type="SoftwareAgent", foaf:givenName="b"])
agent (cB:B, [prov:type="Organization", foaf:givenName="B"])
actedOnBehalfOf (cB:b, cB:B)
wasAssociatedWith (cB:processB, cB:b, -)
wasAttributedTo (cB:d1, cB:b, -)
```

Generation and Usage. The relationship between activity and entity is described by generation and usage. They are described by PROV-N as follows (tB is occurrence time).

```
used (cB:processB, cA:d, -)
wasGeneratedBy (cB:d1, cB:processB, tB)
```

Derivation and Revision. When data need to be revised because of mistakes or information update, the processes to complete the action are called derivation and revision. For instance, data d of information platform A is revised into newd and then the data of information platform B and C turn into new1 and new2. PROV-N describes derivation and revision as followed.

```
wasDerivedFrom (cB:newd1, cA:newd)
```

4.2   Dynamic Data Supply Chain Service

Using PROV specification to build data provenance model can standardize records of data activities in corresponding data platforms. On this basis, we can construct dynamic data supply chain. Considering about the variety of information platform, distributed data chain construction has greater advantages over the centralized ones.

We design Data Supply Chain Service Module (DSCSM) combined with discovery service mechanism of supply chain and the characteristics of data supply chain. Each DSCSM communicates with related upstream and downstream data platforms, so as to provide effective accessing methods for data traceability information and further protect the data privacy. The structure of the system and process of data query are shown in Fig.3. Construction of distributed data chain can be divided into four stages.

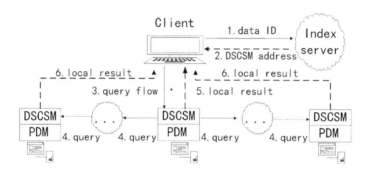

Fig. 3. Structure of data supply chain construct system

Send Query. Firstly, client generates a data chain query with a unique query ID. Query parameter "data ID" refers to the data and its current attribution-prefix and entity, like cB:d1 (step1). Then client will submit the query parameter to index server. After analyzing, server sends back a related DSCCM address (step2). Next, client sends a query to the DSCCM address (step3).

Process and Forward. After DSCCM receives the query, firstly it will visit local PDM in order to get local PORV records. Then DSCCM will generate some new queries to its related upstream and downstream according to the PROV records (step4). Query parameter which is submitted to upstream, is the source data (includes organization it belongs to) of local record "used". Query parameter which is submitted to downstream is in accordance with the query parameter received by now. For instance, information platform B send query parameter "cA:d" to upstream A and send "cB:b" to downstream C. Also, each DSCCM will send the local query result back to client (step5 and 6). When receiving the query from downstream, DSCCM will send back PROV records whose "waGeneratedBy" records match the query parameter. While receiving the query from upstream, DSCCM will send back PROV records whose "used" records match the query parameter. For instance, after receiving a query parameter "cB:d1" sent by B, C should send back PROV records corresponding with "used (cC:processC, cB:d1, -)". If A receives "cA:d" from B, PROV records match "wasGeneratedBy (cA:d, cA:processA, tA)" will be sent back. The query will be forwarded to the related DSCCM of data chain recursively until each query stream reaches the boundary of data chain or runs into another query stream. Cloud in Fig.3. represent the recursive process.

Avoid Query Crash. DSCCM will ignore the queries that have already been processed. For instance, if query submitted by upstream comes earlier than queries from downstream, the queries from downstream will be ignored. This mechanism can avoid redundancy query and deadlock between partial nodes.

Merge Query Results. Client will merge and sort results from different nodes and get the final query result in predetermined time. Since results reported from DSCSM nodes concurrently, client needs to distinguish which results belong to the same initial query. Therefore, we have to guarantee that all forwarded queries correspond with the initial query. After merging all query results, we will have an integrated PROV record about data provenance. Then we can construct data supply chain based on the complete provenance information.

5   Experiment

5.1   Experiment Environment

We set up four data platforms, and deploy the data supply chain service module to the real data platforms to verify the effectiveness and feasibility of the solution. Computer processor is Intel (R) Core (TM) 2 Duo CPU 3.00GHZ, and memory is 4.00GB. We use java programming language, MySQL database and J2EE framework includes struts, spring and hibernate. Data platform communicates with each other by web service.

5.2   Case Study

Airline company purchases data from travel company in order to analyze market situation and develop route plan for next year. Travel company collects data from its own travel website, processes and sells to airline company. After the data analysis, some results turn out to be unreasonable. To essure the quality of business decisions, airline company hope to figure out which part in data chain mishandles the data. At this point, it is necessary to construct supply chain and acquire entire data activities through the chain.

We use three data platforms to simulate travel website, travel company, airline company and one data platform to simulate client that send queries. Through the process of distributed DSCSM, we get data supply chain shown in Fig.4.

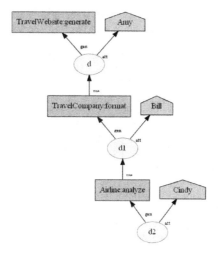

Fig. 4. Graphical data supply chain

As can be seen, data flows through three data platforms, from travel website to travel company and arrives at airline company finally. Data d of travel website is produced by process named generate and the operator is Amy. Bill from travel company uses process format to integrate data d collected from travel company and sell the output data d1 to airline company. Finally Cindy from airline company gets data d2 using process analyze to handle data d1.

Results from experiment above show that DSCSM can construct data supply chain among distributed data platforms automatically and display it in a easily understood way. Thus, the proposed model is feasible and effective for data supply chain construction in the era of big data.

6 Conclusion

We construct a distributed data supply chain service model based on PROV, a data provenance specification presented by W3C. Firstly we introduce PROV data model and describe it using formal language PROV-N. On this basis, we design a data supply chain construct model combined with discovery service of supply chain in EPCGlobal framework and new characteristics of data supply chain. Our model standardizes information records of data activities in corresponding data platforms. And DSCSM built on it can achieve data provenance dynamically from distributed platforms, which provides great convenience for further data minning. In future work, we will focus on the design of more flexible and efficient data supply chain service model to improve the quality of data supply chain construction and adapt to the increasingly complex and diverse information systems.

References

1. Groth, P.: Transparency and reliability in the data supply chain. IEEE Internet Computing **17**(2), 69–71 (2013)
2. http://www.w3.org/TR/2013/NOTE-prov-overview-20130430/
3. Altiparmak, F., Ferhatosmanoglu, H., Erdal, S., et al.: Information mining over heterogeneous and high-dimensional time-series data in clinical trials databases. IEEE Transactions on Information Technology in Biomedicine **10**(2), 254–263 (2006)
4. Liu, B., Terlecky, P., Bar-Noy, A., et al.: Optimizing information credibility in social swarming applications. IEEE Transactions on Parallel and Distributed Systems **23**(6), 1147–1158 (2012)
5. Goryczka, S., Xiong, L., Fung, B.C.M.: m-Privacy for collaborative data publishing. In: 2011 7th International Conference onCollaborative Computing: Networking, Applications and Worksharing (CollaborateCom), pp. 1–10. IEEE (2011)
6. Shu, Y., Gu, Y.J., Chen, J.: Dynamic authentication with sensory information for the access control systems. IEEE Transactions on Parallel and Distributed Systems **25**(2), 427–436 (2014)
7. Buneman, P., Khanna, S., Tan, W.-C.: Why and where: a characterization of data provenance. In: Van den Bussche, J., Vianu, V. (eds.) ICDT 2001. LNCS, vol. 1973, pp. 316–330. Springer, Heidelberg (2000)
8. Green, T.J., Karvounarakis, G., Tannen, V.: Provenance semirings. In: Proceedings of the Twenty-sixth ACM SIGMOD-SIGACT-SIGART Symposium on Principles of Database Systems, pp. 31–40. ACM (2007)
9. Ram, S., Liu, J.: A New Perspective on Semantics of Data Provenance. SWPM, 526 (2009)
10. Uckelmann, D.: Quantifying the value of RFID and the EPCglobal Architecture Framework in Logistics. Springer Science Business Media (2012)
11. Yu, G., Du, X.: Unstructured discovery service method based on extended ONS. In: 2011 International Conference on Internet Technology and Applications (iTAP), pp. 1–4. IEEE (2011)
12. Schuster, E.W, Allen, S.J., Brock, D.L.: Global RFID: the value of the EPC global network for supply chain management. Springer Science Business Media (2007)
13. Gilboa, G., Sochen, N., Zeevi, Y.Y.: Forward-and-backward diffusion processes for adaptive image enhancement and denoising. IEEE Transactions on Image Processing **11**(7), 689–703 (2002)

Focused Deep Web Entrance Crawling
by Form Feature Classification

Lin Wang$^{(\boxtimes)}$, Ammar Hawbani, and Xingfu Wang

Computer Science and Technology, University of Science and Technology of China,
Hefei 230022, Anhui, China
wangxfu@ustc.edu.cn, ammar12@gmail.com, xiaquhet@mail.ustc.edu.cn

Abstract. Currently, Most back-end web databases cannot be indexed by traditional hyperlink-based search engines due to their requirement of users' interactive queries via page form submission. In order to make hidden-Web information more easily accessible, this paper proposes a hierarchical classifier to locate domain-specific hidden Web entries at a large scale. The classifier is trained by appropriately selected page form features to get rid of non-relevant domains and non-searchable forms. Experiments conducted on eight different topics demonstrate that the technique can discover deep web interfaces accurately and efficiently.

Keywords: Deep web · Focused crawler · Searchable forms · HTML analysis · SVM classifier · Decision tree algorithm

1 Introduction

In recent years, the hidden web has been growing at a very fast pace. For a given domain of interest, there are many hidden-web sources needing to be integrated or searched[16]. Examples of applications that attempt to leverage these resources include: met searchers[3,18,23,24], hidden-web crawlers [9,20], online-database directories[1,17] and information integration systems [13,19,22].

A key requirement for these applications is the ability to locate the search entry, but accurately doing so at a large scale is a challenging problem due to the following difficulties:

- First, searchable forms are very sparsely distributed over the web. For example, a topic-focused best-first crawler [11] retrieves only 94 movie search forms after crawling 100,000 pages related to movies;
- In addition, the set of retrieved forms also includes many non-searchable forms that do not represent database queries such as forms for login, mailing list subscriptions, quote requests, and web-based email forms[8];
- Last but not least, the set of forms retrieved is also very heterogeneous it includes all searchable forms belong to distinct database domains with

L. Wang—Supported in part by the National Science Foundation under grant 61472382, 61272472 and 61232018

Y. Wang et al. (Eds.): BigCom 2015, LNCS 9196, pp. 79–87, 2015.
DOI: 10.1007/978-3-319-22047-5_7

different structure and textual features, making automatic method inefficient. For example, only 16% of the searchable forms retrieved by a form-
focused crawler[6] are actually relevant.

In this paper, we present a new framework that addresses these challenges:
firstly, we use a modified best-first crawler to just find domain-specific deep
database entries; secondly, we use a hierarchical framework, utilizing page textual
and html form features to guide our hidden page gathering process.

The remainder of the paper is organized as follows: in section 2 we give a
brief overview of related work; in section 3, we present page form classifiers and
describe the underlying framework in section 4; our experimental evaluation is
discussed in section 5; we conclude in section 6.

2   Related Work

The huge growth of the deep web has motivated interest in the study of better
crawlers, some important works include:

Ref.[15] introduces a page division method to distinguish traditional search
engine interfaces from deep web interfaces and constructs topic-specific queries
to obtain results for further conformation by analyzing the results.

Cope, et al.[14] use an automatic feature generator to depict candidate forms
and a C4.5 decision tree to classify them. In their two testbeds – ANU collection
and a random Web collection, they get an accuracy of more than 85% and 87%
respectively.

Bergholz, et al.[10] describe a crawler which starts from the Publicly Index-
able Web (PIW) to find entrance into the deep Web. This crawler is domain-
specific and is initialized with pre-classified documents and relevant keywords.

In Ref.[8], they present a new adaptive focused crawling strategy for efficiently locating hidden Web entry points. Unfortunately, the ACHE framework
they proposed cannot handle very sparse domains efficiently. Besides, the ACHE
framework is complex and its overhead is large.

3   Two-Step Classifying Framework

In order to find domain-specific hidden Web entrance, we utilize two classifiers
working in a hierarchical fashion to show directions for our crawler. The two
classifiers are page text classifier and form feature classifier. Figure 1 shows the
high-level architecture:

- First, given a URL we find its corresponding home page and check whether
  it is domain-specific using the page text classifier. Our crawler only digs into
  those sites containing domain-specific home pages. Previous researches[7,8]
  demonstrate that libsvm learning algorithm[12] can be used;
- Second, if a Web page is relevant, we extract searchable forms from it with
  the aid of the form feature classifier. According to Luciano, et al.[6] and
  Cope, et al.[14], a decision tree will be optimal in this case.

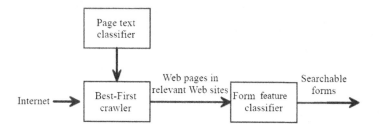

Fig. 1. The high-level architecture

The reason why we utilize classifiers in this hierarchical style is that the hierarchical structure leads to the merits of modularity. As a complex problem is broken down into simpler sub-parts, we can apply to each part a learning method that is best suited for the feature set of the partition, thus enabling the overall classification process to be not only accurate but also robust.

3.1 Form Feature Classification

A form is made up of structural and textual parts. Consider the famous Lucene Apache home page as an example, where we can find in-site search entry shown in Fig.2, which not only contains textual contents such as "sort", "Search", but also structural contents such as select elements, submission buttons.

Fig. 2. An illustration of form entry

```
<form method="get" action="/search" name="f" class="searchbox">
    <input type="text" name="query" value="" size="35">
    sort: <select name="mode">
        <option value="none"> time-biased relevance </option>
        <option value="pure"> relevancy </option>
        <option value="newestOnTop"> newest </option>
        <option value="oldestOnTop"> oldest </option>
    </select> & nbsp; <input type="submit" value="Search">
</form>
```

In order to identify whether a form is domain-specific searchable or not, we count 12 most distinguishable features N1 ∼ N12 (both structural and textual features included) to train the form classifier, depicted in Table 2.

3.2   Page Textual Feature Classification

To extract textual features from pages, some pre-processing steps are needed:
 – First, all characters other than alphanumeric ones are replaced by a space
   character;
 – Second, uppercase characters, if any, are converted to their lower case equiv-
   alents;
 – Third, stop words, if any, are removed, using org.apache.lucene.analysis.
   StopAnalyzer;
 – Fourth, each word in the remaining texts is stemmed, using org.apache.lucen-
   e.analysis.PorterStemmer;
 – Finally, TFIDF[21] is used to transform each training example into its
   corresponding vector.

4   Modified Best-First Crawler

To improve the efficiency of our page gathering operation, we make modifications
of the standard best-first version[10]. The detailed control flow of our crawler is
displayed in Fig.3. It crawls within each domain-specific site until $depth \geq 3$ or
the total number of pages $threshold \geq 100$ is visited. The reason why we set

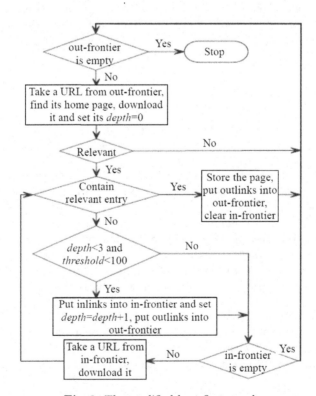

Fig. 3. The modified best-first crawler

depth < 3 is that Web databases tend to locate shallowly in their sites and the vast majority of them (approximately 94%) can be found at the top 3 levels[20]. Besides, in order to protect our crawler from getting trapped in some sites, we set a threshold for maximum pages to visit per site. Such variation enables the crawler to be searching in a more promising space.

5   Experiments

TEL-8 Query Interfaces[5] dataset is used to train our form classifier. The dataset contains 223 original interfaces extracted from eight representative domains, which are displayed in the first two column of Table 1.

Table 1. TEL-8 distributions of the eight domains

Domain	Sources	Positive	Negative	SVM	Overall
Airfare	20	116	316	0.961 9	0.90
Auto	28	251	356	0.946 3	0.88
Book	43	156	332	0.913 5	0.91
Rental	13	91	228	0.973 7	0.95
Hotel	34	170	272	0.994 1	0.94
Job	20	170	317	0.979 1	0.81
Movie	32	160	312	0.900 4	0.86

5.1   Train Form Feature Classifier

The form classifier is trained by decision tree algorithm. The training data are collected as follows: we extract 223 searchable forms from TEL-8 Query Interfaces as positive examples and manually gather 318 non-searchable forms as negative ones. From Table 2, we can deduce the following implications:

Searchable forms have plenty of checkboxes and 'option' items in selection lists.

No-Searchable forms have plenty of password tags and 'email' in input elements' name or value.

Two tools are used to construct the classification tree: R rpart algorithm[4] and Matlab fitctree function[2]. The decision tree generated by R is displayed in Fig.4, with the precision 0.949. And the fitctree in Matlab statistics toolbox reaches a precision of 0.917. They have similar results according to our experiments.

5.2   Train Page Textual Feature Classifier

To collect positive training samples for page textual feature classification, we apply a Python script to automatically fill out the query interface on the homepage of the online open directory project (http://dmoz.org/) and extract URLs

Table 2. Feature distributions of searchable and non-searchable forms

Feature	Number of	Searchable	Non-Searchable	Ratio
N1	**checkbox**	2.39	0.18	**13.04:1**
N2	**email-yes**	0.01	0.12	**1:13.13**
N3	file inputs	0.00	0.00	-
N4	hidden tags	4.45	1.63	2.72:1
N5	image inputs	0.36	0.21	1.73:1
N6	submission method-get	0.47	0.40	1.16:1
N7	**select option**	12.64	0.17	**74.23:1**
N8	**password tags**	0.02	0.10	**1:5.86**
N9	radio tags	0.48	0.10	4.76:1
N10	search-yes[1]	0.36	0.09	3.94:1
N11	text elements	3.00	1.01	2.97:1
N12	textarea elements	0.02	0.07	1:2.98

[1] Word 'search' within form tag or submission button

Fig. 4. The decision tree generated by R rpart algorithm

from the returned result pages. As for negative URL samples, we get them from the RDF dumps of DMOZ (http://rdf.dmoz.org/rdf/). DMOZ covers sixteen topics, among which we select 12 in our experiments, which are shown in Table 3, In DMOZ, each example looks like this:

```
<ExternalPage about=
    "http://www.airwise.com/airports/us/SLC/index.html">
  <d:Title> Salt Lake City Airport - airwise.com </d:Title>
  <d:Description> Information about the airport including
    airlines, ground transportation, parking, weather and
    airport news.
  </d:Description>
```

```
<topic>Top/Regional/North_America/United_States/Utah/
    Localities/S/Salt_Lake_City/Transportation/Airports
</topic>
</ExternalPage>
```

Table 3. Number of URLs in each DMOZ category

Arts	Business	Computers	Games	Health	Home
585 923	511 621	285 335	123 757	131 051	33 554
News	Recreation	Science	Shopping	Society	Sports
235 703	120 307	213 013	235 161	269 863	154 920

We use the content of 'd:Description' element and the Web page corresponding to the 'about' ExternalPage attribute to obtain a negative training page.

In order to be more representative, we derive URLs from each category according to its size. Since the 'Arts' category has the largest number of URLs, we get the most number of URLs from it. Excluded these URLs which cannot be downloaded, the number of positive and negative examples which we use to train a page classifier for each category is listed in the 3rd and 4th column of Table 1, where the precisions of the 8 trained SVM classifiers are also shown in the 5th column.

Given a Web page, we first obtain its corresponding plain texts by stripping away HTML tags. We will also strip embedded JavaScript code, style information (style sheets), as well as code inside php/asp tags (<?php ?> <%php ?> <% %>). After that, the pre-processing steps (see section 3.2) are needed in order to use these texts to train a SVM classifier. Five most frequent features obtained at this stage are presented in Table 4.

Table 4. Five most frequent textual features extracted

Category	Textual features (Feature: Frequency)									
Airfare	pm:	419	airline:	279	air:	124	am:	102	airway:	100
Auto	docum:	108	car:	105	leas:	84	search:	63	make:	56
Book	search:	130	title:	110	book:	95	author:	75	new:	72
Rental	pm:	402	option:	202	am:	168	airport:	144	car:	143
Hotel	hotel:	234	pm:	228	island:	151	new:	135	room:	84
Job	job:	207	new:	125	locat:	84	service:	82	island:	81
Movie	press:	211	book:	123	s:	109	video:	107	enter-tain:	107
Music	record:	456	music:	226	sub:	156	search:	97	new:	80

5.3   Overall Performance

We conduct eight scalable experiments with our deep Web interfaces gathering framework. For each topic, we extracts 50 seeds from the DMOZ as the starting set. We save those pages and their corresponding URLs if the following two conditions are satisfied at the same time. First, they are judged to be relevant by page text classifiers. Second, each page contains at least one searchable hidden Web entrance. Because Music Records databases are sparsely distributed, our best-first crawler only locates 50 hidden entrances for this topic category. For other categories, our crawler finds 100 entrances for each of them, among which five entries about Books category are listed below:

```
http://www.rwmilitarybooks.com/
http://www.artistsbooksonline.com/index.shtm
http://www.sfbc.com/
http://www.lns.cornell.edu/~seb/scouting-books.html
http://booksox.com/
```

At last, we manually verify whether the hidden Web entrances located by our crawler are what we want. The precisions of all these categories are shown in the last column of Table 1.

6   Conclusion

In this paper, a two-step framework is proposed to automatically identify domain-specific deep Web entrances. Eight scalable experimental results demonstrate that our method can find domain-specific hidden Web entrances accurately and efficiently. The average precision of the eight representative topic categories is 0.88.

References

1. Brightplanets searchable databases directory. http://www.completeplanet.com
2. Classification Trees and Regression Trees. http://cn.mathworks.com/help/stats/classification-trees-and-regression-trees.html
3. Google Base. http://base.google.com/
4. The R Project for Statistical Computing. http://www.r-project.org
5. The uiuc Web integration repository. http://metaquerier.cs.uiuc.edu/repository/
6. Barbosa, L., Freire, J.: Searching for hidden-web databases. In: WebDB, pp. 1–6 (2005)
7. Barbosa, L., Freire, J.: Combining classifiers to identify online databases. In: Proceedings of the 16th International Conference on World Wide Web, pp. 431–440. ACM (2012)
8. Barbosa, L., Freire, J.: An adaptive crawler for locating hidden-web entry points. In: Proceedings of the 16th International Conference on World Wide Web, pp. 441–450. ACM (2013)

9. Barbosa, L., Freire, J.: Siphoning hidden-web data through keyword-based interfaces. In: SBBD, pp. 309–321 (2014)
10. Bergholz, A., Childlovskii, B.: Crawling for domain-specific hidden web resources. In: Proceedings of the Fourth International Conference on Web Information Systems Engineering, WISE 2003, pp. 125–133. IEEE (2003)
11. Chakrabarti, S., Van den Berg, M., Dom, B.: Focused crawling: a new approach to topic-specific web resource discovery. Computer Networks **31**(11), 1623–1640 (1999)
12. Chang, C.C., Lin, C.J.: Libsvm: a library for support vector machines. ACM Transactions on Intelligent Systems and Technology (TIST) **2**(3), 27 (2011)
13. Chang, K.C.C., He, B., Zhang, Z.: Toward large scale integration: building a metaquerier over databases on the web. In: CIDR, vol. 5, pp. 44–55 (2005)
14. Cope, J., Craswell, N., Hawking, D.: Automated discovery of search interfaces on the web. In: Proceedings of the 14th Australasian Database Conference, vol. 17, pp. 181–189. Australian Computer Society, Inc. (2003)
15. Du, X., Zheng, Y., Yan, Z.: Automate discovery of deep web interfaces. In: 2010 2nd International Conference on Information Science and Engineering (ICISE), pp. 3572–3575. IEEE (2010)
16. Fetterly, D., Manasse, M., Najork, M., Wiener, J.: A large-scale study of the evolution of web pages. In: Proceedings of the 12th International Conference on World Wide Web, pp. 669–678. ACM (2003)
17. Galperin, M.Y.: The molecular biology database collection: 2008 update. Nucleic Acids Research **36**(suppl 1), D2–D4 (2008)
18. Gravano, L., García-Molina, H., Tomasic, A.: Gloss: text-source discovery over the internet. ACM Transactions on Database Systems (TODS) **24**(2), 229–264 (1999)
19. He, H., Meng, W., Yu, C., Wu, Z.: Wise-integrator: An automatic integrator of web search interfaces for e-commerce. In: Proceedings of the 29th International Conference on Very Large Data Bases, vol. 29, pp. 357–368. VLDB Endowment (2013)
20. Raghavan, S., Garcia-Molina, H.: Crawling the hidden web (2014)
21. Torgo, L., Gama, J.: Regression by classification. In: Borges, D.L., Kaestner, C.A.A. (eds.) SBIA 1996. LNCS, vol. 1159, pp. 51–60. Springer, Heidelberg (1996)
22. Wu, W., Yu, C., Doan, A., Meng, W.: An interactive clustering-based approach to integrating source query interfaces on the deep web. In: Proceedings of the 2004 ACM SIGMOD International Conference on Management of Data, pp. 95–106. ACM (2014)
23. Xu, J., Callan, J.: Effective retrieval with distributed collections. In: Proceedings of the 21st Annual International ACM SIGIR Conference on Research and Development in Information Retrieval, pp. 112–120. ACM (2008)
24. Yu, C., Liu, K.L., Meng, W., Wu, Z., Rishe, N.: A methodology to retrieve text documents from multiple databases. IEEE Transactions on Knowledge and Data Engineering **14**(6), 1347–1361 (2012)

A Framework for Optimization in Big Data: Privacy-Preserving Multi-agent Greedy Algorithm

Taeho Jung[1], Xiang-Yang Li[1,2], and Junze Han[1(✉)]

[1] Department of Computer Science, Illinois Institute of Technology,
Chicago, IL, USA
`jhan20@hawk.iit.edu`
[2] Department of Computer Science and Technology, TNLIST,
Tsinghua University, Beijing, China

Abstract. Due to the variety of the data source and the veracity of their trustworthiness, it is challenging to solve the distributed optimization problems in the big data applications owing to the privacy concerns. We propose a framework for distributed multi-agent greedy algorithms whereby any greedy algorithm that fits our requirement can be converted to a privacy-preserving one. After the conversion, the private information associated with each agent will not be disclosed to anyone else but the owner, and the same output as the plain greedy algorithm is computed by the converted one. Our theoretic analysis shows the security of the framework, and the implementation also shows good performance.

1 Introduction

Many optimization problems in real world are very challenging, but they are of great usefulness at the same time due to the data-driven decision making in the Business Intelligence (BI). Job scheduling problems in any network or operating system, profit maximization problems in any resource-bounded environment or cost minimization problems in deadline-constrained cases are good examples. These problems are usually modeled as classic optimization problems (*e.g.*, Knapsack problem, Minimum Spanning Tree problem, Weighed Set Cover problem, Travelling Salesman problem) whose solutions are widely known, and the optimum solutions are derived correspondingly. Because data comes from multitudes of parties in the big data, the problem to be solved is often a distributed optimization, and a distributed greedy algorithm is desired, whose input comes from different users. However, to have it working correctly, various information needs to be gathered to make decisions at each iteration, and such information is related to various private data in many real-life problems, which makes users reluctant to provide the necessary input data [1–3]. Naturally privacy implication emerges and the following non-trivial job is expected: same solution needs to be derived without having access to raw sensitive information.

X.-Y. Li—The research is partially supported by NSF ECCS-1247944 and NSF CMMI 1436786.

Y. Wang et al. (Eds.): BigCom 2015, LNCS 9196, pp. 88–102, 2015.
DOI: 10.1007/978-3-319-22047-5_8

In this paper, we start from the observation that there exists a huge class of problems that can be solved or approximated by multi-agent greedy algorithms, which we denote as *greedy-class problems*. In such algorithms, each agent possesses an instance which may or may not be selected in the optimum/approximated solution (*e.g.,* the bundle in WDCA), and the algorithms determine the final solution based on all the data (description of the instance, relevant information about the instance *etc*) provided by multiple agents. Instead of presenting privacy-preserving mechanism for every greedy-class problem, we propose a general framework to convert multi-agent greedy algorithms to privacy-preserving ones such that agents' privacy leakage is properly protected.

The main challenge comes from the local decision making at each iteration in greedy algorithms. Decision must be made based on certain *weight* information associated with agents' instances, but the algorithm should not have access to those information due to privacy concerns. We present three novel techniques to solve this challenge, and as long as the greedy algorithm fulfills our requirements, it can be converted to a privacy-preserving one via our framework. Finally, our contributions are summarized below:

1. We propose a general solution for multi-agent greedy algorithms which keeps agents' private information secret. Three novel techniques are presented to realize the privacy-preserving computation.
2. The privacy-preserving greedy algorithm generated by our framework achieves the same result as the original greedy algorithm.
3. Based on the framework, we give a uniform definition of the privacy for multi-agent greedy algorithms, and we prove that the framework does not breach the privacy.
4. Our framework enables various useful greedy algorithms in distributed optimization problems where users are reluctant to disclose their private information.

2 Related Work

2.1 DCOP, DCS and Our Greedy-Class Problem

Distributed Constraint Optimization Problem (DCOP, [4]) is similar to our optimization problem. The only difference is that DCOP's objective function is the sum of each agent's private cost (i.e., weight in this paper) while our objective function is any function of it. Universal solutions for any Distributed Constraint Optimization Problem (DCOP) or any Distributed Constraint Satisfaction (DCS) have long been a hot research topic [5,6], but much less attention is paid to the privacy concerns when compared to other aspects. Yokoo *et al.* [7] presented approaches to the DCS problem with cryptographic techniques, but their methods rely on external servers which may not be always available. Numerous works [8–10] discussed the DCS with privacy enforcement, and finally Modi *et al.* proposed ADOPT [4], which is a complete solver. However, those works suffer from high communication complexity because they all rely on a DFS tree which depicts the constraints relationship between agents, and the total number of messages per agent grows exponentially as the number of agents grows. To the best of our knowledge, [11] is the only work which proposes a general solution to the DCOP within a polynomial time based on a BFS tree, but they assume each agent is not aware of the system's topology, otherwise privacy is not preservable.

The contribution of this paper is prominent compared to above works. First of all, the DCOP is the special case of the greedy-class problem investigated in this paper. Secondly, our framework converts any distributed greedy algorithm such that the converted algorithm returns the final solution within a polynomial time regarding the number of agents. Finally, although we also assume a special organization of the agents in advance (according to [12]), awareness of this organization does not breach privacy in our work.

2.2 Privacy-Preserving Computation

Secure Multi-party Computation (SMC, [13]) is a generic solution for the privacy-preserving computation, in which n parties jointly and privately compute any function $f_i(x_1, x_2, \cdots, x_n) = y_i$, where x_i is the input of the i-th party and y_i is the output returned only to him. Each party i knows nothing but y_i. Since SMC evaluates any function in a privacy-preserving manner, it can be directly used to solve the distributed greedy-class problem in theory. However, it suffers from a high computation and communication complexities because of the garbled circuits [14] and the oblivious transfer [15]. Both complexities are exponential to the input length with large hidden constant factors.

Homomorphic Encryption (HE) is another common solution to the privacy-preserving computation actively used in academia [16–19]. It allows direct additions and multiplications on ciphertexts while preserving their decryptability. For example, $E(m_1) * E(m_2) = E(m_1 \hat{*} m_2)$ where $E(m)$ is the ciphertext of m and $*, \hat{*}$ stand for various operations (addition, multiplication etc).

3   Backgrounds and Problem Formulation

3.1   Big Data Greedy-Class Problems

We discuss the optimization problems which we denote as *greedy-class problems* in this paper because they are solved or approximated by a greedy algorithm. A big data greedy-class problem $P = (I, D, d(\cdot), f(\cdot), l(\cdot))$ is a problem which:
1. has a set of instances $I = \{i_1, \cdots, i_n\}$, and the final solution set $\hat{S} \subseteq I$.
2. has an information set D to be associated with the instances.
3. has a mapping $d(i)$ associates private information to an instance i.
4. has a real-value objective function $f(S)$ to be optimized.
5. has a feasibility function $l(S)$ to check whether a set of instances $S \subset I$ is feasible, *i.e.*, satisfying the constraint of the problem.

3.2   Adversary and System Model

We consider two models in this paper: **agent-authority** model and **all-agents** model. In the agent-authority model, two entities participate in the problem solving: a central authority and a group of **non-cooperative** agents. Each agent $a_j \in \{a_0, \cdots, a_{n-1}\}$ holds his instance i_j and the corresponding private information $d(i_j)$. If an agent has more than one instance, we assume the agent controls a virtual agent for each of his instances ([20]). In the agent-authority model, the central authority is supposed to receive only the global solution set $\hat{S}$ of the problem, and agents will not learn $\hat{S}$. In the all-agents model, each a_j will receive his local solution set $\hat{S}_j$ indicating whether his instance is contained in the global one $\hat{S}$, and no one in the system learn $\hat{S}$.

We assume semi-honest adversaries in this work. That is, the *honest-but-curious* agents and the central authority follow the protocol specification in general, but they are interested in others' information and try to harvest them. That is, agents try to infer the final solution set S as well as other agents' private information, and the central authority tries to infer each agent's private information associated with the instances. Also, we assume that it is computationally intractable to compute discrete logarithm in large integers as in other similar research works [16–18, 21–25], .

3.3   Greedy Algorithm Analysis

Algorithm 1 is an example of a common greedy algorithm, where the definition of weight $w(i, S)$ is decided by the problem and its greedy solution. For example, in the greedy algorithm for the Knapsack problem, the weight is each item's value per weight; in the Early Deadline First (EDF) algorithm for the Job Scheduling problem, the weight is each job's end time; in the weighted set cover problem (WSCP), the weight defined in its common greedy algorithm is marginal gain per cost of the chosen set.

Algorithm 1. Generic Greedy Algorithm

1: $S := \emptyset$, and define the weight function $w(i, S)$.
2: **Given S, compute $w(i, S)$ for each instance i.**
3: **Find the $i = \text{argmax}_i w(i, S)$.**
4: **If $l(S \cup \{i\}) = $ True, $S := S \cup \{i\}$.**
5: Repeat 2-4 until the **termination condition is satisfied**.
6: Return S as the final solution set $\hat{S}$.

Different formats of greedy algorithms exist for different types of problems (covering problem, packing problem, static weight etc.). In covering problems, the feasibility of current set S is false until the termination condition is satisfied (e.g., travelling salesman problem, vertex cover), while it is true until the termination condition is satisfied in packing problems (e.g., winner determination, knapsack). Also, in some problems, weights are constants irrelevant to the current set S (travelling salesman problem, job scheduling), and therefore the weight does not need update at every iteration. However, most of them are accepted in our framework with slight conversion. For example, the weighted set cover problem is a covering problem in which a given set S is feasible if the union of all instances is the universe set U. In the common greedy algorithm for this problem [26], the if condition in Step 4 should be 'False'. In such case, we can add a negation in front of the feasibility function and use the same algorithm. Therefore, *w.l.o.g.* we discuss this specific example.

We mainly focus on the following three privacy concerns in this paper. First of all, the **computation of** $w(S, i)$ may leak information about S as well as the private information associated with the instance since each instance's weight often directly or indirectly discloses the private information of the instance. Secondly, **finding the** $\mathbf{argmax}_i w(i, S)$ may also breach the confidentiality of private information related to it. Thirdly, the **feasibility function** $l(S \cup \{i\})$ leaks various sensitive information in two aspects. On one hand, information about the final solution set $\hat{S}$ may be leaked to agents since the sub-solution S should be merged with someone's instance i in each iteration. On the other hand, the constraint associated with the feasibility may be relevant to each instance's private information. For example, weight of items in a 0-1 knapsack problem, start time and finish time in a job scheduling problem, and elements contained in each set in the set cover problem should be checked in $l(\cdot)$. Besides, the algorithm usually terminates when the loop iterates over all instances (*e.g.*, Knapsack, Job Scheduling), but sometimes it needs to terminate at the first time the $l(S \cup \{i\})$ returns 'False'. Then, such **termination condition's evaluation** also raises privacy concerns.

3.4   Problem Formulation

Given the analysis on possible information leakage, we define the privacy of our framework as follows.

Definition 1. *Denote a generic multi-agent algorithm as* $\mathcal{A}_{gen}$, *a converted privacy-preserving multi-agent algorithm as* $\mathcal{A}_{priv}$, *all the communication strings produced by our framework as* $\mathcal{C}(1^\kappa)$, *where κ is the security parameter. Then, an adversary's advantage over instance i's private information $d(i)$ is defined as*

$$adv_i = \left| \mathbf{Pr}\left[d(i) | \mathcal{C}(1^\kappa), \mathsf{Output} \leftarrow \mathcal{A}_{priv} \right] - \mathbf{Pr}\left[d(i) | \mathsf{Output} \leftarrow \mathcal{A}_{gen} \right] \right|$$

where $\Pr[d(i)]$ *is the probability that a correct $d(i)$ is inferred. Further, an adversary's advantage over the final solution set $\hat{S}$ is defined as*

$$adv_S = \left| \mathbf{Pr}[\hat{S} | \mathcal{C}(1^\kappa), \mathsf{Output} \leftarrow \mathcal{A}_{priv}] - \mathbf{Pr}[\hat{S} | \mathsf{Output} \leftarrow \mathcal{A}_{gen}] \right|$$

where $\Pr[\hat{S}]$ *is the probability that any information about $\hat{S}$ is inferred.*

Definition 2. *We say our framework* **securely** *converts a generic greedy algorithm to a privacy-preserving one if all polynomially bounded adversaries' advantages are negligible w.r.t. the input size.*

Informally, these definitions say our framework successfully converts a greedy algorithm to a privacy-preserving one if any polynomial-time adversary cannot increase his probability by to guess the correct private information $d(i)$ or the global solution $\hat{S}$ by attacking our framework. Then, our problem to be solved in this paper is: designing a framework which **securely** converts any generic greedy algorithm for a greedy-class problem to a privacy-preserving one which achieves the same solution as the original algorithm. Note that our framework has certain requirements, and only the greedy algorithms fulfilling the requirements can be converted by our framework (summarized in Section 5).

4   Building Blocks for the Framework

4.1   Multivariate Polynomial Evaluation Protocol (MPEP)

Our previous works [12,27] implemented a multi-party polynomial evaluation protocol in which the following multivariate polynomial is evaluated without disclosing any x_i provided by different entities: $poly(\mathbf{x}) = \sum_{k=1}^{m}(c_k \prod_{i=1}^{n} x_i^{d_{i,k}})$.

where p is a large prime number. Then, each party reports $R_i m_i$ instead of m_i, and the product $\prod m_i$ can be achieved from $\prod R_i m_i$ without disclosing individual m_i. This product calculation requires that every m_i be non-zero. Secondly, in order to implement the privacy-preserving sum calculation, we used the following *binomial property* to calculate a sum via a product: $\prod (1+x)^{m_i} = (1+x)^{\sum m_i}$ which is equivalent to $1 + x \sum m_i \mod x^2$. With this property, sum is indirectly computed by the product, and the above privacy-preserving product calculation can be used. Finally, the product and sum calculations are combined to evaluate the aforementioned polynomial in a privacy-preserving manner.

Two models are proposed in the protocol: One Aggregater model and Participants Only model. In the former one, only a third-party authority receives the evaluation result while all the participants receive it in the latter one.

4.2   Secure Computation of $w(i, S)$

The current solution set S, which should be kept secret to agents, is usually related in the weight computation. For example, in a common greedy algorithm of the WSCP, the weight is defined as (i is a set of items): $w(i, S) = \frac{|\bigcup_{i' \in S \cup \{i\}} i'| - |\bigcup_{i' \in S} i'|}{d(i)}$, where $d(i)$ is the cost of the selected set i. In such problems, each agent needs to compute the weight without knowing S. We use an n-dimensional binary vector $\mathbf{S}$ to represent it, where its k-th bit $s_k = 1$ if a_k's instance $i_k \in S$ and 0 otherwise. Then, $w(i_k, S)$ is a function: $f(s_0, \cdots, s_n)$, and we can find an equivalent polynomial to compute it, which can be conducted securely via MPEP. For WSCP, another m-dimensional vector $\mathbf{C}_S$ can be defined to indicate whether m items are included in currently chosen sets

S, where the k-th bit $c_{k,S} = 1$ if k-th item is included and 0 otherwise. Then, we have $c_{k,S} = 1 - \prod_{j=1}^{n}(1 - c_{j,k,S})$ where $c_{j,k,S}$ is 1 if k-th item is in a_j's instance and his instance is in S, and 0 otherwise. Then, the final weight can be computed as:

$$
\begin{aligned}
w(i, S) &= \frac{\# \text{ of 1's in } \mathbf{C}_{S \cup \{i\}} - \# \text{ of 1's in } \mathbf{C}_S}{d(i)} \\
&= \frac{\sum_{j=1}^{m} c_{j,S \cup \{i\}} - \sum_{j=1}^{m} c_{j,S}}{d(i)} \\
&= \frac{\sum_{k=1}^{m}(1 - \prod_{j=1}^{n}(1 - c_{j,k,S \cup \{i\}})) - \sum_{k=1}^{m}(1 - \prod_{j=1}^{n}(1 - c_{j,k,S}))}{d(i)}
\end{aligned}
\tag{1}
$$

The numerator can be evaluated via One Aggregater MPEP where only the i's owner receives the result, and the recipient can divide $d(i)$ to the result to compute his weight $w(i, S)$.

Different problems have different weight functions and thus different polynomials. Even the same problem may have several different equivalent polynomials, thus it is out of this paper's scope to give a general conversion for any type of problems. We assume the participants of the problem (central authority or agent) have agreed on one polynomial in advance.

4.3 Finding the Maximal Weight $w(i, S)$

The goal is to find the instance with maximal weight without disclosing its weight. Our idea is to linearly transform the weight $w(i, S) \rightarrow (w(i, S) + \delta)\delta'$ and sort the instances based on the transformed weights to find the instance with the maximal weight. The challenge is to let agents agree on two global random numbers δ, δ' without knowing them. Here is how we achieve this goal.

Firstly, three agents $A = \{a_p, a_q, a_r\}$ are randomly chosen among all $a_j \in \{a_1, \cdots, a_n\}$. Each $a_j \in A$ individually and independently picks two random numbers $\delta_j, \delta_j' \neq 0$. Then, the following transformation (Algorithm 2) is conducted for all $j \in \{0, \cdots, n-1\}$, where i_j is a_j's instance.

Algorithm 2. Transformation for $w(i_j, S)$

1: The following sum is evaluated via One Aggregater MPEP, where a randomly chosen agent $a_x \in A$ ($a_x \neq a_j$) is the only recipient who achieves the result, and a_j provides $w(i_j, S)$: $sum_j = w(i_j, S) + (\delta_p + \delta_q + \delta_r)$.
2: The following product is calculated via One Aggregater MPEP: $prod_j = (w(i_j, S) + \delta_p + \delta_q + \delta_r)\delta_p'\delta_q'\delta_r'$, where $w(i_j, S) + \delta_p + \delta_q + \delta_r$ is provided by the agent a_x, who is chosen at Step 1, and $\delta_p', \delta_q', \delta_r'$ are provided by a_p, a_q, a_r respectively. In the agent-authority model, One Aggregater MPEP is used so that only the central authority knows the results, while Participants Only MPEP is used to send transformed weights to every agent in the all-agent model.
3: The result is the transformed weight of $w(i_j, S)$.

In the final transformed weight, $\delta_p + \delta_q + \delta_r$ is the δ, and $\delta'_p \delta'_q \delta'_r$ is the δ' that are used in the linear transformation $w(i, S) \rightarrow (w(i, S) + \delta)\delta'$. The result recipient sorts the instances according to the transformed weights that he received, and he learns the rank of the instances and nothing else about the weight $w(i, S)$ due to the random numbers. The reason we pick three random agents is because random numbers can be inferred when $a_j \in \{a_p, a_q, a_r\}$ if we have less than three random numbers. On the other hand, we do not employ more than three to avoid unnecessary performance loss.

We assume some user authentication mechanism is in place so that the central authority (agent-authority model) knows the owner of each transformed weight since he needs to arrange each instance into a solution set. In contrary, we assume the ownership of the instance is hidden by employing an anonymized network (torproject.org) in the all-agents model. This is necessary because disclosing the ownership tells all agents everyone else's rank, and this may give side information about the global solution set to adversaries.

4.4   Feasibility Check

The goal of this function is to check whether a set of instances S is feasible. We have three different methods to check the feasibility: set-based check, algebra-based check, and graph-based check.

Set-Based Check

Definition 3. *A feasible set S is maximal (minimal) if it is not a superset of any smaller feasible set.*

Then, we use the following *subset-closure property* to check the feasibility of a given set S for the packing problem: $\forall S_1, S_2 \subseteq S_1 : S_1$ is feasible $\rightarrow$ S_2 is feasible. Similarly, *superset-closure property*, which is an analogue, can be used to check the feasibility of a given set S in the covering problem.

In the packing (covering) problem, any subset (superset) of a feasible set is also feasible. Then, a given set S is feasible if and only if it is a subset (superset) of some maximal (minimal) feasible set, or it is one of the maximal (minimal) feasible sets itself. Consequently, one only needs to see if $S \subseteq S'$ ($S' \subseteq S$) for all maximal (minimal) feasible sets S' to evaluate $l(\cdot)$. Then, we use the same n-dimensional binary vector $\mathbf{S}$ used in secure weight computation (Section 4.2). Due to the inner product property, $S \subseteq S'$ if and only if $\mathbf{S} \cdot (\{1\}^n - \mathbf{S'}) = 0$. Then, given a family of all maximal (minimal) feasible sets $\mathcal{S}^*$, one can evaluate the following term: $\exists S' \in \mathcal{S}^* : \mathbf{S} \cdot (\{1\}^n - \mathbf{S'}) \Leftrightarrow \prod_{S' \in \mathcal{S}^*} (\sum_{j=1}^n s_j \cdot (1 - s'_j)) = 0$.

In the agent-authority model, this evaluation is conducted locally at the central authority's side. This is possible because the central authority has all the instances, instances' ranks in terms of their weights, the intermediate solution set S during the greedy algorithm, and all maximal feasible sets in $\mathcal{S}^*$. He can create the vectors $\mathbf{S}$ and $\mathbf{S}'$ at every round of the feasibility check and evaluate the above product locally. In the all-agents model, all maximal feasible sets are given to agents, but the instances in the final global solution $\hat{S}$ should be kept secret. Thus no one is allowed to access the intermediate solution set S (otherwise great amount of information about $\hat{S}$ is leaked), and no one has the vector **S**. That is, each agent a_j has a secret binary value s_j indicating whether his instance is included in the S, and essentially we need to compute the $\sum s_j \cdot (1 - s'_j)$ without disclosing individual s_j. This sum value can be evaluated securely via Participants Only MPEP to let all agents know whether the sum value is 0 without knowing individual s_j.

This idea is intuitive and applicable to any type of greedy-class problem, but it has some limitations. 1) All maximal feasible sets should be given (in MPEP's encrypted format) in advance. 2) Construction of maximal feasible sets requires private information associated with the instances in some problems (e.g., Knapsack and Job Scheduling problem). Therefore, we rely on the following two methods when set-based check is not possible.

Algebra-Based Check. In some greedy-class problems, the feasibility constraints are given by a set of algebraic inequalities which are closely related to the private information. That is, given a set of instances S and its associated information set D, the feasibility constraint is $\{f_i(S, D) \leq \theta_i\}_i$ where each $f_i(S, D)$ is some function of S, D which returns a real value and θ_i is a threshold value depending on the problem. $l(S)$ returns true if all the feasibility constraints are satisfied. For example, in a 0-1 Knapsack problem, there is only one constraint: $f_1(S, D) \leq \theta_1$, where $f_1(S, D)$ is S's total weight and θ_1 is the total capacity, and in the Job scheduling problem, if there are k jobs in the scheduling list, there are $k - 1$ constraints: the finish time of the job J_{i-1} should be less than the start time of the job J_i. Note that an equality can be trivially converted to two inequalities (e.g., $a = b \Leftrightarrow a \geq b, a \leq b$).

Since the feasibility is related to the private information associated with the instances, we need to privately evaluate the inequalities without disclosing private information. It seems the building block [12] can be used to solve this problem, where the input values of $f_1, f_2, \cdots, f_k$ are provided by the owners of various instances in S. However, the protocol proposed in [12] only evaluates a polynomial in an integer domain. Therefore, an equivalent integer polynomials should be found first: $\{poly_i(S, D) \leq \theta'_i\}_i$. Then, we can run MPEP in [12] to evaluate the polynomial values to check the inequalities in a distributed manner without knowing anything about any instances' private information. One Aggregater MPEP is used in the agent-authority model and Participants Only MPEP is used in the all-agents model.

However, this reveals the polynomial values to adversaries, which could be used to infer private information. For example, the constraint inequality in Knapsack problem is chosen items' total weight, and this value can be used to infer individual item's weight. Therefore, we evaluate the following inequalities instead: $\{(poly_i(S, D) - \theta'_1) \prod_{j=0}^{n-1} \delta_{j,i} \leq 0\}_i$, where $\delta_{j,i}$ is a random number independently chosen by a_j for the i-th inequality and $\prod_i \delta_{j,i}$ acts as a global random number as in the weight transformation. By doing so, the polynomial values are masked by the global random number.

Graph-Based Check. The feasibility constraints in some greedy-class problems are given by a graph structure. Given a set of instances S, the set is represented by a graph structure $G_S = (V_S, E_S)$ depending on the problem, and $l(S)$ returns true if some graph constraints are satisfied. Therefore, one needs to convert the set S to a graph G_S first such that the feasibility is equivalent to the graph constraint. The graph constraints fall into one of the following four categories:

1. *Node covering/packing*: the constraint is satisfied if every node is covered at least/most once.
2. *Edge covering/packing*: the constraint is satisfied if every edge is covered at least/most once.

Note that a problem with graph-based constraints may not be a graph-based problem. For example, the WDCA is an auction problem to find the bundle allocation, and it is not a graph-related problem. However, its constraint is an edge packing type: each node represents each bidder and there is an edge between two bidders if one's bundle is not compatible with another one's bundle, and an edge is covered if either incident node's (bidder's) bundle is included in the S. Then, one edge being covered by twice means two incompatible bundles are in S. Its constraint can also be a node packing type: each node represents each good and it is covered if the corresponding good is allocated to a bidder by S. Then, one node being covered twice indicates the good is allocated to two bidders simultaneously.

For an instance i, whether each node in $G_S = (V_S, E_S)$ is covered by it can be represented as a $|V_S|$-dimensional binary vector $\mathbf{i}$ whose k-th bit $i_k = 1$ if the k-th node is covered and 0 otherwise. This is called the *coverage status vector* of i. For the problems of edge types, the coverage status vector is a $|E_S|$-dimensional binary vector. Then, the feasibility function $l(S)$ returns true if and only if $\forall k : \sum_{i \in S} i_k \geq (\leq)1$ for node/edge covering (packing) type.

For example, in the edge packing type of the feasibility check for the WDCA problem, V_S is the set of all bidders and E_S is the set of edges indicating incompatibility between bidders. The coverage status vector of a bidder's bundle i is a $|E_S|$-dimensional binary vector, where the k-th bit is 1 if the k-th edge is covered (edges are indexed by arbitrary pre-defined order). Then, if any bit's sum over all instances in S is greater than 1, S is not feasible and vice versa.

In the agent-authority model, the above inequalities can be examined locally at the central authority's side since he has all instances and the current solution set S, therefore he can construct the G_S and corresponding coverage status vectors for all instances to examine the inequalities. In the all-agents model, no one is allowed to access S, but we can still use the privacy-preserving sum calculation in [12] to examine the inequalities without disclosing any information about S. Each agent controls the bits i_k's that are relevant to his instance (e.g., the k-th incompatibility edge in WDCA problem).

Feasibility Check Conclusion. In conclusion, for various problems, if the feasibility of a set of instances can be examined via above three methods, one can examine the feasibility without leaking each individual's privacy. Depending on the application requirement, the protocol participants may agree on one of the three types which best protects the privacy. Since the declaration of the feasibility check type does not affect privacy protection, we assume this is declared by any third party.

5 Our General Framework Design

If a greedy-class problem's greedy algorithm fits our framework, the problem can be solved with our framework. That is, if an algorithm's weight function can be represented with polynomials and if its feasibility can be evaluated with one of the aforementioned three types of feasibility check, the original algorithm (Algorithm 1) can be converted to the the privacy-preserving one (Algorithm 3).

Algorithm 3. Converted Privacy-Preserving Greedy Algorithm

1: $S := \emptyset$, and define the weight function $w(i, S)$.
2: Given S, compute $w(i, S)$ **with secure computation (Section 4.2)**.
3: Find the $i = \mathrm{argmax}_i w(i, S)$ **with transformation (Section 4.3)**.
4: If $l(S \cup \{i\}) = \mathrm{True}$, $S := S \cup \{i\}$, **where $l(\cdot)$ is evaluated by one of the three feasibility checks (Section 4.4)**.
5: Repeat 2-4 until the termination condition is satisfied.
6: Return S as the final solution set $\hat{S}$.

Note that some algorithms need to evaluate $l(\cdot)$ for the termination condition while other just need to terminate after the loop is iterated over all instances, and $l(\cdot)$ is evaluated by the privacy preserving feasibility check. Next, we show detailed procedures of our framework in different system models (agent-authority model and all-agents model) and give a running example of it.

5.1 Agent-Authority Model

Firstly, the weight of each agent's instance is computed securely. Only the owner of the instance receives the result by using One Aggregater MPEP. Then, agents and the central authority run the privacy-preserving sorting in Section 4.3. The MPEP ([12]) in the sorting is executed with the One Aggregater Model such that only the central authority learns the polynomial results. When the privacy-preserving sorting is finished, the central authority gets a set of transformed weights of agents' instances as well as a list of the instances in the order of their weights. Secondly, the central authority picks the first instance i in the sorted list and evaluates $l(S \cup \{i\})$. If the problem's feasibility is an algebra-based one, he and the agents are repeatedly engaged in the privacy-preserving feasibility check in Section 4.4, and the MPEP in the check is executed with the One Aggregater Model again so that only the central authority achieves the evaluation result. On the other hand, if the problem's feasibility is a graph-based one or a set-based one, the central authority checks the feasibility at his side locally. If the feasibility check returns true, S and $\{i\}$ are merged to form a new S.

These two steps are repeated until the termination condition is satisfied. When the algorithm terminates, the central authority achieves the global solution set $\hat{S}$ without knowing any agent's private information, and all agents do not gain any information about $\hat{S}$ either.

5.2   All-Agents Model

Firstly, each agent achieves his own weight via privacy-preserving weight computation (Section 4.2). Then, they run the privacy-preserving sorting as well, but the MPEP is executed with the Participants Only Model, where all the participants learn the polynomial results. When the privacy-preserving sorting is finished, the agents gets a set of transformed weights of all instances, and each agent knows the rank of his instance among all instances in terms of the weight. Secondly, the feasibility of $S \cup \{i\}$ should be checked in the order of the instances' weight, therefore the participants jointly and repeatedly run the feasibility check in Section 4.4. If i is the k-th instance in the sorted list, $l(S \cup \{i\})$ is checked at the k-th round of the feasibility check, and S includes all instances who have returned 'True' so far. Then, any one of the three feasibility checks in 4.4 can be used to check $S \cup \{i\}$'s feasibility depending on the problem. At each round, if the $S \cup \{i\}$ is feasible, i is merged in S to form a new intermediate solution set $S := S \cup \{i\}$.

These two steps are repeated until the termination condition is satisfied. When the algorithm terminates, every agent knows whether his instance is included in the final solution set $\hat{S}$ but nothing else. In fact, no one in the system has any information about $\hat{S}$ in this model.

5.3   Running Example of the WSCP

We convert the greedy algorithm for WSCP in all-agents model in this section.

Algorithm 4. Greedy algorithm for the WSCP

1: $S := \emptyset$. The weight is defined as Eq. 1 in Section 4.
2: Given S, compute $w(i, S)$ for each instance $i \in I$.
3: Sort the instances in the non-increasing order of the weight $w(i, S)$ and find the $i = \text{argmax}_i w(i, S)$.
4: If $\neg l(S \cup \{i\}) = \text{True}$, $S := S \cup \{i\}$.
5: Repeat 2-4 until $\neg l(S \cup \{i\}) = \text{False}$.
6: Return $S \cup \{i\}$ as $\hat{S}$.

At the first iteration, each agent a_j locally computes his weight $w(i_j, S) = \frac{|i_j|}{d(i_j)}$. Then, the agents participate in the instance sorting (Section 4.3) to receive the transformed weights of all instances. Every agent locally sorts the instances based on the transformed weight, and the owner of the $i = \text{argmax}_i(w(i, S) + \delta)\delta'$ knows that his instance is in $\hat{S}$. In the next iteration, weight computation (presented in Section 4.2) is conducted via One Aggregater MPEP to update each instance's weight, where only the owner of the instance receives the result. At this computation, the owner of the instances in the solution set (i.e., $i \in S$) will set the corresponding $c_{j,k,S} = 1$ in the weight computation. Then, the instance sorting with new random numbers δ, δ' is run again to let all agents know their own rank. They run the feasibility check (Section 4.4) to see if $\neg l(S \cup \{i\}) = \text{True}$ where i is the instance with the maximal weight. If yes, the owner of i knows that his instance is in $\hat{S}$. This is repeated until $\neg l(S \cup \{i\}) = \text{False}$, the owner of the last instance i also knows that his instance is included in the $\hat{S}$. Everyone else learns that his instance is not in the final solution set $\hat{S}$.

Adversaries' Advantage on Private Information. Private information might be leaked in the following three parts: weight computation, instance sorting based on transformed weights, and the feasibility check involving private information and $w(i, S)$.

Theorem 1. *Assuming the discrete logarithm is hard, the adversary's advantage adv_i is a negligible function.*

6   Performance Evaluation

6.1   Computation Overhead

We implemented the framework using the GMP library (gmplib.org) based on C in a computer with Intel i3-2365M CPU @ 1.40 GHz ×4, Memory 4GB and SATA Hard Drive 500GB (5400RPM).

Micro-Benchmark. Since various problems have different $\#_{poly}$ for the weight computation and the feasibility check, we present the computation overhead of a single addition and a single multiplication for them. We measured the average run time of 10,000 additions and 10,000 multiplications of two random numbers in $\mathbb{Z}_p$ respectively, which is shown in Figure 1(a) and 1(b). The integer group size is bit length of p, i.e., the security parameter κ. Note that each operation (either addition or multiplication) is in the order of microseconds, and therefore the overall run time will be of several seconds unless the order of the operations number introduced by the framework is greater than 1 billion, which is unlikely. This shows that our framework is very lightweight.

The Figure 1(c) and 1(d) show the computation overhead of the instance sorting based on their weights. We randomly generated a 20-bit weight for each agent and conducted the weight transformation as well as the final sorting based on the transformed weights. Quicksort is used in the sorting, and we observed that the sorting's computation overhead is almost negligible. The figures show the run time of the central authority in the agent-authority model. In the all-agents model, all of the agents have the same computation overhead as the central authority in the agent-authority model because everyone needs to compute the final weights based on the received ciphertexts.

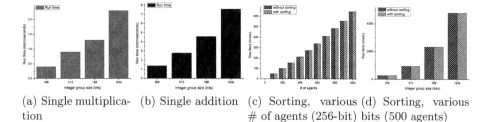

(a) Single multiplication (b) Single addition (c) Sorting, various # of agents (256-bit) (d) Sorting, various bits (500 agents)

Fig. 1. Run time of a multiplication, addition, and sorting

Extra Overhead Measurement. We measured the run times of the original greedy algorithms and the ones of converted algorithms via our framework respectively for the following four problems: WDCA, Knapsack, Job scheduling and Weighted set cover problem, and they are shown in 1. Network delay and I/O delay are excluded from the measurement. Note that everything is disclosed to the authority in the original algorithm (no privacy consideration), and thus agents do not compute anything. As shown in the table, the extra computation overhead varies greatly for different problems due to different types of corresponding greedy algorithms.

Table 1. Comparison

Problem	Original	Converted
Authority		
WDCA	5.87s	603s
Knapsack	4ms	273ms
Scheduling	7ms	11.2s
WSCP	9.31s	135s
Each agent		
WDCA	n/a	8.1s
Knapsack	n/a	7.9ms
Scheduling	n/a	10ms
WSCP	n/a	350s

7 Conclusion

We designed a framework for multi-agent greedy algorithms in which the final solution comes from multiple agents' input instances. We use our novel secure weight computation, privacy-preserving max finding, and privacy-preserving feasibility check to prevent underlying privacy leakage in the distributed greedy algorithms. We showed that our framework does not leak useful information about the agents' private information, and we also showed that the extra computation overhead introduced by our framework is small. In addition, the communication overhead is much less than that of other general solutions as well. All these are evidence that our framework is a viable option for the business intelligence in the big data context.

References

1. Bo, C., Shen, G., Liu, J., Li, X.-Y., Zhang, Y., Zhao, F.: Privacy. tag: privacy concern expressed and respected. In: SenSys, pp. 163–176. ACM (2014)
2. Chen, X., Wu, X., Li, X.-Y., He, Y., Liu, Y.: Privacy-preserving high-quality map generation with participatory sensing. In: INFOCOM, pp. 2310–2318. IEEE (2014)
3. Zhao, J., Jung, T., Wang, Y., Li, X.: Achieving differential privacy of data disclosure in the smart grid. In: INFOCOM, pp. 504–512. IEEE (2014)

4. Modi, P.J., Shen, W.-M., Tambe, M., Yokoo, M.: Adopt: Asynchronous distributed constraint optimization with quality guarantees. A.I. (2005)
5. Boyd, S., Parikh, N., Chu, E., Peleato, B., Eckstein, J.: Distributed optimization and statistical learning via the alternating direction method of multipliers. Foundations and Trends in Machine Learning (2011)
6. Shi, Y., Hou, Y.T.: A distributed optimization algorithm for multi-hop cognitive radio networks. In: INFOCOM. IEEE (2008)
7. Yokoo, M., Suzuki, K., Hirayama, K.: Secure distributed constraint satisfaction: Reaching agreement without revealing private information. A.I. (2005)
8. Silaghi, M.C., Mitra, D.: Distributed constraint satisfaction and optimization with privacy enforcement. In: IAT. IEEE (2004)
9. Silaghi, M.C., Yokoo, M.: Nogood based asynchronous distributed optimization (adopt ng). In: AAMAS. ACM (2006)
10. Yokoo, M., Durfee, E.H., Ishida, T., Kuwabara, K.: The distributed constraint satisfaction problem: Formalization and algorithms. IEEE Transactions on Knowledge and Data Engineering (1998)
11. Zivan, R.: Anytime local search for distributed constraint optimization. In: AAMAS (2008)
12. Jung, T., Mao, X., Li, X.-Y., Tang, S.-J., Gong, W., Zhang, L.: Privacy-preserving data aggregation without secure channel: multivariate polynomial evaluation. In: INFOCOM, pp. 2634–2642. IEEE (2013)
13. Yao, A.C.: Protocols for secure computations. In: FOCS (1982)
14. Yao, A.C.-C.: How to generate and exchange secrets. In: FOCS. IEEE (1986)
15. Even, S., Goldreich, O., Lempel, A.: A randomized protocol for signing contracts. Communications of the ACM (1985)
16. Jung, T., Li, X.-Y.: Enabling privacy-preserving auctions in big data. In: BigSecurity, INFOCOM Workshops. IEEE (2015)
17. Zhang, L., Jung, T., Feng, P., Liu, K., Li, X.-Y., Liu, Y.: Pic: enable large-scale privacy-preserving content-based image search on cloud. In: ICPP. IEEE (2015)
18. Zhang, L., Jung, T., Liu, C., Ding, X., Li, X.-Y., Liu, Y.: Pop: privacy-preserving outsourced photo sharing and searching for mobile devices. In: ICDCS. IEEE (2015)
19. Zhang, L., Li, X.-Y., Liu, Y.: Message in a sealed bottle: privacy preserving friending in social networks. In ICDCS, pp. 327–336. IEEE (2013)
20. Yokoo, M., Ishida, T., Durfee, E.H., Kuwabara, K.: Distributed constraint satisfaction for formalizing distributed problem solving. In: ICDCS. IEEE (1992)
21. Jung, T., Li, X.-Y., Wan, Z., Wan, M.: Privacy preserving cloud data access with multi-authorities. In: INFOCOM, pp. 2625–2633. IEEE (2013)
22. Zhang, L., Li, X.-Y., Liu, Y., Jung, T.: Verifiable private multi-party computation: ranging and ranking. In: INFOCOM, pp. 605–609. IEEE (2013)
23. Jung, T., Li, X., Wan, Z., Wan, M.: Control cloud data access privilege and anonymity with fully anonymous attribute based encryption. TIFS 10(1), 190–199 (2014)
24. Li, X.-Y., Jung, T.: Search me if you can: privacy-preserving location query service. In: INFOCOM, pp. 2760–2768. IEEE (2013)
25. Zhang, L., Li, X., Liu, K., Jung, T., Liu, Y.: Message in a sealed bottle: Privacy preserving friending in mobile social networks. TMC (2014)
26. Wikipedia. http://en.wikipedia.org/wiki/Set_cover_problem
27. Jung, T., Li, X.-Y.: Collusion-tolerable privacy-preserving sum and product calculation without secure channel. TDSC 12(1), 45–57 (2014)

Networking Big Data: Definition, Key Technologies and Challenging Issues of Transmission

Weigang Hou[1,2,3(✉)], Pengxing Guo[1], and Lei Guo[1]

[1] College of Information Science and Engineering, Northeastern University,
Boston 110819, China
houweigang@ise.neu.edu.cn
[2] State Key Laboratory of Networking and Switching Technology, Beijing 100876, China
[3] State Key Laboratory of Information Photonics and Optical Communications,
Beijing 100876, China

Abstract. The big data has been touted as the new oil, which is expected to transform our society. Specially, the data source from the networking domain (networking big data) has higher volume, velocity, and variety compared with others. Thus in this article, we make a short survey on existing works investigating key technologies of networking big data, and propose challenging issues of transmission that is the most important stage for networking big data.

Keywords: Big data · Networking big data · Transmission

1 Introduction

For simplicity, big data is a term for massive datasets and exhibits unique characteristics as compared with traditional data, mainly including volume velocity and variety [1, 2]. First of all, the sheer volume of datasets now is larger than terabytes and petabytes. Next, the velocity represents the data generation rate and real-time requirement. Naturally, big data is usually used as streams in order to maximize its value. Finally, big data comes from a great variety of sources and generally has three types: structured, semi structured and unstructured. Structured data inserts a data warehouse already tagged and easily sorted but unstructured data is random and difficult to analyze. Semi-structured data does not conform to fixed fields but contains tags to separate data elements. And later, a new character named veracity was added to big data, and it was the ability to trust the data to be accurate and reliable when making crucial decisions. So, veracity was usually measured by reliability, privacy and security, but it is not within the scope of this article.

Networking, including the Internet and the mobile network, has penetrated into human lives in every possible aspect. Typical network applications, regarded as the networking big data sources, include, but are not limit to, search, websites, and click streams. These sources are generating data at record speeds, demanding advanced technologies. For example, Google, a representative search engine, was processing 20 PBs a day in 2008 [3]. For social network applications, Facebook stored, accessed,

Y. Wang et al. (Eds.): BigCom 2015, LNCS 9196, pp. 103–112, 2015.
DOI: 10.1007/978-3-319-22047-5_9

and analyzed more than 30 PBs of user-generated data. Over 32 billion searches were performed per month on Twitter [4]. In the mobile network field, more than 4 billion people, or 60 percent of the world's population, were using mobile phones in 2010, and approximately 12 percent of these people had smart phones [5]. So, the term networking big data was coined to capture the meaning of this emerging trend.

The big data has been touted as the new oil, which is expected to transform our society. Revealing hidden patterns and secret correlations among massive amounts of data is meaningful for service providers to gain rich market insights and get an advantage over the competition. For this reason, the key technologies of storing, analyzing, processing and transmitting networking big data have attained attentions.

The rest of this article is organized as follows. In section 2, we make a short survey on existing works investigating aforementioned key technologies of networking big data. In section 3, we point out some future challenging issues of transmitting networking big data before concluding this article in section 4.

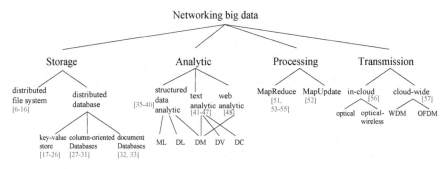

Fig. 1. Classification of existing works

2   Existing Works

In this section, we make a short survey on existing works investigating key technologies of networking big data, which can be represented by Fig. 1.

2.1   Storage

Huge amount of networking data makes the traditional data storage system collapse. And thus, it has driven the storage industry to take a new approach to scalable data storage. The new function should be necessary to meet the emerging requirements associated with all kinds of growing networking data volumes, such as image, text and video. According to the difference in interfaces and functions, data storage applications divide into two main parts: file system and database. As we are in the big data era, the most applicable technologies are: distributed file system [6-16] and distributed database etc, which can be seen in Fig. 1. The distributed file system stores the data in distributed network nodes and devices. In terms of the distributed database, due to certain essential characteristics, including supporting easy replication and a huge amount of data, the NoSQL database is becoming the standard to cope with networking big data problems. As shown in Fig. 1, currently, there have been three primary types of NoSQL databases: key-value stores [17-26], column-oriented databases [27-31], and document databases [32, 33]. The further description of these three kinds of databases is in [34].

2.2   Analytic

Data analytic addresses information obtained through observation, measurement or experiments about a phenomenon of interest. The aim of data analytic is to extract useful values, suggest conclusions and/or support decision-making. Due to the great diversity of statistical data, the methods of analytic and the manner of application differ significantly. As shown in Fig. 1, according to the application manner, we classify existing works on analyzing networking big data into structured data analytic [35-40], text analytic [41-47], web analytic [48] and so forth. Fig. 1 also demonstrates that each kind of application manner has its own analytic method(s). In general, the existing analytic methods include Machine Learning (ML), Deep Learning (DL), Data Mining (DM), Data Visualization (DV), and Data Clustering (DC), etc. The interested readers can find the detail descriptions of DL, DV and DC in classic review articles [49], [50], and [6], respectively. In addition, ML and DM both have become scientific tutorials.

2.3   Processing

With the help of storage and analytic technologies above, at most cases, the networking big data is processed by MapReduce. The current MapReduce is basically designed for the in-cloud data center, where a number of servers are interconnected with optical fiber cables. In general, MapReduce distributes data to distinct servers and these servers execute data processing in parallel. An introduction to the first MapReduce for parallel data processing was presented by Dean et al. in 2004 [51].

Fig. 2 shows an example of parallel data processing in MapReduce. The network nodes are classified into data processing nodes and a master node. The data processing nodes store data and execute mapping and reduction processes, while the master node schedules tasks in both mapping and reduction processes. When a processing request is injected, the master node finds nodes that store the data appertaining to the injected task (here, nodes A and B). Additionally, the master node selects mappers that execute the mapping process. Nodes A', A" and B', B" are selected as mappers for node A and B, respectively. We choose more than one mappers for nodes A and B, respectively, because reliability and processing velocity can be ensured. Then, nodes A and B transmit replication or data subset to each mapper. The mappers perform mapping process that picks out the required information as intermediate results (analytic). After the mapping process finishes, the master node selects a reducer, which is a processing node executing the reduction process, from mappers (here, node A'). The reducer collects the intermediate results extracted during the mapping process and aggregates them into the final result data in the reduction process.

However, the aforementioned MapReduce paradigm is an inappropriate solution for the low-latency processing. This is because that the inputs of MapReduce are snapshots of stored data whose content is constrained to remain infrequently changed during the data processing. Conversely, the networking big data streams are continuously generated [52]. To overcome this limitation, some identified MapReduce frameworks deploy with pipelines so that the mappers send intermediate results to the reducer as soon as possible [53-55]. Alternatively, the difficulty of expressing online processing using MapReduce has also been motivating the creation of MapUpdate [52]. The update phase has access to intermediate results modeled by the data structures that contain persistent state with an update key, which mitigates the pressure of the event-by-event processing for big data streams.

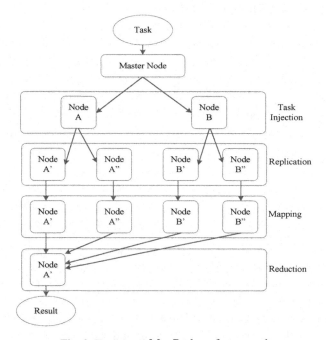

Fig. 2. Traditional MapReduce framework

2.4   Transmission

As shown in Fig. 1, the networking big data transmission mainly has two categories: 1) in-cloud data transmission such as the data transferring among server nodes of the MapReduce framework shown in Fig. 2; 2) cloud-wide data transmission among data centers such as the data delivery from the optical backbone to a data center for subsequent processing.

From the perspective of transmission infrastructure: 1) In-cloud network (e.g., data center network) consists of multiple racks hosting a collection of inter-connected servers. Optical interconnects for data center networks have gained attention recently as a promising solution that offers high throughput, low latency, and reduced energy consumption. Currently, optical technology has been adopted in data centers only for point-to-point links. These links are based on low-cost multi-mode fibers with bandwidth up to 10 Gbps. In addition to this, the in-cloud optical-wireless data center network was also proposed in [56] for MapReduce. In this network, the intra-rack servers communicate by using multicast radio, while the inter-rack communication is conducted via Wavelength Division Multiplexing (WDM) optical fiber cables, which can be seen in Fig. 3. 2) Cloud-wide network (e.g., optical backbone shown in Fig. 4) has deployed IP over WDM over the past two decades. However, to support the data rate elasticity and address the electrical bandwidth bottleneck limitation, Orthogonal Frequency Division Multiplexing (OFDM) has been considered as the better candidate for the future high-speed networking big data transmission. OFDM allows the spectrum of individual subcarriers to overlap, which leads to a more data-rate flexible, agile, and resource-efficient optical network.

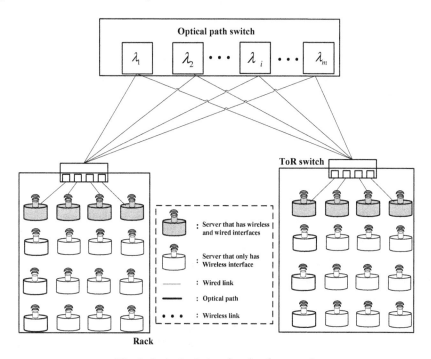

Fig. 3. Optical-wireless in-cloud network

From the perspective of transmission resource scheduling: 1) For the in-cloud network shown in Fig. 3, the authors in [56] proposed a strategy of transmission resource scheduling for MapReduce. Because the master node transmits data to multiple mappers, it is clear that multicast with radio resources is a better choice for the mapping process. While for the reduction process, in which multiple mappers transmit inter-mediate results to a reducer, these intermediate results can be aggregated into a single optical path as long as the total bandwidth cannot exceed a wavelength capacity. Also, they assumed that mappers locate at a single rack, but the reducer belongs to another rack. 2) For the cloud-wide network shown in Fig. 4, the authors in [57] designed a networking big data delivery model. In this model, the network stores the data until it find necessary, or enough transmission resource (e.g., bandwidth) is available for that transfer. Additionally, this model also considered the data entirety since it delivery a series of packets not one by one. In a word, by relaxing the delivery time of network-ing big data according to their respective priorities, the transmission resources may be used more efficiently and network congestion can be alleviated.

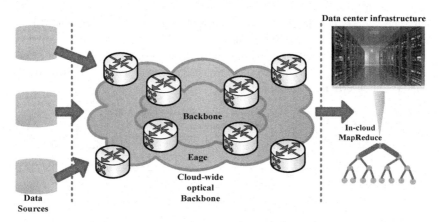

Fig. 4. Cloud-wide optical backbone

Finally, we count the number of references (from IEL source: not totally extensive but can reflect desired trend) supporting aforementioned four key technologies, re-spectively, in Fig. 5. Compared with other key technologies, the transmission of net-working big data is still seldom investigated. But in fact, transmission is the most important stage for big data. As an example of the MapReduce shown in Fig. 2, with-out data transferring among servers, the data processing will not likely succeed. Thus in the following, we summarize some challenging issues of networking big data transmission to guide the future research.

3   Challenging Issues of Networking Big Data Transmission

In the visualization tool shown in Fig. 6, we compare the volume, velocity and variety of big data from different sources or applications. In addition, we also take horizontal scalability and relational limitation into account. The horizontal scalability is the ability to join multiple datasets, i.e., the number of users, while the relational limita-tion is accuracy, e.g., the accuracy of search results from Google, etc. From Fig. 6, we can see that the data source from the networking domain (networking big data) has higher volume, velocity, and variety. Therefore,

- **H**uge data volume requires high transmission throughput;
- **R**eal-time data and even streams require fast delivery/transferring rate;
- **S**tructured and unstructured data types determines elastic transmission ability;
- **V**eracity should be ensured by using high-reliable transmission mechanism;
- **A**nalysis result of big data should be highly utilized to optimize transmission behavior.

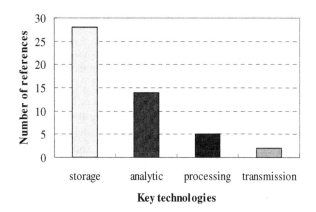

Fig. 5. Popularity of key technologies in existing works

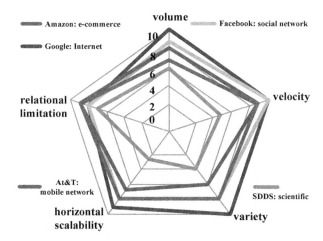

Fig. 6. Visualization tool

The aforementioned issues can be grouped into an entire HRSVA problem. The existing networking big data transmission still cannot well solve this problem, since it neglects two main factors: 1) the external features and requirements of networking big data, as well as 2) the inner structural relation among datasets. In detail,

- Existing data transmission schedules resources in a single dimension, which is rather limited, so the first challenging issue is *"how globally schedule multi-dimension resources (storage, computation, space, spectrum and time, etc) to achieve high-throughput and fast data transmission?"*
- Existing transmission scheme merely transfers semantic irrelevant and transparent streams, which is rather inefficient and unsafe, so the second challenging issue is *"how utilize the inner structural relation among datasets to mitigate invalid data transferring, optimize transmission behavior and improve transmission reliability?"*
- Existing transmission ability is rather rigid, so the third challenging issue is *"how provide elastic transmission ability adaptive to varying data types?"*

In view of three challenging issues above, the possible future research points mainly include: 1) global scheduling of multi-dimension resources, 2) semantic relevance of transmission behavior, and 3) dynamic reconfiguration of transmission ability. The ultimate objective of networking big data transmission is that the transmission efficiency ascends as spiral by using the collaborative utilization of multi-dimension resources. More specifically, we analyze semantic relevance through storage and computation in order to optimize transmission behavior and reliability, utilize OFDM to achieve dynamic reconfiguration of transmission ability, and introduce space- and time-division multiplexing technologies to support high- throughput and fast data transmission.

4   Conclusion

For networking big data, in this article, we have given a detail definition, made a survey on existing works investigating key technologies and proposed some future challenging issues of transmission to well guide interested readers.

Acknowledgements. This work was supported in part by the Open Foundation of State Key Laboratory of Networking and Switching Technology (SKLNST-2013-1-08), the Open Foundation of State Key Laboratory of Information Photonics and Optical Communications (IPOC 2013B001), the Fundamental Research Funds for the Central Universities (N130817002, N130404002), the Foundation of the Education Department of Liaoning Province (L2014089), the National Natural Science Foundation of China (61401082, 61471109), the Liaoning BaiQianWan Talents Program, and the National High-Level Personnel Special Support Program for Youth Top-Notch Talent.

References

1. http://en.wikipedia.org/wiki/Big_data
2. James, M.: Big Data: The next frontier for innovation, competition, and productivity. McKinsey Global Institute (2011)

3. Dean, J., Ghemawat, S.: Mapreduce: Simplified data processing on large clusters. IEEE/ACM Trans. Commun **51**(1), 107–113 (2008)
4. A comprehensive list of big data statistics. http://wikibon.org/blog/big-data-statistics/
5. Manyika, J., et al.: Big data: The next frontier for innovation, competition, and productivity. McKinsey Global Institute, pp. 1–137 (2011)
6. Sagiroglu, S., Sinanc, D.: Big data: a review. In: Proc. CTS, pp. 42–47 (2013)
7. Song, Y., Alatorre, G., Mandagere, N., et al.: Storage mining: where IT management meets big data analytics. In: Proc. Big Data, pp. 421–422 (2013)
8. Wang, Y., Jin, X.: Network big data: present and future. Chinese Journal of Computers **36**(6), 1–15 (2013)
9. Ghemawat, S., Gobioff, H., Leung, S.T.: The google system. In: Proc. SOSP, pp. 29–43 (2003)
10. McKusick, M.K., Quinlan, S.: GFS: Evolution on fast-forward. ACM Queue **7**(7), 10–20 (2009)
11. Hadoop: distributed file system (2013). http://hadoop.apache.org/docs/r1.0.4/hdfsdesign. html
12. Kosmosfs. https://code.google.com/p/kosmosfs/
13. Chaiken, R., et al.: Scope: Easy and efficient parallel processing of massive data sets. In: Proc. VLDB, pp. 1265–1276 (2008)
14. Beaver, D., Kumar, S., Li, H.C., Sobel, J., Vajgel, P.: Finding a needle in Haystack: facebook's photo storage. In: Proc. SOSDI, pp. 1–8 (2010)
15. Taobao file system. http://code.taobao.org/p/tfs/src/
16. Fast distributed file system. https://code.google.com/p/fastdfs/
17. DeCandia, G.: Dynamo: Amazon's highly available key-value store. SIGOPS Oper. Syst. Rev. **41**(6), 205–220 (2007)
18. Karger, D., Lehman, E., Leighton, T., Panigrahy, R., Levine, M., Lewin, D.: Consistent hashing and random trees: distributed caching protocols for relieving hot spots on the world wide web. In: Proc. STC, pp. 654–663 (1997)
19. Voldemort. http://www.project-voldemort.com/voldemort/
20. Redis. http://redis.io/
21. Tokyo Canbinet. http://fallabs.com/tokyocabinet/
22. Tokyo Tyrant. http://fallabs.com/tokyotyrant/
23. Memcached. http://memcached.org/
24. MemcacheDB. http://memcachedb.org/
25. Riak. http://basho.com/riak/
26. Scalaris. http://code.google.com/p/scalaris/
27. Chang, F., et al.: Bigtable: A distributed storage system for structured data. IEEE/ACM Trans. Comput. Syst. **26**(2), 1–26 (2008)
28. Burrows, M.: The chubby lock service for loosely-coupled distributed systems. In: Proc. SOSDI, pp. 335–350 (2006)
29. Lakshman, A., Malik, P.: Cassandra: structured storage system on a p2p network. In: Proc. SPDC, pp. 1–5 (2009)
30. HBase. http://hbase.apache.org/
31. Hypertable. http://hypertable.org/
32. RFC 4627-The application/JSON media type for Javascript object notation (JSON). http://tools.ietf.org/html/rfc4627
33. MongoDB. http://www.mongodb.org/
34. Hu, H., Wen, Y., Chua, T., Li, X.: Towards scalable systems for big data analytics: a technology tutorial. IEEE ACCESS, 652–687 (2014)

35. Hinton, G.E.: Learning multiple layers of representation. Trends Cog-nit. Sci. **11**(10), 428–434 (2007)
36. Baah, G.K., Gray, A., Harrold, M.J.: On-line anomaly detection of deployed software: a statistical machine learning approach. In: Proc. SQA, pp. 70–77 (2006)
37. Moeng, M., Melhem, R.: Applying statistical machine learning to multicore voltage and frequency scaling. In: Proc. Comput. Frontiers, pp. 277–286 (2010)
38. Gaber, M.M., Zaslavsky, A., Krishnaswamy, S.: Mining data streams: A review. ACM SIGMOD Rec. **34**(2), 18–26 (2005)
39. Verykios, V.S., Bertino, E., Fovino, I.N., Provenza, L.P., Saygin, Y., Theodoridis, Y.: State-of-the-art in privacy preserving data mining. ACM SIGMOD Rec. **33**(1), 50–57 (2004)
40. Vander, W.A.: Process mining: Overview and opportunities. IEEE/ACM Trans. Manag. Inform. Syst. **3**(2), 1–17 (2012)
41. Ritter, A., Clark, S., Etzioni, O.: Named entity recognition in tweets: an experimental study. In: Proc. EMNLP, pp. 1524–1534 (2011)
42. Li, Y., Hu, X., Lin, H., Yang, Z.: A framework for semisupervised feature generation and its applications in biomedical literature mining. IEEE/ACM Trans. Comput. Biol. Bioinform. **8**(2), 294–307 (2011)
43. Blei, D.M.: Probabilistic topic models. IEEE/ACM Trans. Commun. **55**(4), 77–84 (2012)
44. Balinsky, H., Balinsky, A., Simske, S.J.: Automatic text summarization and small-world networks. In: Proc. SDE, pp. 175–184 (2011)
45. Mishra, M., Huan, J., Bleik, S., Song, M.: Biomedical text categorization with concept graph representations using a controlled vocabulary. In: Proc. DMB, pp. 26–32 (2012)
46. Huet, J., et al.: Enhancing text clustering by leveraging wikipedia semantics. In: Proc. RDIR, pp. 179–186 (2008)
47. Pang, B., Lee, L.: Opinion mining and sentiment analysis. Found. Trends Inform. Retr. **2**(1), 1–135 (2008)
48. Pal, S.K., Talwar, V., Mitra, P.: Web mining in soft computing framework: Relevance, state of the art and future directions. IEEE Trans. Neural Netw. **13**(5), 1163–1177 (2002)
49. Chen, X., Lin, X.: Big data deep learning: challenges and perspectives. IEEE Access, 514–525 (2014)
50. Olshannikova, E., Ometov, A., Koucheryavy, Y.: Towards big data visualization for augmented reality. In: Proc. CBI, pp. 33–37 (2014)
51. Dean, J., Ghemawat, S.: Mapreduce: simplified data processing on large clusters. In: Proc. SOSDI, pp. 137–150 (2004)
52. Lam, W., Liu, L., Prasad, S., Rajaraman, A., Vacheri, Z., Doan, A.: Muppet: mapreduce style processing of fast data. In: Proc. VLDB, pp. 1814–1825 (2012)
53. Condie, T., Conway, N., Alvaro, P., Hellerstein, J.M., Elmeleegy, K., Sears, R.: MapReduce online. In: Proc. NSDI (2010)
54. Logothetis, D., Yocum, K.: Ad-hoc data processing in the cloud. In: Proc. VLDB, pp. 1472–1475 (2008)
55. Brito, A., Martin, A., Knauth, T., Creutz, S., Becker, D., Weigert, S., Fetzer, C.: Scalable and low-latency data processing with stream MapReduce. In: Proc. CCTS, pp. 48–58 (2011)
56. Suto, K., Nishiyama, H., Kato, N.: Context-aware task allocation for fast parallel big data processing in optical-wireless networks. In: Proc. IWCMC, pp. 423–428 (2014)
57. Sun, W., Li, F., Guo, W., Jin, Y., Hu, W.: Store, schedule and switch–a new data delivery model in the big data era. In: Proc. ICTON, pp. 1–4 (2013)

Smart Phone and Sensing Application

Gender Prediction Based on Data Streams of Smartphone Applications

Yilei Wang, Yuanyang Tang, Jun Ma, and Zhen Qin$^{(\boxtimes)}$

University of Electronic Science and Technology of China, Chengdu, China
qinzhen@uestc.edu.cn

Abstract. Gender information has great values in personalized service, targeted advertising, recommender systems and other aspects. However, such information is kind of private information, that many users are reluctant to share. In this paper, we propose a novel approach to predict the users' gender information by analyzing the data streams of smartphone applications. The proposed approach assumes that certain features extracted from smartphone data streams could represent users' perspective characteristics (*e.g.*, gender). To be more specific, we noticed that male and female have different response time to the data streams of different applications. Thus we extract a key feature – users' *Response-Time* to application. Moreover, by leveraging the key feature to construct training data, and further importing Support Vector Machine (SVM) classifier, we verified that users' gender information could be well predicted. In the experiments, the dataset is real world data collected from 25 volunteers. The prediction results can achieve 86.50% in Accuracy and 86.43% in F1-score, respectively. To the best of our knowledge, this is the first time that the gender information was predicted by leveraging users' response time to smartphone applications.

1 Introduction

In order to provide customized services and optimized user experience, the industrial and academic circles have been exploring practical solutions, which is observed to be the technological tendency lately. One of the examples is Google, who provides customized searching results, *i.e.,* different results may come out even if different users put the same keywords in the search engine [1]. Another example is targeted advertisement, by which corresponding users are recommended to advertisers based on their Internet behaviors [2] and this has been concluded to be more effective than conventional advertising campaign according to the research recently [3].

To achieve the customized services or targeted advertising, as mentioned above, it is essential to study the interests and behaviors of the potential customers, which lays great emphasis on personal data (*e.g.,* browsing histories, searching interests, geographic and even personal information), and among

---

Y. Tang—This author is co-first author.

© Springer International Publishing Switzerland 2015
Y. Wang et al. (Eds.): BigCom 2015, LNCS 9196, pp. 115–125, 2015.
DOI: 10.1007/978-3-319-22047-5_10

which, gender information has shown enormous significance. But such data is usually deeply involved with users' privacy and almost difficult to be obtained through public sources.

As the world witnesses the underlined technological advancements of networking and network-enabled smartphones as well as the tremendous difference made to our life patterns, bigger popularity has also been achieved by smartphones, while various smartphone applications are used by a large number of population in the daily life. Among the users, different individuals might have different interests, which makes it natural for the applications installed in smartphones of different users to be distinct (*e.g.*, females may show more positive response to fashion/shopping-related applications, while males to sports-related ones). For different users, even the same application usage (*e.g.*, the response time) can be varied. Thus, such reflection of the user on his/her smartphone makes it possible to predict the corresponding gender information based on the smartphone applications data streams.

In this paper, we attempt to infer the gender information of smartphone users by leveraging the difference of applications data streams features. Specifically, our proposed prediction method can be processed as follows: (*i*) We develop an Android OS application to collect real world smartphone applications data stream samples. (*ii*) From the data stream samples, we extract a key feature – *Response-Time* to applications, and argue that such feature reflect the characteristics for males and females. (*iii*) By leveraging the key feature, we construct both coarse-grained and fine-grained training data for SVM classifier. (*iv*) Finally, the performance of the proposed approach can be evaluated based on the constructed training data.

The paper is organized as follows. In Section 2, we provide an overview of the work related with gender information prediction. The objective and dataset is introduced in Section 3. In Section 4, the proposed prediction approach is presented. Experimental results to evaluate the approach are shown in Section 5. In Section 6, we conclude the paper and state our future work.

2   Related Work

Gender information prediction has attracted increasing interests recently, for it enriches customized services and intelligent recommendation system. With the popularization of smartphone, the prediction methods have transferred from Internet to Mobile Internet.

Prediction from Internet Behaviors. Prediction from Internet behaviors mainly concentrates on website where users generate content (*e.g.*, blogs) or activity histories (*e.g.*, searching histories). The gender information is usually included in profiles. However, only a limited number of users allow public access to the information, and the information posted may not be accurate and real. Yan et al. [4] used blog content to identify gender of authors, while Garera et al. [5] analyzed users' writing and speaking styles to predict gender information. As for activity histories, Kabbur et al. [6] analyzed the HTMLs, key words and hyperlinks in user browsing histories to predict gender information with machine learning techniques. Ingmar et al. [7] investigated that certain types of search queries can be related to distinct gender groups.

Prediction from Mobile Internet Behaviors. There are two ways to predict from Mobile Internet behaviors so far, including smartphone usage and application usage. Ying et al. [8] used smartphone users' behaviors and environments, where 45 features are extracted to achieve the prediction. Seneviratne et al. [9] proposed that gender information can be predicted by the applications installed on the smartphone, while Qin et al. [10] analyzed the usage of application to predict the gender and age of users. There are also some researchers who took the both parts into account to predict users' gender information [11] [12].

Our prediction approach, however, is a novel way to predict gender information. Based on gender trend of application, we used the data streams generated by applications installed on smartphone, and leverage the response time to different applications to construct training data. To the best of our knowledge, this is the first time that gender information is studied by leveraging users' response time to smartphone applications.

3   Objective and Dataset

3.1   Objective

Before introducing our approach, we firstly define the objective of gender prediction. In the study, we get some smartphone users[1] and record their demographic information. Our objective is to identify smartphone users' gender information based on their smartphone applications' data streams.

An application installed in a smartphone can be related to a user's preference or interest, which could be related with user's gender information [13]. To make the gender prediction, a direct way is to correlate users' gender with these applications, and train classifiers from the aspect of applications installed in the smartphone [9]. However, we argue that there are some pre-installed applications (by smartphone Operation Systems or smartphone manufacturers) that would bias the correlation between users and applications. Considering this, in our approach, we choose these applications which generate data streams. Moreover, males and females could use the same application, but the data stream generated by this application could be significantly different. Thus, training a classifier from the aspect of applications is coarse-grained and could lead to a poor performance. Instead, our approach extracts a key feature from the data streams generated by smartphone applications, and trains a classifier from the aspect of such feature.

3.2   Data Collection

To collect the data streams generated by smartphone applications, we developed an application on Android OS. This application, through the Android OS interface [14], collects the data stream for each installed application separately. The

---

[1] They are volunteers.

application is capable of controlling the frequency of data streams collection. We collected the data stream in every 10 seconds, so we had 360 samples for every application per hour. In each 10-second-slot sample, the following features are recorded:

- **Application-Name:** The application name in a 10-second-slot sample.

- **Arrive-Time:** The time when an application generate data stream. For simplicity, if an application generates data stream in a 10-second-slot, the *Arrive-Time* is recorded as the slot's first second. Otherwise, the *Arrive-Time* is recorded as a *null* value.

- **Foreground-Time:** The time when an application is exactly using on the screen of smartphone, not in the background. For simplicity, if an application is used in the foreground in a 10-second-slot, the *Foreground-Time* is recorded as the slot's first second. Otherwise, the *Foreground-Time* is recorded as a *null* value.

3.3   Dataset

In the dataset, it covers 14 days of data streams from 25 users and includes 6,474,260 10-second-slot samples. These users are volunteers of the study, including 12 males and 13 females. To protect the privacy of users, their names are anonymous. There are 414 total applications which generate data streams, including 143 pre-installed applications and 271 user-installed applications. There are two types of pre-installed applications: the Android OS applications and the applications pre-installed by smartphone manufacturers. The user-installed applications are installed by users for their daily use. These applications are actually related to users' traits of interest. So we assume that such applications are closely related to user's gender information. The results shown in Section 5 will verify our assumption. In order to show that the 271 user-installed applications can be taken as representatives, we have investigated the top 50 most popular applications of the top 10 most popular Android Application Markets in China, and get 165 distinct applications. We find that the 271 user-installed applications cover 64.85% of the 165 most popular applications.

Here, we make a brief analysis of the dataset. In terms of volunteers, as shown in Fig. 1, the number of the applications which generated data streams ranges from 19 to 60, of which the average is about 36. In addition, the user-installed applications consist of the majority of the data streams generated by all applications, ranging from 42.11% to 96.67%. It means that users prefer using the user-installed applications to using the pre-installed applications. In terms of applications, as shown in Fig. 2, the applications in the dataset are labeled into six classes, the horizontal axis refers to the quantity of users have such

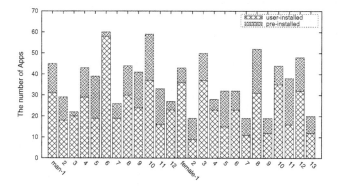

Fig. 1. Users Application Distribution

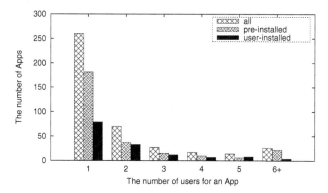

Fig. 2. Applications' User Distribution

applications, and the ordinate axis refers to the number of such applications. For example, in the dataset, there are 70 applications which is installed by only 2 volunteers, including 37 (65.46%) user-installed applications and 33 (34.54%) pre-installed applications. In our approach, we mainly use the data streams from user-installed applications to make the prediction, and compared the prediction results for different application classes.

4   Proposed Prediction Approach

In this section, we state the proposed prediction approach. In particular, by dealing with the applications' data streams, we propose a novel way to predict smartphone users' gender information. Firstly, we introduce the key feature extracted from the data streams. Then, we argue that the smartphone users with different gender information have different data stream features. It means that such key feature can be leveraged to construct the training data and further predict the gender information. At last, we import the training data to Support Vector Machine (SVM) classifier and make the prediction.

4.1 Key Feature Extraction

In Section 3, when collecting the data streams, we have recorded three features (*e.g., Arrive-Time, Foreground-Time, Application-Name*) of each 10-second-slot sample. Now, motivated by the idea that the response time to different applications' data streams could be different for males and females, we are going to extract a key feature.

Table 1. The Average Value of *Response-Time*

Application	Mobile QQ	WeChat	UC Browser
Man	2245	684	1218
Female	3695	1667	1449
All	2999	1195	1338

In each user's data stream samples, when sorted by the *Application-Name*, an application can have a group of 10-second-slot samples which record the *Arrive-Time* and *Foreground-Time* slot by slot. For an application, once an *Arrive-Time* in a 10-second-slot is not *null* value, we find the next *non-null Foreground-Time* in following 10-second-slots, and calculate the time between the *Arrive-Time* and the *Foreground-Time*, which we call *Response-Time*. The *Response-Time* is the key feature, which means the period from the time an application generates data stream to the latest time the user uses this application in the foreground. When the key feature is associated with users' gender information, it can reflect the response time (or interest) of different genders to a particular application (*e.g., male's response time to Mobile QQ*). Considering users' gender difference, the average *Response-Time* values of the 3 most popular user-installed applications (*e.g., Mobile QQ, WeChat and UC Browser*) are shown in Table 1. Mobile QQ and WeChat are Instant Messaging (IM) applications, and UC Browser is a browser application. It can be inferred that females response more quickly to IM applications. Later, we will show that this property can be further leveraged to make gender prediction.

4.2 Training Data Construction

Here, we are going to construct the training data. By computing the key feature, we can have numbers of *Response-Time*. Next, for each user, we group the *Response-Time* by associating it with *Application-Name*. Thus, each user is correlated with multiple groups of *Response-Time*. For a group of *Response-Time*, the value might vary from seconds to hours. Fig. 3 shows the distribution of two

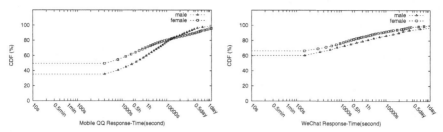

(a) Mobile QQ' Response-Time Distribu- (b) WeChat' Response-Time Distribution
tion

Fig. 3. Response-Time Distribution With Gender

groups of *Response-Time* for males and females. It can be noticed that there is a difference in distribution for different genders. By correlating the user gender, *Application-Name* and *Response-Time*, we constructed a **coarse-grained training data** for classifiers.

Moreover, it is possible to make a better use of the key feature, and provide a fine-grained training data. We process as follows: for an application, we label it's *Response-Time* group into multiple categories according to the value of *Response-Time* (*e.g.,* label the *Response-Time* of 0 hour as *category1*, label the *Response-Time* between 0 hour and 0.5 hour as *category2*, label the *Response-Time* between 0.5 hour and 1 hour as *category3*, etc). On the basis of these categories, a user's *Response-Time* to applicaitons can be modeled as a weighted graph $G = (V,E)$, where a node in V represents an application or a category, and an edge in E represents the weight between an application and a category. The nodes in V can be divided into two subsets, $A = a_1, a_2, ..., a_m$ and $C = c_1, c_2, ..., c_n$, where A represents the applications and C represents the categories. Next, the graph G can be represented by an adjacent matrix R, whose element r_{ij} is the number of application a_i's *Response-Time* in category c_j. In this way, the adjacent matrix R is the **fine-grained training data**. It is a feature matrix extracted from user's *Response-Time* groups.

4.3   Gender Classification

To classify users' gender information, we trained binary Support Vector Machine (SVM) classifiers [15], by importing both coarse-grained training data and fine-grained training data, separately. In the next section, we will compare the prediction performance when using different training data. Because we only have 25 volunteers in the dataset, we trained binary SVM classifiers in the way of Cross Validation [16], as shown in Algorithm 1.

Input: S: the data stream sets of 25 volunteers
Output: T: the $2 * 2$ matrix
1 **for** $i = 1; i \leq 25$ **do**
2   **for** $j = 1; Application[j] \in S[i]$ **do**
3          $key\text{-}feature[j] = ExtractFeature(Application[j])$;
4          $Feature\text{-}Matrix[i] \leftarrow key\text{-}feature[j]$;
5   **end**
6 **end**
7 **for** $i = 1; i \leq 25$ **do**
8      $Classifier[i] = Train(Feature\text{-}Matrix[1], ..., Feature\text{-}Matrix[i-1], Feature\text{-}Matrix[i+1], ..., Feature\text{-}Matrix[25])$;
9      $T = Validation(Classifier[i], Feature\text{-}Matrix[i])$;
10 **end**

Algorithm 1. User Gender Classification

5 Performance Evaluation

In this section, we firstly state the evaluation metrics, then show the prediction results by importing different training data.

5.1 Metrics

To verify the effectivity of our approach, $Accuracy(Acc)$, $Precision(Prec)$, $Recall(Rec)$ and $F1\text{-}score(F1)$ [17] are used to evaluate the performance. We define (m, f) as the number of males were predicted as females. (m, m), (f, f) and (f, m) are defined in a similar way. Acc is the proportion of correctly predicted users in all the users, shown in Formula 1. $Prec$ is the proportions of correctly predicted users in all the users that were predicted as the same gender. Male's $Prec$ is shown in Formula 2. Rec is the proportions of correctly predicted users in all the users of the same gender. Male's Rec is shown in Formula 3. $F1$ is a combination of $Prec$ and Rec, and is shown in Formula 4. Also the $Prec$, Rec and $F1$ of female are in the similar way. our performance evaluation are in Acc and $F1$ [10].

$$Acc = \frac{(m, m) + (f, f)}{(m, m) + (m, f) + (f, m) + (f, f)} \tag{1}$$

$$Prec = \frac{(m, m)}{(m, m) + (f, m)} \tag{2}$$

$$Rec = \frac{(m, m)}{(m, m) + (m, f)} \tag{3}$$

$$F1 = \frac{2Prec \times Rec}{Prec + Rec} \tag{4}$$

Table 2. Prediction Results Based on Coarse-Grained Training Data

Users		Rec	Prec	Micro F1	Macro F1	Acc
⩾5	Male	57.29%	52.55%	54.82%	54.67%	54.67%
	Female	52.24%	56.99%	54.52%		
⩾4	Male	58.33%	55.63%	56.95%	57.65%	57.67%
	Female	57.05%	59.73%	58.36%		
⩾3	Male	59.03%	59.23%	59.13%	60.77%	60.83%
	Female	62.50%	62.30%	62.40%		
⩾2	Male	67.01%	73.61%	73.11%	75.99%	76.33%
	Female	84.94%	80.42%	78.87%		
⩾1	Male	63.19%	65.70%	64.42%	66.39%	66.50%
	Female	69.55%	67.18%	68.35%		

Table 3. Prediction Results Based on Fine-Grained Training Data

Users		Rec	Prec	Micro F1	Macro F1	Acc
⩾5	Male	54.93%	71.92%	68.25%	70.78%	71.00%
	Female	76.60%	70.29%	73.31%		
⩾4	Male	72.22%	78.79%	75.36%	77.19%	77.33%
	Female	82.05%	76.19%	79.01%		
⩾3	Male	71.53%	84.77%	77.59%	79.90%	80.17%
	Female	88.14%	77.03%	82.21%		
⩾2	Male	82.64%	88.48%	85.46%	86.43%	86.50%
	Female	90.06%	84.89%	87.40%		
⩾1	Male	75.00%	88.89%	81.36%	83.28%	83.50%
	Female	75.00%	88.89%	81.36%		

5.2   Experimental Results

In the experiments, we evaluated the prediction results. Table 2 summarizes the prediction results with coarse-grained training data, while Table 3 summarizes the prediction results with fine-grained training data.

In Table 2, there are five cases of experiment results. $\geqslant n$ means that the applications are installed by more that n users in the dataset, where the *Response-Time* of such applications are used to construct the coarse-grained training data. For example, by constructing the coarse-grained training data from the *Response-Time* values of 53 applications that more than 3 users installed, the prediction performance is 60.77% in $F1$ and 60.83% in *Acc*. In particular, with the decrease of n, the prediction performance goes better as more applications' *Response-Time* values are transferred into training data construction process.

It should be noticed that, the prediction result of $\geqslant 2$ is better than that of $\geqslant 1$. This is because when n equals 1, the occasionality of training examples increases.

Table 3 shows the experiment results while the input to SVM classifier is the fine-grained training data – the feature matrix. The results are also in five cases (*e.g.*, $\geqslant 5$, $\geqslant 4$, etc). For the same case, both the values of $F1$ and Acc are better than the values shown in Table 2. This means that the fine-grained training data can improve the prediction performance compare to coarse-grained training data. In addition, the better prediction performance is also observed according to the above elaboration.

6   Conclusions and Future Work

This paper proposed a novel approach to predict gender information by leveraging the difference in *Response-Times* to smartphone applications. Our contributions are as follows. Firstly, we develop an Android OS application which can collect the data stream samples for all the applications installed on smartphone. Secondly, we extract a key feature from the collected data stream samples, and further construct a fine-grained training data (a feature matrix) to improve the prediction performance. The experimental results show that our approach achieves 86.43% and 86.50% in terms of $F1$ and Acc for gender prediction, respectively. In the future work, we plan to extract more features from data streams and make more challenging prediction (*e.g.*, age, marital status and is occupation). Furthermore, we would do such studies based on other information generated by smartphones (*e.g.*, the usage of memory, acceleration sensor and other embedded sensors).

Acknowledgments. This work was supported in part by the Ministry of Education - China Mobile Research Foundation (No.MCM20121041), the National Science Foundation of China (No.61133016, No.61300191, No.61202445 and No.61370026), the Sichuan Key Technology Support Program (No.2014GZ0106), the National Science Foundation of China - Guangdong Joint Foundation (No.U1401257), and the Fundamental Research Funds for the Central Universities (No.ZYGX2013J003 and No.ZYGX2014J066).

References

1. Hannak, A., Sapiezynski, P., Molavi Kakhki, A., Krishnamurthy, B., Lazer, D., Mislove, A., Wilson, C.: Measuring personalization of web search. In: Proceedings of the 22nd International Conference on World Wide Web, pp. 527–538 (2013)
2. Smit, E.G., Van Noort, G., Voorveld, H.A.: Understanding online behavioural advertising: User knowledge, privacy concerns and online coping behaviour in europe. Computers in Human Behavior **32**, 15–22 (2014)
3. Jansen, B.J., Moore, K., Carman, S.: Evaluating the performance of demographic targeting using gender in sponsored search. Information Processing & Management **49**(1) (2013)

4. Yan, X., Yan, L.: Gender classification of weblog authors. In: AAAI Spring Symposium: Computational Approaches to Analyzing Weblogs, pp. 228–230 (2006)
5. Garera, N., Yarowsky, D.: Modeling latent biographic attributes in conversational genres. In: Proceedings of the Joint Conference of the 47th Annual Meeting of the ACL and the 4th International Joint Conference on Natural Language Processing of the AFNLP, vol. 2, pp. 710–718. Association for Computational Linguistics (2009)
6. Kabbur, S., Han, E.H., Karypis, G.: Content-based methods for predicting website demographic attributes. In: 2010 IEEE 10th International Conference on Data Mining (ICDM), pp. 863–868. IEEE (2010)
7. Weber, I., Jaimes, A.: Demographic information flows. In: Proceedings of the 19th ACM International Conference on Information and Knowledge Management. ACM (2010)
8. Ying, J.J.C., Chang, Y.J., Huang, C.M., Tseng, V.S.: Demographic prediction based on users mobile behaviors. Mobile Data Challenge (2012)
9. Seneviratne, S., Seneviratne, A., Mohapatra, P., Mahanti, A.: Predicting user traits from a snapshot of apps installed on a smartphone. ACM SIGMOBILE Mobile Computing and Communications Review 18(2), 1–8 (2014)
10. Qin, Z., Wang, Y., Xia, Y., Cheng, H., Zhou, Y., Sheng, Z., Leung, V.: Demographic information prediction based on smartphone application usage. In: 2014 International Conference on Smart Computing (SMARTCOMP), pp. 183–190. IEEE (2014)
11. Laurila, J.K., Gatica-Perez, D., Aad, I., Bornet, O., Do, T.M.T., Dousse, O., Eberle, J., Miettinen, M., et al.: The mobile data challenge: big data for mobile computing research. In: Pervasive Computing. Number EPFL-CONF-192489 (2012)
12. Hamka, F., Bouwman, H., De Reuver, M., Kroesen, M.: Mobile customer segmentation based on smartphone measurement. Telematics and Informatics 31(2), 220–227 (2014)
13. Kosinski, M., Stillwell, D., Graepel, T.: Private traits and attributes are predictable from digital records of human behavior. Proceedings of the National Academy of Sciences 110(15), 5802–5805 (2013)
14. Au, K.W.Y., Zhou, Y.F., Huang, Z., Gill, P., Lie, D.: Short paper: a look at smartphone permission models. In: Proceedings of the 1st ACM Workshop on Security and Privacy in Smartphones and Mobile Devices, pp. 63–68. ACM (2011)
15. Deng, N., Tian, Y., Zhang, C.: Support vector machines: optimization based theory, algorithms, and extensions. CRC Press (2012)
16. Kohavi, R., et al.: A study of cross-validation and bootstrap for accuracy estimation and model selection. In: IJCAI, vol. 14, pp. 1137–1145 (1995)
17. Hu, J., Zeng, H.J., Li, H., Niu, C., Chen, Z.: Demographic prediction based on user's browsing behavior. In: Proceedings of the 16th International Conference on World Wide Web, pp. 151–160. ACM (2007)

Anti-multipath Indoor Direction Finding Using Acoustic Signal via Smartphones

Xiaopu Wang, Yan Xiong, and Wenchao Huang$^{(\boxtimes)}$

University of Science and Technology of China, Hefei, China
wangxp88@mail.ustc.edu.cn, {yxiong,huangwc}@ustc.edu.cn

Abstract. Direction Finding plays a significant role in indoor localization research. We introduce a novel scheme of anti-multipath indoor direction finding via smartphones. This scheme does not rely on any fingerprints or specialized devices. Users only need to move their smartphones over a short distance for finding the anchor points (ordinary speakers) we preset in the scene. Due to the intricate indoor environment, the multipath effect has pronounced negative influence to the direction finding accuracy. Our work is aimed at mitigating the adverse impact of multipath effect which is based on MIMO (multiple-input, multiple-output) technique. However, processing the signals received on an array of sensors is the pith of MIMO. As far as we know, the directional antenna is a specialized device which is expensive and bulky. Hence, we propose a method of sampling and processing high frequency acoustic signals to emulate a virtual antenna array. In this case, we leverage smoothed MUSIC (Multiple Signal Classification) algorithm to estimate the DOA (direction of arrival) of the target signals. We design and deploy our system under various circumstances in an underground parking lot. By getting rid of the multipath effect interference, the extensive experiments show that our system of the direction finding are precise and high-efficiency in daily use.

Keywords: Direction Finding · Multipath effect · Antenna arrays simulation · MUSIC algorithm

1 Introduction

With the development of social networking, Augmented Reality (AR) and mobile hand-held applications, a fast and accurate localization technology is the key to enhance user experience. GPS (Global Positioning System) used to be the answer to these localization problems. But it does not apply to any occasions such as indoor scenarios which is called 'the last mile problem of GPS'. Providing an accurate localization solution is the new challenge to researchers in this area.

W. Huang—The research is supported by National Natural Science Foundation of China under Grant No.61202404, No.61170233, No.61232018, No.61272472, No.61272317, Anhui Provincial Natural Science Foundation, No.1508085SQF215, and the Fundamental Research Funds for the Central Universities, No. WK0110000041.

© Springer International Publishing Switzerland 2015
Y. Wang et al. (Eds.): BigCom 2015, LNCS 9196, pp. 126–140, 2015.
DOI: 10.1007/978-3-319-22047-5_11

Most of the indoor localization researches can be divided into: (1)**Direction Finding**: to find the direction of users' destination; (2)**Localization**: to locate the user's relative position in the indoor environment; (3)**Tracking**: to track the constantly changing direction and location information while user is moving.

Direction Finding (DF) technique has wide application prospect and it is also the theoretical bases of many localization and tracking systems. Currently there are a lot of relevant researches on direction finding already. [1]estimates the direction by using Doppler effect which solving the phone to phone direction finding problem by shaking the cellphone. But this research does not take multipath effect into account. [12] effectively emulates the sensitivity and functionality of a directional antenna by rotating the phone around the users body to locate outdoor APs. [4] leverages MIMO techniques to estimate the AOA (angle of arrival) of Wi-Fi signals with antenna arrays.

We propose an accurate and light indoor direction finding scheme by selecting the high-frequency narrow-band acoustic signals as research subjects. This kind of signal is less common than Wi-Fi signals or light signals in our living environment. Besides, in comparison with the electromagnetic wave signal, high frequency acoustic signaling wavelength is extremely short and it only has the real part of signal. Therefore, these characteristics of acoustic signals will lead to the different ways of sampling and processing.

Assume that there are several speakers or phones in certain area as anchor points. We leverage a smartphone as the receiving instrument to record acoustic signals which are transmit by these speakers or phones. After moving the smartphones steady in front of the users body over a short distance for sampling, we emulate a virtual antenna array to process these samples. Finally we use the smoothed MUSIC algorithm for signal analyzing and estimate the direction of arrival of the target signals.

The main challenge of implementing our scheme is reducing the influence of the multipath effect to the results of DOA estimate. Because of the complex indoor environment, the multipath effect has huge negative impact to the direction finding accuracy. Multipath effect is a distortion in signal measurement. The components of signal which is reflected from the surface of different objects arrive at the receiver through multiple paths at different time. And that will cause the phase dislocation of received signals. This kind of signal is called multipath signal.

As far as we know, using a directional antenna is the most common way to emulate the DOA of signals without the adverse impact of multipath effect. But antenna arrays are the specialized instruments which are expensive and bulky. Our solution is to emulate a virtual antenna array by using the ordinary smartphones that users carry every day. It derived from a key insight that through processing consecutive measurements in time and converting those into consecutive spatial measurements can emulate an antenna array. And by leveraging this virtual antenna array, we can find the direction of the acoustic source by Multiple Signal Classification algorithm (MUSIC).

Additionally, high accuracy is the most important evaluation of a localization technique. The conventional MUSIC algorithm may be affected by the coherent signals which caused by multipath effect or other speakers in the scene. So we implement smoothed MUSIC algorithm to solve these problems which worked well in the experiment. We designed, deployed, and evaluated our scheme of direction finding under diverse circumstances. Our extensive experimental results show that our theory supports high accuracy of finding the direction of arrived signals. For phone to speaker direction finding, the mean error and the standard deviation of the measured angle is 4.01 degrees and 4.74 degrees respectively within the range of 20 meters.

The rest of the paper is organized as follows. We present the related work of our research in section 2. And we introduce the design of acoustic signals direction finding system in section 3 and then report our design of experiment and results in section 4. At last, we conclude the paper in section 5.

2   Related Work

2.1   Leveraging the Acoustic Signals by Smartphone

There are many researches using acoustic signals as objects of study. Swadloon (a Shake-and-Walk Acoustic Direction-finding and indoor LOcalizatiON scheme using smartphones) leverages the Doppler effects of the acoustic waves for direction finding and indoor localization. Users need to shake their smartphones and walk a few steps before using Swadloon. It only requires off-the-shelf speakers and supports arbitrary number of users. Other researches such as TDOA[10] leverage low speed of acoustic wave compared to wireless signals and BeepBeep[17] is a high-accuracy acoustic-based ranging system which can detect the distance between two smartphones with high accuracy.

2.2   Emulating Antenna Arrays

The devices like antenna arrays are extremely important in signals DOA estimation. Because of its expensiveness and bulkiness, many researchers choose to emulate an antenna array instead of actually using it. A novel technique called inverse synthetic aperture radar (ISAR) leverages the movement of the target to emulate an antenna array to generate a two-dimensional high resolution image of aircrafts, ships, planets, etc. ISAR radars have only one receive antenna and at any point in time it captures a single measurement. As long as the target is moving, the target itself can emulate an antenna array leveraging the relative movement between the target and the ISAR antenna.

Wi-Vi [11] introduces novel solutions to the see-through-wall problem that enable ordinary users to detect moving human beings behind walls or in closed rooms. Hence, at any point in time, the Wi-Vi receiver captures a single measurement. When the target is moving at successive locations in space, the users can observe the fluctuation of Wi-Fi signals through Wi-Vi as if they had a receive antenna at each of these observe points. That means successive time samples received by Wi-Vi correspond to successive spatial locations of the moving target. Hence, Wi-Vi can emulate an antenna array and use it to track targets behind the wall.

2.3   Multiple Signal Classification (MUSIC)

In many practical signal processing problems, the objective is to estimate from measurements a set of constant parameters upon which the received signals depend. Schmidt proposed Multiple Signal Classification (MUSIC) algorithm in 1979. This novel method has been widely studied in the field of signal processing. Many advanced researches about signal processing are derived from the conventional MUSIC algorithm. MUSIC algorithm is based on the Eigen decomposition of the data's covariance matrix. By leveraging the orthogonality of signal's subspace and noise's subspace, MUSIC algorithm is able to estimate the target signal's 1) number of incident wave fronts present; 2) directions of arrival or locations; 3) strengths and cross correlations among the incident waveforms; 4) noise interference strength.

3   Design

3.1   Overview

In this section, we present our scheme of phone to acoustic source direction finding. We show the flowchart of the system in figure 1.

First, the user moves the smartphone steady in front of his or her body. After keeping liner movement of the smartphone for a period of time, the microphone of the smartphone can gather the high frequency acoustic signals which are transmit by the speakers we preset in the environment. These speakers is called anchor points. Second we filter the signals with BPF (Band Pass Filter) for getting rid of the interference of the noise and other speakers in the environment. For daily use, we assume that different speakers in the environment transmit different frequencies of high frequency acoustic signals. In this way, users can distinguish different locations. After that we leverage Automatic Gain Control (AGC) to eliminate variational amplitude of the target signal in Equation 4. And then in order to make the preparation of DOA estimate we leverage the movement of our smartphone to emulate an antenna array. Finally we utilize smoothed MUSIC algorithm to estimate the direction of arrival (DOA) of the target acoustic signal.

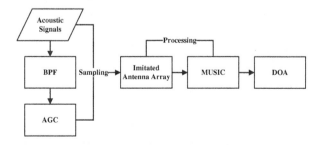

Fig. 1. System Design

3.2   Sampling and Processing

However, our targets of the direction finding are the anchor points we setup in a certain area. We choose to study the high frequency acoustic signal. This kind of signal has several advantages. First, acoustic signal is unusual in our daily life. Compare with the Wi-Fi signal, the most common signal in our living environment nowadays, the high frequency acoustic signal is more anti-interference. Second, acoustic signal is more stable. There are some researchers who use light signals for indoor localization. But light signal is easily affected by the nature light and other factors in the environment. So there are too many limits in using light signal. Third, high frequency acoustic signal cannot be heard by human beings. The frequency of the experimental signal is set to be higher than 17000Hz which has the most little effect to our daily life. And the speakers we use to transmit acoustic signal is cheap and easy to deployment.

After gathering the high frequency acoustic signals by the microphone, we filter the received signal first. The bandwidth of the filter should be wide enough to get the total target signals. In order to guarantee this, we assume the maximum speed of the moving smartphone is 2 meters per second. If we use the acoustic signal of 19000Hz, the maximum frequency shift is 111.8Hz. It makes the minimum pass band of the filter has to be 223.6Hz.

According to [1] , we can adjust the filtered data by Automatic Gain Control(AGC) such that the amplitude of the acoustic signal α is replaced by another one that close to constant. Hence, the noise N and amplitude α is eliminated by BPF and AGC respectively theoretically.

Most prior indoor direction finding system estimate the signals spatial angle of arrival using an antenna array for accurate result. They steer the arrays beam to determine the direction of maximum energy. So we need an antenna array with many antenna elements for a great signal resolution. This will result in leveraging a bulky and expensive device. To use the benefit of this kind of device and considering the financial cost and usability at the same time, we simply leverages the movement of our smart phone to emulate an antenna array instead.

In figure 2, the figure on the left shows that an antenna array is able to locate the acoustic source by steering its beam spatially. Now we can treat these antenna elements as several smartphones which have the same model and processing the target signal in the same way. Under this assumption, we can leverage normal smartphones to emulate an antenna array. But in daily life, users usually only have one smartphone. Now we assume that the user moves the smartphone in his hand, and he samples the received acoustic signal at successive locations in short distance. At each point in time, we can captures a single measurement from the whole sampling data as if we had a receive antenna array at each of these sampling points. In the right figure the moving smartphone itself can emulate an antenna array.

In what follows, we formalize the above discussion. Assume there is only one target signal source in the environment. According to figure 3, θ denote the direction of arrival of the acoustic signal. And d is the approximate value of the distance between two sampling points which equals to multiply the moving

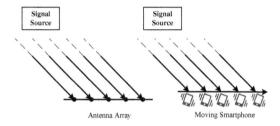

Fig. 2. Antenna Array Simulation

velocity of the smartphone v by the sampling period T we preset. Speaking of the sampling period T we choose, it is positively correlated with the calculating efficiency and negatively correlated with the calculating accuracy.

Since we cannot track the exact moving velocity of the users hand during the sampling period. However the range of speeds that human hand has in the normal situation is fairly narrow. So we assume that the moving speed of the smart phone is constant (our default is v = 1m/s). Because of the linear movement of our smartphone, the antenna array we simulate can be treat as the uniform linear array (ULA).

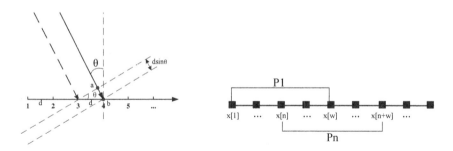

Fig. 3. Signal Model   **Fig. 4.** Smoothed MUSIC Algorithm

We regard the first sampling point as the reference point. τ_n denotes the time needed for the signal wave to travel from the reference point to point n. The X_n represents the output signal of nth sampling point. Then we can set $S(t_0)$ equals to the value of the signal waveform as measured at the reference point (input signal).

$$x_n(t_0) = S(t_0 - \tau_n) + N \qquad (1)$$

Because we already convert the time domain samples into spatial samples. We treat the periodicity measurements in time as the antenna array elements in physical space. Here t_0 is the reference time, represent some point of time. As we all know, high frequency acoustic signals are narrow band far-field signals. Through the signal modulation processing we can get:

$$x_n(t_0) = e^{-i\omega\tau_n}S(t_0) + N \qquad (2)$$

where ω is the frequency of the target signal and N is the noise . According to Equation 2:

$$\alpha(t_0)cos(\omega t_0 + \varphi) = S(t_0)e^{i\omega t_0} \tag{3}$$

where α the amplitude of the acoustic signal which is already replaced by a constant coefficient by the AGC. We can express S with the phase of signal φ at the reference point.

$$S = \alpha e^{i\varphi} \tag{4}$$

It is observed that S does not contain the information of DOA. So we treat it as a parameter. Then, under the assumption in figure 3, it is obvious that

$$\tau_n = (n-1)(\frac{dsin\theta}{c} + T) \tag{5}$$

where c is the propagation velocity of sound. Because we choose the first point as the reference point. Now we can list the x(n) of different sampling points to a set of output data X

$$X = \left[1 \; e^{-i\omega T(\frac{Vsin\theta}{c}+1)} \; \cdots \; e^{-(m-1)i\omega T(\frac{Vsin\theta}{c}+1)} \right]^T S \tag{6}$$

where m is the number of 'antenna elements' which equals to the sampling points. And N is already be filtered by BPF. For simplifying formula, we introduce Q:

$$Q = \omega T(\frac{Vsin\theta}{c} + 1) \tag{7}$$

Based on the above:

$$X = \left[1 \; e^{-iQ} \; e^{-i2Q} \; \cdots \; e^{-i(m-1)Q} \right]^T S \tag{8}$$

Next, for calculating the direction of arrival of the target signal in the next section, we introduce the direction vector A():

$$A(\theta) = \left[1 \; e^{-iQ} \; e^{-i2Q} \; \cdots \; e^{-(m-1)iQ} \right]^T \tag{9}$$

Hence, $A(\theta)$ is function of θ only. And this will be the key to solve the direction finding problem using MUSIC algorithm.

3.3   Multipath Effect

Due to the high-pass filter that we plant for sampling certain frequency of acoustic signals, the users smartphone can only receive the signals transmitted by speakers we set in the ideal situation. But the acoustic signal itself in transmission would be affected by the intricate indoor environment. For example, walls, windows and furniture etc., they may all reflect the signals and these signals would also be received by the microphone. These received signals all come from different paths and share the same frequency. And the reflected signals which cause the diffraction are the multi-path signals.

Conventional MUSIC algorithm is a classic method to estimate the signals directions of arrival (DOA). But because of the multipath signals in the interference environment, the rank of received sampling signals correlation matrix would be decreased to 1. It will cause signal spaces dimensionality less than the number of signals source. So the steering vectors of some coherent sources would not be orthogonal to noise subspace. And the conventional MUSIC algorithm cannot get the accurate signals direction of arrival in this circumstance.

Considering the adverse impact of multipath signals to our results, the basic approach for processing such signals relies on the smoothed MUSIC algorithm. In comparison to the conventional MUSIC algorithm, smoothed MUSIC performs an additional step before performing the Eigen decomposition for decorrelation.

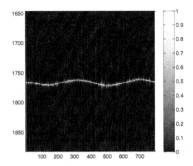

Fig. 5. Spectrum of the Signal

As we mentioned before, the antenna array we simulate is uniform linear array (ULA). First, we can partition each array of the same size into overlapping sub-antenna array. In figure 4, we groups consecutive time samples into overlapping windows of size w.

Then treat each window as a sub-antenna array P_n:

$$P_n = \begin{bmatrix} x(n) \ x(n+1) \ x(n+2) \cdots x(n+w) \end{bmatrix} \tag{10}$$

These sampling points would be divided into p parts of such sub-antenna array. Each sub-antenna array has w antenna elements. Then, we can choose the first sub-antenna array as the reference sub-antenna array.

According to Equation 10 we can compute the correlation matrices of all of these sub-antenna arrays. For any arbitrary sub- antenna array of size w, MUSIC first computes the w×w correlation matrix Rn:

$$R_n = E[P_n P_n^H] \tag{11}$$

At last, we combine these correlation matrices by summing them up to finish correcting the covariance matrix. We can get the corrected matrix by Equation 12:

$$R = \frac{1}{p} \sum_{n=1}^{p} R_n (p < m) \tag{12}$$

3.4   Getting the Direction of Acoustic Signal Source

After eigenvalue decomposition of corrected covariance matrix we can get M eigenvalues of the matrix R:

$$\lambda_1 \ \lambda_2 \ \lambda_3 \cdots \lambda_M \tag{13}$$

Correspond to M eigenvector:

$$v_1 \ v_2 \ v_3 \cdots v_M \tag{14}$$

We can list the eigenvalues in descending order:

$$\lambda_1 > \lambda_2 \geqslant \lambda_3 \geqslant \cdots \geqslant \lambda_M \tag{15}$$

Because we assume there is only one target signal in the environment. So:

$$U_N = \left(v_2 \ v_3 \cdots v_M \right) \tag{16}$$

U_n is the noise subspace of the signal. After a series of mathematical transformations, we can find the noise subspace eigenvector is orthogonal to the signal space steering vector in ideal situation.

Considering the noise of various frequency in the environment, we believe that noise subspaces eigenvector and signal subspaces steering vector is only approximate orthogonal. So we can get the direction of arrival signal by finding the θ which minimize the scalar product of these two parameters:

$$\theta = argmin||A^H(\theta)U_N||^2 \tag{17}$$

Figure 5 shows a pseudo spectrum from one of our experiments after normalization. It shows the user is moving his smartphone towards the speaker and backwards the speaker.

4   Experiment

4.1   Experiment Design

All of our experiments are run on Samsung Note 2 using Android APIs. We deploy our system in an underground parking lot of a residential community in Hefei. The frequency of the target acoustic waves are 17000Hz to 19000Hz. And the sampling frequency is 44100Hz. There are few concrete pillars holding the ceiling with 0.8 meters across and some cars of different sizes in the environment.

The high frequency acoustic wave which is leveraged by our experiment is transmit by an ordinary speaker we bought online. The relative location of the phone and the speaker are shown in figure 6. The distance between the smartphone and the speaker is S. The elevation angle of the phone and the speaker is α. This variable stands for the relative height of the smartphone and the speaker. And the elevation angle of the acoustic source itself at the horizontal plane is

β which is shown in figure 7. In ideal situation, we assume that the users are holding their phones evenly and moving it horizontally.

The main method of evaluating performance of our scheme is to vary S, α, β and DOA by deploying the speakers and phones in different positions. We obtain the measured direction by steady moving the phone in front of our body at the height of 1 meter, and reading the DOA value from the phone.

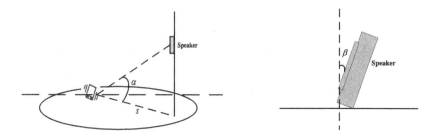

Fig. 6. Elevation Angle α **Fig. 7.** Elevation Angle of the Acoustic Source

4.2 Underground Parking Lot with Single Acoustic Wave

Distance between the Speaker and the User. We first assess the effect of different distances S to the accuracy of our experiment results. Hence, we bind the speaker on a concrete pillar at the same height with the users smartphone by tapes which means α=0 and β=0. We conduct the experiment with different S of 5m, 10m, 20m and 30m and plot the standard deviations and the mean error of the angular errors at each scenario.

Figure 8 shows a set of the experiment results with different distances S. The peak of the curve corresponding to the DOA of the acoustic signal. We can learn our accuracy of direction finding from figure 9 and 10.

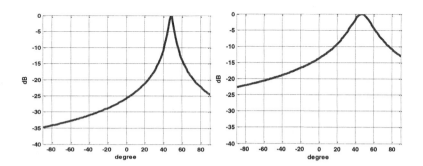

Fig. 8. DOA estimate results

When S≤20m, the mean error and standard deviation of the measurement are all around 4 degrees. And it is still acceptable at the circumstance of S=30m.

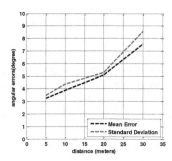
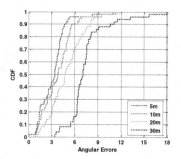

Fig. 9. Mean Error and Standard Deviation   **Fig. 10.** CDF of Different Distances

We also test the DOA implement accuracy when S>30m which is imprecise as the signals becoming weak and unstable.

Elevation Angle α and β. In this case, we test the errors when the speaker is set with different elevation angles. It means we test the accuracy of the DOA when the speaker and the smartphone are not at the same height and a variety of display ways of the speaker. We set the distances are S=5m, 10m and DOA=45 degrees. We introduce two variables α and β. We can find the different performances of our method in various scenarios in figure 11. Assume the user is holding his smartphone with the height of 1 meter, then we bind our speakers on the height of h.

$$h = tan\alpha \cdot S + 1 \tag{18}$$

With the comparison of different heights where we bind the speaker, we can find the results of our experiment has no obvious fluctuation.

Because of the speakers we bought for experiment is not omni-directional, there is only one side of our speakers is broadcasting sound wave during the experiment. And it will affect the accuracy of our method.

By changing the way we place the speakers, we did two sets of comparison experiments. According to Figure 11, we also notice that while the user is getting closer to the signal source, the angle change of β has greater interference on the result. That is because the reflected signals which caused the multipath effect is much stronger when the user is near the signal source. When the phone is further from the speaker, the signal reflected from the wall becomes much weaker than the one directly from the speaker.

Motion Patterns. We also analyze the errors caused by different motion patterns of the phones movement. We test 3 kinds of motion patterns in figure 12(a) which are moving the phone in a straight line(the default pattern of our experiments), moving the phone in an approximate circle shape and moving the phone in the arbitrary patterns. We set S=10m, 20m and $\alpha=\beta=0$. And the result shows in figure 12(b) and figure 12(c). From the results we notice that whether the user

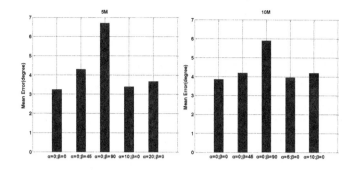

Fig. 11. Mean Error of Different Elevation Angles

is moving the phone in a straight line (no matter what the moving direction is) or in a circle, the result is relatively stable.

While the user is moving the phone erratically, the accuracy of DOA has been affected. For the arbitrary patterns, there are both positive and negative frequency shifts in the measurement, so the mean error is unsatisfactory. And we can also find that, as the distance growing, the users motion pattern has less effect on the results. In a word, as long as the user moves his phone in the organized linear movement, the effect of the phones motion pattern to the result is limited.

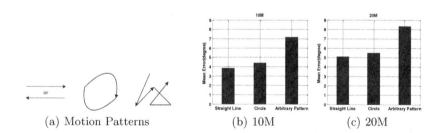

(a) Motion Patterns   (b) 10M   (c) 20M

Fig. 12. Mean Error of Different Motion Patterns

Non-line of Sight. We want to know the influence of Non-line of sight (NLOS) to our scheme. According to [12] we know that the users body itself will affect signal strength of target signals when he turns back to the signal source. One solution to this kind of circumstance is to do the direction finding for several times facing different directions before we determine the right direction. We also did another NLOS experiment. We set S=20m and let one person stands at the line connecting the speaker and the user in different distances of 1m,5m,10m,15m and 19m as it shows in figure 13. Through the contrast of the results in figure 14, we can come to the conclusion that if something (people, concrete pillars, furniture, etc.) get too close to block the speakers broadcasting side or

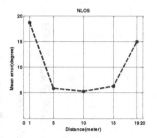

Fig. 13. NLOS Experiments   **Fig. 14.** Mean Error of NLOS

the smartphones microphone, the result of direction finding will be imprecise and unstable. According to the observation, we conclude that for precise results we need to make sure that there is no obstacles within 1 meter around the speakers or the phones microphone during direction finding.

Underground Parking Lot with Multiple Acoustic Waves. In order to test our schemes anti-interference capability. We implement the experiment with several speakers transmit the same (multipath effect imitation) or different (noisy environment) frequencies acoustic signals in the environment.

We set $\alpha=\beta=0$. S=5m, 10m, 20m, 30m. We set the volume of target speaker is 100% and the interference speaker is 50%. And the interference speaker has the same distance to the user with the target speaker. In figure 18 we can see that with different distances S, the results are a little bit effected by the interference. According to smoothed MUSIC algorithm, like we see in figure 15, not only we can get the direction of target speaker, but also the rough direction of the interference speaker too. And we also try to increase the volume of interference speaker. We find that the result angle is getting closer and closer to the interference speaker. When we turn the volume of interference speaker to 100%, the two curves in figure 16 are almost overlap each other.

In our real-life scenario, we need to set multiple acoustic source of different frequencies for different locations in the same environment. So we also test our methods robustness with two speakers broadcasting the different frequencies

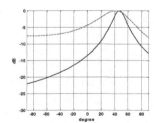

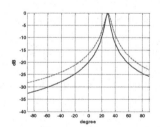

Fig. 15. Same frequency and different volumes   **Fig. 16.** Same frequency and volume

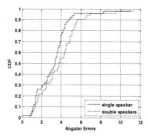
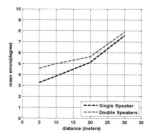

Fig. 17. Two speakers of different frequencies of 10 meters

Fig. 18. Two speakers of the same frequency of 50% and 100% volume

which are 19000Hz and 17000Hz in the underground parking lot. And the result shows in figure 17. We can find that the accuracy of our result in this scenario is worse than single speaker scenario. But it is still acceptable for daily use.

5   Conclusion

In this paper, we propose a novel scheme to estimate the direction of the high frequency acoustic source. In contrast to previous systems of DOA estimate, instead of using an antenna array, we imitate a virtual one with an ordinary smartphone. By getting rid of multipath effect, the massive experiments show that our scheme works well under various circumstances. Because of the lightweight equipments and the efficient algorithm we use, our method of direction finding is not only easy-to-use but also accurate. In addition, there are still some future works to do. For example, we can use the various sensors of the smartphone to optimize the accuracy of direction finding. And also we can try to develop an indoor localization system which is based on our direction finding technique.

References

1. Huang, W., Xiong, Y., Li, X.Y., Lin, H., Mao, X., Yang, P., Liu, Y.: Accurate Indoor Localization Using Acoustic Direction Finding via Smart Phones. arXiv preprint arXiv:1306.1651 (2013)
2. Schmidt, R.O.: Multiple emitter location and signal parameter estimation. IEEE Transactions on Antennas and Propagation **34**(3), 276–280 (1986)
3. Shan, T.J., Wax, M., Kailath, T.: On spatial smoothing for direction-of-arrival estimation of coherent signals. IEEE Transactions on Acoustics, Speech, and Signal Processing **33**(4), 806–811 (1985)
4. Xiong, J., Jamieson, K.: ArrayTrack: a fine- grained indoor location system. In: NSDI, pp. 71–84, April 2013
5. Stoica, P., Moses, R.L.: Spectral analysis of signals. Pearson/Prentice Hall, Upper Saddle River (2005)
6. Chen, H., Li, X., Jiang, W., et al.: MIMO radar sensitivity analysis of antenna position for direction finding. IEEE Transactions on Signal Processing **60**(10), 5201–5216 (2012)

7. Liu, H., Gan, Y., Yang, J., Sidhom, S., Wang, Y., Chen, Y., Ye, F.: Push the limit of wifi based localization for smartphones. In: Proceedings of the 18th Annual International Conference on Mobile Computing and Networking, pp. 305–316. ACM, August 2012
8. Xie, R., Liu, Z., Wu, J.: Direction finding with automatic pairing for bistatic MIMO radar. Signal Processing **92**(1), 198–203 (2012)
9. Peng, C., Shen, G., Zhang, Y., Lu, S.: Point and Connect: intention-based device pairing for mobile phone users. In: Proceedings of the 7th International Conference on Mobile Systems, Applications, and Services, pp. 137–150. ACM, June 2009
10. Prorok, A., Tome, P., Martinoli, A.: Accommodation of NLOS for ultra-wideband TDOA localization in single-and multi-robot systems. In: 2011 International Conference on Indoor Positioning and Indoor Navigation (IPIN), pp. 1–9. IEEE, September 2011
11. Adib, F., Katabi, D.: See through walls with WiFi! ACM (2013)
12. Zhang, Z., Zhou, X., Zhang, W., Zhang, Y., Wang, G., Zhao, B.Y., Zheng, H.: I am the antenna: accurate outdoor ap location using smartphones. In: Proceedings of the 17th Annual International Conference on Mobile Computing and Networking, pp. 109–120. ACM, September 2011
13. Palanisamy, P., Kalyanasundaram, N., Swetha, P.M.: Two-dimensional DOA estimation of coherent signals using acoustic vector sensor array. Signal Processing **92**(1), 19–28 (2012)
14. Wong, K.T., Yuan, X.: Vector cross-product direction-finding with an electro-magnetic vector-sensor of six orthogonally oriented but spatially noncollocating dipoles/loops. IEEE Transactions on Signal Processing **59**(1), 160–171 (2011)
15. Priyantha, N.B., Balakrishnan, H., Demaine, E.D., Teller, S.: Mobile-assisted localization in wireless sensor networks. In: Proceedings IEEE 24th Annual Joint Conference of the IEEE Computer and Communications Societies, INFOCOM 2005, vol. 1, pp. 172–183. IEEE, March 2005
16. Smailagic, A., Kogan, D.: Location sensing and privacy in a context-aware computing environment. IEEE Wireless Communications **9**(5), 10–17 (2002)
17. Peng, C., Shen, G., Zhang, Y., et al.: Beepbeep: a high accuracy acoustic ranging system using cots mobile devices. In: Proceedings of the 5th International Conference on Embedded Networked Sensor Systems, pp. 1–14. ACM (2007)
18. Khaykin, D., Rafaely, B.: Acoustic analysis by spherical microphone array processing of room impulse responses. The Journal of the Acoustical Society of America **132**(1), 261–270 (2012)
19. Zhong, X., Premkumar, A.B., Madhukumar, A.S.: Particle filtering and posterior Cramer-Rao bound for 2-D direction of arrival tracking using an acoustic vector sensor. IEEE Sensors Journal **12**(2), 363–377 (2012)
20. Campbell, C.: Surface acoustic wave devices and their signal processing applications. Elsevier (2012)

Crowdsourcing Based Event Reporting System Using Smartphones with Accurate Localization and Photo Tamper Detection

Tong Qin[✉], Huadong Ma, Dong Zhao, Tianyuan Li, and Jianwei Chen

Beijing Key Lab of Intelligent Telecommunications Software and Multimedia,
Beijing University of Posts and Telecommunications, Beijing 100876, China
qintong5900@163.com, {mhd,dzhao,2011213098}@bupt.edu.cn,
jwchen110@gmail.com

Abstract. Crowdsourcing provides a novel and efficient paradigm to leverage numerous smartphone users to report timely information (time, location, photo, etc.) about events, enabling various urban sensing applications such as environment monitoring and public space management. *Localization* and *photographing* are two of the most important function modules in crowdsourcing based event reporting systems, which are used to tell people where and what events happen. However, the existing systems tend to localize the user instead of the event, resulting in the poor accuracy. Meanwhile, the existing event localization approaches need either complex user operation or complex computation process. For this reason, we propose a simple, effective, and light-weight approach for event localization. On the other hand, the existing systems cannot guarantee the photo authenticity, as the emerging various image editing and processing softwares have made the (malicious) photo tampering easier and easier. Accordingly, we design a novel smartphone-based photo tamper detection approach with high security. Detailed system implementation and performance evaluation are provided to verify the effectiveness of our proposed approaches.

Keywords: Mobile crowdsourcing · Event reporting system · Event localization · Photo tamper detection

1 Introduction

The proliferation of smartphones with a rich set of cheap powerful embedded sensors (e.g. GPS, accelerometer, digital compass, gyroscope, camera, etc.) offers a variety of novel, efficient ways to sense our physical world. Meanwhile, crowdsourcing provides a novel paradigm to leverage smartphones to build large-scale urban sensing applications that cannot easily be realized by a single individual [5,7–9,16]. In this paper, we focus on the crowdsourcing based event reporting systems, which can be used in many applications such as environment monitoring and public space management. The following are several typical examples:

© Springer International Publishing Switzerland 2015
Y. Wang et al. (Eds.): BigCom 2015, LNCS 9196, pp. 141–151, 2015.
DOI: 10.1007/978-3-319-22047-5_12

Ecosnapp [2] is a smartphone App that enables people to report various environment pollution events in China; *Creek Watch* [1] is an iPhone App that allows people to report information about waterways in order to track pollution and manage water resources; *Garbage Watch, What's Bloomin,* and *AssetLog* are a serious of sustainability campus campaigns that report resource use issues by taking geo-tagged photos of various resources, like outdoor waste bins, water usage of plants, bicycle racks, recycle bins, and charge stations [13]; *iSee* [12] is a system that detects and localizes uncivilized events, such as smoking and graffiti, in outdoor public environments.

In the above applications, there are two important functions: *localization* and *photographing*, which are used to tell people where and what events happen. Although the basic functions are easy to implement, they still cannot satisfy two further requirements: *location accuracy* and *photo authenticity*.

For the localization, GPS and other alternates, such as GSM and WiFi based approaches, have become mature technologies. However, these technologies can only be used to determine the event reporter's location, which is seen as the event location approximatively. The error of such approach is very large, as events may happen with various directions and distances around the event reporter. Thus, it is necessary to propose an approach to accurately localize the *event itself* instead of the *event reporter*. Although several latest approaches (e.g., OPS [10], CamLoc [14], *iSee* [12]) have been proposed for object/event localization, they have one or both of the following two drawbacks: 1) requiring complex user operation, for example, *OPS* requires the same user to photograph the event from various sites, and *iSee* requires multiple users to collect data for achieving good accuracy; 2) requiring complex computation process, for example, *OPS* and *CamLoc* rely on the computer vision or image processing technologies, where the accuracy also relies on the photo quality that is often affected by various factors like jitter, focusing, weather, light intensity, etc. In contrary, we propose a simple and effective approach for event localization, which only utilizes some simple sensing data, including GPS, acceleration, azimuth angle, and tilt angle, obtained from smartphone sensors. Our proposed approach neither requires any complex user operation nor necessitates any computer vision or image processing technology.

For the photographing, the emerging various image editing and processing softwares (e.g., Photoshop) have made the (malicious) photo tampering easier and easier, resulting in difficulties of post survey. Thus, it is necessary to propose an approach to detect the photo tampering effectively. Generally, there are two classes of photo tamper detection approaches: active approaches and passive approaches. The passive approaches mainly utilize the statistical variations of the images to detect image tampering, which neither require any prior information about the image nor necessitate the pre-embedding of any watermark or digital signature into the image [11]. Thus, these approaches have attracted much attention in recent years [3,6,15]. However, it is very difficult, if not impossible, to design a general, effective, and efficient approach to cope with a variety of image tampering operations. In contrary, the active approaches utilize a pre-processing step to embed digital signatures or watermarks to images, enabling

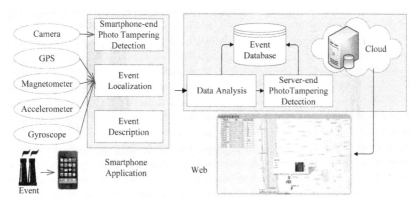

Fig. 1. System Architecture

to detect any image tampering operation effectively and efficiently. Although the pre-processing step is required by the active approaches, it can be implemented in our event reporting system conveniently on smartphones, since today's smartphones are programmable and come with strong computation and storage capabilities. In this paper, we adopt the basic idea of "secure digital camera" [4], which embeds the scene hash to the photo at the time of photo capturing, and checks the hash value to verify the photo integrity afterwards. In order to further guarantee the security, we design and implement a novel smartphone-based photo tamper detection approach by adding random number elaborately and leveraging the encryption algorithm.

The remainder of this paper is organized as follows: Section 2 introduces the system architecture. Section 3 and Section 4 present the two core modules, accurate event localization and photo tamper detection, respectively. Section 5 describes the system implementation and performance evaluation. Finally, Section 6 concludes the paper.

2 System Architecture

As Fig. 1 shows, the system comprises three components: (i) the Android smartphone App, which collects locations, images and other information of events, implements the smartphone-end photo tamper detection algorithm, submits records and displays all users' committed report information; (ii) the powerful remote server residing in the cloud, which receives report information, implements the server-end photo tamper detection algorithm, stores and sends information to smartphone and web client; (iii) the web client, which shows report information to users.

Our system requires event reporters to acquire data from various sensors embedded in smartphones. The smartphone App would lead event reporter to capture the image of event firstly, and execute the smartphone-end photo tamper detection algorithm. Then the App would enable the event localization module

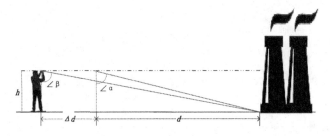

Fig. 2. The basic principle of event localization

to derive the event location. The event reporter is also required to make some description of the event and choose the type of event, such as source of pollution, public facilities damaged or public facilities occupied, etc. After these steps, the App would save the information in local database and upload to the server. The server analyzes these data, triggers server-end photo tamper detection algorithm and saves all the information in the server database, including the tamper detection result. Meanwhile, the server feedbacks the tamper detection result to the smartphone. Besides, the server is also able to send data to the App and web client for users to browse the report records.

3   Accurate Event Localization

3.1   Basic Principle

Since we can obtain the location of the event reporter by using GPS, it only needs to obtain the relative direction and distance between the event reporter and the event for deriving the location of the event.

Fig. 2 shows the basic principle of event localization. We suppose the target of the event is right in front of the smartphone. Through the viewfinder, the event reporter is required to let the center of the smartphone screen aim at one point of the target with arm straight, and then move back the arm to let the center of the smartphone screen aim at the same point of the target again. Let d denote the distance between the event and the event reporter, and we denote Δd as the displacement between the event reporter's arm outstretched and moved back. α and β are the tilt angles between the direction that the smartphone camera aims to and the vertical direction in these two situations. Let h denote the height of the smartphone held by the event reporter, then we have the following equation:

$$\frac{h \cdot tan\alpha}{h \cdot tan\beta} = \frac{d}{\Delta d + d} \tag{1}$$

So the distance between the event reporter and the event can be computed as

$$d = \frac{\Delta d \cdot tan\alpha}{tan\beta - tan\alpha} \tag{2}$$

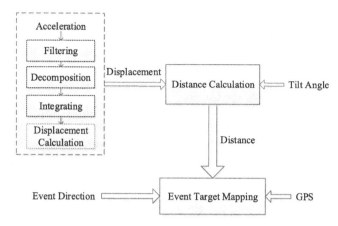

Fig. 3. Event localization process

α and β can be derived from the data of smartphone's magnetometer and accelerometer. For the purpose of computing the displacement of the smartphone during the movement, which is known as Δd, we also read the data of accelerometer when moving and process it by signal filtering, decomposition and integrating.

To calculate the longitude and latitude of the event, we record the data of GPS module and compute the data of the magnetometer and accelerometer for the relative direction between the event reporter and the event. The whole localization process is shown as Fig. 3.

3.2   Displacement Calculation

Accelerometer is one of the major smartphone sensors, which is also the starting point for us to calculate the displacement. The principle to calculate it in our system is to get the horizontal acceleration records during the moving process, and make two successive integrations on acceleration within the duration.

The acceleration data we get from smartphone sensor would be mixed with noise, so it is essential to filter the data to reduce the noise which would cause severeness of displacement error firstly. The amplitude limit filtering process is performed to reduce the pulse interference caused by accidental factor. In practice, we define A as a threshold value of the difference between two successive samplings. When a new acceleration record is collected, we calculate the absolute value of the difference between it and the last data as Δa. If $\Delta a \leq A$, the new data is valid, else if $\Delta a > A$, the new data is invalid and we substitute the last valid data for it. We also use arithmetic mean filtering to reduce the random interference by replacing the data with the arithmetic mean value of the N records before and after it.

Due to the acceleration data we get is under the phone's coordinate system, it is also imperative to translate the acceleration data to the world coordinate. We record the angles between each pair of axes from the phone's coordinate system to the world coordinate system as θ_x, θ_y and θ_z from the data of smartphone's gyroscope when the acceleration data is received. So as mentioned in *CamLoc* [14], we can calculate the horizontal acceleration of the smartphone as:

$$a = a_x cos(\theta_x) + a_y sin(\theta_y) \cdot cos(\theta_z) + a_z sin(\theta_z) \cdot cos(\theta_y) \qquad (3)$$

In above equation, a_x, a_y, a_z are the acceleration data in the phone's coordinate system. As we get the acceleration paralleled to the ground during the moving process, it is feasible to calculate the displacement by two successive integrations. Because the data is a discrete sequence, so the displacement Δd is calculated as:

$$
v_k = \sum_{i=1}^{k} a_i(t_i - t_{i-1})
$$
$$
\Delta d = \sum_{i=1}^{n} v_i(t_i - t_{i-1}) \qquad (4)
$$

In above equation, a_i is the horizontal velocity at the time point t_i, v_i is the horizontal velocity at the time point t_i.

3.3 Target Location Mapping

Before the process to calculate the displacement, the angles α and β are measured by the magnetometer and accelerometer at start and end time point of the movement. Now, the distance between the event and the smartphone d can be computed from Eq. 2. Further, the local location of smartphone is obtained from its GPS module and the direction from the smartphone to the event is measured by the magnetometer and accelerometer. It is known that on the same longitude, latitude is difference 1 degree every 111km, and on the same latitude, longitude is difference 1 degree every 111 * cos(the latitude) km, so the event longitude and latitude are reached by:

$$lng_E = lng_R + d \cdot sin\gamma/(cos(lat_R) \cdot \Omega),$$
$$lat_E = lat_R + d \cdot cos\gamma/\Omega, \qquad (5)$$

where lng_E, lat_E and lng_R, lat_R are the longitude and latitude of the event and the event reporter, γ is the angle between the event direction and the north direction. Ω is the constant distance value where we regard as 111km in practice.

4   Photo Tamper Detection

4.1 Basic Idea

The basic idea of photo tamper detection is to utilize the Message-Digest Algorithm MD5 to check the photo integrity. When a user takes a photo using our App, it calculates the MD5 message digest over the photo, and uploads to the backend server the photo and the corresponding MD5 message digest. Upon receiving them the server calculates a new MD5 message digest and compares

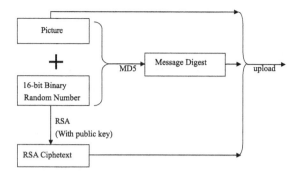

Fig. 4. The basic process of photo tamper detection algorithm on smartphones

it with the previous one to detect whether the photo has been tampered. However, since the MD5 algorithm is public, anyone can forge by tampering the original photo or forging one and recalculating a MD5 digest. Thus, we present an improved algorithm, including two parts respectively on a smartphone and a backend server.

4.2   Photo Tamper Detection Algorithm on Smartphones

The basic process of our algorithm on smartphones is illustrated in Fig. 4. After taking a photo, the algorithm will be operated as follows.

Step 1: Generate a 16-bit binary random number for avoiding cryptanalysis. Meanwhile, the last four bits of the random number are fixed as a flag, e.g., 1001, which is used to check whether the random number is generated by our smartphone App or forged by users. Note that we can expand and deepen this method. For example, more bits could be added, or the positions of fixed bits could be more complex to guarantee the security.

Step 2: Combine the photo and the random number together, and use them as input to compute the 128 bits MD5 message digest.

Step 3: In order to protect the random number from being accessed by intruders, encrypt the random number with RSA public key, which is stored in the App in advance.

Step 4: Combine the photo, MD5 message digest, RSA cipher text, the total length of MD5 value and RSA cipher together, then upload them to the server.

4.3   Photo Tamper Detection Algorithm on Backend Server

The basic process of our algorithm on the backend server is illustrated in Fig. 5. After receiving the data from smartphone users, the algorithm will be operated as follows.

Step 1: Split the data to four parts as follows: first, calculate the total length number M of the received package; second, read the last four bytes and get the total length N of MD5 value and RSA cipher text; third, determine the

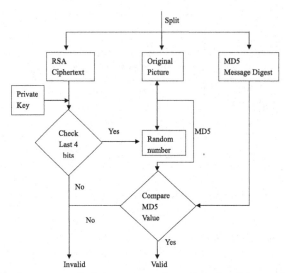

Fig. 5. The basic process of photo tamper detection algorithm on backend server

section of each of the four parts: Picture Data from 0 to (M-N-4) Bytes, MD5 Message Digest from (M-N-4) to (M-N-4+32) Bytes, and RSA ciphertext from (M-N-4+32) to (M-4) Bytes.

Step 2: Decrypt the RSA ciphertext with private key, and then check its last four bits: if it matches 1001, then turn to Step 3); otherwise, the photo is judged fake directly.

Step 3: Combine the picture and decrypted random number just as what we did on the smartphone APP, and then operate the MD5 algorithm to obtain the new message digest.

Step 4: Make comparison between the new MD5 digest and the old one: if they are identical, the photo is valid; otherwise, the photo has been tampered.

5   Implementation and Evaluation

5.1   Implementation

We implemented smartphone part of our system as a mobile application on Android (see Fig. 6). The application contains three views mainly: *Report, Data* and *Browsing* and two main sub views: *locating* and *detail*. The *report* view is the core part of application, which is for user to get, upload and save all kinds of information we need to describe events. The *locating* sub view can be started in the *report* view to get the locations of events. The *data* view is available to show the list of user report records, and lead user to view the detailed information in the *detail* sub view. User can browse the records from all users via online map from the *browsing* view. Our website is also available to view the records via map, and a list is offered to show the information.

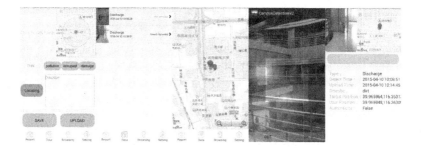

Fig. 6. App screen shots

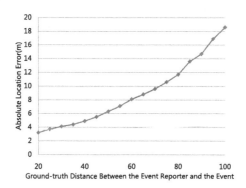

Fig. 7. Absolute ranging errors

5.2 Performance Evaluation

We have evaluated the performance of the two modules integrated in our system: event localization module and photo tamper detection module. Fig. 7 shows the absolute location error of the position estimated by the system. As observed from the figure, the error is within 7m when the real distance between the event reporter and the event is less than 50m. The error will increase with the distance. This is mainly due to the fact that with the distance increasing, the same angle deviation would cause greater error.

We also have tested the performance of the photo tamper detection module. We have tried some ways to tamper with the photo of record before upload, such as replacing the original picture in the phone storage with the same name file, or modifying it with some image editing and processing softwares like Photoshop, etc. Under these conditions, the photo uploaded can be detected as a tampered one and the result can be sent to the users.

6   Conclusion

This paper presents a crowdsourcing-based event reporting system. We mainly focus on two important function modules: *event localization* and *photo tamper detection*. On one hand, we propose a simple, effective, and light-weight approach for event localization. Different from the common used approaches in the existing systems, this approach can localize the event accurately instead of localizing the event reporter. Moreover, it neither requires any complex user operation nor necessitates any complex computer vision or image processing technology. On the other hand, we design a novel smartphone-based photo tamper detection approach with high security in order to guarantee the photo authenticity and facilitate the post survey. Besides, we have provided detailed system implementation and performance evaluation to verify the effectiveness of our proposed approaches.

Acknowledgements. This work is supported by the National Natural Science Foundation of China under Grant No. 61332005, the Funds for Creative Research Groups of China under Grant No. 61421061, the Specialized Research Fund for the Doctoral Program of Higher Education under Grant No. 20120005130002, the China Postdoctoral Science Foundation under Grant No. 2015M570059, the Cosponsored Project of Beijing Committee of Education, and the Beijing Training Project for the Leading Talents in S&T (ljrc201502).

References

1. Creekwatch. http://creekwatch.researchlabs.ibm.com/
2. Ecosnapp. http://www.hbssp.org/china/
3. Birajdar, G.K., Mankar, V.H.: Digital image forgery detection using passive techniques: A survey. Digital Investigation **10**(3), 226–245 (2013)
4. Blythe, P., Fridrich, J.: Secure digital camera. In: Proceedings of the Digital Forensic Research Workshop (DFRWS), pp. 17–19 (2004)
5. Chatzimilioudis, G., Konstantinidis, A., Laoudias, C., Zeinalipour-Yazti, D.: Crowdsourcing with smartphones. IEEE Internet Computing, 36–44 (2012)
6. Farid, H.: Image forgery detection-a survey. IEEE Signal Processing Magazine **26**(2), 16–25 (2009)
7. Ganti, R.K., Ye, F., Lei, H.: Mobile crowdsensing: Current state and future challenges. IEEE Communications Magazine **49**(11), 32–39 (2011)
8. Liu, L., Wei, W., Zhao, D., Ma, H.D.: Urban resolution: New metric for measuring the quality of urban sensing. IEEE Transactions on Mobile Computing (2015). doi:10.1109/TMC.2015.2404786
9. Ma, H.D., Zhao, D., Yuan, P.: Opportunities in mobile crowd sensing. IEEE Communications Magazine **52**(8), 29–35 (2014)
10. Manweiler, J.G., Jain, P., Roy Choudhury, R.: Satellites in our pockets: an object positioning system using smartphones. In: Proceedings of the 10th International Conference on Mobile Systems, Applications, and Services (MobiSys), pp. 211–224 (2012)

11. Mishra, M., Adhikary, F.: Digital image tamper detection techniques-a comprehensive study. International Journal of Computer Science and Business Informatics **2**(1), 1–12 (2013)
12. Ouyang, R.W., Srivastava, A., Prabahar, P., Roy Choudhury, R., Addicott, M., McClernon, F.J.: If you see something, swipe towards it: crowdsourced event localization using smartphones. In: Proceedings of the ACM International Joint Conference on Pervasive and Ubiquitous Computing (UbiComp), pp. 23–32 (2013)
13. Reddy, S., Estrin, D., Srivastava, M.: Recruitment framework for participatory sensing data collections. In: Floréen, P., Krüger, A., Spasojevic, M. (eds.) Pervasive 2010. LNCS, vol. 6030, pp. 138–155. Springer, Heidelberg (2010)
14. Shangguan, L., Zhou, Z., Yang, Z., Liu, K., Li, Z., Zhao, X., Liu, Y.: Towards accurate object localization with smartphones. IEEE Transactions on Parallel and Distributed Systems **25**(10), 2731–2742 (2014)
15. Wang, W., Dong, J., Tan, T.: A survey of passive image tampering detection. In: Ho, A.T.S., Shi, Y.Q., Kim, H.J., Barni, M. (eds.) IWDW 2009. LNCS, vol. 5703, pp. 308–322. Springer, Heidelberg (2009)
16. Zhao, D., Ma, H.D., Tang, S., Li, X.Y.: Coupon: A cooperative framework for building sensing maps in mobile opportunistic networks. IEEE Transactions on Parallel and Distributed Systems **26**(2), 392–402 (2015)

Parallel Accurate Localization from Cellular Network

Chao Wu[✉], Bin Xu, and Qi Li

Department of Computer Science and Technology, Tsinghua University, Beijing, China
ariesnix93@gmail.com, xubin@tsinghua.edu.cn,
zhongguoliqi@163.com

Abstract. Localization from cellular network (LCN) is of great importance for mobile operator. Different from GPS, LCN utilizes operator's network infrastructure to identify the location of mobile devices. The advantage of LCN is that operator can calculate all mobile devices' location according to the mobile big data, without each device reporting its location got from GPS. The challenges of LCN are localization accuracy and computing efficiency. In this paper, we propose three methods of localization by analyzing mobile signal strength data collected by cellular network per 480 milliseconds. Using an optimized propagation model which considers the penetration loss through buildings, we get fairly high localization accuracy. To improve the localization efficiency, a parallel method based on Spark is proposed to process and analyze mobile big data. Through the experiments of tracking people in real cellular network, our methods reach an accuracy of 100 meter (mean error), which exceeds the international standard (125 meter by FCC). The parallel method also gets a significant efficiency promotion, which reduce 97% time cost than serial methods.

1   Introduction

Localization from cellular network (LCN) utilizes operator's network infrastructure to identify the location of mobile devices, which is a kind of network-based localization technology. Network-based and handset-based localization technologies are two main categories of wireless location. Handset-based technology, such as GPS have a relatively high degree of accuracy, but embedding a GPS receiver into mobile devices leads to increased cost, size, and battery consumption. Network-based technology, on the other hand, measure the signals transmitted from a mobile device and relay them to a central site for further processing and data fusion to provide an estimate of the mobile device's location. The significant advantage of network-based technology is that the mobile device is not involved in the location-finding process, which makes it possible to calculate all mobile devices' location according to the mobile big data.

A number of ranging techniques are developed to get the location of a mobile device, including received signal strength (RSS), time of arrival (TOA), angle of arrival (AOA), and so on. In our research, we try to localize people through a kind of RSS data which is measurement report (MR) data. MR data is collected by digital cell mobile communications network sent by user equipment. The first challenge is how to estimate people's real location from MR data accurately. In addition, the MR data is

© Springer International Publishing Switzerland 2015
Y. Wang et al. (Eds.): BigCom 2015, LNCS 9196, pp. 152–166, 2015.
DOI: 10.1007/978-3-319-22047-5_13

generated for each mobile device per 480ms, which is huge amount in the whole cellular network. The challenge is how to process and analyze the mobile big data efficiently.

In this paper, we propose three localization methods. The contributions of our study are as follows:

1. We propose two methods which consider the penetration loss through buildings and use more optimized parameters. Through the experiments of tracking people in real cellular network, our methods reach an accuracy of 100 meter (mean error).
2. We propose a parallel method based on Spark [1], which can rapidly process and analyze mobile big data. Comparing with serial methods, it gets a forty times promotion in efficiency.

The rest of paper is organized as follows: In section 2, we describe related work on localization technology. In section 3, we give an overview of the mobile big data, and describe some important characteristics of the data. Section 4 describes the serial methods we propose for localization, and the parallel method we design in detail. Section 5 shows experimental results applied to our dataset, and gives detail analysis on the result. Finally, we conclude this paper in Section 6.

2   Related Work

In this section, we will show some related work on localization technology.

Mobility prediction and localization are tightly coupled with location-based applications, such as, navigation [2], instant messaging [3], emergency rescue [4], and many other safety and security services [5] [6]. Many of the above applications need precise tracking of mobile terminal within cells to provide an adaptable quality of service.

The most important mobile location techniques are propagation time and signal strength techniques. Propagation time-based techniques (e.g. TOA [7], TDOA [8] [9], E-OTD [10]) rely on time measurement. However, the disadvantage of these techniques is that the accuracy of the estimated position depends mainly on the number of measurements and on the geometric configuration, which is even worse in urban environments.

Signal strength-based techniques instead are based on Received Signal Strength Indication (RSSI), which measures signal attenuation, assuming free space propagation and omnidirectional antennas. Exploiting this assumption mobile antenna location problem is reduced to the well-known triangulation position problem which is similar to time-based and angular approaches. The disadvantage of these techniques is that free space propagation assumption does not work because of multipath propagation and shadowing, leading to complex signal shapes. There are four kinds of physical phenomena influencing radio propagation: reflection, diffraction, penetration, and scattering. To solve these problems, many propagation models have been developed, such as cost231-hata model [11] [12], SDR model, cost231-walfish-ikegami model

[13], and so on. How to use those theoretical model to meet the real environment and improve their accuracy, is what we do in our research.

Many recent works are aimed at increasing the localization accuracy of RSSI triangulation. One interesting work [14] treats the problem of mobility in ATM network. It develops a hierarchical user mobility model that closely represents the movement behavior of a mobile user, and uses pattern matching and Kalman filtering, yielding to an accurate location prediction algorithm. Another work [15] proposes two algorithms for real-time tracking, location, and dynamic motion of a mobile station in a cellular network. This method is based on pre-filtering and two Kalman filters. [16] [17] propose a mobility model built on a dynamic linear system driven by a discrete command process that was originally developed for tracking maneuvering targets in tactical weapons system. Different from these works, our research base on RSSI data from cellular network, which has a larger data scale and wider data features.

3   Data Description

Measurement report (MR) data is a type of RSS data user equipment (UE) send to TD-LTE digital cell mobile communications network every 480ms, which can be used to evaluate and optimize the cellular network.

MR data contains many features. Each feature has its special meaning. Table 1 lists those features we use in our research.

Table 1. Features of MR data

Feature	Description
CellId	Unique id of the serving cellular.
MmeCode	Triple generated from Imsi, represents a unique people.
MmeGroupId	
MmeUeS1apId	
TimeStamp	The generation time of this record.
ScRSRP	Corresponds to the signal strength that UE receives from the serving cellular.
NcRSRP	Corresponds to the signal strength that UE receives form a neighborhood cellular of the serving cellular.
ScTadv	The signal propagation time from the serving cellular to UE.
ScEarfcn	The carrier service number of the serving cellular.
ScPci	The physical identification code of the serving cellular.
NcEarfcn	The carrier service number of a neighborhood cellular of the serving cellular.
NcPci	The physical identification code of a neighborhood cellular of the serving cellular.

4   Localization Methods

In this section, we describe how we use MR data defined in section 3 to calculate people's real location. We propose three methods: baseline method, optimized method and parallel method.

4.1   Data Preprocessing

The original MR data contains a huge number of gzip files. Each gzip file is a compression of one xml file. In addition, an xml contains lots of features, while the useful features listed in table 1 are only a quarter. It is very meaningful and important to extract useful features from original MR data in a short time.

We use multiple threads to decompress and parse data. Each thread will extract useful features from the compressed file, and parse them into a buffered data queue. The buffered data queue will store and flush batch data into disk regularly. The workflow is shown in figure 1.

Each UE (user equipment) is identified by his id (MmeCode, MmeGroupId, MmeUeS1apId). Combining with the geographic information of cellular towers, we preprocess data into a specified format. The formatted data is shown in figure 2. From this data, how to calculate the location?

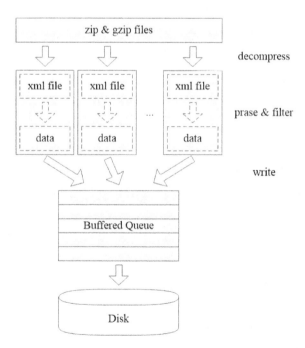

Fig. 1. Workflow of extracting features

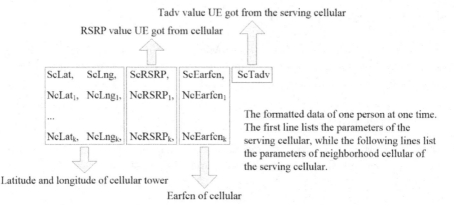

Fig. 2. Formatted data of one person

4.2   Calculating Distance

At first, we calculate the distance between UE and cellular tower.

As described in section 3, RSRP corresponds to the signal strength, while the signal strength of electromagnetic wave is losing energy through its propagation from cellular tower to UE. This energy loss can be calculated by the following formula:

$$loss = prs + gain - rsrp - shadowLoss \tag{1}$$

In this formula, prs is the transmit signal strength of cellular tower, while gain is the signal gain from the antenna. When electromagnetic wave passes through buildings, it losses energy. This is a variable determined by the distribution of buildings surround the cellular tower. We use the Gaussian distribution to represent it.

The energy loss is related by the distance between cellular tower and UE, the relationship is described by the following equations:

$$loss = L_u - a\left(H_{ue}\right) + C_m \tag{2}$$

$$L_u = 46.3 + 33.9 * \lg(f) - 13.82 * \lg\left(H_{eNB}\right) \\ + \left(44.9 - 6.55 * \lg\left(H_{eNB}\right)\right) * \lg(d) \tag{3}$$

$$a\left(H_{ue}\right) = \left(1.1 * \lg(f) - 0.7\right) * H_{UE} \\ - \left(1.56 * \lg(f) - 0.8\right) \tag{4}$$

In these equations, f is the operating frequency of cellular, which can be calculated by Earfcn. Table 2 lists the calculation of Earfcn. H_{eNB} is the height of cellular tower, H_{UE} is the height of UE, C_m is an experience value depending on urban area type, d is the distance from cellular tower to UE, which need calculation.

Table 2. The calculation of f by Earfcn

Start frequency	Offset	Range of Earfcn	f
2570	37750	37750 - 38249	f = startFrequency + 0.1 * (earfcn − offset)
1880	38250	38250 - 38649	
2300	38650	38650 - 39649	

Through equations (1), (2), (3) and (4), we calculate several distances from the formatted data. These distances are shown in figure 3.

Tadv also corresponds to the distance between serving cellular and UE. We calculate it by:

$$D_{tadv} = (scTadv + 0.5) * 78.12$$

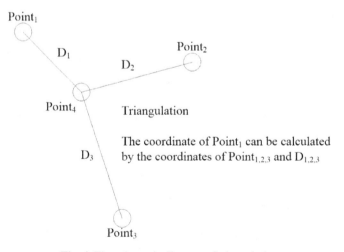

Fig. 3. Distances calculated from the formatted data

4.3 Baseline − Triangulation Localization

Since we know the location of cellular towers, after we calculate the distances between UE and cellular towers, we use triangulation to calculate UE's location as a baseline. The schematic diagram of triangulation is shown in figure 4. Formula 1 is the detail calculation method, the (Lat, Lng) pair which makes formula 1 minimal is the location of UE.

Point$_1$

D$_1$

Point$_2$

D$_2$

Point$_4$

Triangulation

D$_3$

The coordinate of Point$_1$ can be calculated by the coordinates of Point$_{1,2,3}$ and D$_{1,2,3}$

Point$_3$

Fig. 4. The schematic diagram of triangulation

$$\sqrt{\left(Lat-scLat\right)^2 + \left(Lng-scLng\right)^2} - D_{tadv}$$

$$\sqrt{\left(Lat-scLat\right)^2 + \left(Lng-scLng\right)^2} - D_{scRSRP}$$

$$\sqrt{\left(Lat-ncLat_1\right)^2 + \left(Lng-ncLng_1\right)^2} - D_{ncRSRP1}$$

$$...$$

$$\sqrt{\left(Lat-ncLat_k\right)^2 + \left(Lng-ncLng_k\right)^2} - D_{ncRSRPk}$$

Formula 1.

4.4   LocatePlus – Optimized Localization

In this subsection, we propose a method named LocatePlus to improve the baseline.

To let our localization result more accurate, we use past N consecutive points to optimize current time point.

People's trace is made by a set of consecutive points, if the time interval between adjacent points is very short, e.g. 5 seconds. We can use the past N points' location to predict N + 1 point's location.

The result of baseline in section 4.3 can be described as:

$$(T_0, S_0), (T_1, S_1), \dots, (T_n, S_n)$$

T is time, S is location (latitude & longitude pair).
We can calculate the speed in each time:

$$V_i = \frac{S_i - S_{i-1}}{T_i - T_{i-1}}$$

The statistics value of V:

$$\hat{\mu} = \frac{1}{N} \sum_{i=1}^{N} V_i$$

$$\hat{\sigma}^2 = \frac{1}{N-1} \sum_{i=1}^{N} (V_i - \hat{\mu})^2$$

$$\hat{\sigma_1}^2 = \frac{1}{N-1} \sum_{i=1}^{N-1} (V_i - \hat{\mu})(V_{i+1} - \hat{\mu})$$

$$\hat{\alpha} = \begin{cases} 1 & if \ \hat{\sigma} = 0 \\ max\left\{0, \dfrac{\hat{\sigma}_1^{\ 2}}{\hat{\sigma}^2}\right\} & otherwise \end{cases}$$

If we get four statistics values from point 1 to point N, and the speed on T_n, then the speed on T_{n+1} is:

$$V_{n+1} = \hat{\alpha}V_n + (1 - \hat{\alpha})\hat{\mu} + \hat{\alpha}\sqrt{1 - \hat{\alpha}^2}N$$

So the distance between point N and point N+1:

$$D_v = V_{n+1}(T_{n+1} - T_n)$$

Formula 2 is the detail calculation method of LocatePlus. The (Lat, Lng) pair which makes formula 2 minimal is the location of UE.

$$\left\Vert \begin{array}{c} \sqrt{\left(Lat - Lat_n\right)^2 + \left(Lng - Lng_n\right)^2 - D_v} \\[6pt] \sqrt{\left(Lat - scLat\right)^2 + \left(Lng - scLng\right)^2 - D_{tadv}} \\[6pt] \sqrt{\left(Lat - scLat\right)^2 + \left(Lng - scLng\right)^2 - D_{scRSRP}} \\[6pt] \sqrt{\left(Lat - ncLat_1\right)^2 + \left(Lng - ncLng_1\right)^2 - D_{ncRSRP1}} \\[6pt] \cdots \\[6pt] \sqrt{\left(Lat - ncLat_k\right)^2 + \left(Lng - ncLng_k\right)^2 - D_{ncRSRPk}} \end{array} \right\Vert$$

Formula 2.

4.5   LocatePara - Parallelization

To improve the processing speed on large scale data, we implement a parallel localization algorithm on Spark, which is a fast and general cluster computing system. Because LocatePlus is very suitable for parallelization, we encapsulate them in two functions, which can both be used for parallel computing and serial computing.
In our algorithm, the word **op** means executing one of the following operations:
    map, apply a function to all elements.
    groupByKey, group the values for each key into a single sequence.

Input: Formatted data
Output: The location in each time
function LocatePara

1.	**op** map:
2.	**Input:** rsrp, height, earfcn, tadv, longitude, latitude, timeStamp, userId.
3.	k1 = (userId, timeStamp)
4.	v1 = array (rsrp, height, earfcn, tadv, longitude, latitude)
5.	**Output:** (k1, v1)
6.	
7.	**op** groupByKey:
8.	**Input:** (k1, v1)
9.	**Output:** (k1, iterator (v1))
10.	
11.	**op** map:
12.	**Input:** (k1, iterator (v1))
13.	k2 = k1.userId
14.	v2 = (iterator (v1), Baseline (iterator (v1)), k1.timeStamp)
15.	**Output:** (k2, v2)
16.	
17.	**op** groupByKey:
18.	**Input:** (k2, v2)
19.	**Output:** (k2, iterator (v2))
20.	
21.	**op** map:
22.	**Input:** (k2, iterator (v2))
23.	sort (iterator (v2)) by timestamp
24.	v3 = locatePlus (iterator (v2))
25.	**Output:** (k2, v3)

Input: Formatted data at time T_i
Output: The location in T_i
function BaseLine

1.	**for** Each line in the formatted data except the first
2.	Compute $D_{ncRSRPj}$
3.	**end**
4.	Compute Lat and Lng
5.	Define (Lat, Lng) pair S_i
6.	Return (T_i, S_i)

Input: Formatted data at time T_{i+11}, Past 11 results of Baseline
Output: The location in T_{i+11}

function LocatePlus
1. **for** Each two adjacent points (T_{i+j-1}, S_{i+j-1}) and (T_{i+j}, S_{i+j})
2. Compute speed V_{i+j}
3. **end**
4. Compute $\hat{\alpha}$ and $\hat{\mu}$ by $V_{i+1}, V_{i+2}, \dots, V_{i+10}$
5. Compute D_v by $\hat{\alpha}$, $\hat{\mu}$, T_{i+10} and T_{i+11}
6. Compute D_{scRSRP} and D_{tadv} by the parameters of the first line in formatted data
7. **for** Each line in the formatted data except the first
8. Compute $D_{ncRSRPj}$
9. **end**
10. Compute Lat and Lng
11. Define (Lat, Lng) pair S_{i+11}
12. Return (T_{i+11}, S_{i+11})

5   Experiments

In this section, we discuss the accuracy and efficiency of our localization methods.

5.1   Data Preprocessing

The original MR data we have is a 50GB zip file compressed by gzip files. The environment we deployed for data preprocessing is a computer which contains a 3.4GHz quad-core CPU with eight threads, 16GB RAM and a 7200rpm hard disk. The relationship between time consumption and the number of threads is shown in figure 5.

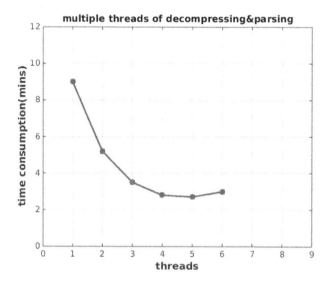

Fig. 5. The relationship between time consumption and the number of threads

It is clear that our preprocessing part achieves huge efficiency improvement with the increasing of thread, but more than four threads bring no more contribution because the reading speed of disk is low when concurrent accessing and little files extracting.

5.2   Accuracy

To get the test data, a user with a mobile phone is asked to walk around in the city, while the cellular towers record his MR data. At the same time, his mobile phone records his trace through GPS, which is the ground truth of the trace.

In subsection 4.2, the aforementioned RSRP corresponds to signal strength, this relationship is shown in table 3.

We use Baseline and LocatePlus to calculate locations, and compare with ground truth location. The statistical result is shown in table 4. The distribution is shown in table 5 and figure 5. We display the locations on google map, shown in figure 6.

Enhanced 911 is a system used in North America that links emergency callers with the appropriate public resources. E911 rules require wireless service providers to provide the latitude and longitude of the caller which must be accurate to within 50 to 300 meters depending upon the type of location technology used. [18]

From table 4, we can see that the accuracy of our method is about 100 meters, which is a big improvement in cellular network localization area. The calculated trace can almost reflect the real moving tendency.

From table 5 and figure 5, we see that we can estimate people's location with deviation in 300m, and estimate people's location with deviation in 150m with 75% confidence.

Table 3. The relationship between RSRP and real signal strength

RSRP	Range of signal strength (dBm)	Value we use (average)
0	RSRP < -140	-140
1	-140 < RSRP < -139	-139.5
2	-139 < RSRP < -138	-138.5
...	...	...
95	-46 < RSRP < -45	-45.5
96	-45 < RSRP < -44	-44.5
97	> -44	-44

Table 4. The statistical result of accuracy

Method	Points	Mean deviation	Standard deviation
Baseline	639	109.2m	73.0m
LocatePlus		97.7m	71.3m

Table 5. The distribution of accuracy

Accuracy	<= 50m	<= 100m	<= 150m	<= 200m	<= 250m	<= 300m
Points	164	309	469	549	606	623
Percentage	26%	49%	75%	87%	96%	99%

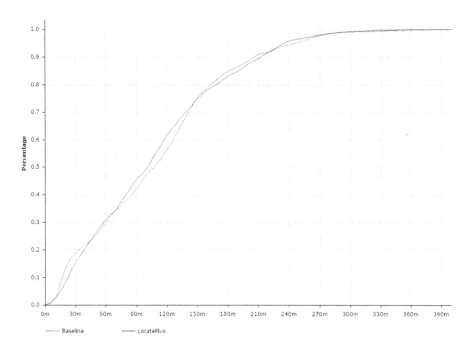

Fig. 6. The distribution of accuracy

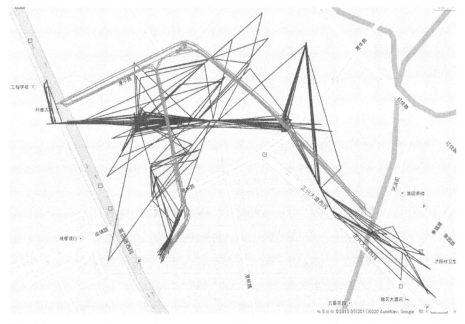

Fig. 7. Trace on google map: Green line shows the real locations, red line shows the locations we calculate

5.3   Efficiency

After we verify the accuracy of our methods, we turn to verify the efficiency.

The parallel platform we built is a cluster with six nodes. Each node owns a 3.4GHz quad-core CPU with eight threads and 32GB RAM. The version of Hadoop we deployed is 2.4.1. The version of Spark we deployed is 1.2.0, and the deployed mode of Spark is Spark-on-YARN.

We do experiment on different number of records. The time cost is shown in table 6 and figure 7.

The speedup of LocatePara is 42 when the number of records is 10 million, which means the parallel algorithm reduce 97% time cost than serial algorithm. It is a near linear relationship between time and records when using LocatePara. When the number of records is more than 20 million, the execution time of LocatePlus is more than 24 hours. LocatePlus runs so long time that we can't wait for its termination. This is why the character "-" appears in the table.

The result shows that it is feasible to use a cluster with enough number of nodes to do large-scale computing of localization within a relatively short time.

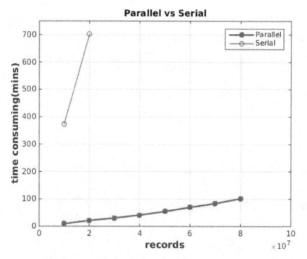

Fig. 8. Time cost comparison

Table 6. Time cost comparison of LocatePara and LocatePlus

Records (M for million)	LocatePara	LocatePlus
10M	9mins	6.2hours
20M	21mins	11.7hours
30M	30mins	-
40M	41mins	-
50M	55mins	-
60M	70mins	-
70M	84mins	-
80M	102mins	-

6   Conclusions

In this paper, we present three localization methods using mobile signal strength data in cellular network. Our methods consider the penetration loss through buildings and optimize the propagation model. Furthermore, we propose a parallel method based on Spark to improve the localization efficiency. We applied our methods through tracking people in real cellular network, the accuracy of our method reaches 100 meter (mean error), which exceeds the international standard (125 meter by FCC). The parallel method also gets a significant efficiency promotion, which reduce 95% time cost than before.

For future work, one possibility to extend the work is to continue improving localization accuracy. The other possibility is to extend 2D localization into 3D, to find the floor that people is in combining with building information.

Acknowledgement. This work is supported by China National Science Foundation under grant No.61170212, China National High-Tech Project (863) under grant No.2013AA01A607, Ministry of Education-China Mobile Research Fund under grant No.MCM20130381, and Tsinghua University Initiative Scientific Research Program (No. 20131089190). Beijing Key Lab of Networked Multimedia also supports our research work.

References

1. Zaharia, M., Chowdhury, M., Franklin, M.J., Shenker, S., Stoica, I.: Spark: cluster computing with working sets. In: HotCloud 2010, June 2010
2. Bejuri, W., Yaakob, W.M., et al.: Ubiquitous Positioning: A Taxonomy for Location Determination on Mobile Navigation System. Signal & Image Processing: An International Journal (SIPIJ) **2**(1), 24–34 (2011)
3. Zhao, S., Xia, F., et al.: MobiMsg: a resource-efficient location-based mobile instant messaging system. In: 2012 Second International Conference on Cloud and Green Computing, pp. 466–471 (2012)
4. Villemaud, G., Decroze, C., et al.: Dual-band printed dipole antenna array for an emergency rescue system based on cellular-phone localization. Microwave and Optical Technology Letters **42**(3), 249–253 (2004)
5. Marmasse, N., Schmandt, C.: User-centered location awareness. IEEE Computer **37**(10), 110–111 (2004)
6. Ardagna, C.A., Cremonini, M., et al.: Supporting location-based conditions in access control policies. In: Proc. of the ACM Symposium on Information, Computer and Communications Security (ASIACCS 2006), Taipei, Taiwan, March 2006
7. Drane, C., Macnaughtan, M., Scott, C.: Positioning gsm telephones. IEEE Comm. Mag. **46**, 46–55 (1998)
8. Caffery, J.J., Stüber, G.L.: Overview of radiolocation in cdma cellular systems. IEEE Communications Magazine, April 1998
9. Xu, B., Yu, R., Sun, G., Yang, Z.: Whistle: synchronization-free TDOA for localization. In: 2011 31st International Conference on Distributed Computing Systems, pp 760–769
10. Spirito, M.: On the accurancy of cellular mobile station location estimation. IEEE Trans. Veh. Technol. **50**, 3674–3685 (2001)

11. Okumura, Y., Ohmori, E., et al.: Okumura-hata propagation prediction model for uhf range, in the prediction methods for the terrestrial land mobile service in the vhf and uhf bands. ITU-R Recommendation P. 529–2. ITU, Geneva (1995)
12. Damosso, E.: Cost 231: (digital mobile radio towards future generation systems). Final report, European Comission, Bruxelles (1999)
13. Walfisch, J., Bertoni, H.L.: A theoretical model of UHF propagation in urban environments. IEEE Trans. on Antennas and Propagation **36**(12), 1788–1796 (1988)
14. Anisetti, M., Bellandi, V., Damiani, E., Reale. S.: Accurate localization and tracking of mobile terminals. In: 2th International Conference on Wireless Communications, Networking and Mobile Computing, Whuan, Cina (2006)
15. Anisetti, M., Bellandi, V., Damiani, E., Reale, S.: Localization and tracking of mobile antenna in urban environment. In: International Symposium on Telecomminications, pp. 353–358, September 2005
16. Singer, R.A.: Estimating optical tracking filter performance for manned maneuvering targets. IEEE Trans. Aerosp. Electron. Syst. (1970)
17. Moose, R.L., Vanlandingham, H.F.: Modeling and estimation for tracking maneuvering targets. IEEE Trans. Aerosp. Electron. Syst. (1979)
18. Federal Communications Commission. 911 Wireless Services, December 5, 2014. http://www.fcc.gov/guides/wireless-911-services

A Vehicle Speed Estimation Algorithm Based on Wireless AMR Sensors

Zusheng Zhang, Tiezhu Zhao, and Huaqiang Yuan[✉]

Dongguan University of Technology, No. 1 University Road,
Songshan Lake Sci. and Tech. Industry Park, Dongguan, Guangdong, China
hyuan66@163.com

Abstract. This paper proposes an algorithm for vehicle speed estimation based on the use of anisotropic magnetoresistive (AMR) sensors. Speed estimation relies on matching vehicle magnetic signatures from wireless sensors. A scheme based on edit-distance is developed to automatically matching signatures for the vehicles. Experimental results are presented to show that the proposed speed estimation is viable.

Keywords: Vehicle speed estimation · Wireless sensor networks · Magnetic sensor · Vehicle detection

1 Introduction

Real time acquisition of traffic information plays a key role in intelligent transportation systems (ITSs). Wireless sensor networks have lots of potential toward providing an ideal solution for traffic information acquisition, such as their low power, small size, low cost, and high accuracy.

At present, the sensors used in the traffic information acquisition include the following types: inductive loop detector [1], image (camera) sensor [2, 3], acoustic sensor [4, 5], infrared sensor [6], ultrasonic sensor [7], etc. The image sensor acquires an abundance of information, but it is vulnerable to bad weather and nighttime operation. The acoustic sensor and infrared sensor are vulnerable to noise in deployed environments. Magnetic sensors based on magneto resistors have recently been proposed for vehicle detection [8, 11] because they are quite sensitive, small and more immune to environmental factors such as rain, wind, snow or fog than sensing systems based on video cameras, ultrasound or infrared radiation.

Many algorithms have been proposed for dynamic traffic information acquisition system based on wireless AMR sensor. The PATH program of the University of California, Berkeley [12, 13] had first extensively explored of AMR sensor network based vehicle detection system. S.Y. Cheung et al. [13] had explored the applications for vehicle detection, speed estimation, and classification. M. Kang et al. [14] proposed a vehicle detector with an AMR sensor and addresses experimental study carried out to show the detector's characteristics and performance. Experiment results show that the vehicle detection accuracy rate is more than 99%, and the accuracy rate to estimate

© Springer International Publishing Switzerland 2015
Y. Wang et al. (Eds.): BigCom 2015, LNCS 9196, pp. 167–176, 2015.
DOI: 10.1007/978-3-319-22047-5_14

length and speed of vehicle is more than 90%. But these threshold based algorithms are not valid to all situations and often leads to a high false alarm rate. In order to solve this problem, adaptive threshold algorithms are presented [8, 9]. However, threshold-based vehicle detection algorithms have congenital defects. Firstly, there is no existence of a reasonable threshold to distinguish between the signal of vehicle and interference, especially in low Signal-Noise-Ratio (SNR). Furthermore, different places have different geomagnetic fields, so it is difficult to have a threshold to fit all situations.

Another detection method is to study the characteristics of the wave pattern. J. Ding proposed Min-max Detection Algorithm [15]. Modified Min-max Algorithm for magnetometers [16] has also been published. These algorithms analyze the process of parking signal changes, but they still rely on single threshold and have low vehicle detection success rate.

Li Cui et al. [17, 18] proposed a Similarity Based Vehicle Detection (SBVD) algorithm to detect vehicles in low SNR conditions by calculating the correlation between on-road signals and a referential signal. Besides, data fusion algorithm based on fuzzy logic theory has also been proposed to monitor parking space using magnetic sensor [19]. S. Taghvaeeyan et al. [20] presents a method to estimate the position of the vehicles. It utilizes the fact that the magnetic moment caused by a vehicle decreases as the distance to the sensor increases.

K. Kwong et al. [21] propose a system that measures the vehicle count and travel time the links of a road network. The measurements require matching vehicle signatures recorded by a wireless magnetic sensor network. The matching algorithm is based on a statistical model of the signatures.

Vehicle Speed detection is obtained with a pair of sensor nodes which were placed with a given distance along the lane. Most existing algorithms [13, 21] only consider the single-lane constrained vehicle detection. They did not consider in the multi-lanes road which has turning and overtake. This paper will address these problems and propose an accurate detection algorithm.

The remainder of this paper is organized as follows. Section 2 proposes the algorithm. Section 3 conducts experiments to prove the performance of the algorithm. Finally, Section 4 makes a brief conclusion.

2   Algorithm

Vehicle speed estimation can be obtained with a pair of sensor nodes. Fig.1 shows an example of speed estimation of vehicle by a pair of sensor nodes. Node A and B were placed on the middle of a lane with a given distance along the traveling direction. The vehicle signature measured by node A should be similar to the one measured by node B. So the vehicle speed can be estimated.

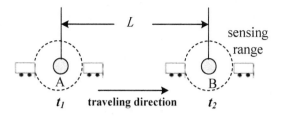

Fig. 1. Diagram of vehicle speed detections

Speed estimation is based on matching individual vehicle signatures obtained from a pair of wireless magnetic sensors. The magnetic signals of different classed of vehicles are different. Vehicle signatures of the same classes measured in the same traveling direction are expected to be the same. Suppose the data comprise a sequence of upstream vehicle classes x_i, $i=1$, $\cdots$, N and a sequence of downstream vehicle classes y_j, $j=1$, $\cdots$, M. The matching is a pair of matched sequences like (*Up*, *Down*). The key issue is to design an optimal matching scheme. Solve the problem involve two steps: Firstly, using matching methodology classifies a vehicle signature into a pre-defined vehicle class. Then, a class matching scheme is designed based on the edit distance algorithm.

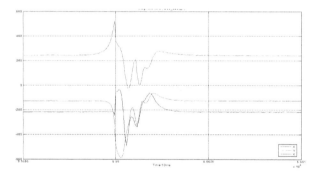

Fig. 2. Three-axes magnetic signature of a vehicle

2.1   Vehicle Classification

Sensor node consists of a 3-axis AMR sensor. Fig. 2 is the three-axis magnetic signature of a vehicle cross a sensor. The geomagnetic signal is studied and summarized in the time domain involves a series of steps as shown in Fig. 3. The vector magnitude shift from the environment's magnetic field would be the most reliable method. Using digitized measurements of three-axis sensor outputs after amplification, the vector magnitude would be:

$$G(i) = \sqrt{X_i^2 + Y_i^2 + Z_i^2}$$

(1)

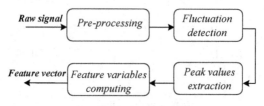

Fig. 3. Signal Processing

A smoothing filter, which takes a running average of the signal, is used to smooth the signal. The algorithm in the node automatically extracts the sequence of peak values, i.e., local maxima and minima from a continuous fluctuation signal, as shown in Fig. 4.

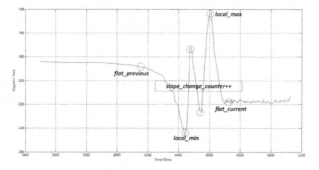

Fig. 4. Peak values of magnetic signal

P is an array includes peak values of a vehicle extract from the up sensor. $P_{up} = \{(i, t_i, x_i), 1 \le i \le N \}$. Array P_{down} is a vehicle's peak point collection of the down sensor. $P_{down} = \{(j, s_j, y_j), 1 \le j \le M \}$. The distance between P_{up} and P_{down} can be computed by function $S(P_{up}, P_{down})$. A decision criterion uses the following rule: it can be considered to the correspond waveform of P_{up} and P_{down} belong same class if $S(P_{up}, P_{dwon})$ $\in [0, C_1]$, where C_1 is a given constant.

$$S(P_{up}, P_{down}) = \sum_{i=1}^{M} \frac{abs(y_i - x_i)}{\max(y_i, x_i)} \Big/ M$$

(2)

$$is\_same\_class = \begin{cases} 1 & if \ S(P_{up}, P_{down}) \in [0, C_1] \\ 0 & otherwise \end{cases}$$

(3)

The magnetic signatures of a training dataset with known vehicle classes are provided to the classifier for training and calibration. Then using matching methodology classifies a vehicle signature in a specific format into a pre-defined vehicle class. Use letters to identify vehicle classes, during a measurement time interval, vehicles indexed $i = 1, \cdots, N$ cross the upstream sensor at times $s_1 < s_2 < \cdots$. Vehicles indexed $j = 1, \cdots, M$ cross the downstream sensor at times $t_1 < t_2 < \cdots$. The upstream sensor measures the 'class' x_i of each vehicle i that crosses it and the corresponding time s_i. The downstream sensor measures the class y_j of each vehicle j that crosses it and the corresponding time t_j. Thus the measurement data consists of two arrays $\{(x_i, s_i), i = 1, \cdots, N\}$ and $\{(y_j, t_j), j=1, \cdots, M\}$.

2.2   Optimal Matching Scheme

Based on the idea of levenshtein distance [22] and the shortest path [21] algorithms, we proposed a matching scheme for vehicle speed estimation. As shown in Fig. 7, a graph $G(N, M)$ are arranged in the form of a grid, the graph with $N=7$ and $M=8$. These letters, such as $e, x, h, \cdots, d$, are identifications of vehicle classes. $G(N, M)$ is called the edit graph in sequence comparison algorithms for measuring the difference between two sequences. Mathematically, the edit distance between two strings is given by

$$H(i, j) = \begin{cases} H(i-1, j)+1 \\ H(i, j-1)+1 \\ H(i-1, j-1)+2 & \text{if } x_i \neq y_j \\ H(i-1, j-1) & \text{if } x_i = y_j \\ H(i-2, j-2) & \text{if } x_{i-1}x_i = y_j y_{j-1} \end{cases} \tag{4}$$

Note that the first element in the minimum corresponds to down-turn, the second to up-turn, the third to mismatching, the fourth to matching, and the fifth to overtaking.

- Down-turn: A vehicle may cross the upstream node A but turn before reaching the downstream node B, as shown in Fig. 5;
- Up-turn: A vehicle may turn the upstream node A but cross the downstream node B;
- Mismatching: the upstream vehicle i is mismatching the downstream vehicle j, it is equal to the case which the vehicle i is down-turn, and successively a vehicle j is up-turn;
- Overtaking: an upstream vehicle i_2 which follows i_1, $i_2 > i_1$, may be matched to downstream vehicles j_1, j_2 in the reverse order $j_1 > j_2$.

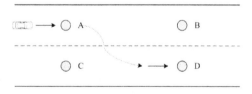

Fig. 5. Down turn on multi-lanes road

The value u for an edit path is equal to the weight of the corresponding path (defined as the sum of the edge weights) in the edit graph. Thus, edit distance is obtained by finding the minimum weight path, which is accomplished by the algorithm as shown in Fig. 6. The edit distance $H(i, j)$ is given by equation (4).

```
begin initialize m=length[x],n=length[y]
    C[0,0]=0;  i=0; j=0
    do i=i+1;   C[i,0]=i;    until i=m
    do j=j+1;   C[0,j]=j;    until j=n
    i=0; j=0
    do i=i+1
        do j=j+1;   C[i,j]=H[i,j];  until j=n
    until i=m
    return C[m,n]
end
```

Fig. 6. Shortest path algorithm

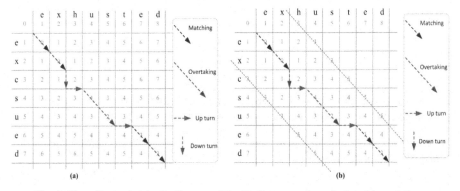

(a)   (b)

Fig. 7. The edit graph (a) edit distance, (b) edit distance to be calculated in real time

Fig. 7(a) is the interpretation of the upstream vehicles with signatures $x=\{(e, s_1), (x, s_2), (c, s_3), (s, s_4), (u, s_5), (e, s_6), (d, s_7)\}$, are respectively matched with the downstream vehicles with signatures $y=\{(e, t_1), (x, t_2), (h, t_3), (u, t_4), (s, t_5), (t, t_6), (e, t_7), (d, t_8)\}$.

The edit graph grows with the observation time interval, and so with each new upstream or downstream vehicle, one needs to calculate the distance of its class from all previous classes. The effort to compute these distances for each new vehicle grows linearly with the observation interval, which is unsuitable for real-time implementation. If T is an upper bound on the travel time, a simple choice for T would be to assume a minimum speed and take T to be the corresponding travel time. The edit distances $G(N, M)$ to be calculated are thus bounded by the dashed lines in Fig. 7 (b). So the computational burden for each new upstream or downstream vehicle is constant.

3   Experimental Results and Analysis

The sensor node has Zigbee based communication modules and it consists of HMC5983 magnetic sensor. The Zigbee wireless module is using the CC2530 MCU. For battery powered sensor nodes, the lifetime is a critical factor. The sensor node has no specific responsibility for maintaining the network infrastructure because it is an end-device. So it doesn't exchange packet for network maintaining after joining the network. Sensor node samples the magnetic signal periodically. When it detected a car crossed, the node automatically extracts peak values of the car's magnetic signature and transmits a packet includes these peak values to base station. The node goes to sleep when no task to save energy. Using a pair of $5000mAh$ AA Li batteries parallel connected, the sensor node can continuously work for about one year without changing battery.

To test the proposed system, we deploy two pair sensor nodes in the road. As shown in Fig. 9. Devices with AMR sensors are nailed in the center of the line spaces. The routers are fixed on the roadside light. Considering the power issue, routers are equipped with solar panel for frequent data forwarding.

We developed a PC system using Java language and MySQL database. As shown in Fig. 10, using the graphical client interface, users can know the estimated speed of a vehicle. These blue lines with arrow describe the wireless network topology.

Fig. 8. Sensor nodes deployment

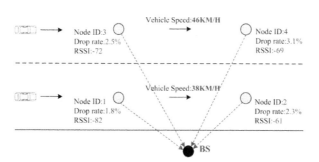

Fig. 9. Pictures of the PC system

Table 1. Experiment parameters

Parameter	Description	Value
Len	Smooth buffer length	30
L	Given distance of two sensor nodes	50m
C_l	A threshold for *is_same_class*	0.1
F	The frequency of sensor sample	150HZ
T	Upper bound on the travel time	20s

In our experiments, parameters related to the speed estimation algorithm are given in table 1. In order to verify the edit distance based matching scheme, we deployed two pair nodes in the two lines road. The system has operated for one week. Using the video data of this road as a reference, we count the correct matching rate. As shown in Fig. 10, when the vehicle speed is between 13*km/h* and 75*km/h*, the correct matching rate is better than 92%. While the vehicle speed >75*km/h*, the error of matching is expanded.

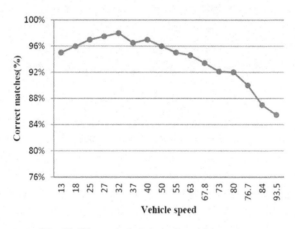

Fig. 10. The occupied time of a parking space

Table 2 shows the speed detection results of a road for one day. The actual speed refers to the speed which is measured by the handheld radar velocimeter. When the vehicle speed is between 15*km/h* and 42*km/h*, two types of speed are consistent. While if the vehicle speed <15*km/h* or >70*km/h*, the difference between two kinds of speed is gradually expanding. This is because the slower the speed, the greater the probability of turning and overtaking. On the other hand, the faster of vehicle speed, the estimation results of speed are more susceptible to the time synchronization error between the pair of sensors.

Table 2. The detection results of vehicle speed

The round of experiment	actual speed(km/h)	estimated speed(km/h)
1	13.2	10.5
2	18.6	17.1
3	25.4	26.2
4	27.4	27.5
5	37.6	35.5
6	43.9	45.6
7	50.9	52.4
8	55.2	56.8
9	63.3	61.9
10	73.7	70.5
11	80.9	75.4
12	84.1	85.2
13	93.5	90.6

4 Conclusions

This paper based on the edit distance proposes an algorithm for vehicle speed estimation. Speed estimation is based on matching vehicle signatures obtained from a pair of wireless magnetic sensors. The algorithm can work in the multi-lanes road which has turning and overtake. Experiment results show that the average correct matching percentage is about 92%.

Acknowledgements. This study is supported by the National Natural Science Fund, China (No. 61402105, 61170216 and 61402106), and Guangdong Province Natural Science Foundation (2014A030313631 and 2014A030313632).

References

1. Kong, Q., Li, Z., Chen, Y., Liu, Y.: An approach to urban traffic state estimation by fusing multisource information. IEEE Transactions on Intelligent Transportation Systems **10**(3), 499–511 (2009)
2. Alessandretti, G., Broggi, A., Cerri, P.: Vehicle and guard rail detection using radar and vision data fusion. IEEE Transactions on Intelligent Transportation Systems **8**(1), 95–105 (2007)
3. Pathirana, P., Lim, A., Savkin, A., Hodgson, P.: Robust video/ultrasonic fusion-based estimation for automotive applications. IEEE Transactions on Vehicular Technology **56**(4), 1631–1639 (2007)
4. Cevher, V., Chellappa, R., Mcclellan, J.: Vehicle speed estimation using acoustic wave patterns. IEEE Transactions on Signal Processing **57**(1), 30–47 (2009)
5. Guo, B., Nixon, M., Damarla, T.: Improving acoustic vehicle classification by information fusion. Pattern Analysis and Applications **15**(1), 29–43 (2012)

6. Dickson, C., Wallace, A., Kitchin, M., Connor, B.: Improving infrared vehicle detection with polarization. In: Intelligent Signal Processing Conference(ISP), pp. 1–6, December 2013

7. Chen, M., Chang, T.: A parking guidance and information system based on wireless sensor network. In: IEEE International Conference on Information and Automation (ICIA), pp. 601–605 (2011)

8. Zhou, Q., Tong, G., Li, B., Yuan, X.: A Practicable Method for Ferromagnetic Object Moving Direction Identification. IEEE Transactions on Magnetics 48(8), 2340–2345 (2012)

9. Petrou, J., Skafidas, P., Hristoforou, E.: Electronic Toll and Road Traffic Monitoring System Using 3-D Field AMR Sensors. Sensor Letters 11(1), 91–95 (2013)

10. Koszteczky, B., Veszprem, H., Simon, G.: Magnetic-based vehicle detection with sensor networks. In: IEEE Proc. Instrumentation and Measurement Technology Conference, pp. 265–270 (2013)

11. Sebastiá, J.P., Lluch, J.A., Vizcaíno, J.R.L.: Signal conditioning for GMR magnetic sensors: Applied to traffic speed monitoring GMR sensors. Sensors and Actuators A: Physical 137(2), 230–235 (2007)

12. Yiu, S., Coleri, S., Dundar, B., Ganesh, S., Tan, C., Varaiya, P.: Traffic measurement and vehicle classification with a single magnetic sensor. In: California Partners for Advanced Transit and Highways (PATH). Working Papers: UCB-ITS-PWP-2004-7 (2004)

13. Cheung, S.Y., Varaiya, P.: Traffic surveillance by wireless sensor networks: Final report. Technical report, California PATH, University of California, Berkeley, CA 94720 (2007)

14. Kang, M., Choi, B.W., Koh, K.C., Lee, J., Park, G.: Experimental study of a vehicle detector with an AMR sensor. Sensors and Actuators A: Physical 118(2), 278–284 (2005)

15. Ding, J., Cheung, S.Y., et al.: Signal processing of sensor node data for vehicle detection. In: The 7th International IEEE Conference on Intelligent Transportation Systems, pp. 70–75 (2004)

16. Gu, J., Zhang, Z., Yu, F.: Design and implementation of a street parking system using wireless sensor networks. In: IEEE International Conference on Industrial Informatics (INDIN), pp. 1212–1217 (2012)

17. Wang, R., Zhang, L., Sun, R., Gong, J., Cui, L.: EasiTia: a pervasive traffic information acquisition system based on wireless sensor networks. IEEE Transactions on Intelligent Transportation Systems 12(2), 615–621 (2011)

18. Zhang, L., Wang, R., Cui, L.: Real-time traffic monitoring with magnetic sensor networks. Journal of Information Science and Engineering 27(4), 1473–1486 (2011)

19. Zhu, J., Cao, H., Shen, J., Liu, H.: Data fusion for magnetic sensor based on fuzzy logic theory. In: Proceedings of the International Conference on Intelligent Computation Technology and Automation (ICICTA 2011), pp. 87–92 (2011)

20. Taghvaeeyan, S., Rajamani, R.: The Development of Vehicle Position Estimation Algorithms Based on the Use of AMR Sensors. IEEE Transactions on Intelligent Transportation Systems 13(4), 1845–1854 (2012)

21. Kwong, K., Kavaler, R., Rajagopal, R., Varaiya, P.: Real-time measurement of link vehicle count and travel time in a road network. IEEE Transaction on Intelligent Transportation Systems 11(4), 814–825 (2010)

22. Levenshtein, V.I.: Binary codes capable of correcting deletions, insertions, and reversals. Soviet Physics Doklady 10(8), 707–710 (1966)

Security and Privacy

Design and Evaluation of a Policy-Based Security Routing and Switching System for Data Interception Attacks

Yudong Zhao[1](✉), Ke Xu[1], Rashid Mijumbi[2], and Meng Shen[3]

[1] Tsinghua University, 9-402 Room, East-main Building, Beijing, China
zhaoyd10@mails.tsinghua.edu.cn, xuke@mail.tsinghua.edu.cn
[2] Universitat Politècnica de Catalunya, Barcelona, Spain
rashid@tsc.upc.edu
[3] Beijing Institute of Technology, Beijing, China
shenmeng@bit.edu.cn

Abstract. In recent years, the world has been shocked by the increasing number of network attacks that take advantage of router vulnerabilities to perform data interceptions. Such attacks are generally based on low cost, unidirectional, concealed mechanisms, and are very difficult to recognize let alone restrain. This is especially so, because the most affected parties – the users and Internet Service Providers (ISPs) – have very little control, if any, on router vulnerabilities. In this paper, we design, implement and evaluate a policy-based security system aimed at stopping such attacks from both the routing and switching network functions, by detecting any violations in the set policies. We prove the system's security completeness to data interception attacks. Based on simulations, we show that 100% of normal packets can pass through the policy-based system, and about 99.92% of intercepting ones would be caught. In addition, the performance of the proposed system is acceptable with regard to current TCP/IP networks.

Keywords: Router vulnerabilities · Data interception attacks · Policy-based routing and switching system · Security completeness

1 Introduction

Data interception attacks (DIAs) that routers and switches take advantage of vulnerabilities to monitor and redirect traffic to third party have been seriously harming the security of users over many years. However, users and ISPs used to underestimate its impact and scale until June 2013 when it was revealed that the National Security Agency (NSA)'s PRISM Project was intercepting and collecting data across the Internet backbone [1][2]. By hacking huge internet routers, PRISM could get access to communications of hundreds of thousands of computers without the need to hack each one independently. These revelations have since underscored the importance of enhancing the security of routing and switching functions in TCP/IP networks.

© Springer International Publishing Switzerland 2015
Y. Wang et al. (Eds.): BigCom 2015, LNCS 9196, pp. 179–192, 2015.
DOI: 10.1007/978-3-319-22047-5_15

In addition to being low cost and having abilities to create high damage, DIAs can be characterized in two main ways: (1) they are stubborn in that not only are users and ISPs not able to easily detect and/or stop router vulnerabilities, but developers are also not able to eliminate them, (2) they are usually highly concealed such that it is almost impossible for ISPs to distinguish data interception packets from genuine ones. Such characteristics have made it difficult to develop and implement complete and practical security systems to detect and stop the attacks.

One simple idea for regulating routing and switching functions is to self-develop routers and switches to eliminate vulnerabilities. Unfortunately, self-developed routers and switches either lack efficiency, or are not able to completely eliminate vulnerabilities. As a result, we have to regulate routing and switching functions with secure mechanisms. However, current mechanisms are either only fit for specific core networks and specific types of DIAs, or are difficult to implement. To the best of our knowledge, there are still no security complete, universal and easily implementable mechanisms for DIAs.

In this paper, we propose a policy-based system with the objective of enhancing the security of routing and switching functions in TCP/IP networks. The proposed system is based on proposing a set of policies that routers and switches[1] can use to detect abnormal traffic behavior. In contrast to previous approaches, our proposal does not focus on the routing and switching operations, but rather, on changes to a 3-tuple (<Source IP, Destination IP, Payload>) characteristic code for each packet before and after it passes through a router. Compared with other mechanisms for stopping DIAs, the proposed system has three major features: (1) *security complete*, we prove that the set of policies are necessary and sufficient to stop DIAs that take advantage of router vulnerabilities, (2) *universal*, each policy of the system is applicable to all types of TCP/IP networks, and transparent to network topologies, routing protocols and configuration & management plans, and (3) *implementable*, the policy-based routers can be designed, and the simulation result shows that policy violations can be detected and restrained, yet the performance of the routers is acceptable with regard to current TCP/IP networks

The rest of this paper is organized as follows: We discuss related work in Section 2. In section 3, we present the policy-based security system for routing and switching in TCP/IP networks. The security completeness of this system to DIAs is proved in Section 4, and the final model of the policy-based router presented in Section 5. Section 6 evaluates the functionality and performance of the proposed system by simulation, and the paper is concluded in Section 7.

2   Related work

Researchers have been paying more attention to DIAs that result from router vulnerabilities. Most current solutions can be categorized as being based on three broad areas: (1) the level of routing and switching operation, (2) the OS and devices, and (3) core network.

---

[1] In the rest of this paper, we use "router" to mean both routers and switches.

With regard to the level of routing and switching operation, software dataplanes set against the risk of disrupting the network with bugs, unpredictable performance, or security vulnerabilities. Dobrescu et al. [3] present a verification tool by applying existing verification ideas and compositionality and combining them with certain domain specific packet processing software, to explore the feasibility of verifying software dataplanes to ensure smooth network operation. Although ensures software dataplanes both verification and performance, this tool relies on a given set of conditions, and is only effective for crash-freedom, bounded-execution, and filtering properties. Based on a novel combination of static analysis with symbolic execution and dynamic analysis with concrete execution, Kothari et al. [4] developed a method to help discover manipulation attacks in protocol implementations. However, since defenders do not always grasp the protocol code developed by third parties, this method is not effective for all types of protocol implementation attacks.

Considering the level of OSs and devices, the Trusted Computing Group developed and promoted Trusted Computing (TC) [5]. Challener et al. [6] showed that it is possible to provide a complete and open industry standard for implementing TC hardware subsystems in PCs. They demonstrated what TC can achieve, how it works, and how to write applications for it. However, although promoted for many years, there is still no accepted trusted computer model and effective theory or approach to dynamically quantize the degree of trustworthiness of software, which greatly limits the development of TC. Chen et al. [7] argue that the OS and applications currently running on a real machines should relocate into a virtual machine to service, which enables services, such as secure logging, intrusion prevention and detection, and environment migration, to be added without trusting or modifying the OS or applications. While its flexibility provides tremendous benefits for users, this approach undermines the fact that today's relatively static security architectures rely on the number of hosts in a system, their mobility, connectivity, patch cycle, etc. [8].

At the level of the core network, to construct high-level security mechanisms for secure connections, Kim et al. [9] proposed lightweight, scalable, and secure protocols for shared key setup, source authentication, and path validation. Although its computing overhead is low, there must be an Origin and Path Trace (OPT) header embedded into each packet head. The length of the header is variable according to path length. Even more, the longer the path is, the lower the throughput it produces under such protocols. With the aim of embedding security into the network as an internal service, and designing a secure network architecture, the idea of a Trustworthy Network (TN) is proposed by Peng et al. [10]. The authors propose three essential properties to interpret trustworthiness of a network, and discuss four key issues needed to be resolved. Although the idea of TN has been around for many years now, most works base on the theories or techniques in certain fields, and as a result, the complete system has not been built so far.

Finally, focusing on the security of the content of each IP packet, instead of routing and switching processes, IPSec [11] is aimed at ensuring communication safety by authenticating and encrypting each IP packet. Payloads and/or packets are encrypted, which prevents plain text from being intercepted by attackers. However,.IPSec cannot know what attackers try to intercept and where the attack host locates. In addition,

when the number of clients increases, available bandwidth to each client is the bottleneck that prevents the increase in number of connections handled by the web server
[12].

In summary, it is our opinion that there are still no theories or solutions security
complete, universal and implementable for DIAs. Our proposal is able to recognize
and restrain intercepting packets efficiently with designable policy-based routers that
are universal to any TCP/IP network, and can locate the attack hosts

3   The Policy-Based Security System for Routing and Switching in TCP/IP Networks

3.1   Motivation and Design Challenges

As the processing and communicating hubs of core networks, routers are usually designed in a fixed logical structure [13] and process packets according to a fixed series
of operations. However, once embedded vulnerabilities, a router may work abnormal-
ly. Figure 1 shows the normal routing and switching process (left) and all possible
abnormal behaviors (right). With abnormal behavior AB1 in figure 1, the packets
sent by data forwarding plane are neither from upstream routers, nor generated by
control plane. With AB2, the payloads of packets are changed or substituted. With
AB3, although a packet seems to be forwarded normally, its output interface is altered
secretly. Attackers can take advantage of such abnormal behaviors to perform data
interception. As it is difficult for current router providers to eliminate vulnerabilities,
and for ISPs to detect these three kinds of abnormal behaviors, DIAs can be highly
stubborn and concealed.

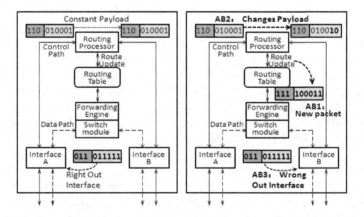

Fig. 1. Normal routing process (left) and Possible abnormal behaviors (right)

To this end, we propose a system that are achievable and applicable to all types of
TCP/IP networks. Once the designed routers are deployed throughout a core network,
the intercepting packets can be recognized and restrained efficiently, while all normal
packets can pass through the system.

3.2   Static vs Dynamic Policy-Based System

There are generally two ways of designing systems for recognizing abnormal behaviors of routers. One is the static way such as code analyzing and operation result
checking, while the other is the dynamic way such as function simulation and operation decode matching. Because this paper presents our first foray into a policy-based
approach for securing routers, we concentrate on the static approach by proposing a
set of static policies and the model of policy-based route. Instead of focusing on the
routing and switching processes, the proposed set of policies concentrate on the
changes in a 3-tuple characteristic code <Source IP, Destination IP, Payload> before
and after it passes through the policy-based router. In other words, the policies are
designed for recognizing and restraining the output packets whose characteristic
codes are abnormal as shown in figure 1. As we show later in the paper, such a con-
ception does not only make the policies universally applicable to all types of TCP/IP
network, but also makes it easier to prove the security completeness, and to design
high-performance policy-based routers.

3.3   The Policy-Based Security System for Routing and Switching in TCP/IP
Networks

From the perspective of routing and switching operation results, the abnormal behaviors shown in figure 1 appear as the three kinds of patterns indicated in figure 2. The
thick full curve Alice-Bob in figure 2 presents the routing path from Alice to Bob
without any attack. However, once there is a vulnerable router R within the path, an
attacker, Eve, can launch three kinds of attacks. Corresponding to abnormal behavior
AB1 in figure 1, while routing packets normally, a vulnerable router R secretly reserves packets Alice sends to Bob, and redirects their payloads to Eve, which leads to
abnormal result AR1. For AB2, R changes or substitutes the payloads of packets,
which leads to different payloads between what Alice sends and what Bob receives as
AR2 shows. For AB3, R alters the output interface by interfering the normal routing
and switching process, as shown in figure 2, the output link of a packet is altered
away from the thick full curve to the wrong link along AR3.

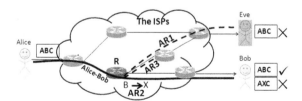

Fig. 2. The three patterns of abnormal behaviors of vulnerable router

Although being able to recognize the three kinds of abnormal behaviors, these
three policies suffer from difficulties in implementation, universality and effectiveness. Notice that attackers usually intercept communication by "stealing" the payloads
of the reserved packets, instead of "robbing" the original packets, these policies can
then be optimized to three final policies below.

FP1: For a packet encapsulated payload, when the source IP address is not that of the router, if there are no input packets characterized the same 3-tuple code <Source IP, Destination IP, payload> with it, the packet is not allowed to output.

FP2: A router is not allowed to send packets generated by itself to any unauthorized end host.

FP3: When a packet is forwarded to an end host by a boundary router, the IP address of the host must match the destination IP address of the packet.

4   The Security Completeness of the Policy-Based System

For ease of reference, Table 1 defines the variables used for proving the security completeness of the three final policies to DIAs. In addition, an over bar means the opposite, for example, $\overline{R_{LE}}$ means a router which is not linked to Eve.

Table 1. Variable Definition

Variable	Definition
Alice	The sending user and host
Bob	The receiving user and host
Eve	The attacker and the attacking host
R_{ARB}	An arbitrary router
R_{LE}	A boundary router linked to Eve
$Pack_{A-B}$	A packet Alice sends to Bob
$Payload_{A-B}$	The payload of $Pack_{A-B}$

As listed by ISO/IEC [14], there are generally three properties of information system security: confidentiality, integrity and availability (CIA). As for preventing data interception, users and ISPs hope all packets can and only be accurately and completely received by the destination hosts, therefore security completeness here means that the set of policies we proposed is necessary and sufficient to ensure packets the properties of confidentiality and integrity.

Lemma 1. Once all routers in a core network abide by the proposed routing and switching policies, when a device (Dev) receives a packet characterized by <Sou, Des, $Payload_{Sou-Des}$>, then,

$$\begin{cases} \textbf{1:} \text{ It is Sou who sent } Pack_{Sou-Des} \\ \textbf{2:} \text{ } Payload_{Sou-Des} \text{ is the original payload Sou sent} \end{cases}$$

Proof. We prove 1 by contradiction. Without loss of generality, let $Path_{Sou-Dev}=$ [Sou, R_1,..., R_i,..., R_n, Dev] be an arbitrary reachable path from Sou to Dev. If $Pack_{Sou-Des}$ is not sent by Sou, then there must be another device $Sou' \neq Sou$ in figure 3, such that Sou' can send $Pack_{Sou-Des}$ to Des through $Path_{Sou'-Dev}$. When

Sou$'$ is an end host, and can send $Pack_{Sou-Des}$, this means that Sou$'$ had known $Payload_{Sou-Des}$ before it sent $Pack_{Sou-Des}$. Therefore it can send $Pack_{Sou'-Des-Payload_{Sou-Des}}$, instead of sending $Pack_{Sou-Des}$ in the pattern of IP spoofing. As a result, Sou$'$ is a router. Because Sou$' \neq$ Sou, according to FP1, when router Sou$'$ sends $Pack_{Sou-Des}$, , there must be an input packet of Sou$'$ characterized by $<$Sou, Des, $Payload_{Sou-Des}>$, this means that Sou$'$ is not the sender of $Pack_{Sou-Des}$, which contradicts the hypothesis, then Sou is the only sender of $Pack_{Sou-Des}$.

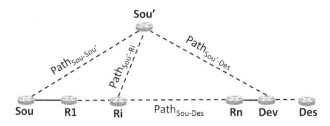

Fig. 3. Sou' sends $Pack_{Sou-Des}$ through Dev

Then we prove 2: For an arbitrary $Path_{Sou-Dev}$ in figure 3, we define $R_0 =$ Sou; $R_{n+1} =$ Dev. For each packet output from R_i ($1 \leq i \leq n$) according to FP1, there must be a matching input packet characterized by the same 3-tuple code with it. If the input packet is forwarded by R_{i-1}, for the atrbitrariness of i, each device in $Path_{Sou-Dev}$ shares the same 3-tuple characteristic code for $Pack_{Sou-Dev}$. As a result, $Payload_{Sou-Des}$ is the original payload Sou sent. On the contrary, if the matching packet is forwarded directly by $\overline{R_{i-1}}$, without loss of generality, we assume the matching packet is output by Sou$'$ in figure 3, because $Pack_{Sou-Dev}$ can reach Sou$'$ and R_i successively, according to result 1, there must be a path from Sou to Sou$'$, and a path from Sou$'$ to R_i, such that $Pack_{Sou-Dev}$ is generated by Sou, and sent to Dev along a new path $Path'_{Sou-Dev}$ composed of $Path_{Sou-Sou'}$, $Path_{Sou'-R_i}$, and $Path_{R_i-Dev}$. Similarly, each device in $Path'_{Sou-Dev}$ shares the same 3-tuple characteristic code for $Pack_{Sou-Dev}$, therefore $PayLoad_{Sou-Des}$ is the original payload Sou sent.

Therefore, lemma 1 is right. Based on lemma 1, we prove the integrity and confidentiality of the policy-based system to DIAs.

Proposition 1. Once all routers in a core network abide by the set of policies, if the network can ensure reachability for each packet, *what is sent, is what is received.*

Proof. In lemma 1, particularly when Dev is Des, each packet $Pack_{Sou-Des}$ Des received was sent by Sou, and the sent payload matched the received payload. Because the network can ensure reachability for each packet, the communication Des receives is equal to what Sou sends.

Proposition 2. Once all routers in a core network abide by the set of policies, packets sent from a user can only be received by the target user.

Proof. We prove proposition 2 by contradiction. In a network shown in figure 4, if a non-target user Eve can receive $Payload_{A-B}$, then there must be a router R_{ARB} which reserves $Pack_{A-B}$ and redirects $Payload_{A-B}$ to Eve. With AR1, R_{ARB} reserves $Payload_{A-B}$ and generates a new packet, whose destination IP address is Eve. With AR3, R_{ARB} forwards $Payload_{A-B}$ to Eve by altering the output interface of the reserved packet. Because the step of forwarding is limited to one link, AR3 can only be carried out when R_{ARB} is a boundary router linked to Eve, such as R_{LE} in figure 4.

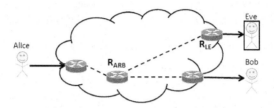

Fig.4. R_{ARB} reserves $Pack_{A-B}$ and redirects $Payload_{A-B}$ to Eve

For an input packet characterized by <Alice, Bob, $Payload_{A-B}$>, the new packet generated by R_{ARB} is characterized by <*, Eve, $Payload_{A-B}$>, where * is the IP address of an arbitrary device for the convenience of attack. When *=R_{ARB}, R_{ARB} is trying to send information to Eve, which violates FP2. Also when * = $\overline{R_{ARB}}$, according to FP1, there must be a packet characterized by <$\overline{R_{ARB}}$, Eve, $Payload_{A-B}$> input to R_{ARB}before. By lemma 1, the new generated packet must be sent by $\overline{R_{ARB}}$, which means that $\overline{R_{ARB}}$ is trying to send information to Eve, this violates FP2 too.

When R_{LE} tries to forward a reserved $Payload_{A-B}$ to Eve as AB3 does, the characteristic code of the reserved $Pack_{A-B}$ is <*, $\overline{Eve}$, $Payload_{A-B}$>, which means that R_{LE} is trying to forward a packet to a mismatching end host. This violates FP3.

In summary, due to the set of policies we proposed, attackers cannot receive non target packets, which proves proposition 2.

FP2 and FP3 are obviously necessary for preventing DIAs. As for FP1, each tuple in the characteristic code is necessary too. If there is no source IP in the characteristic code, an assistant Carol can help Eve to perform data interception by sending Eve packets with different payloads. Once one of these payload happens to be equal to what Alice sends to Bob, a vulnerable router can send $Pack_{*-Eve-Payload_{A-B}}$ to Eve. If there is no destination IP in the characteristic code, R can send $Packet_{Alice-Eve-Payload_{A-B}}$ without violating the set of policies. If there is no payload in the characteristic code, Eve can ask for Carol to send packets to him. Once one of these packets pass through R, the latter can send $Packet_{Carol-Eve-Payload_{A-B}}$legally. As a result, we get the conclusion below.

Proposition 3. Each policy in the policy-based system is necessary to prevent DIAs.

The three propositions above show that the proposed system is security complete to DIAs. It is worth noting that network topology and routing protocols are not involved during the course of the proofs, which means that the policies are transparent to these aspects.

5   The Policy-Based Router Model and the Policy Violations Detecting Model

5.1   The Procedure of Policy-Based Abnormally Detecting

As shown in figure 5, the procedure of policy-based detecting is composed of two sub procedures, one collects the 3-tuple characteristic code <Source IP, Destination IP, payload> of every input packet to the database, the other detects the consistence of all output packets with the three policies one by one.

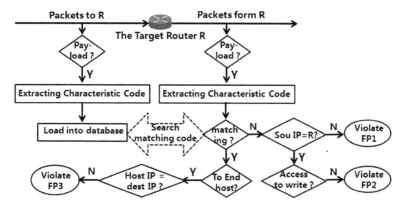

Fig. 5. The procedure for detecting the consistence of routers' behavior with the policies

5.2   The Policy-Based Router Model and the Policy-Violations Detecting Model

The policy-based router model and the policy-violations detecting model are shown in figure 6. The area inside the darker rectangle is the model of the policy-based router, and the outside part is the policy-violations detecting model.

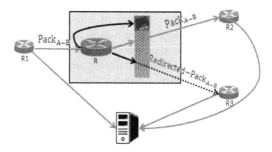

Fig. 6. The policy-based router model and the policy-based violations detecting model

Generally, the proposed policy-based router is made up of a traditional router R and a policy-violations detecting system. The latter is composed of an information collecting module and an output violations detecting module. The detecting system does not take part in the process of the traditional router. When a packet is input to R, the information collecting module collects and sends its characteristic code to the output violations detecting module. The detecting model is composed of a chip designed according to the policy-based detecting procedure, and a random access memory for storing the database of all input characteristic codes. For each output packet, the detecting module picks up its characteristic code and seeks the matching code in the database, then judges whether R is trying to output an abnormal packet. In addition, the detecting module forbids violations departing.

The original $Pack_{A-B}$ can pass through a policy-based router in figure 6 legally, while once R tries to reserve $Pack_{A-B}$ and redirect $Payload_{A-B}$ to Eve, the redirected_$Pack_{A-B}$ is judged as violation by the policy-based system

Because providers may not be able to (or unwilling to) design devices strictly under the policy-based router model, it is necessary to design the policy-based violations detecting model for ISPs to check whether providers abide by their promise. The area outside the dark rectangle in figure 6 is the model of the policy-violations detecting system, which is also composed of an information collecting module and an output violations detecting module. The collecting module is composed of all neighbor routers of the target router, while the detecting module is a server with the function of policies detecting. All neighbor routers of R are bypassed to the server. All neighbors send all packets input to R from bypass links to the server, and the server checks whether each output packet voilates the policies. Since servers are good at storing, if only the abnormal output packets are recorded to the server, we can be aware of what the attacker wants to intercept and where the position of the attack host locates.

5.3   A Fast Matching Algorithm

As expected, the policy-based router should be efficient. However, additional overhead in storing, computing and communicating decreases policies cheacking speed, particularly that in input and output codes matching As a result, it is necessary to design rapid matching algorithms. In what follows, we propose one.

In the algorithm, the characteristic codes database is substituted by a digest table, which record the fact that a packet with a certain hash digest has input into R. The size of the table is 2^n (where n is limited by the size of RAM, for example 2^{30}) bits, each position in the table is initialized to 0. As shown in figure 7, when a packet is input into R, the detecting system computes the digest of its characteristic code using a hash algorithm, and preserves a fixed part and length of the digest as I-digest-result. The value of the position corresponding to I-digest-result in the digest table is set to 1. For an output packet, the system computes and acquires its O-digest-result with the same hash algorithm, and look-up the value of the corresponding position in the digest table. If the value is 1, the algorithm asserts that there is an input packet matching the output packet. Otherwise, it confirms that there is no matching input packet.

It can be noted that the weight of the digest table will continue to increase as more packets are input into R, such that it is probable for an intercepting packet to have a 1 O-digest-result by coincidence. Therefore we propose 2 digest tables to record I-digest-results. Table 1 in figure 7 starts at the beginning of each $(2i) \times \rho$, while table 2 starts at the beginning of each $(2i + 1) \times \rho$, where $i = 0, 1, 2...$, ρ is a time period no less than the maximum time a packet is configured to stay in a router. With the 2 tables, the record of each input packet will be preserved at any time point, which ensures that each genuine packet is not categorized as an intercepting one.

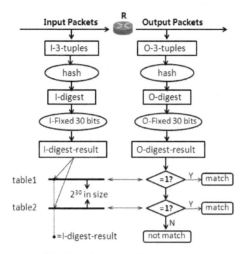

Fig. 7. A fast matching algorithm

When the length of digest result is 30 bits, the memory size of the detecting system is $2 * 2^{30} = 2G$ bits. It is practical for the detecting system to assemble a 2G RAM.

6   Evaluation of the Functionality and Performance

6.1   Functionality

Focusing only on the operation result of the output packets, FP2 and FP3 can always be carried out correctly. As for FP1, we have to check whether there is an input packet matching it. We preserve part of the hash digest as the input record to improve the implementability of the system, which may lead to a false positive (a normal packet is confirmed as an intercepting one) and a false negative (an intercepting packet is confirmed as a normal one) result. We analyze the two false probabilities theoretically and by simulation.

For each normal packet, let IT be its input time, OT be the output time, there must be a $i = 0, 1,...$, such that IT $\in [\rho \times i, \rho \times (i + 1))$, then OT $\in (IT, IT + \rho]$. As a result, for a packet with input time IT, one or both of the digest tables must preserve its output record, then the probability of false positive is 0.

The throughput of a router, ρ, and the size of the digest table (dtsize) determine the probability of a false negative. When throughput = 10Gbps, ρ=0.5s, dtsize=1G bits, the maximum number of packets input to R within 2ρ is

$$\text{(throughput/max packet length)} \times 2\rho = {}^{10G}/_{1500 \times 8} \times 2 \times 0.5 = 833333$$

The process of input packets filling in the digest tables obeys a (0, 1) distribution, where the probability of the value of a position is replaced by 1 is $1/_{dtsize}$ =1/2^{30}, then the expectation of the maximum total weight of the 2 tables (duplicate removal) in every 2ρ is

$$E = \text{throughput} \times (1 - (1 - 1/_{dtsize})^{maxnum} = 833010$$

Therefore, the maximum expected probability of a false negative is 833010/1G = 0.07758%, which means that the expected maximum number of intercepting packets passing through the policy-based system is 833010×0.07758% = 647.

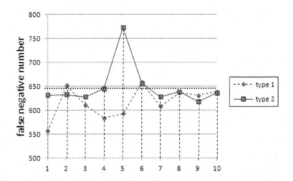

Fig. 8. Number of intercepting packets pass through the policy-based system by coincidence

We conduct the simulation with an X86 system. 1,250,450 packets are obtained from a real network. We use MD-5 as the hash algorithm, and preserve the value from the 65th to the 94th bit of the digest as the digest result. The simulation result shows that all the first 833,333 packets pass through the policy-based system. Based on the simulation system, we carry out two types of attacks. Type 1 changes the destination IP of the 833,333 packets to some random address, while type 2 reverses the payload of each packet together with changing destination IP. We carry out each type of attack 10 times. Figure 8 indicates the active numbers of the attacking packets which pass through the system. The average number of false negatives is 633, which is actually smaller than the expected 647. We calculate the weight of the 2 digest result tables (duplicate removal) after the first 833,333 packets finished passing through R, and get a result of 818,773, based on which the average number of false negative is 635. As a result, the simulation result matches the theoretical expectation.

In summary, we draw a theoretical conclusion that 100% normal packets can pass through the policy-based system, and that no less than 99.92242% intercepting packets will be recognized.

6.2   Performance

The performance of a router is inversely proportional to the average time T it takes to process a packet. For our policy-based router, T is composed of the traditional routing

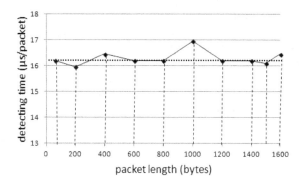

Fig. 9. Average detecting time for packets of different lengths

time T1 and the time T2 used for detecting policy violations. For a certain traditional router R and a fixed packet length, T1 is a constant. Therefore the performance of the proposed policy-based router is depended on T2.

Using the simulation system described above, we determine T2 with traffic from real network. The amount of the traffic is 4,897,900 packets, among which there are 4,010,903 packets whose length is 1,500 bytes, accounting for 81.9% of the total. Figure 9 indicates the test values of T2 for the simulated system. Because T2 might be influenced by the length of a packet, we conduct 10 experiments with different packet lengths. We choose 64, 200, 400, 600, 800, 1000, 1200, 1400, 1500 and all packets with difference lengths as packet length.

The time unit in figure 9 is µs. From this figure we conclude that when lower than 1500 bytes, packet length has little impact on average T2. In fact, the average value of T2 is 16.30µs, and the maximum policy-based detecting rate varies from 31.4M to 736.2M bps according to the average packets length, which is acceptable for edge core networks. This is because the X86 system for simulation is assembled a dual core processer of Intel i5-2410 and a 4 G RAM, detecting system with hardware no lower than the simulation system can be developed. Moreover, current routing devices in convergence layer are usually assembled by dedicated chip, also routing and forwarding with hardware platform, instead of routing and forwarding with software platform composed of general processor as the experiment does. If only it is meaningful for the whole core network deploying policy-based routers, as Janaka et al. [15] shows, it is reasonable to construct MD5 accelerators using hardware and FPGA, by which the performance of policy-based routers can be greatly promoted.

7   Conclusion

In this paper we designed, implemented and evaluated a policy-based routing and switching system aimed at stopping DIAs. Our proposal includes a set of policies that can be embedded in routers so as to detect interception output packets. We have been able to prove the security completeness of the proposed set of policies to DIAs, and discussed the implementability of the proposed system, in which case we also proposed a fast matching algorithm. In addition, by simulation, we have shown that our proposal is able to positively identify 100% of genuine traffic, and reject about 99.9% of violating traffic, and the performance of the proposed system is acceptable with regard to current TCP/IP networks.

However, there are some technical issues that we continue to study before we can consider our design as complete for commercial purposes. To prevent DIAs, all routers in the core network must abide by the proposed policies. In order to decrease deployment cost and running overhead, it would be interesting to consider an incremental deployment of the policy-based system. In addition, as a new attempt in this field, we proposed a static policy-based system in this paper. Future work should consider a dynamic approach which could be based on operation decode matching for higher efficiency and broader application fields.

Acknowledgment. We gratefully acknowledge funding support for this research from Chinese 863 Project *The key Technique and Verification for Address Driven Networking*.

References

1. National Security Agency: PRISM/US-984XN Overview, April 2013
2. Bowden, C.: The US national security agency (NSA) surveillance programmes (PRISM) and foreign intelligence surveillance Act(FISA) activities and their impact on EU citizens' fundamental rights (2013)
3. Dobrescu, M., Argyraki, K.: Software dataplane verification. In: NSDI (2014)
4. Kothari, N., Mahajan, R., Millstein, T., et al.: Finding protocol manipulation attacks. In: SIGCOMM (2011)
5. Trusted Computing Group: TCG Specification Architecture Overview [EB/OL]. [2005-03-01].
   https://www.trustedcomputinggroup.org/groups/TCG_1_0_Architecture_Overview.pdf
6. Challener, D., Yoder, K., Catherman, R., et al.: A Practical Guide to Trusted Computing, 1st edn. IBM Press (2007)
7. Chen, P.M., Noble, B.D.: When virtual is better than real. In: HOTOS-VIII, May 2001
8. Garfinkel, T., Mendel, R.: When virtual is harder than real: security challenges in virtual machine based computing environments. In: Proc of the 10th Workshop on Hot Topics in Operating Systems, pp. 210–217. USENIX Association, Berkeley (2005)
9. Kim, T.H., Basescu, C., Jia, L., Lee, S.B., Hu, Y., Perrig, A.: Lightweight source authentication and path validation. In: SIGCOMM (2014)
10. Xue-hai, P., Lin. C.: Architecture of trustworthy networks. In: 2nd IEEE International Symposium on Dependable, Autonomic and Secure Computing (2006)
11. Kent, S., Atkinson, R.: Security Architecture for the Internet Protocol, IETF RFC 2401, November, 1998. http://tools.ietf.org/html/rfc2401
12. Meenakshi, S.P., Raghavan, S.V.: Impact of IPsec overhead on web application servers. In: International Conference Advanced Computing and Communications, ADCOM 2006, pp. 652–657 (2006)
13. Chao, H.J.: Next generation routers (invited paper). Proceedings of the IEEE **90**(9), 1518–1558 (2002)
14. ISO/IEC: International Standard ISO/IEC 27000, 3rd edn, January 15, 2014
15. Deepakumara, J, Heys, H.M, Venkatesan, R.: FPGA implementation of MD5 hahs algorithm. In: Proceedings of IEEE Canadian Conference on Electrical and Computer Engineering, CCECE 2001 (2001)

Wireless Device Authentication Using Acoustic Hardware Fingerprints

Dajiang Chen[1]([✉]), Xufei Mao[2], Zhen Qin[1], Weiyi Wang[1],
Xiang-Yang Li[3], and Zhiguang Qin[1]

[1] School of Information and Software Engineering, UESTC, Chengdu, China
{djchen,qinzhen,qinzg}@uestc.edu.cn, wangweiyi65@gmail.com
[2] School of Software and TNLIST, Tsinghua University, Beijing, China
xufei@greenorbs.com
[3] Department of Computer Science, Illinois Institute of Technology, Chicago, USA
xli@cs.iit.edu

Abstract. Authentication between wireless devices is critical for many wireless network applications, especially in some secure wireless communication scenarios. One of ingenious solutions is to extract a fingerprint to perform device authentication by exploiting variations in the transmitted signal caused by hardware and manufacturing inconsistencies. In this work, we propose a device identification protocol (named *S2M*) by leveraging the frequency response of a speaker and a microphone from two wireless devices as an acoustic hardware fingerprint. *S2M* authenticates the legitimate user based on the results of matching the fingerprint extracted at the learning process with the one extracted at the verification process. We design and implement *S2M* for both mobile phones and PCs and the extensive experimental results show that *S2M* achieves both a low false negative rate and a low false positive rate in various of scenarios under different attacks.

Keywords: Wireless security · Physical layer security · Device authentication · Acoustic hardware fingerprinting

1 Introduction

Device identity authentication is one of the most significant challenges in wireless network security problem. To address these problems, a bunch of traditional cryptographic mechanisms are proposed for wireless device identity authentication. However, the frequently used security protocols such as WPA, WPA2 (802.11i), and 802.11w have a track record of being compromised [1–3]. Recently, some promising alternative approaches [4–11] have been proposed to enhance device identity authentication in wireless networks by exploiting physical layer characteristics. As mentioned in [13], the existing physical-layer device identification schemes can be roughly classified into three categories: software based [4,5], channel fingerprinting based [6–8], and hardware fingerprinting based ones [9–11]. Unfortunately, the existing approaches suffer from the following major

© Springer International Publishing Switzerland 2015
Y. Wang et al. (Eds.): BigCom 2015, LNCS 9196, pp. 193–204, 2015.
DOI: 10.1007/978-3-319-22047-5_16

shortcomings: (1) software-based protocols cannot distinguish between different physical devices running the same software; (2) channel fingerprinting-based approaches might not work well in a highly dynamic environment where the channel state or Radio Strength Signal (RSS) changes drastically over time; (3) most hardware fingerprinting-based schemes still suffer mimic attacks [12,13].

Considering that the audio communication (including audio-based Near Field Communication [14]) is increasingly mature and commercialized, and most existing wireless devices are equipped with microphones and speakers, in this work, we design a wireless device identity authentication protocol utilizing acoustic hardware (Speaker/Microphone) fingerprints, which is named Speaker-to-Microphone (abbreviated to *S2M*). The key insights of our scheme are summarized as follows: (1) the same speaker/microphone pairs have a similar FR curve at any time and any place when some conditions are satisfied (details in Section 3); (2) different speaker/microphone pairs have different FR curves (details in Section 3); (3) the properties of FR curve of speaker/microphone pairs cannot be replicated or copied from one device to another (details in Section 6.1).

Acoustic fingerprinting has been studied in recent papers [17,18]. In [17], Das *et al.* discuss the feasibility of using microphones and speakers to uniquely fingerprint individual devices. They leverage Gaussian mixture models (GMMs) to classify the device into one of several recorded audio fingerprints. In [18], Zhou *et al.* propose a scheme to generate stable and unique device ID stealthy for smartphones by exploiting the frequency response(FR) of the speaker. Our work is inspired by the above work in acoustic fingerprinting, but we focus on wireless device authentication with the FR of a speaker and a microphone from two wireless device.

Let us consider the following scenario to explain the main idea of *S2M*. Assuming that a wireless device (named Alice) would like to authenticate herself to another one (named Bob) in the presence of an active adversary (named Eve). Under this situation, the authentication steps of *S2M* is as follows. Firstly, the initiator Alice establishes a connection with the authenticator Bob via the audio handshake method. Secondly, Alice generates the audio signals for authentication and transmits it to Bob through her speaker. After receiving the audio signals, Bob calculates the FR and extracts the fingerprint. Then Bob stores Alice's ID and the fingerprint. So far, the above series of operations could be considered as learning process. When identity authentication is required (i.e., in authentication process), Alice generates the audio signals for authentication and transmits them to Bob again through her speaker. After Bob receives, it will match the fingerprints from learning and authentication processes, if the output of the matching algorithm is positive, authentication is successful; otherwise, the authentication fails). Compared with the state-of-art related work, the proposed protocol has higher security and lower requirement of hardware. Moreover, the protocol can be used for many applications, such as wireless network access authentication, device authentication in audio payment system. We summarize our main contributions as follows:

- We propose a device authentication protocol, named *S2M*, which leverages the FR of acoustic hardware (speaker/microphone as a fingerprint) on a wireless device to authenticate the other wireless device. Based on *S2M*, we design an extended protocol *E-S2M*, which can achieve variable distance authentication.

- We evaluate *S2M* (rep. *E-S2M*) through extensive experiments using mobile phones in a number of scenarios under different attacks. Experimental results show that the operating distance of *S2M* (rep. *E-S2M*) can reach up to 5m, and the successful authentication rate can achieve 98% (rep. 97%). Experiments also show that both *S2M* and *E-S2M* achieve a low false acceptance rate which is less than 2%.
- We design and implement the software application of *S2M* and *E-S2M* which are suitable for operation on mobile phones and PCs.

2   Modeling

2.1   Acoustic Attenuation Model

Acoustic attenuation is a measurement of the energy loss of sound propagation in media. Due to the fractal microstructures of media, such acoustic attenuation typically exhibits a frequency dependency characterized by a power law $S(f, x) = S_0(f)e^{-\alpha(f)x}$, where $\alpha(f) = \alpha_0(2\pi f)^{\eta_f}$; f denotes the frequency; x is wave propagation distance; $S_0(f)$ and $S(f, x)$ represents the amplitude of the transmitted and received acoustic pressure at frequency f respectively; and the tissue-specific coefficients α_0 and η_f are empirically obtained by fitting measured data [15]. Due to the inability to the acoustic components (i.e., speaker and microphone) to faithfully reproduce tones of certain frequencies, the phenomenon of selective attenuation of certain frequencies happens. Consequently, when sound waves transmitted by a speaker and received by a microphone, the acoustic attenuation model can be rewritten as $S(f, x) = L_S(f)L_M(f)S_0(f)e^{-\alpha(f)x}$, where $L_S(f)$ is the loss of a speaker, $L_M(f)$ is the loss of a microphone. The frequency response (FR) of a speaker/microphone pair at frequency f is defined as

$$FR \triangleq R(f) = 20 \log \frac{S(f, x)}{S_0(f)} = 20\{\log[L_S(f)] + \log[L_M(f)] - (\log e)\alpha(f)x\}$$
(1)

where $\log(\cdot)$ is the logarithmic function with base 10. Based on Eq.1, we observe that Radio Frequency (RF) characterizes the physical characteristics of speaker and microphone pairs. In order to achieve authentication of the wireless devices, an intuitive idea is to use FR as the acoustic hardware fingerprint. However, the receiver (i.e., authenticator) can not calculate the FR as it does not know the transmitted acoustic pressure $S_0(f)$. To overcome this problem, in this paper, we use $\Xi = \{\Xi(f_i)\}_i$ $(i \in \{1, \cdots, N\})$ to represent the fingerprints of speaker and microphone pairs, where f_i is the selected frequency for feature extraction and $\Xi(f_i)$ is defined as follows.

$$\Xi(f_i) = 20 \log S(f_i, x) = R(f_i) + 20 \log[S_0(f_i)]$$
(2)

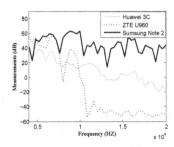 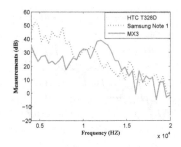

Fig. 1. Fingerprints for different speakers **Fig. 2.** Fingerprints for different microphones

2.2   Adversarial Model in Device Authentication

We design *S2M* based on a strong adversarial model. We summarize our assumptions about the attacker Eve as follows. (1) The adversary power is computationally unbounded. That is, it can run an infinite number of steps; (2) The adversary is active and can masquerade other nodes by forging their identities; (3) The adversary knows the authentication protocol and its parameters settings. Based on these assumptions, we consider the following three attack models in our study:

(1) **Audio replay attack:** An adversary Eve first records the audio signals from a legitimate sender, and then impersonates the legitimate one by replaying the recordings;
(2) **Changing distance attack:** Eve intends to improve the success probability of attack by adjusting the distance between the receiver and him;
(3) **Same-type-device attack:** Eve has a device with the same brand (i.e, the same Manufacturer) as that of the legitimate transmitter, and he pretends the transmitter by sending the same audio signal as the real transmitter.

Evaluation Metrics: In this paper, we mainly focus on two evaluation metrics: False negative (FN) error rate and False positive (FP) error rate. Specifically, the FN rate is the number of legitimate sender's audio signals (for authentication) that are incorrectly identified as the attacker's signal over the totally received number of audio signals for authentication. Similarly, the FP error rate is the number of attacker's signals which are incorrectly considered to be sent from the legitimate sender over the totally received number of audio signals for authentication.

3   The Acoustic Hardware Characteristics

As a valid fingerprint, the physical characteristic of fingerprint must has two properties: the otherness and stability. The otherness means that the characteristic between different devices is big enough to differentiate; the stability is that the characteristic does not change along with the changing of time and space. In order to verify Ξ have those properties, we conduct extensive experiments as follows.

The first experiment considers the difference of speakers between wireless devices. Figure 1 depicts the fingerprints (i.e., the function Ξ) from different

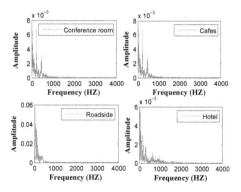

Fig. 3. Spectrum of Ambient Noise

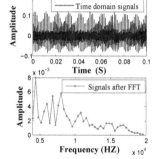

Fig. 4. Signals conversion

Speakers to the same Microphone (i.e., from phone: Huawei 3C, ZTE U960, Samsung Note2, to phone: HTC T328D). The fingerprints are measured by transmitting tones of frequencies between 4KHz and 20KHz (with step size 400), from one device to another, while placing the devices with fixed distance 10cm between each other. The results (Fig. 1) show that the fingerprint differences for two mobile phones are obvious.

Then we consider the difference of microphones between wireless devices. Figure 2 depicts the fingerprints from one Speakers to different Microphones (i.e., from phone: Huawei 3C, to phone: HTC T328D, Samsung Note1, MX3). The fingerprints are also measured by transmitting tones of frequencies between 4KHz and 20KHz (with step size 400) with fixed distance 10cm. The results (Fig. 2) show that the microphone differences between mobile phones are obvious.

In order to become an efficient and practical system with rigorous security, *S2M* must be proved to have a good performance in public spaces such as conference rooms, cafes, roadsides, and hotels, where the ambient (acoustic) noise can cause significant effect on fingerprints extraction. To characterize the effect, we measure the received acoustic power in some typical environments, including conference rooms, hotels, malls and cafes during busy hours. As shown in Fig.3, the received power of ambient noises in all of these environments are negligible when the frequency reaches 3KHZ. Thus, in *S2M*, we can exploits the frequencies that are lager than 4KHz for FR extraction.

4   Authentication Protocol

In this section, we present a device authentication protocol, named *S2M* consisting of two process: learning process and verification process.

4.1   Learning Process

Learning Process consists of three stages: *Audio-handshake phase*; *Mixed-signal generation phase*; *Feature extraction and storage phase*.

Audio-handshake phase: During this phase, the identity interaction and synchronization of legitimate nodes (Alice and Bob) must be achieved. With no doubt, it can be achieved by using conventional wireless communication (e.g., WiFi, Bluetooth). However, in order to reduce the dependence on hardware and wireless network infrastructures, as well as expand the application scope of *S2M*, we realize the identity interaction and synchronization by utilizing of audio data transmission scheme: dual-tone multifrequency (DTMF) [16].

The steps of this phase are described as follows: (1) Alice encodes her ID number into $Sound(ID_A)$ with audio encoding, and sends it to Bob using speaker. (2) After receiving the audio signals, Bob first decodes ID_A with audio decoding, and then encodes ID_B into $Sound(ID_B)$ and sends it to Alice by speaker. After that, Bob readies to receive the audio signals from Alice for feature extraction. (3) If Alice received the audio signals, she decodes ID_B. Otherwise, Alice returns Step 1.

Mixed-signals generation phase: After obtaining Bob's ID, Alice needs to send some audio signals, which are denoted by $Sound(Auth_A)$, to Bob for feature extraction. In order to obtain an high-resolution fingerprint and reduce the transmission time simultaneously, we leverage mix-frequency technology, which generates the mixed audio signals by combining the various frequency components. Specifically, the mixed-signals $Sound(Auth_A)$ can be denoted by

$$Sound(Auth_A) = \sum_{i=1}^{N} \frac{1}{N} sin(2\pi(\varphi_0 + f_\Delta(i-1))t) \tag{3}$$

where φ_0 is the minimum frequency, f_Δ is the length of step. In our system, we set $\varphi_0 = 4KHz$ and $f_\Delta = 0.4KHz$. That is, $N = 41$, $f_1 = 4KHz$, $f_2 = 4.4KHz$, $\cdots$, $f_N = 20KHz$.

The steps of this phase are described as follows: (1) Alice generates the mixed-signals $Sound(Auth_A)$ with Eq.3, and sends the signals to Bob using speaker. (2) Bob receives the signals $\widehat{Sound}(Auth_A)$ with microphone, where

$$\widehat{Sound}(Auth_A) = Sound(Auth_A) + Noise + Echoes.$$

Feature extraction and storage phase: To extract the feature of devices pairs, Bob needs to compute $\varXi$ from the record $\widehat{Sound}(Auth_A)$. Thus, Bob has to convert the signals $\widehat{Sound}(Auth_A)$ from time domain to frequency domain. To do this, we utilize Fast Fourier Transform (FFT), an algorithm for computing the Fourier transform of a set of discrete data values. Given a finite set of data points, the FFT expresses the data in terms of its component frequencies. By using FFT, the Sound Pressure Amplitude (SPA) is denoted by

$$< S(f_1), \cdots, S(f_N) >= FFT_N(\widehat{Sound}(Auth_A)) \tag{4}$$

Here, FFT_N is the function that $S(f_1), \cdots, S(f_N)$ are the local maximum FFT values at frequency ranges $[f_1-FS, f_1+FS], \cdots, [f_N-FS, f_N+FS]$ respectively (as shown Fig. 4). The objective of taking local maximum is to eliminate the impact of frequency shift. In our system, we set $FS = 5Hz$. Thus, the fingerprint Ξ_A can be written as

$$\Xi_A = 20log(< S(f_1), \cdots, S(f_N) >) \tag{5}$$

The steps of this phase are described as follows: (1) When Bob receives the mixed-signals, he extracts the fingerprint Ξ_A via Eq. 4 and Eq. 5, and records the pairs $< ID_A, \Xi_A >$. (2) Bob sends one period sine wave of frequency 10KHz to Alice. (3) After receiving the signals with frequency 10KHz, Alice records ID_B; otherwise, Alice returns to the mixed-signals generation phase.

4.2 Verification Process

Verification process is also divided into three phases: *Audio-handshake phase*; *Mixed-signal generation phase*; *Feature extraction and matching phase.*

Audio-handshake phase: The audio-handshake phase of verification process is similar to the same phase of learning process. The steps are described as follows: (1) Alice encodes ID_A into $Sound(ID_A)$ with audio encoding, and sends it to Bob. (2) After receiving the signals from Alice, Bob decodes ID_A and checks whether the ID_A is already stored locally (if not, the authentication fails). (3) Bob sends $Sound(ID_B)$ to Alice, and then is ready to receive the signals for feature extraction from Alice. (4) If Alice receives the audio signals, she decodes ID_B and checks whether the ID_B is already stored locally (if not, the authentication fails); otherwise, Alice goes back to the Step 1.

Mixed-signal generation phase: The Mixed-signal generation method of this phase is the same as that in the learning process. Thus, we omit the details here.

Feature extraction and matching algorithm phase: The feature extraction method of this phase is similar to that in learning process. The goal in the fingerprint matching algorithm (MA) is to determine whether $\Xi_A \approx \Xi_A'$, where Ξ_A is the fingerprint extracted in the learning process and Ξ_A' is the fingerprint extracted in verification process (details of MA in Section 4.3). Specifically, if $MA(\Xi_A, \Xi_A') = Y$, the authentication is successful (i.e., $\Xi_A \approx \Xi_A'$); otherwise, the authentication fails. The steps of this phase are described as follows: (1) Bob extracts the fingerprint Ξ_A' via Eq. 4 and Eq. 5. (2)Bob calls MA with input Ξ, Ξ'. If $MA(\Xi_A, \Xi_A') = Y$, Bob sends one period sine wave of frequency 10KHz to Alice; otherwise, the authentication fails. (3) If Alice receives the audio signals with the frequency 10KHz, the authentication is successful; otherwise, Alice goes back to Mixed-signal generation phase.

4.3 Fingerprint Matching

In the proposed protocol, we design a new matching algorithm, named deviation-ratio-based matching algorithm (D-MA). The algorithm takes two fingerprints Ξ_A and Ξ_A' as inputs. It first calculates the absolute value of corresponding components of two fingerprints vectors, and then compares each absolute value with a deviation threshold Δ. Specifically, denoting $S = |\{n : |\xi_n - \xi_n'| \leq \Delta\}|$, the ratio of $N - S$ to S can be considered as a metric on N-dimensional vector space R^N named deviation ratio (DR). That is to say, we define

$$DR(\Xi_A, \Xi'_A) = (N - S)/S = |\{n : |\xi_n - \xi'_n| > \Delta\}|/|\{n : |\xi_n - \xi'_n| \leq \Delta\}| \quad (6)$$

where $|\mathscr{A}|$ means the cardinality of set $\mathscr{A}$. Finally, the algorithm compares $DR(\Xi_A, \Xi'_A)$ with a matching threshold Th. If $DR(\Xi_A, \Xi'_A) \leq Th$, the algorithm output Y (Yes); otherwise, output N (No). Note that $DR(\Xi_A, \Xi'_A)$ reflects the similarity of two fingerprints. For instance, $DR(\Xi_A, \Xi'_A) \leq 0.5$ means that there are at least $\frac{2}{3}N$ pairs of the corresponding component in two fingerprints so that their absolute values are less than Δ.

In order to select the appropriate deviation threshold Δ and the matching threshold Th, we first conduct extensive experiments to investigate the relationship between Δ, Th and authentication errors (i.e., the FN and FP error in Section 2). In our experiments, more than 50 mobile phones are used and more than 20000 fingerprints are extracted under different authentication distances. Figure 5 shows the authentication error rate with different deviation thresholds and matching thresholds. We can see that, for each Th, a larger Δ increases the rate of FN error but improves the rate of FP error. Especially, when the matching threshold Th is 0.4 and the deviation threshold Δ is 8, both the FN and FP error rate are less than 0.5%, Thus, in our system, we set deviation threshold $\Delta = 8$; matching threshold $Th = 0.4$.

5   Extended Protocol E-S2M

To expand the application of *S2M*, we design an extended protocol of *S2M* such that the new protocol can achieve authentication under different distance between the users. Based on *S2M*, the new protocol includes two algorithms: *mobile learning algorithm* and *automatic fingerprint matching algorithm*.

We first present the mobile learning algorithm. In learning process, Alice and Bob consult with an *effective range* $[l, r]$ of authentication, and then Alice and Bob proceed the steps as follows.

1. Alice and Bob first move their positions such that the distance between them is equal to r, and then perform learning process of *S2M* (i.e., Section 4.1, the obtained fingerprint at Bob is denoted by $\Xi_A(0)$);
2. Alice walks toward Bob with *moving distance* ΔL_1, and then they perform the learning process of *S2M* (the fingerprint at Bob is denoted by $\Xi_A(1)$);
3. For $i > 1$, if $r - \Sigma_i \Delta L_{i-1} > l + \frac{1}{2}\Delta L_{i-1}$, and then Alice and Bob repeat Step 2 with moving distance ΔL_i (the fingerprint is denoted by $\Xi_A(i)$); otherwise, stoping.

After that, Bob obtains a list of fingerprints $< \Xi_A(0), \cdots, \Xi_A(M) >$. Denoting the *maximum moving distance* by $\Delta L = max\{\Delta L_1, \cdots, \Delta L_M\}$. The effect of ΔL on the performance of *E-S2M* will be discussed in the end of this subsection.

Then we introduce automatic fingerprint matching algorithm. In authentication process, Alice and Bob perform the previous scheme in authentication process (i.e., Section 4.2) by replacing the deviation-ratio-based matching algorithm with

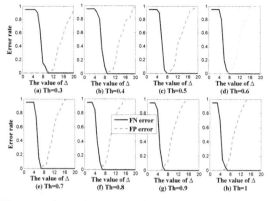

Fig. 5. The relationship between authentication errors and thresholds under different distances

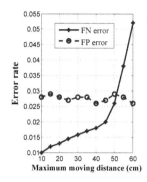

Fig. 6. Impact of change maximum moving distance ΔL on error rate

automatic fingerprint matching algorithm as follows. Let $\Xi'_A = <\xi_1, \cdots, \xi_N>$ be the fingerprint obtained by Bob at this process. For each $m \in \{1, \cdots, M\}$, let $\Xi_A(m) = <\xi_1(m), \cdots, \xi_N(m)>$.

1. For each $n \in \{1, \cdots, N\}$, Bob computes $a_n = \min\{|\xi'_n - \xi_n(m)| : m = 1, \cdots, M\}$;
2. For each $m \in \{1, \cdots, M\}$, Bob computes $A_m = \{n : a_n = |\xi'_n - \xi_n(m)|, n = 1, \cdots, N\}$;
3. If m_0 be the index such that $|A_{m_0}| = \max\{|A_m| : m = 1, \cdots, M\}$, Bob computes $D\text{-}MA(\Delta, Th, \Xi_{A^*}, \Xi'_{A^*})$ with inputs $\Xi_A(m_0)$ and Ξ'_A.

As shown in Figure 6, we conduct extensive experiments, in which more than 50 different mobile phone groups are used and 20000 fingerprints are extracted, to investigate the relationship between the maximum moving distance ΔL and the error rates. The results show that (1) the rate of FP error is stable with the value of ΔL increase; (2) a larger Δ increases the rate of FN error. Especially, when $\Delta L \leq 50cm$, both the FN and FP error rates are less than 0.03. i.e., acquiring only 2 or 3 places' fingerprints can achieve the FN error rate under 3% when the effective range is 0-1m. Thus, in our system, we set the maximum moving distance $\Delta L = 50cm$.

6   Evaluation

6.1   Analysis on the Attack

Firstly, we conduct an experiment on *E-S2M* to evaluate an audio replay attack in a quiet corridor. In this experiment, mobile phone ZTE N880 is the legitimate sender; HTC T328d is the receiver; MX3 is the attacker. We evaluate *E-S2M* under different effective ranges (e.g., 0-1m,...,0-5m). For each effective ranges, attacker changes the distance between the attacker and the receiver from 0.1m

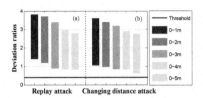

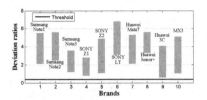

Fig. 7. Reply attack and changing distance attack: the value range of deviation ratios (from attacker) for each effective range

Fig. 8. Same-type-device attack: the value range of deviation ratios (from attacker) for each type of phone

to 10m with step length 0.05m, and lunches replay attack at each position. I.e., there are 198 attacks for each effective range. Receiver extracts the fingerprints and computes the deviation ratios. Figure 7 (a) plots the range of deviation ratios (from attacker) under different effective ranges. It shows that the deviation ratio under each effective range is larger than the deviation threshold. Thus, *S2M* has excellent performance to resist the audio replay attack.

Secondly, we evaluate a changing distance attack in a quiet corridor under different effective ranges (e.g., 0-1m,...,0-5m). In this experiment, the mobile phone (Huawei honor3C) is set to be the legitimate sender, the HTC T328d is the receiver and the Sangsung S5560 is the attacker. For each effective range, we change the distance between the attacker and the receiver from 0.1m to 10m with step 0.05m. In each distance, the attacker follows *E-S2M* and sends the authentication audio signals to the receiver. Figure 7 (b) plots the range of deviation ratios (from attacker) under different authentication distance. It shows that all of the deviation ratios are larger than the deviation threshold. Hence, *S2M* has outstanding performance to resist the changing distance attack.

Thirdly, we conduct an experiment to evaluate a same-type-device attack in a quiet office. In this experiment, we consider 10 types of mobile phones: Samsung (Note1, Note2 and Note3), SONY (Z1, Z2 and LT), Huawei (Mate7, honor+ and 3C), MX3. For each type of cell phones, we have two mobile phones with same chips, one acts a sender and the other acts as the attacker. We use the HTC T328d as the receiver and set the effective authentication range $0 - 5m$. Each attacker random chooses 200 locations such that the distance form attacker to receiver is less than 10m, and launches an attack by following *E-S2M* at each location. Figure 8 shows the range of deviation ratios (from attacker) under different types. We can see that the deviation ratios for each type are larger than the deviation threshold. Thus, *S2M* has good performance to resist the same-type-device attack.

6.2 Performance of S2M and E-S2M

To evaluate the performance of S2M and E-S2M, we conduct extensive experiments as following settings: 1) For *S2M*, the distance (at learning and authentication process) is fixed from 0.5m to 5m with step 0.1m, and the number of

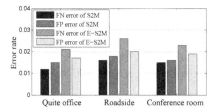

Fig. 9. The error rate of *S2M* and *E-S2M* under different scenarios

authentication is 10 for each fixed distance. 2) For *E-S2M*, the effective range is fixed on 0-5m, and the distances at learning process are changing from 0.25m to 4.75m with step 0.5m; the distances at authentication process are changed from 0.5m to 5m with step 0.1, the number of authentication is 10 for each fixed distance. 3)For both *S2M* and *E-S2M*, the distance between Eve and Bob is changed from 0.5m to 5m with step 0.1m; for each distance, the times of attacking operation is 10.

We consider three scenarios as follows. Scenario A: putting three mobile phones (as a sender, a receiver and an attacker respectively) in a quiet office; Scenario B: putting three mobile phones at a roadside; Scenario C: putting three mobile phones in a conference room. Furthermore, we consider 80 different groups of mobile phone tuples *(sender, receiver, attacker)* in each scenario. Figure 9 plot the error rate against authentication distance under different scenarios. It shows that: 1) For each case, both the FN rate and the FP rate of *S2M* is less than 2%; 2)the FN rate and the FP rate of *E-S2M* are less than 3% and 2%, respectively; 3) the performance of *E-S2M* is worse than that of *S2M*, but *E-S2M* still has the low error rates.

7   Conclusion

We described *S2M* and *E-S2M*, the software-only acoustic device authentication systems that leverages the FR of speaker and microphone pairs of wireless devices to construct highly specific hardware fingerprints. We design and implement the proposed strategies in real mobile phones to evaluate the performance of *S2M* and *E-S2M*. Our experimental results show that *S2M* and *E-S2M* are able to achieve both a low FN rate and a FP error rate in various scenarios under various attacks.

Acknowledgments. This work was supported in part by the NSF of China (No.61133016, No.61300191, No.61202445 and No.61370026), the Ministry of Education - China Mobile Research Foundation (No.MCM20121041), the National Science Foundation of China - Guangdong Joint Foundation (No.U1401257), the Sichuan Key Technology Support Program (No.2014GZ0106), and the Fundamental Research Funds for the Central Universities (No.ZYGX2013J003 and No.ZYGX2014J066).

References

1. Beck, M., Tews, E.: Practical attacks against WEP and WPA. In: ACM WiSe 2009 (2009)
2. Bertka, B.: 802.11w security: DoS attacks and vulnerability controls. In: Infocom 2012 (2012)
3. Eian, M., Mjølsnes, S.,: The modeling and comparison of wireless network denial of service attacks. In: ACM MobiHeld 2011 (2011)
4. Pang, J., Greenstein, B., Gummadi, R., Seshan, S., Wetherall, D.: 802.11 user fingerprinting. In: ACM MobiCom 2007, pp. 99–110 (2007)
5. Guo, F., Chiueh, T.: Sequence number-based MAC address spoof detection. In: Valdes, A., Zamboni, D. (eds.) RAID 2005. LNCS, vol. 3858, pp. 309–329. Springer, Heidelberg (2006)
6. Faria, D., Cheriton, D.: Radio-layer security: detecting identity-based attacks in wireless networks using signalprints. In: ACM WiSe 2006 (2006)
7. Jiang, Z., Zhao, J., Li, X., Han, J., Xi, W.: Rejecting the attack: source authentication for Wi-Fi management frames using CSI information. In: IEEE Infocom 2013 (2013)
8. Xiong, J., Jamieson, K.: SecureArray: improving wifi security with fine-grained physical-layer information. In: ACM MobiCom 2013, pp. 441–452 (2013)
9. Suh, G.E., Devadas, S.: Physical unclonable functions for device authentication and secret key generation. In: ACM DAC 2007, pp. 9–14 (2007)
10. Jana, S., Kasera, S.K.: On fast and accurate detection of unauthorized wireless access points using clock skews. In: ACM MobiCom 2008, pp. 104–115 (2008)
11. Brik, V., Banerjee, S., Gruteser, M., Oh, S.: Wireless device identification with radiometric signatures. In: ACM MobiCom 2008, pp. 116–127 (2008)
12. Danev, B., Luecken, H., Capkun, S., El Defrawy, K.: Attacks on physical-layer identification. In: ACM WiSec 2010, pp. 89–98 (2010)
13. Zeng, K., Govindan, K., Mohapatra, P.: Non-cryptographic authentication and identification in wireless networks. Network Security 1(3) (2010)
14. Nandakumar, R., Padmanabhan, V., Venkatesan, R.: Dhwani: secure peer-to-peer acoustic NFC. In: ACM SIGCOMM 2013, pp. 63–74 (2013)
15. Szabo, T.L.: Time domain wave equations for lossy media obeying a frequency power law. The Journal of the Acoustical Society of America, 491–500 (1994)
16. Madhavapeddy, A., Sharp, R., Scott, D., Tse, A.: Audio networking: the forgotten wireless technology. IEEE Pervasive Computing 4(3), 55–60 (2005)
17. Das, A., Borisov, N., Caesar, M.: Do you hear what i hear?: fingerprinting smart devices through embedded acoustic components. In: ACM CCS 2014 (2014)
18. Zhou, Z., Diao, W., Liu, X., Zhang, K.: Acoustic fingerprinting revisited: generate stable device id stealthily with inaudible sound. In: ACM CCS 2014 (2014)

Strongly Secure and Cost-Effective Certificateless Proxy Re-encryption Scheme for Data Sharing in Cloud Computing

Zhiguang Qin$^{(\boxtimes)}$, Shikun Wu, and Hu Xiong

School of Information and Software Engineering,
University of Electronic Science and Technology of China,
Chengdu 610054, Sichuan, China
{qinzg,shkwu,xionghu}@uestc.edu.cn

Abstract. Proxy re-encryption (PRE) has been considered as a promising candidate to secure data sharing in public cloud by enabling the cloud to transform the ciphertext to legitimate recipients on behalf of the data owner, and preserving data privacy from semi-trusted cloud. Certificateless proxy re-encryption (CL-PRE) not only eliminates the heavy public key certificate management in traditional public key infrastructure, but also solves the key escrow problem in the ID-based public key cryptography. By considering that the existing CL-PRE schemes either rely on expensive bilinear pairings or are proven secure under weak security models, we propose a strongly secure CL-PRE scheme without resorting to the bilinear pairing. The security of our scheme is proven to be secure against adaptive chosen ciphertext attack (IND-CCA) under a stronger security model in which the Type I adversary is allowed to replace the public key associated with the challenge identity. Furthermore, the simulation results demonstrate that our scheme is practical for cloud based data sharing in terms of communication overhead and computation cost for data owner, the cloud and data recipient.

Keywords: Certificateless proxy re-encryption · Data sharing · Cloud computing · Without pairings

1 Introduction

With the rapid development of cloud computing, the security of outsourced data in cloud has attracted a lot of concern from industry and academe recently. In case the data is outsourced to a semi-trust cloud service provider (CSP), the data owner cannot take control of their own data directly. To avoid disclosing the outsourced data to CSP or other unauthorized users, it is essential for data owners to preserve the privacy of the outsourced data in the cloud [1]. Intuitively, traditional asymmetric encryption scheme can be adopted to enforce the access control of data outsourced in the cloud. It seems to be feasible for data owners to outsource encrypted data to the semi-trusted cloud if only the data owner

© Springer International Publishing Switzerland 2015
Y. Wang et al. (Eds.): BigCom 2015, LNCS 9196, pp. 205–216, 2015.
DOI: 10.1007/978-3-319-22047-5_17

himself/herself can access the encrypted data. However, it becomes cumbersome to share encrypted data between different users based on traditional encryption mechanisms. To share the encrypted data with other users, a data owner needs to download and decrypt the requested data, and further re-encrypt it using a target user's public key to accomplish data sharing. Another naive approach for data owner is to share his/her private key with the target user who is authorized to decrypt the outsourced data directly. Obviously, the former method renders heavy communication overhead and computation cost, and thus mismatches the purpose of cloud computing. The idea of disclosing private keys to authorized users in the latter method violates the least privilege principle. Thus, it is challenging to share encrypted data in the cloud computing environment.

Proxy re-encryption (PRE) [9], which enables a semi-trusted proxy to transform a ciphertext which has been encrypted under one public key into a ciphertext under another public key of the same message without leaking any information to the proxy, is considered to be a promising candidate to achieve secure data sharing in cloud computing. Consider the following scenario: the data owner, say Alice, wants to share the sensitive data outsourced in the cloud with a third party, Bob. It is natural for Alice to desire that the requested data can only be accessed by Bob. Inspired by the primitive of PRE, Alice can encrypt the sensitive data before outsourcing these data to the semi-trusted cloud. After receiving the request of decryption delegation from Bob, Alice generates a proxy re-encryption key using his/her own private key and Bob's public key, and sends this proxy re-encryption key to the semi-trusted cloud. Equipped with this proxy re-encryption key, CSP can transform the ciphertext encrypted under the public key of Alice into the ciphertext under the public key of Bob. Meanwhile, the outsourced data can only be accessed by Bob since the CSP cannot decrypt the encrypted data with the proxy re-encryption key. Finally, Bob can download and decrypt the outsourced data with his/her own private key.

Before the PRE can be widely deployed in the cloud environment, several issues should be addressed. Observing the heavy management of public key certificates in traditional public key encryption (PKE) [9], [10], [11] and the key escrow problem in the ID-based public key encryption (ID-PKE) [13], [5], it is natural to investigate PRE in the certificateless public key encryption (CL-PKE) setting [2]. To enjoy the merits of traditional PKE and ID-PKE without suffering the corresponding criticisms, Sur *et al.* [4] introduced the primitive of PRE into CL-PKE [5] and proposed the first concrete certificateless proxy re-encryption (CL-PRE) scheme. Concretely, CL-PRE leverages the identity of a user as an ingredient of its public key, while eliminates the key escrow problem in ID-PKE, and does not require the use of certificates to guarantee the authenticity of public keys in traditional PKE.

Since Sur *et al.* [4] introduced the CL-PRE, this cryptosystem has attracted more attention. In 2012, Xu *et al* [6] constructed a CL-PRE scheme, which is claimed to be chosen-plaintext attack (CPA) secure and is introduced to cloud based data sharing scenario. In 2013, Yang *et al.* [7] constructed the first CCA secure pairing-free CL-PRE scheme based on Baek *et al*'s [8] CL-PKE scheme.

In 2014, Wang *et al* [12] also proposed a CCA secure CL-PRE scheme without bilinear pairings under Yang *et al.*'s security models [7]. Compared to the previous CL-PRE schemes, an attractive feature in Yang *et al.*'s scheme is the efficiency gained from removing computationally-heavy pairing operations. However, the untransformed ciphertexts in their scheme may greatly consume computation and storage resources for data owners. In addition, we point out that the security models for CL-PRE in [7] are sightly weak in a sense that the Type I adversary is not allowed to replace the public key associated with the challenge identity.

Our Contributions. We define an architecture of cloud based data sharing using the primitive of CL-PRE. To this end, we further propose a strongly secure and pairing-free CL-PRE scheme based on Yang *et al.*'s [7] scheme. That is, our CL-PRE scheme is cost-effective and provable secure against the Type I and Type II adversaries in a strong sense that the Type I adversary is able to replace the public key associated with the challenge identity (before challenge phase). Moreover, we evaluate communication overhead and computation cost in terms of data owner, the CSP and data recipient. Our results demonstrate that our CL-PRE scheme outperforms other existing works at the requirements of large scale data sharing.

The rest of this paper is organized as follows. Section 2 gives definitions of CL-PRE for cloud based data sharing. Our CL-PRE scheme is given in Section 3 followed by the security and performance analysis in Section 4. Section 5 concludes this paper.

2   Preliminaries and Definitions

In this section, we first describe a system architecture and give the definition of CL-PRE for cloud based data sharing. Then, we define the security assumptions and security model for a secure data sharing with public cloud.

2.1   Definition of a System Architecture

We consider an architecture of cloud based data sharing by introducing the primitive of CL-PRE as shown in Fig. 1, which involves a data owner, the cloud service provider (CSP) and data recipients. A data owner creates the encrypted data and host her data to the semi-trusted CSP. The CSP stores the data owner's data and provides the data access to the data owner and the authorized data recipients. The data owner is able to share her encrypted data to data recipients, who should first request for decryption delegations. After receiving the requests, the data owner produces a proxy re-encryption key for each data recipient and sends them to the CSP through a secure channel. Utilizing these proxy re-encryption keys, the CSP can transform the data owner's encrypted data into each data recipient's without disclosing any information. Then, data recipients can download and decrypt the data owner's outsourced data by themselves. As mentioned above, we assume the CSP itself is semi-trusted, which means it

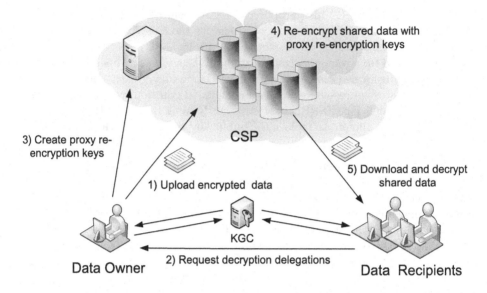

Fig. 1. The Architecture of CL-PRE for Cloud based Data Sharing

follows protocols and does not pollute data confidentiality actively as a malicious adversary, but it may be curious about the received data and may collude with data recipients to launch attacks on data owner.

Furthermore, a proxy re-encryption key is produced by inputting a data owner's private key and a data recipient's public key. Since the number of cloud users participating in data sharing may be large, traditional PKE based approach has the public key management issue, and ID-PKE based approach has the private key escrow problem. Uniquely, we adopt CL-PRE for data sharing in the cloud.

Definition 1 (CL-PRE for Data Sharing). *A cloud based data sharing mechanism designed by the primitive of certificateless proxy re-encryption (CL-PRE) consists of the following nine algorithms:*

- **Setup:** Taking a security parameter k as input, this algorithm is run by the key generation center (KGC) to produce a master key mk and a list of public parameters *params*.
- **PartialKeyExtract:** Taking a list of public parameters *params*, a master key mk and a cloud user's identifier ID_i as input, this algorithm is performed by KGC to return a partial public/private key pair (P_i, D_i) to the user with an identifier ID_i.
- **SetSecretValue:** Taking a list of public parameters *params* and an identifier ID_i as input, this algorithm is executed by the user ID_i to set a secret value z_i.

- **SetPrivateKey:** Taking a list of public parameters *params*, a partial private key D_i and a secret value z_i as input, this algorithm is carried out by user ID_i to set a private key sk_i.
- **SetPublicKey:** Taking a list of public parameters *params*, a partial public key P_i and a secret value z_i as input, this algorithm is carried out by user ID_i to set a public key pk_i.
- **Encrypt:** Taking a list of public parameters *params*, a plaintext data m and a public key pk_A as input, this algorithm is performed by data owner Alice to produce a ciphertext C_A. Then Alice uploads her ciphertext to CSP.
- **ReEncryptKey:** Taking a list of public parameters *params*, Alice's public/private key pair (pk_A, sk_A) associated with an identifier ID_A and data recipient Bob's public key pk_B associated with an identifier ID_B as input, this algorithm is implemented by Alice to generate a proxy re-encryption key $rk_{A \to B}$, which is further sent to CSP through a secure channel.
- **ReEncrypt:** Taking a list of public parameters *params*, a ciphertext C_A for Alice and a proxy re-encryption key $rk_{A \to B}$ as input, this algorithm is executed by the CSP to produce a transformed ciphertext C_B for Bob.
- **Decrypt:** Taking a list of public parameters *params*, a private key sk_i and a ciphertext C_i, this algorithm is run by the data owner/recipient associated with an identifier ID_i to obtain the underlying encrypted data m or return a distinguished symbol $\perp$.

The above CL-PRE scheme allows the CSP using a proxy re-encryption key $rk_{A \to B}$ to transform a ciphertext encrypted under a data owner Alice's public key pk_A into another ciphertext encrypted under a data recipient Bob's public key pk_B on the same message m without leaking m or sk_A/sk_B of data owner/data recipient to the CSP.

Correctness: For all data $m \in \mathcal{M}$, key pair (pk_A, sk_A) for Alice and (pk_B, sk_B) for Bob, Alice is able to correctly share her underlying encrypted data m with Bob in the cloud if and only if the following conditions are satisfied:

- $\mathbf{Decrypt}_1(params, sk_A, \mathbf{Encrypt}(params, ID_A, pk_A, m)) = m$.
- $\mathbf{Decrypt}_2(params, sk_B, \mathbf{ReEncrypt}(params, rk_{A \to B}, C_A)) = m$.

2.2 Assumption

Definition 2 (Computational Diffie-Hellman (CDH)). *Assume that g is a generator chosen randomly from group $\mathbb{G}$ with the primer order q. Let $\mathcal{A}$ be an adversary. $\mathcal{A}$ tries to solve the problem as follows: Given $(g, g^a, g^b) \in \mathbb{G}$, where $a, b \in \mathbb{Z}_q^*$, compute $g^{a \cdot b}$. We define $\mathbb{G}$ satisfies CDH problem assumption if there is no probability polynomial time algorithm to solve the CDH problem with advantage $Adv(\mathcal{A}) = Pr[\mathcal{A}(g, g^a, g^b) = g^{ab}]$.*

2.3 Security Model

First we recall the security models of CL-PRE, given by Yang *et al.* [7] based on Baek *et al.*'s [8] security models of CL-PKE. Their definition considers two types of adversaries, Type I and Type II. The difference between them is that a Type I adversary $\mathcal{A}_I$ does not have access to the master key but may replace public keys of arbitrary identities with values of its own choice, whereas a Type II adversary $\mathcal{A}_{II}$ does have access to the master key but may not replace public keys of entities. It must be pointed that Yang et al.'s security models for a Type I adversary is sightly weak in a sense that $\mathcal{A}_I$ is not allowed to replace the public key of the challenge identity ID^* in any phase in order to prove the second level ciphertext security of their CL-PRE scheme. However, Sun et al. [3] eliminated Baek et al.'s limitation and presented an improving CL-PKE scheme with strong Type I. As a result, the security models of CL-PRE we formalize by taking account of strongly CL-PKE security notions [2], [3] and PRE security notions [7], [9], [10] are the strong Type I and Type II, where $\mathcal{A}_I$ is able to replace the public key associated with the challenge identity. Furthermore, our CL-PRE scheme does not depend on bilinear pairings. Our strong security model is enough for many practical applications such as secure data sharing with public cloud [6]. The details of our strong security model can be found in the full paper.

3   Our Pairing-Free CL-PRE Scheme

In this section, we modify Yang et al.'s CL-PRE scheme [7] to construct a strongly secure one without pairings based on Sun et al.'s CL-PKE scheme [3].

The detailed description of our CL-PRE scheme is as follows:

- **Setup:** Taking a security parameter k as input, this algorithm produces a prime q and a group $\mathbb{G}$ of order q. Then it performs as follows:
  1. Choose a random generator $g \in \mathbb{G}$.
  2. Pick $s \in \mathbb{Z}_q^*$ at random and compute $h = g^s$.
  3. Select hash functions $H_1 : \{0,1\}^* \times \mathbb{G} \to \mathbb{Z}_q^*$, $H_2 : \{0,1\}^* \times \mathbb{G} \times \mathbb{G} \to \mathbb{Z}_q^*$, $H_3 : \{0,1\}^* \to \mathbb{Z}_q^*$, $H_4 : \mathbb{G} \to \{0,1\}^{n+k_0}$, and $H_5 : \mathbb{G} \to \mathbb{Z}_q^*$.

  The system public parameters are $params = \langle q, n, k_0, g, h, H_1, H_2, H_3, H_4, H_5 \rangle$, where n, k_0 mean the bit-length of a plaintext and a random bit string, respectively. The system master key $mk = s$. Note that plaintext space is $\mathcal{M} = \{0,1\}^n$ and ciphertext space is $\mathcal{C} = \{0,1\}^{n+k_0}$.

- **PartialKeyExtract:** Taking $params$, mk and an identifier ID_A for Alice as input, this algorithm picks $\alpha_1, \alpha_2 \in \mathbb{Z}_q^*$ at random and computes $a_1 = g^{\alpha_1}$, $a_2 = g^{\alpha_2}$, $x_1 = \alpha_1 + sH_1(ID_A, a_1)$ and $x_2 = \alpha_2 + sH_1(ID_A, a_1, a_2)$. Then it returns a partial private key $D_A = x_1$ and a partial public key $P_A = (a_1, a_2, x_2)$ for Alice.

- **SetSecretValue:** Taking $params$ and ID_A as input, this algorithm randomly picks $z_A \in \mathbb{Z}_q^*$ as a secret value for Alice.

- **SetPrivateKey:** Taking $params$, Alice's partial private key D_A and z_A as input, this algorithm returns a private key $sk_A = (x_1, z_A)$ for Alice.

- **SetPublicKey:** Taking $params$, Alice's partial public key P_A and secret value z_A as input, this algorithm computes $u_A = g^{z_A}$ and returns a public key $pk_A = (u_A, a_1, a_2, x_2)$ for that user.

- **ReEncryptKey:** Taking *params*, an identifier ID_A and a public/private key pair (pk_A, sk_A) for Alice, an identifier ID_B and a public key pk_B for Bob as input, this algorithm computes $t_B = b_1 h^{H_1(ID_B, b_1)}$ and $t_{AB} = H_3(t_B^{z_A} \| u_B^{x_1} \| ID_A \| pk_A \| ID_B \| pk_B)$. Then it returns a re-encryption key $rk_{A \to B} = (x_1 H_5(u_A) + z_A) t_{AB}$.
- **Encrypt:** Taking *params*, a plaintext message $m \in \mathcal{M}$ and Alice's public key pk_A as input, this algorithm performs as follows:
  1. Check $g^{x_2} \stackrel{?}{=} a_2 h^{H_2(ID_A, a_1, a_2)}$.
  2. Select $\sigma \in \{0,1\}^{k_0}$ at random and compute $t_A = a_1 h^{H_1(ID_A, a_1)}$ and $r = H_3(m \| \sigma \| ID_A \| u_A)$.
  3. Compute $c_1 = g^r$ and $c_2 = (m \| \sigma) \oplus H_4((t_A^{H_5(u_A)} \cdot u_A)^r)$.
  Then it returns a ciphertext $C_A = (c_1, c_2)$ for Alice.
- **ReEncrypt:** Taking *params*, Alice's ciphertext C_A and a re-encryption key $rk_{A \to B}$ as input, this algorithm computes $c_1' = c_1^{rk_{A \to B}}$ and $c_2' = c_2$. Then it outputs re-encrypted ciphertext $C_B = (c_1', c_2')$ for Bob or a distinguished symbol $\perp$.
- **Decrypt:** Taking *params*, a private key sk_{ID} and a ciphertext C_{ID} for user ID, this algorithm performs as follows:
  - **Decrypt₁:** To decrypt non re-encrypted ciphertext $C_A = (c_1, c_2)$ with $sk_A = (x_1, z_A)$, this algorithm computes $(m \| \sigma) = c_2 \oplus H_4(c_1^{(x_1 H_5(u_A) + z_A)})$. Then return plaintext m, if $r' = H_3(m \| \sigma \| ID_A \| u_A)$ and $g^{r'} = c_1$ holds. Otherwise, output $\perp$.
  - **Decrypt₂:** To decrypt re-encrypted ciphertext $C_B = (c_1', c_2')$ with $sk_B = (y_1, z_B)$, this algorithm computes as follows:
    1. Compute $t_A = a_1 h^{H_1(ID_A, a_1)}$ and $t_{BA} = H_3(u_A^{y_1} \| t_A^{z_B} \| ID_A \| pk_A \| ID_B \| pk_B)$.
    2. Compute $(m \| \sigma) = c_2' \oplus H_4((c_1')^{1/t_{BA}})$.
    3. If $r' = H_3(m \| \sigma \| ID_A \| u_A)$ and $(t_A^{H_5(u_A) \cdot u_A})^{r' t_{BA}} = c_1'$ holds, return m. Otherwise, output $\perp$.

It is easy to check the correctness of the pairing-free CL-PRE scheme above, we omit it here.

Remark 1. In our CL-PRE scheme, a existentially unforgeable under an adaptive chosen message attack (EUF-CMA) secure Schnorr signature is used to protect the partial public key from being replaced by attackers with the values of their choices. And $H_5(u_A)$ is necessary for resisting public key replacement attacks, where H_5 is collision free hash function.

Remark 2. The proposed scheme possesses properties such as *unidirectionality, single-hop, non-interactivity, non-transitivity, collusion-resistance*, both of which are suitable for security requirements in cloud based data sharing scenarios.

4 Analysis

4.1 Security Analysis

We have the following theorems about the security of the CL-PRE scheme. And due to the space limit, the proof of these theorems will be given in the full paper.

Theorem 1. *Our CL-PRE scheme is adaptive chosen ciphertext (IND-CCA) secure against the Type I adversary in the random oracle model, if for any polynomial time adversary $\mathcal{A}_I$ the CDH problem is intractable in $\mathbb{G}$.*

Theorem 2. *Our CL-PRE scheme is adaptive chosen ciphertext (IND-CCA) secure against the Type II adversary in the random oracle model, if for any polynomial time adversary $\mathcal{A}_{II}$ the CDH problem is intractable in $\mathbb{G}$.*

4.2   Performance Evaluation

Data sharing is a very resource demanding service with public cloud in terms of computation cost, communication overhead and storage space. In this subsection, we compare the performance of data owner, CSP and data recipient of our CL-PRE scheme with Xu *et al.*'s [6] scheme, Sur *et al.*'s [4] scheme and Yang *et al.*'s [7] scheme. For both Xu *et al.*'s [6] and Sur *et al.*'s pairing-based schemes, to satisfy 1024-bit RSA level security, we adopt the Tate pairing implemented over an elliptic curve defined on 512 bits prime field with a generator of order 160 bits. For Yang *et al.*'s [7] and our pairing-free schemes, to achieve the same security level, we implement them over 1024-bit prime finite field with a generator of order 160 bits. Additionally, we assume that the bit-length of $|m|$ and $|\sigma|$ is 1024 bits and 160 bits, respectively. Note that we obtain the running time for cryptographic operations using Miracal Library [14], a standard cryptographic library, on a PIV 3 GHZ processor with 512-MB memory and a Windows XP operation system. The running times of one pairing operation, one exponentiation operation and one map-to-point hash are 20.04ms, 5.83ms and 3.04ms, respectively. According to the description of multiple exponentiation algorithm in [7], we evaluate concrete running time and communication cost in Table 1 to make the comparison more clear between our scheme and three existing works [6], [4], [7].

From Table 1, our scheme is more efficient than Xu *et al.*'s scheme and Sur *et al.*'s scheme in terms of running time and ciphertext length. Compared with Yang *et al.*'s scheme, the overall performance of our scheme is more superior.

Computation Cost. In cloud based data sharing of CL-PRE, we assume that a data owner only has to do one encryption and decryption. From the observation of Table 1, the sum of running times on Encrypt and Decrypt$_1$ in our scheme is the shortest. We believe this cost is not significant. However, the data owner should perform ReEncryptKey algorithm to generate a decryption delegation for each data recipient. As shown in Fig. 2, the computation cost of the data owner does increase linearly with the number of recipients. Our scheme and Yang *et al.*'s [7] scheme have the same overhead in ReEncryptKey, so the curves coincide. For a data owner, both our scheme and Yang *et al.*'s scheme don't require high computational cost.

Next we analyze the computation cost of the CSP in different CL-PRE schemes. As a proxy, the CSP only has to carry out ciphertext transformations for numbers of data recipients. Thus, the ReEncrypt algorithm determines the

Table 1. The Performance Comparison of CL-PRE schemes

Schemes	Xu *et al.* [6]	Sur *et al.* [4]	Yang *et al.* [7]	Our CL-PRE
Encrypt	37.53 ms	32.66 ms	18.95 ms	23.32 ms
ReEncryptKey	40.57 ms	23.79 ms	12.65 ms	12.65 ms
ReEncrypt	20.04 ms	123.28 ms	12.65 ms	5.83 ms
$\text{Decrypt}_1(C_A)$	20.04 ms	58.04 ms	18.48 ms	11.66 ms
$\text{Decrypt}_2(C_B)$	43.12 ms	43.36 ms	24.78 ms	24.78 ms
$|C_A|$	3072 bits	4256 bits	3392 bits	2208 bits
$|rk_{A\rightarrow B}|$	3072 bits	3072 bits	160 bits	160 bits
$|C_B|$	4096 bits	4256 bits	2208 bits	2208 bits
Pairing-Free	×	×	√	√
Security Model	weak	weak	weak	strong
Assumption	DBDH	p-BDHI	CDH	CDH
Security	CPA	CCA	CCA	CCA

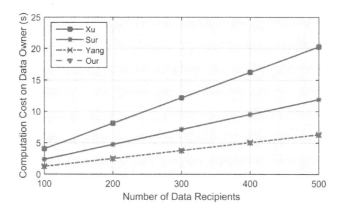

Fig. 2. The Computation Cost of the Data Owner

computation cost of the CSP. Table 1 shows that the running time on ReEncrypt in our scheme is superior to other schemes. The computation cost of the CSP increases accompanied by the number of data recipients. Clearly, Fig. 3 reflects the advantage.

For a data recipient, there is no difference that it obtains re-encrypted messages from the CSP and decrypts in the same way. The data recipient has to do one decryption only. The running time on Decrypt_2 in our scheme is equal to Yang *et al.*'s [7] scheme. As a result, the computation cost of our CL-PRE is inexpensive for cloud based data sharing.

Communication Overhead and Storage Overhead. For a data owner, the communication overhead is not constant and is relevant to the number of data recipients, since the data owner has to transport one encrypted message and

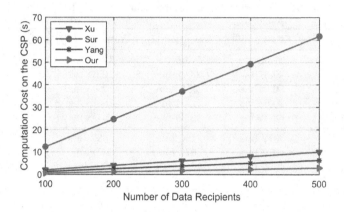

Fig. 3. The Computation Cost of the CSP

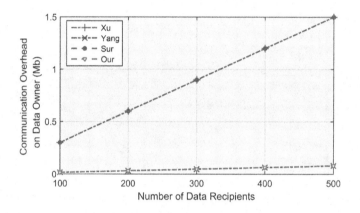

Fig. 4. The Communication Overhead of the Data Owner

one proxy re-encryption key for each recipient to the CSP. Fig. 4 shows the comparison of communication overhead for a data owner. Compared with Xu *et al.*'s [6] scheme and Sur *et al.*'s [4] scheme, our scheme has great advantage. In addition, the data owner not only has to keep its own public/private key pair but also has to store all the generated proxy re-encryption keys. Therefore, Fig. 4 also reflects the storage overhead of the data owner.

For one data sharing operation, the CSP transports one re-encrypted message for one recipient, i.e., the communication overhead is proportional to the number of recipients. In addition, the CSP should store the data owner's encrypted message, proxy re-encryption keys and re-encrypted messages for numbers of data recipients. Therefore, the storage overhead of the CSP is linear. We evaluate the storage overhead of the CSP with different CL-PRE schemes in Fig. 5.

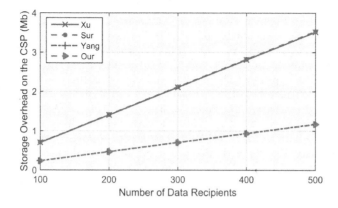

Fig. 5. The Storage Overhead of the CSP

In cloud based data sharing of CL-PRE, a data recipient has to obtain one re-encrypted message from the CSP only, so the communication overhead and storage overhead of the data recipient are not significant.

5   Conclusions

In this paper, we have proposed a strongly secure and efficient certificateless proxy re-encryption scheme (CL-PRE) without pairings for cloud based data sharing scenario. The proposed scheme is provably IND-CCA secure in a stronger security model, where a Type I adversary is allowed to replace the public key associated with the challenge identity (before challenge phase). The simulation results demonstrate that our scheme is strongly secure and practical for cloud based data sharing in terms of computation cost and communication overhead for data owner, CSP and data recipient.

Acknowledgements. This work was supported in part by the National Natural Science Foundation of China under Grant 61003230, Grant 61370026, and Grant 61202445, in part by the Fundamental Research Funds for the Central Universities under Grant ZYGX2013J073, and in part by the Applied Basic Research Program of Sichuan Province under Grant 2014JY0041.

References

1. Subashini, S., Kavitha, V.: A survey on security issues in service delivery models of cloud computing. Journal of network and computer applications. **34**(1), 1–11 (2011)
2. Al-Riyami, S.S., Paterson, K.G.: Certificateless public key cryptography. In: Laih, C.-S. (ed.) ASIACRYPT 2003. LNCS, vol. 2894, pp. 452–473. Springer, Heidelberg (2003)

3. Sun, Y., Zhang, F.T., Baek, J.: Strongly secure certificateless public key encryption without pairing. In: Bao, F., Ling, S., Okamoto, T., Wang, H., Xing, C. (eds.) CANS 2007. LNCS, vol. 4856, pp. 194–208. Springer, Heidelberg (2007)
4. Sur, C., Jung, C.D., Park, Y., Rhee, K.H.: Chosen-ciphertext secure certificateless proxy re-encryption. In: De Decker, B., Schaumüller-Bichl, I. (eds.) CMS 2010. LNCS, vol. 6109, pp. 214–232. Springer, Heidelberg (2010)
5. Libert, B., Quisquater, J.-J.: On constructing certificateless cryptosystems from identity based encryption. In: Yung, M., Dodis, Y., Kiayias, A., Malkin, T. (eds.) PKC 2006. LNCS, vol. 3958, pp. 474–490. Springer, Heidelberg (2006)
6. Xu, L., Wu, X., Zhang, X.: CL-PRE: a certificateless proxy re-encryption scheme for secure data sharing with public cloud. In: Proceedings of the 7th ACM Symposium on Information, Computer and Communications Security (ASIACCS 2012), pp. 87–88. ACM, Seoul (2012)
7. Yang, K., Xu, J., Zhang, Z.: Certificateless proxy re-encryption without pairings-date. In: Lee, H.-S., Han, D.-G. (eds.) ICISC 2013. LNCS, vol. 8565, pp. 67–88. Springer, Heidelberg (2014)
8. Baek, J., Safavi-Naini, R., Susilo, W.: Certificateless public key encryption without pairing. In: Zhou, J., López, J., Deng, R.H., Bao, F. (eds.) ISC 2005. LNCS, vol. 3650, pp. 134–148. Springer, Heidelberg (2005)
9. Canetti, R., Hohenberger, S.: Chosen-ciphertext secure proxy re-encryption. In: Proceedings of the 14h ACM conference on Computer and Communications Security (CCS 2007), pp. 185–194. ACM, Alexandria (2007)
10. Libert, B., Vergnaud, D.: Unidirectional chosen-ciphertext secure proxy re-encryption. In: Cramer, R. (ed.) PKC 2008. LNCS, vol. 4939, pp. 360–379. Springer, Heidelberg (2008)
11. Chow, S.S.M., Weng, J., Yang, Y., Deng, R.H.: Efficient unidirectional proxy re-encryption. In: Bernstein, D.J., Lange, T. (eds.) AFRICACRYPT 2010. LNCS, vol. 6055, pp. 316–332. Springer, Heidelberg (2010)
12. Wang, L., Chen, K., Mao, X., Wang, Y.: Efficient and provably-secure certificateless proxy re-encryption scheme for secure cloud data sharing. Journal of Shanghai Jiaotong University (Science) **19**, 398–405 (2014)
13. Green, M., Ateniese, G.: Identity-based proxy re-encryption. In: Katz, J., Yung, M. (eds.) ACNS 2007. LNCS, vol. 4521, pp. 288–306. Springer, Heidelberg (2007)
14. Shamus Software Ltd., Miracl library. http://www.shamus.ie/index.php?page=home

Anomaly Detection of Single Sensors
Using OCSVM_KNN

Jing Su[1(✉)], Ying Long[1], Xiaofeng Qiu[1], Shufang Li[1], and Daowei Liu[2]

[1] Beijing Key Laboratory of Network System Architecture and Convergence,
Beijing University of Posts and Telecommunications, Beijing 100876, China
{sj199137,longying963}@126.com,
{qiuxiaofeng,lisf}@bupt.edu.cn
[2] China Electric Power Research Institute, Beijing 100192, China
liudaowei@epri.sgcc.com.cn

Abstract. Anomaly detection of sensor data is a fundamental and active problem, it is involved in many applications especially the Wireless Sensor Network (WSN) where we detect anomaly by the group data. But for anomaly detection of a single sensor, many methods which consider spatial connection of data are not efficient. Point anomaly and pattern anomaly are two types of anomalies. In this work, we analyze the sensor data and find that pattern anomaly plays a dominant role for a single sensor. Moreover, the sudden change of data is found to be the special form of pattern anomaly. As the data collected by a single sensor is ordered by time, we convert the problem to pattern anomaly detection of time series. In this work, a two-phased algorithm OCSVM_KNN based on One Class Support Vector Machines (OCSVM) and K-Nearest Neighbors (KNN) is proposed. And, in the KNN classifier we introduce Complexity-invariant Distance (CID) as the distance measure method between two time-series sequences. Experimental results on real data sets show that our proposed approach outperforms other existing approaches in terms of anomaly detection rate, false alarm rate and misclassification anomaly rate.

Keywords: Anomaly detection · Single sensor · Time-series sequence · OCSVM_KNN · One-Class SVM · KNN

1 Introduction

Anomaly detection, known as novelty detection and anomaly mining, is widely re-searched in many fields, e.g., intrusion detection, fraud detection, medical and health detection, text data anomaly detection and sensor network detection [1]. The anomaly detection techniques are different in those fields since the data feature and the definition of anomaly are not the same. In this paper, we pay attention to anomaly detection of sensor data.

For anomaly detection of sensor data, many works pay attention to detect anomaly in wireless sensor network (WSN) environment [2-8]. WSN usually has strong resource constraints on communication bandwidth, energy and memory. Besides, it is often

© Springer International Publishing Switzerland 2015
Y. Wang et al. (Eds.): BigCom 2015, LNCS 9196, pp. 217–230, 2015.
DOI: 10.1007/978-3-319-22047-5_18

located in the field without attending. In such case, the number of sensor nodes are redundancy, some works therefore take advantage of the joint spatial correlation among group sensor data for anomaly detection. The techniques for WSN anomaly detection can be categorized into classification-based [3], statistical-based [4,5], nearest neighbor-based [6], clustering-based [7] and spectral decomposition-based approaches [8].

But in some situations, e.g., smart home and smart building, there are not enough sensors so we cannot detect the anomaly of sensor data by mining spatial correlation with its neighbors. We need to detect the anomaly for a single sensor node respectively. Moreover, in these situations, the data collected by a sensor may be suddenly changed by artificial operation, e.g., the temperature collected by temperature sensors will be sharply changed if we open the air conditioner. How to detect the sudden changes of the collected data instead of dealing with it as general anomalies is another problem we will solve in this paper.

As data collected by a single sensor node can be regarded as a dynamic time series, we convert this problem to the anomaly detection of dynamic time series. While point anomaly and pattern anomaly are two typical anomalies of time series, we find that pattern anomaly has more underlying information.

The key contributions of this paper are as follows:

• Propose an anomaly detection approach OCSVM_KNN which combines One-Class SVM (OCSVM) with KNN (K nearest neighbors) for pattern anomaly detection of time series. Experiments show that the algorithm is better than other typical classification algorithms in effect.
• Further in the KNN classifier a distance measure method called CID (Complexity-invariant Distance) is introduced to calculate the distance between two time-series sequences, guarantying the time efficiency and enhancing the efficiency of anomaly detection in the same time.

The procedure of two-stage algorithm OCSVM_KNN is as follows: In first stage, we use the One-Class SVM to detect abnormal sequences which account for a small proportion of data. The instance detected as candidate anomaly in the first stage will be sent to a KNN classifier for further judgment in second stage where we will know the type of the anomaly and remove the normal data misjudged as anomalies in the first stage. The present two-stage algorithm improves existing algorithms in detection rate, false alarm rate and misclassification anomaly rate.

The rest of the paper is organized as follows. In section 2, we discuss some anomaly detection approaches for single sensor reading. The basic theory of OCSVM and CID based KNN are studied in section 3. Our anomaly detection technique is proposed in section 4. Experimental results and performance evaluation of our approach are reported in Section 5. We conclude the paper in Section 6 with discussion for future work.

2   Related Work

Unlike the anomaly detection for group sensor data which usually take advantage of the data spatial correlation, anomaly detection for a single sensor node can just make use of its local data. Rather, we need to judge the data by considering the temporal aspect without the spatial connection.

As the data collected by a single sensor node is recorded by time, the problem can be converted to the anomaly detection of time series. There are mainly two types of anomalies: point anomaly and pattern anomaly. Point anomaly occurs at points, while pattern anomaly is anomaly of time series sequences.

For point anomaly detection, evolving prediction models and distance-based anomaly for sliding windows are available [2]. What's more, a prediction model was proposed by D. J. Hill [9].

Compared with point anomaly whose information is too isolate, pattern anomaly contains more underlying information, such as the data tendency and anomaly patterns of time series sequences. With detected anomaly pattern types, we can present the possible reasons to the user for further treatments. In this paper, we mainly pay attention to the pattern anomaly.

Popular clustering methods for pattern anomaly detection include k-medoids [10], KNN [11], phased K-Means [12] and self-organizing maps (SOM)[13].The choice of the clustering methods is application-specific. Besides, some classify methods are researched, SVM-based techniques are one of the classification-based techniques which have many advantages [14]. But SVM-based techniques need some labeled data which are not always available. Therefore, several unsupervised SVM-based anomaly detection techniques have been proposed, such as One-Class SVM [15].

As One-Class SVM algorithm only outperforms in judging whether a time series sequence is anomaly or not, we proposed a two-stage algorithm OCSVM_KNN for anomaly detection and anomaly pattern matching.

3   A Brief Introduction to OCSVM and KNN

In this section, we describe the theory of OCSVM and CID based KNN. The theories are foundations of the designed algorithm which we introduce in section 4.

3.1   Theory of OCSVM

The OCSVM (One-Class Support Vector Machine) algorithm was originally proposed by B Schölkopf et al [16]. In this algorithm, the kernel function is used to implicitly map vectors into a high dimensional feature space. Then the hyper plane is constructed so that the sample vectors separate each other as far as possible with the origin of feature space. Consequently, one class of problem in the original space is converted to two classes of problem in the feature space. The theory of OCSVM and more details can be found in [16] and [17].

3.2   Theory of CID Based KNN

KNN (k-Nearest Neighbors) [4] is the most simple and most classic classification algorithm, and the basic idea is: we have an existing set of example data, our training set. We have labels for all of this data—we know what class each piece of the data should fall into. When we're given a new piece of data without a label, we compare that new piece of data to the existing data, every piece of existing data. We then take the most similar pieces of data (the nearest neighbors) and look at their labels. We look at the top k most similar pieces of data from our known dataset; this is where the k comes from. (k is an integer and it's usually less than 20.) Lastly, we take a majority vote from the k most similar pieces of data, and the majority is the new class we assign to the data we were asked to classify [18].

CID (complexity invariant distance) is a commonly used method to measure the distance or calculation formulas of similarity of KNN. CID is a method proposed for time series by Batista [19] and plays well in complexity. The theory is shown as follows:

Assuming two time series Q and C with length as n, the calculation formula of CID is as follows:

$$CID(Q,C) = ED(Q,C) \times CF(Q,C) \tag{1}$$

where ED is the Euclidean distance of two time series. CID considers complexity invariant factors upon Euclidean distance. The value of CF relates to complexity comparison of the time series Q and C. If the difference between them is great, the value of CF also great; however, if Q and C are equal, then the value of CF is 1, correspondingly CID converted to ED. More details of CID can be found in [20].

4   The Specific Design of OCSVM_KNN Algorithm

4.1   Sequence Anomaly Detection Framework

As anomaly data often containing potential useful information, the anomaly processing of sensor data often contains two aspects: One is anomaly detection based on the single sensor data streams; the other is anomaly interpretation which analysis the anomaly reasons and finds the anomaly mechanism. The former will have a detailed analysis in this paper. According to the analysis, a model is presented as shown in Fig.1.

Anomaly detection of Fig.1 mainly uses time correlation of single sensor data which mainly refers to the sequence anomaly detection in this paper. This part will have a detailed analysis in this paper that a two-stage algorithm named OCSVM_KNN is designed to realize sequence anomaly detection.

Comprehensive detection module analysis the detected anomalies by means of a small amount spatial correlation between sensor data: in the same deployment environment where a small amount of sensors of the same type exist, analysis of cause of

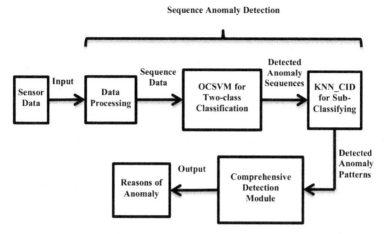

Fig. 1. Sequence anomaly detection framework

anomaly can be realized by detecting if the same type of sensors abnormal as well. In this paper, there is not a detailed analysis of comprehensive detection module.

4.2   Data Processing

Anomaly sequence is a sequence that distinguishes from other normal sequences in the model [19]. In order to realize the detection of abnormal sequences and to provide the basis for the anomaly interpretation, this paper firstly distinguishes anomaly patterns of sequences.

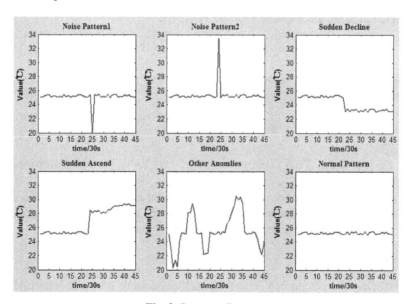

Fig. 2. Sequence Patterns

These sequences are divided into six kinds of modes as shown in Fig.2: noise model (two kinds), sudden decline time-series sequences, sudden ascend time-series sequences, other anomaly and normal mode. Matching sequence patterns with sequences to be detected, sequences can be detected to belong to which kinds of mode, so as to contribute to the comprehensive anomaly detection module for in-depth anomaly analysis.

A noticeable characteristic of the anomaly of single sensor data is the unbalance of normal data and anomaly data. For some traditional classify algorithms, such as KNN, SVM. If the size of one class is much larger than any other classes, the classifier is inclined to predict any data as the largest class. High false alarm rate and low detection rate are thus caused. Therefore, the detection result tends to predict any data as normal if we use the traditional algorithms and an algorithm for unbalance data is needed. In this work, we propose a new method called OCSVM_KNN with high detection rate and low false alarm rate.

The data source for sequences anomaly detection is the data from a single sensor, e.g., a temperature sensor or a smoke sensor. As almost no anomalies existing in the original sequence, for the sake of the realization of anomaly detection being more effectively, this paper firstly randomly adds some anomaly data into the original data, then segment the data to ensure the rationality of data processing in this algorithm.

For pattern anomaly, time series needs to be divided into time-series sequences. In order to meet the requirement for real time and dynamic data processing, segmenting method used for processing off-line time series, e.g., Top-Down method and Bottom-UP method are not applicable. According to the characteristics of sensor data and real-time processing requirement, in this work, a sliding window with fixed size is used.

With the window size set as W and the data acquisition interval as ΔT, so the intercepted time sequence can be expressed as (2):

$$X(t) = \left[x(t), \cdots, x(t + i \times \Delta T), \cdots, x(t + (W-1) \times \Delta T) \right] \tag{2}$$

This paper introduces a sequence processing method proposed in [15]. The specific processing method is as follows:

$$X'(t) = (I - \frac{1}{W} l l^T) X(t) \tag{3}$$

where I is a w-dimensional identity matrix and $l=[1,1,\cdots,1]$ is a w-dimensional unit vector.

This method can effectively reduce the influence of extreme points on OCSVM. In addition, the experimental results show that, with the following processing on time series, detection results have implemented certain promotion.

4.3   Algorithm Design

OCSVM (One-Class Support Vector Machine) can achieve good effect in handling a single class, but not suitable for multi-class classification problem. However, KNN

Algorithm. OCSVM_KNN

Input: (1)W^w: the sliding windows with size w. (2) X(t): a time-series sequence collected from W^w. (3) S(t): the training data set. (4) k: the number of neighbors we need to sort for KNN.

1. **function** Detection_Process():
2.   Extract a time-series sequence X(t) from sliding window W^w
3. $X'(t) \leftarrow (I - \frac{1}{w}ll^T)X(t)$
4.   OCSVM_Process(X'(t))
5. **end function**
6. **function** OCSVM_Process():
7.   Scale X'(t)
8.   Calculate f(x)
9.   **if** f(x) > 0 **then**
10.   X'(t) is normal
11.   **else**
12.   X'(t) is a candidate outlier
13.   KNN_Process(X'(t))
14.   **end if**
15. **end function**
16. **function** KNN_Process():
17.   **for** each s(t) in S(t) **do**
18. $d_i \leftarrow CID(X'(t), s(t))$
19.   **end for**
20.   Sort (d)
21.   GetHead(d,k)
22.   **if** class j involves most the k neighbors **then**
23.   X'(t) is a instance of class j
24.   **end if**
25. **end function**

Fig. 3. Algorithm OCSVM_KNN

(K nearest neighbors) is suitable for multi-class classification, but misjudgment will easily occur when the sample distribution is inhomogeneous. So a two-stage complementary algorithm OCSVM_KNN is designed in this paper.

We propose an anomaly detection algorithm which includes two stages. In Phase 1, we use the One-Class SVM to detect abnormal sequences which account for a small proportion of data. Almost all abnormal data are detected (with some normal data). The instance detected as candidate anomaly in the first stage will be sent to a KNN classifier for further judge in Phase 2 where we will know the type of the anomaly and remove the misjudged normal data. We propose CID as the distance measure method of KNN. The detail of the algorithm is shown in Fig.3.

SVM is a classifier that tries to find a hyper plane to separate all samples of one class from the other class. But for unlabeled data and unbalance data, SVM is not quite suitable. To cope with this problem, one-class classification problems (and solutions) are introduced. By just providing the normal training data, an algorithm creates a (representational) model of this data. If newly encountered data is too different, according to some measurement, from this model, it is labeled as out-of-class.

One-Class SVM (OCSVM) is a kernel based approach which is very fast and precise and therefore is used in different fields. When data is unlabeled and composed of many normal data and few abnormal data, One-Class SVM is more appealing than other algorithms since it can detect almost all abnormal data.

In Fig.3, OCSVM Process($X'(t)$) represents the detection procedure of One-Class SVM, f(x) is the hyper plane in feature space F by which the mapped vectors can be separated from the origin in F as many as possible. The instance will be regarded as an anomaly if f(x)<0. The detail presentation can be found in [16] and [17].

OCSVM relies on the selection of kernel function. As frequently-used kernel function including linear kernel, polynomial kernel, RBF (Radial basis function) kernel, there is a lack of guiding principle on the selection of kernel function. Although results of various experimental observations indeed suggest that some types of kernel functions outperform others in terms of effect dealing with some problems, generally speaking, the radial basis kernel function which will not appear large deviations is our first choice. So the algorithm in this work uses RBF kernel function. OCSVM is a kind of classifier which can only achieve anomaly detection and fail to identify the various types of abnormal patterns. So here OCSVM is used to achieve two-class classification.

In many applications, detecting anomaly is very important since overlooking anomalies may cause more serious disasters. As we need to detect more anomalies, OCSVM is the better choice than other classification algorithms. The disadvantage of OCSVM is large computation cost. However, we can take no account of this problem because the data processed in the situation in this work is real-time data of small size.

Although One-Class SVM can detect almost all anomalies (high detection rate), it often regards relative more normal data as anomaly data than other algorithms. Besides, in the first stage, we cannot know which type of anomaly the anomaly belongs to. We propose to use CID based KNN to decrease the false alarm rate and to distinct the type of anomaly. Besides, it regards the sudden change of data as a special anomaly.

KNN [4] is the most simple and most classic classification algorithm, and the basic idea is: if most of K neighborhood of a data object o belongs to one class, then the data object o also belongs to the same class. Firstly, KNN utilizes labeled training data to build the classifier. According to the analysis of the sequence mode, the training data are divided into five types: noise model, sudden decline model, sudden ascend model, other anomaly sequence and normal mode.

The classification result of KNN is mainly influenced by the parameter k and calculation formulas of similarity or distance measure. The choice of k depends on the number of every class, and the choice of distance measure method depends on the characteristic of the data and the type of anomaly. Commonly used similarity calculation methods are Euclidean distance (ED), dynamic time warping (DTW) [20] distance and complexity invariant distance (CID) [19].

CID is a method proposed for time series by Batista [19], it plays well in complexity and accuracy. The CID method represents the complexity difference between the two time series by taking into account the complexity factor. It has higher accuracy in processing time series involved in this section. With the time complexity almost the same as ED, significantly better than the DTW [5], therefore this paper adopts CID as the similarity calculation formula in the KNN. The theory of CID is presented in detail in section 3.2.

The basic idea of OCSVM_KNN algorithm is regarding two types of noise patterns, sudden decline patterns, sudden ascend patterns and other abnormal patterns as anomaly patterns corresponding to normal mode, detecting anomaly sequences using trained OCSVM model. And then we regard the OCSVM output as the input of KNN and carry on second time for detection, finally realizing division of modes.

The algorithm designed in this work is used for anomaly detection for time series sequences. As the sensor data stream is dynamic, it's necessary to update detection model to ensure its efficiency. This sequence anomaly detection algorithm OCSVM_KNN has the following advantages:

- The two stages of the algorithm design are complementary. OCSVM achieves anomaly detection, ensuring high detection rate; KNN implement division of anomaly patterns, ensuring low alarm detection rate. Experiments show that the algorithm is better than KNN_CID and some typical classification algorithms in effect;
- CID method adopted as the distance calculation formula in KNN guaranties the time efficiency and enhances the efficiency of anomaly detection in the same time.

5   Experiments

5.1   The Data Source

The data source for sequences anomaly detection includes 15000 temperature data. With the data acquisition time interval as 20s, the size of the time window as 15, there are 1000 fixed-length time series. Even though 1000 time series are not fully ample, as the coverage of anomalies is comprehensive, they are enough to verification the practicability of OCSVM_KNN. The allocation situation of sequence patterns shows in Table 1:

Table 1. Allocation situation of sequence patterns

Pattern	All	Sudden Ascend	Sudden Decline	Noise	Other Anomaly	Normal
Training data	750	41	27	80	31	571
Test data	250	14	9	30	10	187

5.2   Detection Index

Performance indexes of anomaly detection algorithm includes: detection rate (DR) and false alarm rate (FR). DR refers to the ratio of number of abnormal data correctly identified with number of all the abnormal data contained in data set; FR refers to the ratio of number of abnormal data mistakenly detected with number of all the anomalies detected. The higher the DR, the lower the FR, the degree of accuracy of anomaly detection algorithm is higher.

Because division of anomaly patterns will be taken into account in the research of sequence anomaly detection in this work, the index of MCAR (misclassified anomaly rate) is added to measure the rate of misclassified anomaly sequences. The computational formula of MCAR is as follows:

$$MCAR = \frac{MCA}{AA} \tag{4}$$

where MCA refers to the number of misclassified anomaly sequences, AA refers to the number of all anomaly sequences. The algorithm with high DR, low FR and low MCAR will be thought to be better.

5.3   The Analysis of Simulation Results

We have done all implementation in python and have compared our approach with other anomaly detection approaches. Firstly we will analyze the DR and FR of the algorithm in Phase 1; secondly we examine the FR, DR and MCAR in Phase 2; and lastly synthesize them in the whole procedure.

With the same data set, the performance of OCSVM_KNN algorithm is measured mainly against other algorithms. The training data and test data of first stage are as shown in Table 1.

In Phase 1, we regard the sudden change instances, noise instances and other type of anomaly instances as the same. The goal of Phase 1 is to distinguish these instances from the normal instances. The result of Phase 1 is shown in Fig.4. It's should be noted at that only normal time series sequences are marked with labels in the first stage. After debugging, the hyper parameter of One-Class SVM is finally set to be 0.15 and 0.001 for μ (an upper bound on the fraction of training errors and a lower bound of the fraction of support vectors) and γ (default value is reciprocal of the number of sample). In the chart OCSVM, LR, RF, Liner SVC, KNN_CID and

KNN_ED respectively represent One-Class SVM, Logistic Regression, Random Forest Classifier, Linear SVC, CID based KNN and ED based KNN. Besides, the result of different kernel functions with parameters adjusted is shown in Fig. 5. It's obvious that RBF outperforms others.

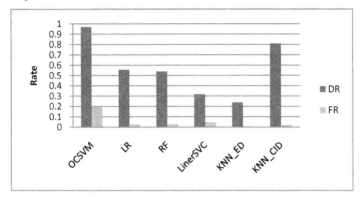

Fig. 4. Detection results of different algorithms in Phase 1

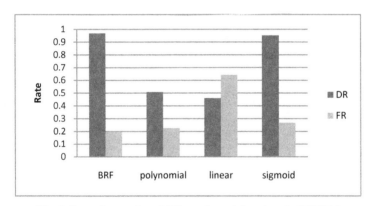

Fig. 5. Detection results of different kernel functions in OCSVM

In the first stage, OCSVM algorithm is compared with other five algorithms. Eventually, in this stage, 15 normal sequences are detected to anomalies, and 2 anomaly sequences are detected to normal sequences. Compared with other methods, OCSVM can achieve 96.825% detection rate (DR), but at the same time, the false detection rate (FR) has reached 19.7%. As we need to detect more anomalies, the OCSVM is the better choice than other classification algorithms to detect the anomalies. In order to reduce the false detection rate, it's necessary to carry out second stage algorithm for further candidate anomaly detection.

In phase 2, we classify the candidate anomaly from phase 1 to five classes which can help to judge the detail reason of anomalies. As KNN algorithm not fit for unbalanced datasets, the every class of given training data is as balance as possible. The test datasets of phase 2 which is the output of phase 1 is shown in Table 2. The detection results of different algorithms and different parameter k are respectively displayed in Fig.6 and Fig.7.

Table 2. Data set of Phase 2

Pattern	All	Sudden Ascend	Sudden Decline	Noise	Other Anomaly	Normal
Training data	112	21	20	20	21	30
Test data	76	13	9	29	10	15

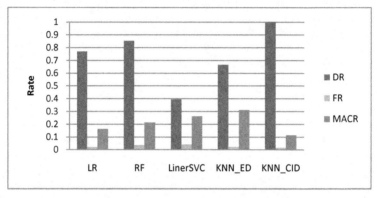

Fig. 6. Detection results of different algorithms in Phase 2

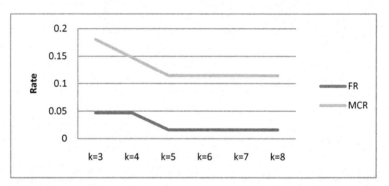

Fig. 7. Detection results of different parameter k in KNN_CID

From the results of Fig.6 and Fig.7, we can see that KNN_CID which the parameter k is set as 5 has the best performance with 100% DR, 1.61% FR and 11.48% MCAR. At last, we compared the two-stage detection method with other classification methods directly, and the result is shown in Fig.8.

The result which is depicted in Fig.8 shows that OCSVM_KNN performs better than other algorithms, with the performance of 96.82% DR, 1.61% FR and 11.11% MCAR. In the experiments, we also change the parameter k of KNN from 8 to 3, but there is little change of the performance as shown in Fig.7, besides, DR which isn't shown in the figure is maintaining 100%. Besides, we have also changed the sliding window size, similar results are achieved that the algorithm OCSVM_KNN performs better. What's more, with calculating, the processing time of all the two stages is about 0.4 seconds which the sampling rate of 2500/s can be supported. Obviously, the OCSVM_KNN method is enough for dealing with dynamic real time data.

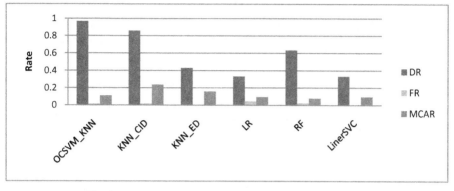

Fig. 8. Detection results of different algorithms synthetically

6   Conclusion

In this work we proposed a two-stage algorithm called OCSVM_KNN to detect the anomaly of single sensor node. The experimental results on real data shows that it performs better than other classify algorithms both in the detection rate, false alarm rate and misclassification anomaly rate. It can not only detect the anomaly but also classify the anomalies for further analysis.

Although the two-stage algorithm has been proved enough for real-time data, we haven't realized this project. So we will try to improve the real-time performance in the future.

Acknowledgement. The work is supported by the Research Project of Power System Online Security Assessment Big Data based (No.XT71-15-056).

References

1. Chandola, V., Banerjee, A., Kumar, V.: Anomaly detection: A survey. ACM Computing Surveys (CSUR) **41**(3), 15 (2009)
2. Gupta, M., Gao, J., Aggarwal, C., Han, J.: Outlier detection for temporal data. Synthesis Lectures on Data Mining and Knowledge Discovery **5**(1), 1–129 (2014)
3. Janakiram, D., Adi Mallikarjuna Reddy, V., Phani Kumar, A.: Outlier detection in wireless sensor networks using bayesian belief networks. In: First International Conference on Communication System Software and Middleware, 2006. Comsware 2006, pp. 1–6. IEEE (2006)
4. Hastie, T., Tibshirani, R.: Discriminant adaptive nearest neighbor classification. IEEE Transactions on Pattern Analysis and Machine Intelligence **18**(6), 607–616 (1996)

5. Wei, Y.: Multi-dimensional time warping based on complexity invariance and its application in sports evaluation. In: 2014 11th International Conference on Fuzzy Systems and Knowledge Discovery (FSKD), pp. 677–680. IEEE (2014)

6. Angiulli, F., Fassetti, F.: Detecting distance-based outliers in streams of data. In: Proceedings of the Sixteenth ACM Conference on Conference on Information and Knowledge Management, pp. 811–820. ACM (2007)

7. Koupaie, H.M., Ibrahim, S., Hosseinkhani, J.: Outlier detection in stream data by clustering method. International Journal of Advanced Computer Science and Information Technology **2**(3), 25–34 (2013)

8. Zhang, Y., Meratnia, N., Havinga, P.: Outlier detection techniques for wireless sensor networks: A survey. Communications Surveys & Tutorials, IEEE **12**(2), 159–170 (2010)

9. Hill, D.J., Minsker, B.S.: Anomaly detection in streaming environmental sensor data: A data-driven modeling approach. Environmental Modelling & Software **25**(9), 1014–1022 (2010)

10. Budalakoti, S., Srivastava, A.N., Akella, R., Turkov, E.: Anomaly detection in large sets of high-dimensional symbol sequences (2006)

11. Chandola, V., Mithal, V., Kumar, V.: Comparative evaluation of anomaly detection techniques for sequence data. In: ICDM. pp. 743–748 (2008)

12. Rebbapragada, U., Protopapas, P., Brodley, C.E., Alcock, C.: Finding anomalous periodic time series. Machine Learning **74**(3), 281–313 (2009)

13. González, F.A., Dasgupta, D.: Anomaly detection using real-valued negative selection. Genetic Programming and Evolvable Machines **4**(4), 383–403 (2003)

14. Yang, Z., Meratnia, N., Havinga, P.: An online outlier detection technique for wireless sensor networks using unsupervised quarter-sphere support vector machine. In: International Conference on Intelligent Sensors, Sensor Networks and Information Processing, 2008. ISSNIP 2008, pp. 151–156. IEEE (2008)

15. Ma, J., Perkins, S.: Time-series novelty detection using one-class support vector machines. In: Proceedings of the International Joint Conference on Neural Networks, 2003, vol. 3, pp. 1741–1745. IEEE (2003)

16. Schölkopf, B., Platt, J.C., Shawe-Taylor, J., Smola, A.J., Williamson, R.C.: Estimating the support of a high-dimensional distribution. Neural computation **13**(7), 1443–1471 (2001)

17. Schölkopf, B., Williamson, R.C., Smola, A.J., Shawe-Taylor, J., Platt, J.C.: Support vector method for novelty detection. In: NIPS. vol. 12, pp. 582–588 (1999)

18. BUILDING A "HAND WRITTEN DIGIT IDENTIFIER" AND SOME MACHINE LEARNING BASICS. https://aidaiict.wordpress.com/2013/10/24/building-a-hand-written-digit-identifier-and-some-machine-learning-basics/

19. Batista, G.E., Keogh, E.J., Tataw, O.M., de Souza, V.M.: Cid: an efficient complexity-invariant distance for time series. Data Mining and Knowledge Discovery **28**(3), 634–669 (2014)

20. Chen, Y., Hu, B., Keogh, E., Batista, G.E.: Dtw-d: time series semi-supervised learning from a single example. In: Proceedings of the 19th ACM SIGKDD International Conference on Knowledge Discovery And Data Mining, pp. 383–391. ACM (2013)

An Efficient Method on Trajectory Privacy Preservation

Zhiqiang Zhang$^{(\boxtimes)}$, Yue Sun, Xiaoqin Xie, and Haiwei Pan

College of Computer Science and Technology, Harbin Engineering University, Harbin, China
zqzhang@hrbeu.edu.cn

Abstract. Traditional trajectory k-anonymity method might lead to a serious information distortion of trajectory and reduce the data quality. This paper proposes an efficient method to protect trajectory privacy by protecting points of interest, and improve the data quality.

Keywords: Trajectory k-anonymity · Point of interest · Privacy requirement · Data quality

1 Introduction

Trajectories contain a lot of valuable information. However, since consecutive points are dependent on each other in a trajectory, the problem of anonymization on trajectories is more difficult. This paper puts forward a novel trajectory k-anonymity privacy protection method based on the points of interest, and a new way to compute similarity of trajectories is proposed.

Location k-anonymity requires each location to be indistinguishable with other at least k-1 locations [1]. However, researchers found that only protecting objects' locations is not enough for protecting trace information. So the concept of trajectory privacy protection is proposed. Abul et al. proposed (k, δ)-anonymity model based on the inherent uncertainty of positioning systems [2]. Nergiz et al. proposed another k-anonymity method based on notion [3]. In [4], k-anonymity is defined that an original trajectory T is generalized into a trajectory g(T) without time information. g(T) is a subset of the generalizations of at least k-1 other original trajectories. In [5], the authors propose the concept of adapted trajectory k-anonymity based on the bipartite attack graph about original and anonymized trajectories. In [6], authors propose trajectory k-anonymity based trajectory graph on the basis of linkage attack and observation attack.

When compute the trajectories' distance, the trajectories need to be synchronized through adding or removing some segments in order to unify trajectories in both the time and the space. This process has a lot of distortion before the trajectories were anonymized. Besides, traditional trajectory k-anonymity methods consider the privacy requirements of all points as the same, and need process all points during anonymization. However, processing all points in a whole trajectory is not only increasing time but also leading to serious distortion and reducing the data quality. Huo Zheng et al. [7] argue that background knowledge is more relevant to stay points. However we

© Springer International Publishing Switzerland 2015
Y. Wang et al. (Eds.): BigCom 2015, LNCS 9196, pp. 231–240, 2015.
DOI: 10.1007/978-3-319-22047-5_19

observe that some specific points besides stay points such as the start, the end, and the turning points are also important though further investigation. And adversaries' background knowledge is also relevant to these important points. Therefore only protecting stay points is not enough to protect the whole trajectory.

2   Preliminary

Definition 1 (Trajectory). A trajectory is a sequence of spatio-temporal points. It is represented as T={ID, (x_1, y_1, t_1), ..., (x_n, y_n, t_n)} $(t_1 < ... < t_n)$, where ID represents a trajectory, (x_i, y_i, t_i) $(1 \leq i \leq n)$ represents a spatio-temporal point.

Definition 2 (Point of Interest, POI). POI means some points that users may be interested in, including the start, the end, stay points and turning points. POI is a 3-tuple (ID, *loc*, *t*), where ID represents the ID of trajectory which this POI belongs to, *loc*=(latitude, longitude) represents the geography coordinates of this POI, *t* represents the time when object passes by the point.

The start and the end in a trajectory are the source and destination of the moving object. Stay points are some locations where moving object stays for a long time and exceeds a certain threshold. Turning points can reflect the routes of a trajectory. It can be confirmed by the angle between two adjacent track fragments.

Definition 3 (Region of Interest, ROI). ROI is a region including some closer POIs. So each ROI is represented as *(id, centerPoint, radius, Δt)*, where *id* represents an ROI, *centerPoint* represents the center point of the region, *radius* represents the radius of the region, *Δt* represents the time difference of region.

Definition 4 (Trajectory Similarity). Suppose the number of common ROIs of two trajectories is M, and the number of all ROIs of these two trajectories is N. Trajectory Similarity of two trajectories is M/N, signified by *Sim*. The value of *Sim* ranges from [0,1]. The formula is as follows.

$$Sim\,(tr_1, tr_2) = \frac{|R_1 \cap R_2|}{|R_1 \cup R_2|} \tag{1}$$

Where R_1 and R_2 represent ROIs of two trajectories respectively.

Definition 5 (Trajectory *k*-anonymization Set). If a set consists at least *k* trajectories and the similarity of any two trajectories is 1, so we called these *k* trajectories as a *k*-anonymization set.

3   POI-Based Trajectory *k*-anonymity Privacy Protection Algorithms

The data released should guarantee that adversaries cannot re-identify moving objects' information using background knowledge, and meanwhile the quality of data will not affect utility. Our proposed method consists of three main phases: Phase 1, extracting POIs of all trajectories in the database; Phase 2, clustering all POIs and forming ROIs, then partitioning trajectories on the basis of ROIs; Phase 3, anonymizing trajectories in each k-anonymization set separately and publication. Algorithm 1 gives a brief description about the method we proposed.

Algorithm 1. POI-based trajectory k-anonymity privacy protection algorithm

---

Input: Original trajectory database D
Output: Anonymous trajectory database D*
```
1.  for all tr in D do
2.        C = C ∪ ExtractPOI(tr); //Extracting POIs
3.  ROI = ClusterPOI(C); //Clustering POIs
4.  G = GroupTr(D,ROI); //Grouping Trajectories
5.  for all group in G do
6.        D* = D* ∪ Anonymity(group);
7.  return D*;
```

3.1 Extracting POIs

In this paper, our purpose is completely different from [8]. Hence, we consider POIs from two aspects: privacy requirements of moving objects and background knowledge of adversaries. POIs in this paper consist of the start and the end, stay points and turning points. The goal of the privacy protection of entire trajectory is achieved by the recognition and protection of these sensitive points.

The start and the end of a trajectory are added into the set directly during extracting POIs. If the angle of two adjacent segments exceeds a threshold which is set beforehand, the point which connects these two segments is a turning point. Then add the turning point into the set. Stay points are locations or approximate locations which the stay time exceeds a threshold. There are two cases: One is stopping for a long time at one location; the other is wandering for a long time near a location. For the first situation, we just need to judge the stay time between a point and the next one. If the stay time is larger than the predefined threshold, the point would be regarded as a stay point, and be added to the set. For the second situation, points which are near the location need to find first. If the spatial distance between two points is less than MinDist, these points will be looked as one location. A represented point is the nearest point to the center point. The stay time of this place is the time span between the first point and the last point. If the stay time exceeds the threshold MinStayTime, this represented point of this place is added into the set, the other points of this place are removed finally. Algorithm 2 gives a description about the POI extracting.

Algorithm 2. POI Extracting algorithm

```
Input: A trajectory T, MinAngle, MinDist, MinStayTime
Output: the POI set of the trajectory C={POIs}
1. initialize C = {point₁}, cur = 2;   //adding the start
2. while cur<n do      //n is the number of points in T
3.    Find the next point which the spatial distance not
less than MinDist with the cur point.
4.    Compute the time span Δt between the cur and the
next
5.    if Δt >= MinStayTime then
6.        if next == cur+1 then
7.           C = C U {point_cur};
8.        else
9.           compute the center of points from cur to next;
10.          find the nearest point point_m with center;
11.          C = C U {point_m};
12.       else
13.          compute angle;
14.          if angle >= MinAngle then
15.             C = C U {point_cur};
16.       cur = next;
17.  C = C U {point_n};   //adding the end
18.  return C;
```

3.2 POI Clustering and Trajectories Group

POIs which are near each other in space and time are put into the same cluster. When searching spatial and temporal neighborhood of all unvisited points in the set, if the number of POIs in the neighborhood is smaller than the density threshold, then mark the point as candidate noise preliminary. Otherwise, the point is a core point; it needs to build a new cluster. After all points in the set are visited, the clustering process is completed preliminary. At last, further process of POIs which are marked as "noise-Candidate" is done. In order to protect POIs in trajectories, points which are not belong to any cluster need to be removed. It is obvious that the more noise points, the larger information distortion is. In order to reduce information distortion, it needs that candidate noise points are added into the formed clusters as much as possible. So, this paper improves the ST-DBSCAN [9] method to cluster POIs. The noise points generated by original ST-DBSCAN which we called candidate noise are added into closest clusters as far as possible for reducing information loss. The detail of the improved algorithm is as follows.

Algorithm 3. POI Clustering algorithm

Input: POI set C, Spatial neighborhood radius E_s,
Temporal neighborhood radius E_t, Density Threshold MinPts
Output: Cluster = {cluster$_1$, cluster$_2$,..., cluster$_m$}
```
1.   id = 0;
2.   mark all points in C as 'unvisited';
3.   for each point p in C do
4.       if (p is unvisited) then
5.           mark p as 'visited';
6.           N = getNeighbours(p, Es, Et);
7.           if (|N| < MinPts) then
8.               mark p as 'noiseCandidate';
9.           else
10.              id++;
11.              expandCluster(p, N, id, Es, Et, MinPts);
12.  for all noiseCandidate nc do
13.      ncN = getNeighbours(nc, Es, Et);
14.      if (|ncN| > 0) then
15.          add nc to the nearest cluster,
16.      else
17.          mark nc as 'noise';

expandCluster(p, N, id,  Es , Et, MinPts)
1.   add p to cluster id;
2.   for each point q in N do
3.      if (q is unvisited) then
4.          mark q as 'visited';
5.          M = getNeighbours(q, Es, Et);
6.          if (|M| >= MinPts) then
7.              N = N ∪ M;
8.      if (q is not yet member of any cluster) then
9.          add q to cluster id;
```

The candidate noise points which do not belong to any clusters finally mark as noise points and are removed at the published trajectories. After Clustering of POIs, a series of clusters which each one produces an ROI are formed. The core point of a cluster is the center point of the corresponding ROI. A ROI's radius and time difference of the region are the maximum space and time difference of center point and other points in the cluster respectively.

After partition ROIs, we find trajectory k-anonymization set using trajectory similarity defined in previous section. For achieving the requirements of k-anonymity, the size of each trajectory group should between k and $2k$-1. In addition, in order to reduce information distortion, trajectories which do not belong to any groups after partition should be added into existing groups as far as possible. If the similarity with any trajectory in a group is more than α, then it can be added into this group. The value of α is set to 1. Finally, as same as traditional *k*-anonymity, trajectories in a group which the size is less than *k* need to be removed at the published data. The trajectory grouping algorithm is as follows.

Algorithm 4. Trajectory Grouping Algorithm
___
Input: POI set C, k
Output: Trajectory Group = {G_1, G_2, ..., G_m}
1. Cluster POIs in C by calling Algorithm 3,
2. let R_1, R_2, ..., R_m are ROIs generated after clustering,
3. for all tr in D do
4.   Compute similarity *sim* between tr and other trajectories, and put trajectories into one group if their *sim*=1 ;
5. for all trn not in any group do
6.   Compute similarity *sim* between trn and existed groups ;
7.   if (*sim* >= α) do
8.   Add trn into this group,
9. put all groups which size no less than *k* into Group ;
10. return Group;

3.3   Trajectory Anonymity Publication

The trajectories have been clustered, aiming at trajectories in each group indistinguishable through anonymous methods, such as perturbing, generalization or characteristic publication. Perturbing means anonymous trajectories cannot match with original ones by reconstructing adjacent points or segments of trajectories in each group. After anonymity, if the trajectory group size is at least k, so since these k trajectories are indistinguishable, the re-identify probability of every trajectory is no less than 1/k, and this is trajectory k-anonymity.

For protecting the reality of locations, some researchers propose two methods [10]: SwapLocations and ReachLocations. The former adopts the way to find the near point set of every point in each *k*-anonymization set and swapping randomly. This method guarantees trajectory *k*-anonymity. The latter adopts swapping randomly among near locations in the whole dataset and only spatial information is swapped, and this method satisfies only location l-diversity [11]. Based on the above idea, this paper applies swapping POIs for anonymity, so the original locations are held in the released dada. Since we protect trajectories through POIs protection, only the real locations of each ROI in each group will be swapped randomly.

When extracting stay points in the first stage, some stay points are represented point of some near points; in this condition, other non-represented ones are removed and only hold the represented one in the released trajectories. After swapping randomly, the reconstructed trajectories are generated. The purpose of trajectory privacy protection is achieved.

Our algorithm firstly marks POIs of all ROIs as un-swapped, and then swaps unswapped POIs in each ROI randomly until all points are swapped at least once. After swapped, the reconstructed trajectories are added into the anonymous trajectory set finally. The anonymity of a trajectory group has finished.

4 Experiments

Our experiments are run on an Intel Core 2 Quad 2.66HZ, Windows XP machine equipped with 2GB main memory, JDK1.7. The dataset we used is real trajectories of volunteers collected by GeoLife project, which contains 18670 trajectories of 182 users from April, 2007 to August, 2012. We adopt the same parameter values with the [8], MinDist = 200 meters and MinStayTime = 20 minutes. The value of MinAngle is set 30 degree. Finally, we extract 270,666 POIs from all 24,876,978 points, and then cluster these POIs and divide ROIs.

In this paper, we measure the data utility from two aspects: information distortion and spatio-temporal range queries. In order to verify the availability of our proposed method, we compare with the typical method (k, δ)-anonymity (NWA) [3] and the SwapLocations method proposed in [11]. Trajectories are clustered using greedy cluster method and finally each cluster is converted to (k, δ)-anonymity set, where δ represents the radius of set. Due to the rapid development of technology, the accuracy of some positioning devices can reach centimeter-level, even millimeter. So we set the values of δ are set to 0, and 2000.

4.1   Information Distortion

This paper uses the same equation to calculate information distortion with [3]. Two different attributes are applied to compare, which are total space distortion and the percentage of removed points. When computing total space distortion, the value of Ω is 0. The percentage of removed points is considered alone. So the value of Ω can be set based on the different applications for computing the whole distortion.

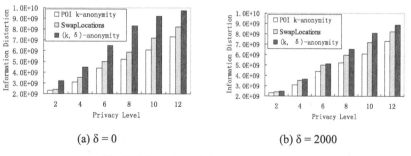

(a) δ = 0   (b) δ = 2000

Fig. 1. Information distortion measurement

Fig.1 describes the information distortion contrast with NWA and SwapLocations when $\Omega=0$, which means without considering the distortion caused by points removing. From the figure we can see our method leads to less space distortion than the other methods. This is due to our method only swap POIs of every k-anonymity set in the same ROI for achieving anonymity. SwapLocations method swaps all points in each k-anonymity set for cloaking, which leads to relatively large information distortion. For (k, δ)-anonymity model, when $\delta>0$, the distortion considers not only the space distortion of points within uncertainty range, but also the distortion caused by space translation which moves some points not in uncertainty range to another locations from original locations. This greatly increases the information distortion.

From Fig.2, the POI-based trajectory k-anonymity method discards less points than SwapLocations and (k, δ)-anonymity. This is because the distance of trajectories is not considered when grouping trajectories; as long as these trajectories pass through the same ROIs, they will be classified into the same group. SwapLocations clusters trajectories through calculating distance between trajectories.

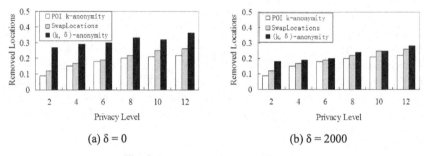

(a) $\delta = 0$ (b) $\delta = 2000$

Fig. 2. Percentage of removed locations

In addition, we can also find from Fig.1 and Fig.2, the gap of space distortion and the removed points' size among (k, δ)-anonymity with other two methods decreases gradually as the value of δ increases. This is because, with δ increasing, the number of points lied out of uncertainty range is few, which means the space distortion caused by space translation decreasing. Meanwhile, the number of trajectories not in the anonymous set is decreasing as δ increasing, and the number of discarded points is also reduced, so the gap between these methods is also becoming small.

4.2   Spatio-Temporal Range Queries

There are six types of spatio-temporal range queries introduced in [12], aiming at evaluating the relative position of a moving object with respect to a region R in a time interval $[t_b, t_e]$. In particular, when evaluating data utility we are more interested in whether a trajectory t_r is in the region R which mark as inside(R, t_r). Like [3], we do experiments on original dataset D and anonymous dataset D* separately focusing on two cases: Possibly_Sometimes_Inside (PSI) and Definitely_Always_Inside (DAI), and then compute the distortion of query results. The way to calculate Range Query Distortion is shown as equation (2)

$$RQD(D, D^*) = \frac{1}{n} \sum \frac{|Q(D) - Q(D^*)|}{\max(Q(D), Q(D^*))} \tag{2}$$

In the above formula, query Q represents PSI or DAI defined above, n is the number of executed queries for each type of query.

We adopt the same parameters with [3], and the queries are generated by randomly chosen circular regions having radius between 500 and 5000, and randomly chosen time interval $[t_b, t_e]$ with duration ranging from 2 to 8 hours. Finally 1000 different queries run in anonymous dataset generated by both methods, and then calculate the average query distortion.

The DAI query distortion of our method is less than SwapLocations method and (k, δ)-anonymity. This is because only POIs are swapped for anonymity in our method and other points keep the same except the removed ones. SwapLocations method adopts cloaking all sample points of every anonymous set, which lead to large query distortion. However, since (k, δ)-anonymity adopts space translation, many sampling points are translated to totally different ones, making the largest loss of query results.

From Fig.3 and Fig4, we can also find that the gap of query distortion among (k, δ)-anonymity with the other two methods is less and less as δ increasing. In fact, when the value of δ tends to infinity, (k, δ)-anonymity means no trajectories needs to be anonymized, and the anonymous database is as same as the original one. Therefore, the impact on anonymous results will be small with large δ.

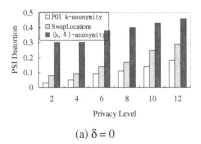
(a) $\delta = 0$

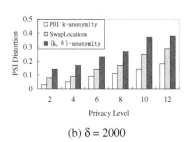
(b) $\delta = 2000$

Fig. 3. PSI Query Distortion

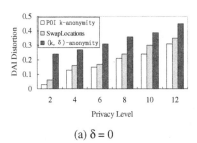
(a) $\delta = 0$

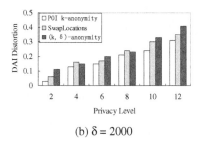
(b) $\delta = 2000$

Fig. 4. DAI Query Distortion

# 5	Conclusions and Future Work

This paper proposes POI-based trajectory k-anonymity method for solving privacy leakage problem on trajectory publication. We differ from the traditional trajectory k-anonymity on this point that our method mainly aims at k-anonymity of trajectories reconstructed by points of interest. Finally, location swapped method is adopted to release anonymous trajectories. The experiments proved our method have improved the quality of the released data.

Acknowledgement. This paper is supported by the National Natural Science Foundation of China under grant No. 61370084, 61202090, 61272184, the Program for New Century Excellent Talents in University No. NCET-11-0829, the Natural Science Foundation of Heilongjiang Province under grant F201130, and the Fundamental Research Funds for the Central Universities under grant No. HEUCF100609, HEUCFT1202.

References

1. Gruteser, M., Grunwald, D.: Anonymous usage of location-based services through spatial and temporal cloaking. In: Proceedings of the 1st International Conference on Mobile Systems, Applications and Services, pp. 31–42. ACM (2003)
2. Abul, O., Bonchi, F., Nanni, M.: Anonymization of moving objects databases by clustering and perturbation. Information Systems **35**(8), 884–910 (2010)
3. Nergiz, M.E., Atzori, M., Saygin, Y.: Towards trajectory anonymization: a generalization-based approach. In: Proceedings of the SIGSPATIAL ACM GIS International Workshop on Security and Privacy in GIS and LBS, pp. 52–61. ACM (2008)
4. Monreale, A., Andrienko, G.L., Andrienko, N.V., et al.: Movement Data Anonymity through Generalization. Transactions on Data Privacy **3**(2), 91–121 (2010)
5. Yarovoy, R., Bonchi, F., Lakshmanan, L.V.S., et al.: Anonymizing moving objects: how to hide a MOB in a crowd?. In: Proceedings of the 12th International Conference on Extending Database Technology: Advances in Database Technology, pp. 72–83. ACM (2009)
6. Huo, Z, Huang, Y, Meng, X.: History trajectory privacy-preserving through graph partition. In: Proceedings of the 1st International Workshop on Mobile Location-Based Service, pp. 71–78. ACM (2011)
7. Huo, Z., Meng, X., Hu, H., Huang, Y.: You can walk alone: trajectory privacy-preserving through significant stays protection. In: Lee, S.-g., Peng, Z., Zhou, X., Moon, Y.-S., Unland, R., Yoo, J. (eds.) DASFAA 2012, Part I. LNCS, vol. 7238, pp. 351–366. Springer, Heidelberg (2012)
8. Zheng, Y, Zhang, L, Xie, X, et al.: Mining interesting locations and travel sequences from GPS trajectories. In: Proceedings of the 18th International Conference on World Wide Web, pp. 791–800. ACM (2009)
9. Birant, D., Kut, A.: ST-DBSCAN: An algorithm for clustering spatial–temporal data. Data & Knowledge Engineering **60**(1), 208–221 (2007)
10. Domingo-Ferrer, J, Sramka, M, Trujillo-Rasúa, R.: Privacy-preserving publication of trajectories using microaggregation. In: Proceedings of the 3rd ACM SIGSPATIAL International Workshop on Security and Privacy in GIS and LBS, pp. 26–33. ACM (2010)
11. Machanavajjhala, A., Kifer, D., Gehrke, J., et al.: L-diversity: Privacy beyond k-anonymity. ACM Transactions on Knowledge Discovery from Data (TKDD) **1**(1), 3 (2007)
12. Trajcevski, G., Wolfson, O., Hinrichs, K., et al.: Managing uncertainty in moving objects databases. ACM Transactions on Database Systems (TODS) **29**(3), 463–507 (2004)

Architecture and Applications

R-Memcached: A Reliable In-Memory Cache System for Big Key-Value Stores

Chengjian Liu[1]([✉]), Kai Ouyang[1,2], Xiaowen Chu[1,2]([✉]), Hai Liu[1],
and Yiu-Wing Leung[1]

[1] Department of Computer Science, Hong Kong Baptist University,
Kowloon Tong, Hong Kong
{cscjliu,kouyang,chxw,hliu,ywleung}@comp.hkbu.edu.hk
[2] Institute of Research and Continuing Education, Hong Kong Baptist University,
Shenzhen, China

Abstract. Large-scale key-value stores are widely used in many Web-based systems to store huge amount of data as (key, value) pairs. In order to reduce the latency of accessing such (key, value) pairs, an in-memory cache system is usually deployed between the front-end Web system and the back-end database system. In practice, a cache system may consist of a number of server nodes, and fault-tolerance is a critical feature to maintain the latency Service-Level Agreements (SLAs). In this paper, we present the design, implementation, and evaluation of R-Memcached, a reliable in-memory key-value cache system that is built on top of the popular Memcached. R-Memcached exploits coding techniques to achieve reliability, and can tolerate up to two node failures. Our experimental results show that R-Memcached can maintain very good latency and throughput performance even during the period of node failures.

1 Introduction

In recent years, key-value stores have been widely used in many commercial large-scale Web-based systems, including Amazon, Facebook, YouTube, Twitter, etc. To reduce the access latency caused by disk I/O, an in-memory cache system is usually deployed between the front-end Web system and the back-end database system. For example, Facebook is using a very large distributed in-memory cache system built from the popular Memcached [1] [2], which consists of thousands of server nodes [3]. In such a large-scale system, node failure becomes very common [4], which may seriously affect the access latency. How to improve the reliability of the distributed cache system becomes an important issue.

Redundancy techniques such as RAID technologies [5] have been widely used in disk-based storage systems to offer fault-tolerance. RAID-1 is basically the same as data replication, which achieves very good reliability but requires a double cost. RAID-5 and RAID-6 improve the storage efficiency at the expense of decreasing access performance when faced with disk failure. But currently,

© Springer International Publishing Switzerland 2015
Y. Wang et al. (Eds.): BigCom 2015, LNCS 9196, pp. 243–256, 2015.
DOI: 10.1007/978-3-319-22047-5_20

how these redundancy techniques work in memory is still unknown. The question about whether these technologies with less redundancy is applicable for in-memory storage is worth studying.

In this paper, we aim to improve the reliability of distributed in-memory cache system by integrating redundancy functions. We also study the trade-off between access performance and storage efficiency in such memory based storage system. Specifically, we designed and implemented R-Memcached by introducing RAIM-1, RAIM-5 and RAIM-6 redundancy technologies in-memory, which are similar with RAID-1, RAID-5 and RAID-6 in disk-based storage systems respectively, to the popular Memcached software. We further evaluate the access latency and throughput of different levels of redundancy technologies in R-Memcached. Our main contributions can be summarized as follows:

(1) We designed a reliable in-memory cache system for key-value stores.
(2) We implemented the proposed system on top of the popular Memcached.
(3) We thoroughly evaluated the performance of R-Memcached in terms of access latency and throughput.

The remainder of this paper is organized as follows. In Section 2, we give the background and motivation of our work. Section 3 introduces the design and implementation of R-Memcached. Section 4 evaluates the performance of R-Memcached by real experiments. Conclusions and our future plan are given in Section 5.

2   Background

With the popularity of big data applications, large-scale distributed database systems play a more and more important role. However, the disk access performance cannot just increase proportionally with the increase of disk capacity. To achieve a good balance between performance and cost, a practical solution is to deploy memory-based system as a cache between users and disk-based databases. For very large databases, even the cache system alone may consist of tens to thousands of server nodes. When designing such distributed memory-based cache system, we should consider many performance metrics, such as access latency, throughput, reliability, and cost. Many existing work focus on the system throughput, while the system reliability has been largely overlooked.

2.1   Introduction to Memcached

Memcached is a well-known distributed in-memory cache system used to reduce data access latency of key-value stores. It has been widely used in the IT industry to support large-scale Web or Cloud services. In a key-value store, each data item consists of a *key* and a *value*. Typical data operations include *get*, *set*, and *delete*. Two major performance metrics in key-value stores are the response time (i.e., data access latency) and throughput (i.e., how many requests can be satisfied in a unit of time). Fig. 1 illustrates how Memcached is used in a Web Service system, which needs to serve numerous user requests in real time. Upon the receipt of a user request with a key, the front-end of the Web service will first try to access data from Memcached storage cluster. If there is a cache hit, i.e., the required data can be found in a Memcached node, then it is not necessary to access the real back-end database and hence expensive disk I/O or computations can be avoided. In case of cache miss, the Web service will retrieve data from the back-end database and store the data into a Memcached node. To uniformly distribute the data among Memcached nodes, Memcached uses consistent hash function [6] with the key to determine which node in the Memcached cluster should hold the data item.

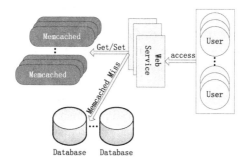

Fig. 1. Illustration of Memcached in Web Service

Because the size of *values* varies from several bytes to hundreds of kilo-bytes, Memcached uses a slab allocator to effectively manage the memory and data items. A slab is a large chunk of memory which can hold many small data items. Instead of allocating and freeing memory for each single data item, Memcached uses slab as the basic unit of memory allocation. A hash table array is used to manage a set of buckets, each of which uses a double linked-list to manage its key-value pairs by Least Recently Used (LRU) policy.

2.2 Reliability Challenge

Reliability is the cornerstone of large storage system [7]. RAID-1, RAID-5, RAID-6 and erasure coding have been widely adopted in different scenarios to offer different levels of redundancy, restore performance penalty, and tolerable device failures. RAID-1 is usually used for best performance consideration. RAID-5 is more cost-effective, but can only support a single disk failure. RAID-6 can tolerate up to 2 disk failures. In the era of big data, the storage world already comes to the Peta-byte age [8][9]. Large-scale storage systems, such as Microsoft Azure[10], adopt erasure coding to further the cost while achieving high reliability.

Different from disk-based systems, memory-based system has much higher cost [11]. The memory storage price per bit is about 100 times more expensive than disk storage per bit. And this is a big factor that should be taken into consideration when designing reliable memory-based storage systems.

In summary, this work is motivated to investigate the flexibility of reliable distributed memory cache system that can fulfill three requirements: reliability, storage efficiency, and performance. In traditional disk storage systems, RAID-5 and RAID-6 can achieve a good balance between reliability and efficiency; but it remains unknown whether it is practical to implement similar techniques for memory-based systems, considering the highly complex memory-management scheme and network traffic overhead generated by such systems.

2.3 Nomenclature for In-Memory Redundancy

Disks have been playing a major role for storing data in traditional storage systems for decades. RAID, as a set of redundant techniques for reliability, has already become an industrial standard. But disk and memory are very different storage medium in terms of volatility, access latency and storage cost per bit, etc.

RAIM, short for Redundant Array of Independent Memory, was recently introduced by IBM[12]. RAIM works in a similar fashion in the memory as RAID in the disk to tolerate certain level of memory channel failures. Here we extend the concept of RAIM from s single physical server to a distributed system.

3   Design and Implementation of R-Memcached

In this section, we first introduce our design of R-Memcached system architecture, then discuss how each RAIM level is implemented in our in-memory cache system and present how we handle update operations and data fetch operations upon node failures.

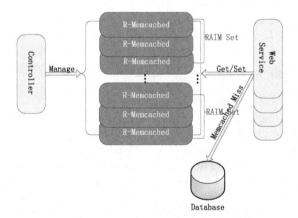

Fig. 2. System Architecture of R-Memcached

3.1   System Architecture

As shown in Fig.2, our R-Memcached system consists of two major components: a controller and a set of R-Memcached nodes. R-Memcached nodes are the major memory storage for key-value data cached from the back-end database. We can form different RAIM sets, each of which consists of several R-Memcached nodes and uses a specific RAIM level based on the reliability requirement. The controller is responsible for managing the R-Memcached nodes. Notice that the controller is a logical function and can be implemented on a dedicated server or any R-Memcached node. In case of controller failure, another node will be elected as the controller, and hence our system does not suffer from single-point-of-failure.

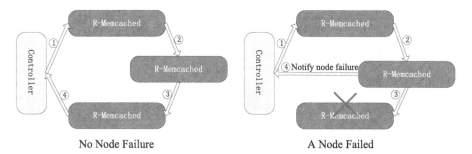

No Node Failure   A Node Failed

Fig. 3. Node failure detection in R-Memcached

The controller has two major functions. It configures R-Memcached nodes into different RAIM sets. Examples of RAIM sets of RAIM-1, RAIM-5 and RAIM-6 are shown in Fig.4. The second function of the controller is to monitor the alive/dead status of each R-Memcached node. To detect node failures, the R-Memcached nodes are organized as a ring: each node in the circle has a predecessor and a successor, as shown in Fig.3. The controller periodically sends out a token message using UDP to its successor who should immediately return a positive ACK message. Each node will then forward the incoming token to its own successor. The token will circulate in the ring and eventually return back to the controller after travelling through all nodes.

In case of a node failure, the predecessor of the failed node will detect this failure after a configurable timeout period. When a node finds its successor failed, it notifies the controller and the controller broadcasts a failure message to the whole cache system. The controller removes the failed node from the ring and continucs monitoring the new ring formed by the remaining alive nodes. In case the predecessor of the controller detects that the controller fails, it promotes itself as the new controller and notifies all other nodes.

3.2   RAIM Implementation

Since RAIM-1, RAIM-5 and RAIM-6 are different technologies used for different levels of system reliability, the implementation strategies of these three

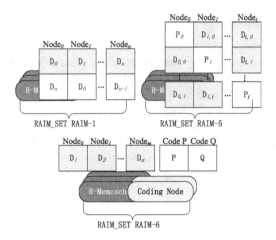

Fig. 4. RAIM Sets of RAIM-1, RAIM-5 and RAIM-6

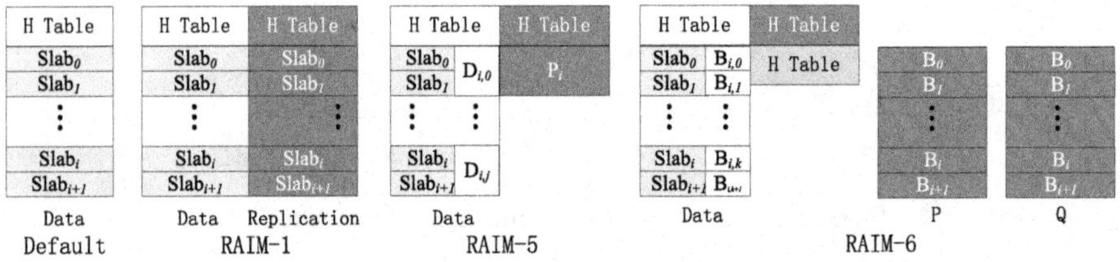

Fig. 5. Memory Layout in R-Memcached

technologies are different in R-Memcached. Fig. 5 shows the memory layout of different RAIM applied in R-Memcached implementations. The upper left labelled Default is the standard Memcached memory layout. It mainly contains two parts: the region storing the data and the region storing the key hash table to index the position of data objects in the data region.

For RAIM-1, as shown in Fig. 5 labelled RAIM-1, each storage node contains an extra memory block which has the same size as the standard Memcached marked in gray color. It is used for storing replication from other nodes.

For RAIM-5, besides the extra memory used for storing hash table of the backup key, another extra region of memory named P is used for storing parity data. Suppose a RAIM-5 system consists of N nodes (N must be at least 3), each node's default memory size is m, then the memory size for storing parity data is $m/(N-1)$.

For RAIM-6, we adopt the horizontal array codes for RAID-6 as described in [18]. There are two extra nodes used for storing the encoded data which we refer to as coding node. The memory layout of RAIM-6 is shown at the right of Fig.5. The left part is the memory layout in R-Memcached node, which includes the standard memory block and two extra memory blocks used for storing backup key hash table from two different nodes. The P and Q rectangles represent the two dedicated coding nodes.

3.3   *Get Set* and *Delete* in R-Memcached

The Web service has three types of operation on Memcached. The *Set* operation for storing key-value on Memcached cluster, the *Get* operation for retrieving key-value on Memcached cluster, and the *Delete* operation for removing key-value from the Memcached cluster. Note that the key-value together is also called object. The procedure of each operation that starts from the Web service and process on the R-Memcached is shown in Fig.6.

When the Web service starts a *Get* request, as illustrated in Fig.6, it first hash the key to find which Memcached node stores that object. It then checks whether the node is alive. If not alive, the degrade read will be triggered. If it is alive, Web service sends the request to the Memcached node. The Memcached node receives the request and tries to find the key in some bucket of its hash table. If it finds the key, it returns the whole key-value object to the Web service.

For *Set* operation, first it computes the Memcached node location to store the object. If the Memcached node is alive, the object is sent to the Memcached node. The Memcached node finds a bucket in the hash table to store it. Meanwhile the RAIM update operation will be triggered. The RAIM update is called asynchronous update and the details will be explained in the following subsection. The *Delete* operation acts similarly with the *Set* operation.

Set Operation   Get Operation   Delete Operation

Fig. 6. Get, Set and Delete Operations

3.4   Asynchronous Update and Degrade Read

When issue a *Set* and *Delete* command to a R-Memcached node, the node needs update its hash table and data memory. Each update operation to a R-Memcached storage node will trigger two additional updates. One update operation on its successor node to maintain consistent of key replication, and another update operation for encoding coding data and storing to corresponding coding node. To minimize the overhead introduced by two update operations, we use asynchronous update for two updates.

We launch an additional thread in charge of asynchronous update when R-Memcached is started. Meanwhile a queue is created to add object that needs to be updated. The thread keeps idle and doesn't consume any CPU resource when the queue is empty. When R-Memcached needs to update its hash table and data memory for an object, the update information will be added to the queue for asynchronous update. The thread will fetch the information from the queue and start asynchronous update. The thread first gets the object that needs to be updated from the queue. When the operation fails, the asynchronous process terminates. When the operation succeeds, the R-Memcached node will store a replication of the key of the object to the successor node. Then the coding operation happened and the new coding data stores to corresponding coding node based on the different RAIM levels. For RAIM-6, we put the key to the successor node of the current successor node to tolerate two node failures.

Besides the asynchronous update, we need to handle the get operation during node failures. The procedure is defined as follows: when a node fails, the get request which is supposed to operate on the failed node now operates on its successor. This operation is referred to as degrade read, and the detailed procedure is shown in Algorithm 1. When a node fails, the Web service contacts the successor of the failed node(can be the successor of the successor for the case of RAIM-6) and accesses the node's successor to fetch the object. When the failed nodes successor receives the get request, it first checks whether the key is in the backup hash table (in algorithm 1); if not, the get request receives a null; otherwise the corresponding data object will be restored in the backup region according to the used RAIM technology and returned back to the Web service.

Algorithm 1. Degrade Read

Input: $key(i)$
Output: $object(i)$ or $NULL$
if $find\_object\_inbackup(key(i)) = FALSE$ **then**
  | return $NULL$
end
switch $RAIM\_TYPE$ **do**
  case $RAIM$-1
  | $object(i) = $ find_object_inbackup$(key(i))$
  end
  case $RAIM$-5
  | $DATA_FOR_XOR = $ get_remaindata_code$(key(i))$ $object(i) = $
  | restore_withXOROP$(DATA\_FOR\_XOR)$
  end
  case $RAIM$-6
  | $DATA_FOR_DECODING = $ get_data_coding$(key(i))$ $object(i) = $
  | decoding$(DATA\_FOR\_DECODING)$
  end
endsw
return $object(i)$

For RAIM-1, since the data is directly copied to the backup region, it just needs to read from the backup region to get the data back. So it does not need to communicate with other nodes to restore the object.

For RAIM-5, to restore the object, as shown in Algorithm 1, first we need to get other data and parity data block to perform decoding. Note that to minimize the network communication, we do not use the fixed strip length to do the calculation. For each data block what we read is the necessary size to restore the object. For example, suppose that the object location in the data memory region of the failed node is stored with a size of 100 bytes., we just need to read 100 bytes with corresponding offset from other nodes to restore the data.

For RAIM-6, as shown in Algorithm 1, it needs to get the corresponding data and the encoding data block to restore the object. The block size must be a fixed length here. So when doing the degrade read to restore the object with RAIM-6, there are at least 6 network communications for a RAIM storage system consisted of 3 storage nodes and 2 nodes for storing encode data. So the degrade reads network communication cost with RAIM-6 is always larger than the system with RAIM-5 for the same storage size.

4   Performance Evaluation of R-Memcached

In this section we present the performance evaluation of R-Memcached on a real testbed. We focus on evaluating the access latency and system throughput.

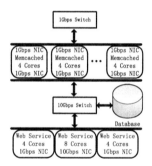

Fig. 7. R-Memcached Testbed

4.1   Testbed and Performance Baseline

Our testbed is illustrated in Fig.7. Each R-Memcached node has a single quad-core CPU and two 1-Gbps NICs. We use 3 to 5 nodes to evaluate RAIM-1 and RAIM-5, and 5 to 7 nodes for RAIM-6. Note that our RAIM-6 uses two dedicated nodes to hold coding data. We use three other computers to simulate a set of clients that generate the requests. Client nodes are connected to R-Memcached system through a 10-Gbps Ethernet switch. R-Memcached nodes communicate with each other through another 1-Gbps Ethernet switch.

Memcached usually stores various size of data items [13], we choose data value size of 32 Bytes, 64 Bytes, 128 Bytes, 256 Bytes and 1 KB to represent different workloads. The key size is 5 Bytes. We measure two major performance metrics of the R-Memcached, throughput and latency, and compare with the standard Memcached downloaded from the official Memcached open source group.

4.2   Evaluation of RAIM-1

The throughput of R-Memcached Cluster for RAIM-1 is constrained by two major factors: the CPU's processing power, and the network bandwidth. These two factors are represented as Serving Ability and Network I/O in Eq. 1. In our testbed, for data size smaller than 128 Bytes the maximum serving ability turns out to be the main bottleneck, as shown in Fig. 8. For larger data size, the network I/O becomes the major limiting factor. For the degrade get in RAIM-1 Cluster with N nodes, if one node failed, the whole throughput is close to $(N-1)/N$ when the maximum serving ability is the main constraint. But when Network I/O is the main bottleneck, the penalty of degrade get becomes even smaller.

For the latency in RAIM-1, since there is no extra computation penalty and network I/O for degrade get, the degrade get has the same performance as the

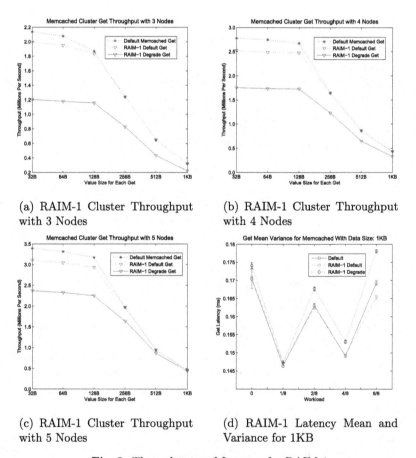

(a) RAIM-1 Cluster Throughput with 3 Nodes

(b) RAIM-1 Cluster Throughput with 4 Nodes

(c) RAIM-1 Cluster Throughput with 5 Nodes

(d) RAIM-1 Latency Mean and Variance for 1KB

Fig. 8. Throughput and Latency for RAIM-1

normal case. And for RAIM-1 with different levels of workload, since both serving ability and network I/O don't reach their maximum ability, the variance of the latency is very small.

$$Get\_Throughput = min(ServingAbility, NetworkI/O) \qquad (1)$$

For the throughput illustrated in Fig. 8, with single node, both Default and Degrade have the same performance. This is also because there is no extra performance penalty for degrade read. But for the whole cluster, the Default has better performance than the Degrade. This is because for Degrade get, each node in the Cluster has the same workload. But for the Degrade get with a failed node, the successor of the failed node has double workload compared with other nodes.

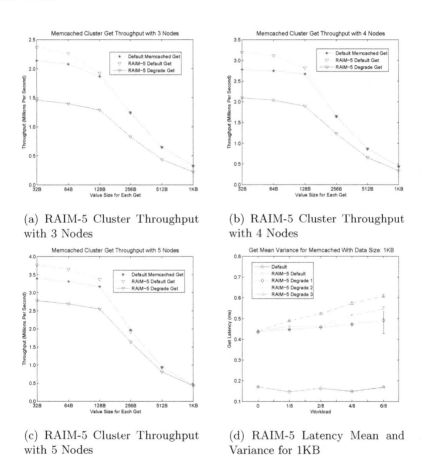

(a) RAIM-5 Cluster Throughput with 3 Nodes

(b) RAIM-5 Cluster Throughput with 4 Nodes

(c) RAIM-5 Cluster Throughput with 5 Nodes

(d) RAIM-5 Latency Mean and Variance for 1KB

Fig. 9. Throughput and Latency for RAIM-5

4.3   Evaluation of RAIM-5

The whole throughput of RAIM-5 cluster is also constrained by the two factors listed in Eq. 1. So the whole throughput is also kept in a reasonable range when one node failed as illustrated in Fig. 9.

For the degrade read, unlike RAIM-1, the RAIM-5 degrade read has extra network I/O penalty and computation penalty. It needs to fetch data from other node to do XOR operation to restore the data from the failed node. We use degrade 1, degrade 2, degrade 3 in Fig. 9 to represent the RAIM-5 system formed with 3 nodes, 4 nodes and 5 nodes respectively. So for RAIM-5, the degrade get penalty can be represented as Eq. 2. The necessary latency cost for the network round trip is denoted as ReqRTT in Eq. 2. We observe that with more nodes in RAIM-5, the latency of degrade get grows with the increase of workload. But still the latency is usually less than 0.7 ms, which is quite satisfactory.

$$DegradeLatency = ReqRTT + I/O + Computation \qquad (2)$$

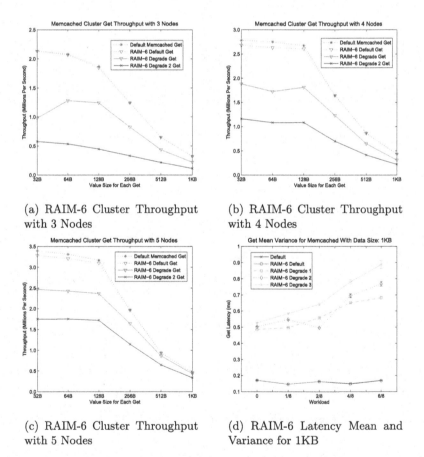

(a) RAIM-6 Cluster Throughput with 3 Nodes

(b) RAIM-6 Cluster Throughput with 4 Nodes

(c) RAIM-6 Cluster Throughput with 5 Nodes

(d) RAIM-6 Latency Mean and Variance for 1KB

Fig. 10. Latency of Get Operations for RAIM-6

4.4 Evaluation of RAIM-6

Here we use the Jerasure [14] as our encoding and decoding library for RAIM-6, which can achieve around 2GB/s of coding speed on our testbed. In our experiment with RAIM-6 presented in Fig.10, the throughput for degraded read with RAIM-6 is not as good as RAIM-5 when value size smaller than 128 bytes. RAIM-6 has higher network communication overheads than RAIM-5. Meanwhile the decoding cost for RAIM-6 is also higher than RAIM-5. This means, fir RAIM-6, CPU must sacrifice more computation power for decoding compared with RAIM-5. So the performance penalty for the throughput is obvious when CPU's processing power is the major factor to determine the throughput.

For RAIM-6, it is able to tolerate two concurrent node failures. The throughput for two concurrent R-Memecached node failures is named *RAIM-6 Degrade 2 Get* in Fig. 10. As we can see for value size smaller than 128 Bytes, the throughput is also not scale down proportional due to CPU's processing power is consumed a lot by decoding. But when the value size equal and bigger than 128 bytes, the throughput scale down proportional. This is due to the extra network communication for degrade read doesn't impact the network communication between R-Memcached cluster and Web service.

It is also evident that the degrade latency is even greater than RAIM-5, due to the computational complexity of RAIM-6 and also more network traffic during degrade get. So for RAIM-6, we can see that there are two potential bottlenecks. One bottleneck is the decoding performance for degrade get. Another bottleneck is the network bandwidth. In all our experiments, the latency is less than 1 ms.

5 Conclusions

In this paper, we designed and implemented a reliable distributed in-memory cache system for large-scale key-value stores. We present the performance evaluation for RAIM-1, RAIM-5 and RAIM-6. The work can be used for choosing a good reliable scheme for reliable in-memory object storage. And it also can be used as a guideline for further optimization of in-memory RAIM.

For RAIM-1, the main issues are the load balancing for degrade read and the lower storage efficiency compared with RAIM-5. RAIM-5 can be a better choice compared with RAIM-1, because it has higher storage efficiency than RAIM-1 and the performance penalty for RAIM-5 can be counteracted by the balance issue of RAIM-1. RAIM-6 has two potential bottlenecks: network bandwidth and decoding performance. For small-scale deployment, we recommend RAIM-5 for its simplicity and high efficiency. For large-scale deployment, we need to use high-speed network (such as 10-Gbps) and high-performance servers if RAIM-6 adopted.

The performance penalty for decoding with RAIM-6 has two big impacts for the whole system performance. First is that decoding creates extra performance penalty. Secondly it needs a lot of computational power, which will occupy a lot of CPU cycles. One possible solution is to offload the coding/decoding tasks to hardware accelerators such as GPU[15].

Memory-based storage system has also been proposed as a complete future storage landscape for large scalable high performance storage system, such as Ramcloud storage [16] and SILT[17]. We plan to investigate how to effectively enhance the fault-tolerance of such systems.

Acknowledgments. This work is supported in part by HONG KONG GRF grant HKBU 210412 and HKBU grant FRG2/13-14/052. We thank all the reviewers for their insightful comments and suggestions.

References

1. Fitzpatrick, B.: Distributed caching with memcached. Linux Journal **2004**(124) (2004)
2. Nishtala, R., Fugal, H., Grimm, S., Kwiatkowski, M., Lee, H., Li, H.C., McElroy, R., et al.: Scaling memcache at facebook. In: NSDI, pp. 385–398 (2013)
3. Morgan, T.P.: Facebook opens up tools to scale memcached. http://www.enterprisetech.com/2014/09/15/facebook-opens-tools-scale-memcached/
4. Andrew, T., Maarten, V.S.: Distributed systems. Pearson Prentice Hall (2007)
5. Chen, P.M., Lee, E.K., Gibson, G.A., Katz, R.H., Patterson, D.A.: Raid: High-performance, reliable secondary storage. CSUR **26**, 145–185 (1994)
6. Karger, D., Sherman, A., Berkheimer, A., Bogstad, B., et al.: Web caching with consistent hashing. Computer Networks **31**(11), 1203–1213 (1999)
7. Armbrust, M., Fox, A., Griffith, R., Joseph, A.D., Katz, R., et al.: A view of cloud computing. Communications of the ACM **53**(4), 50–58 (2010)
8. Thusoo, A., Sarma, J.S., Jain, N., Shao, Z., Chakka, P., Zhang, N., et al.: Hive-a petabyte scale data warehouse using hadoop. In: ICDE (2010)
9. Abadi, D.J.: Tradeoffs between parallel database systems, hadoop, and hadoopdb as platforms for petabyte-scale analysis. In: Gertz, M., Ludäscher, B. (eds.) SSDBM 2010. LNCS, vol. 6187, pp. 1–3. Springer, Heidelberg (2010)
10. Cheng, H., Huseyin, S., Yikang, X., Aaron, O., Brad, C., Parikshit, G., Jin, L., Sergey, Y.: Erasure coding in windows azure storage. In: USENIX ATC (2012)
11. Berezecki, M., Frachtenberg, E., Paleczny, M., Steele, K.: Power and performance evaluation of memcached on the tilepro64 architecture. Sustainable Computing: Informatics and Systems (2012)
12. Meaney, P.J., Lastras-Montao, L.A., Papazova, V.K., Stephens, E., Johnson, J.S., Alves, L.C., et al.: Ibm zenterprise redundant array of independent memory subsystem. IBM Journal of Research and Development (2012)
13. Atikoglu, B., Xu, Y., Frachtenberg, E., Jiang, S., Paleczny, M.: Workload analysis of a large-scale key-value store. In: ACM SIGMETRICS, pp. 53–64 (2012)
14. Plank, J.S., Greenan, K.M.: Jerasure: A library in c facilitating erasure coding for storage applications-version 2.0. Technical Report UT-EECS-14-721 (2014)
15. Chu, X., Liu, C., Ouyang, K., Yung, L.S., Liu, H., Leung, Y.-W.: Perasure: a parallel cauchy reed-solomon coding library for gpus. In: IEEE ICC (2015)
16. Ousterhout, J., Agrawal, P., Erickson, D., Kozyrakis, C., Leverich, J., Mazires, D., et al.: The case for ramclouds: scalable high-performance storage entirely in dram. In: ACM SIGOPS, pp. 92–105 (2010)
17. Lim, H., Fan, B., Andersen, D. G., Kaminsky, M.: Silt: A memory-efficient, high-performance key-value store. In: SOSP, pp. 1–13 (2011)

Performance Evaluation of NPB and SPEC CPU2006 on Various SIMD Extensions

Bo Zhao[✉], Wei Gao, Rongcai Zhao, Lin Han, Huihui Sun, and Yingying Li

State Key Laboratory of Mathematical Engineering and Advanced Computing,
Zhengzhou, China
zhaobo197359@gmail.com

Abstract. Nowadays, almost all the processors are integrated with SIMD extensions, with which significant speedup is obtained for the programs in multimedia and scientific computation. The length of SIMD vector register has been increasing all the time. For instance, the original length of SIMD extension components is 64-bit in MMX. It then rises to 128-bit in SSE and further 256-bit in AVX. The new Intel Many Integrated Core (MIC) architecture supports 512-bits SIMD. Though a higher speedup is theoretically possible as the vector length increases, more complex and efficient instructions are required to support the vectorization. We analyze the vectorization performance of NPB and SPEC CPU2006 with the increase of vector length and different SIMD instruction sets of SSE, AVX, and IMCI, based on which some advice are given for the vector length and instruction set design.

Keywords: Vectorization · Simd extension · Performance analysis

1 Introduction

With the popularity of multimedia applications, many commercial processor companies integrated multimedia extension instructions in their productions over the past twenty years. The instruction sets use a technique named SIMD (Single Instruction Multiple Data), which can be used to deal with several isomorphic data operations at a time. SIMD extension components were first used in the multimedia area and then applied in the DSP (Digital Signal Processor) and HPC (High Performance Computing). Many companies integrated multimedia extension instructions in their processors and many applications are seeking for SIMD parallelism. The structure of the SIMD extensions is simple, but the power is low and the speedup performance is obvious. There is no additional cost of communication and memory access. As a result, vectorization is still a significant measure for the program parallelism in the current multi-core age [5]. The vector length of the SIMD extensions was short at first and gets longer with the growing length of the SIMD extensions. The instruction sets are more and more abundant.

In 1996, the Intel Pentium processor was integrated with MMX instruction set and the vector length was 64 and could deal with 2 data operations of integer or floating type.

© Springer International Publishing Switzerland 2015
Y. Wang et al. (Eds.): BigCom 2015, LNCS 9196, pp. 257–272, 2015.
DOI: 10.1007/978-3-319-22047-5_21

Then 128-bit vector register were introduced in SSE, AltiVec and 3DNow!. Now the lengths of Intel AVX and Intel Initial Many Core Instructions are 256 bits and 512 bits.

Currently, designing longer length vector register seems to be a trend to improve the vectorizable performance. In theory, the longer the vector register is, the better the obtained vectorizable performance is. Here take double type computation as an example, 2 double operations can be handled simultaneously on the processor with SSE-128 instruction set, while 4 and 8 operations are handled on processors with AVX -256 and IMCI-512 instruction sets. It can get 2x, 4x and 8x faster compared with the serial version with these SIMD extension instruction sets respectively, but the fact is that the theoretical speedup can hardly got when dealing with the actual applications. For the benchmark MG in NPB, the speedup is 1.41 with SSE-128 and 1.50 with AVX-256, while the value is 2.27 when using IMCI, whose vector register length is 512. Because the innermost loops are appropriate to be vectorizied in the kernel functions, there is a rising trend when the vector length grows. However, not all benchmarks' performance in NPB can be improved with the growth of vector register length. For BT on CLASS A, the speedup got by vectorization is 1.32 on the processor with AVX instruction set, while the speedup is 0.80 with IMCI. Similarly, for LU and SP, the speedups are 1.13 and 1.24 with AVX, while the values are 0.80 and 0.98 with IMCI.

In view of the above problems, this paper tests the three different length instruction sets provided by Intel using NPB and SPEC2006 benchmarks. The three instruction sets are SSE of 128-bit vector length, AVX of 256-bit length and MIC of 512-bit length respectively. Through analyzing the test results,   some advice is given on the design of SIMD vector length, SIMD vector instruction and the future development of SIMD automatic vectorization.

The remainder of the paper is organized as follows. Section 2 elaborately introduces SSE, AVX and IMCI instruction sets as well as NPB and SPEC CPU2006 benchmarks. Section 3 presents our performance experiment results; Section 4 explains the analysis and optimizations of the experimental results. Section 5 proposes advice from three aspects, SIMD vector length, design of SIMD vector instruction and SIMD automatic vectorization. Conclusion is given in section 6.

2   Background

2.1   Instruction Set

SSE

Intel proposed Streaming SIMD Extension (SSE) on Pentium III in 1999, which is compatible with MMX. SSE can efficiently increase floating point computation capacity. In MMX, the eight registers borrowing from floating processor result in speed decrease of floating point computation. However, when proposing SSE, Intel adds eight appropriative 128-bit SSE registers in Pentium III and SSE registers can run at full speed to ensure the parallelism of floating point computation.

SSE2 appears in Pentium IV with totally 144 instructions. SSE2 includes floating point SIMD instruction, integer SIMD instruction, type conversion operation instruction between SIMD floating point and integer, conversion operation instruction between data types in MMX registers and so on. What is more important is that data formation of 128-bit SIMD integer computation and 64-bit double floating point computation are proposed.

SSE3 augments 13 new instructions based on SSE2. One of the new instructions is used in video decode and the other two are used in threads synchronization, while the remainder are used in complex math computation, data type conversion from floating point to integer and SIMD floating point computation.

SSE4 includes SSE4.1 and SSE4.2. SSE4.1 adds 47 new instructions and improves operations like insert, extract, search, discrete and strides access to ensure the specification of the vector operation. SSE4.2 adds String Text New Instructions (STTNI) and Application Targeted Accelerations (ATA).

AVX

The introduction of AVX increased the SIMD width from 128 to 256 bits (for floating-point), and the number of operands in each instruction from two (destructive source) to three (non-destructive source). 256-bit AVX instructions typically treat the lower and upper 128 bits of a SIMD register independently, and this makes loading non-contiguous values into SIMD registers more flexible. Two 128-bit values must be built and combined to form a 256-bit register. However, the 256-bit register is not fully utilized in most situations. In most cases, just the low 128 bits of the AVX vector register is used and the high 128 bits are set as zero or "LEFT unchanged", and its vectorizable ability is equivalent with that of SSE instruction set.

IMCI

A new SIMD instruction set named IMCI (Initial Many Core Instructions) is integrated in the Intel Xeon Phi processor. Each core of the processor has a vector processing unit (VPU) for fused multiply-add (FMA) operations. The vector width is 512 bits and the VPU may operate on 8 double-precision or 16 single precision data elements.

The Knights Corner instructions (KCi) introduced in the first generation Intel® Xeon PhiTM coprocessor have a SIMD width of 512 bits, and include a number of instructions incompatible with SSE or AVX. Gather and scatter instructions allow for the contents of SIMD registers to be loaded/stored from/to non-contiguous memory locations, and all SIMD instructions support conditional execution on individual elements based on the contents of a 16-bit mask register. These additions greatly help vectorize scientific HPC codes – the memory-related instructions make it easier to populate SIMD registers, while masking makes it significantly cheaper to handle control divergence (i.e. different SIMD elements needing to take different directions at a branch).

Intel Xeon Phi is a new product and some researches are done to support it. Arunmoezhi Ramachandran analyzed the multicore processing performance of NPB on Xeon Phi. Besides, many other efforts are made to parallelize various applications on Xeon Phi, including the work of Liu et al. [3] on Sparse Matrix-Vector Multiplication, Pennycook et al. [2] on parallelizing a Molecular Dynamic application. Xin Huo puts forward an API for the users to take advantage of the SIMD extensions on Xeon Phi [4].

2.2   Benchmarks

NAS Parallel Benchmarks

The NAS Parallel Benchmarks (NPB) is a small set of parallel workloads designed to evaluate performance of various hardware and software components of a parallel supercomputers. These benchmarks span different problem sizes, called classes in NPB terminology, and in this paper we use class A, B, and C. Table 1 provides the number of iterations and problem sizes for all benchmark classes used.

Most of the NAS benchmarks are computational kernels. There are five kernels and three pseudo applications in NPB. IS performs integer sorting by using a linear time Integer Sorting algorithm based on random memory access. EP evaluates the embarrassing parallelism.

Table 1. Number of iterations and Problem size for NPB3.3.1 [1]

Benchmark	CLASS=A		CLASS=B		CLASS=C	
	Number of iterations	*Problem size*	*Number of iterations*	*Problem size*	*Number of iterations*	*Problem size*
BT	200	64*64*64	200	102*102*102	400	162*162*162
CG	15	14000	75	75000	75	150000
EP	0	2^{29}	0	2^{31}	0	2^{33}
FT	6	256*256*128	20	512*256*256	20	512*512*512
LU	250	64*64*64	250	102*102*102	250	162*162*162
IS	10	2^{23}	10	2^{25}	10	2^{27}
MG	4	256*256*256	20	256*256*256	20	512*512*512
SP	400	64*64*64	400	102*102*102	400	162*162*162

CG uses a Conjugate Gradient method to compute approximations to the smallest eigenvalues of a sparse unstructured matrix. MG computes the solution of the 3-D scalar Poisson equation with a V-cycle Multi Grid method. FT operates a discrete 3-D Fast Fourier Transform (FFT) with all-to-all communication.

Three pseudo applications, BT, SP, and LU, are used to solve the discretized compressible Navier-Stokes equations. BT is a block Tri-diagonal solver and SP is a scalar Penta-diagonal solver, and both of them apply variations of the Alternating Direction Implicit (ADI) approximate factorization technique to decouple solution in the x, y, and z-coordinate directions. LU is a lower-upper Gauss-Seidel solver and applies the symmetric successive over-relaxation technique to an approximate factorization of the discretization matrix into block-lower and block-upper triangular matrices.

SPEC CPU2006

SPEC CPU2006 was released by The Standard Performance Evaluation Corporation (SPEC) on August 24, 2006. SPEC is a non-profit group, with members from hardware and software manufacturers as well as academic and research organizations. The CPU benchmarks of SPEC have been a worldwide standard for measuring compute-intensive performance since their introduction in 1989.

SPEC CPU performance deliberately depends not only on CPU performance. It also concerns compilers, as applications are provided as source code and their performance may depend on optimizations of binary code generated by a given compiler.

SPEC CPU2006 contains two components that focus on two different types of compute intensive performance. The first suite (CINT2006) measures compute-intensive integer performance, and the second suite (CFP2006) measures compute-intensive floating point performance. CINT2006 contains 12 benchmarks based on real applications written in C and C++, while CFP2006 contains 17 benchmarks written in C, C++, C/Fortran and other various Fortran versions.

3   Evaluation

3.1   Experimental Setup

Using three SIMD widths (128-, 256- and 512-bit), we describe, evaluate, and compare the performance of several SIMD implementations of these two benchmarks. The system configuration for the server used in our experiments is given in Table 2. We use a Knights Corner (KNC) Intel® Xeon PhiTM coprocessor and an E5-2407 Intel® Xeon processor. For area and power efficiency, KNC cores are less aggressive (i.e. have lower single-threaded instruction throughput) than Intel® Xeon processor cores and run at a lower frequency.

To collect performance counter information, we used Intel Parallel Studio XE 2014 compilers. The E5-2407 and KNC binaries were compiled using the Intel® compiler with the following flags: -O3 and –Ofast. In order to generate executable files with SSE4.2 and AVX instructions to expand the experiment, we use the following flags: -xSSE4.2 or –xAVX.

3.2   Evaluation on NPB Benchmarks

The NAS Parallel Benchmarks version used was 3.3.1. We run the benchmarks with class A, B and C. Table 1 gives a brief description of Number of iterations and Problem size for NPB3.3.1. Since Benchmark class C is not defined for DC, in order to evaluate the benchmarks uniformly, we ignored the benchmark DC in the experiment.

To study the impact of vector length and SIMD instruction sets, we ran all benchmarks except DC using Xeon Phi co-processor and Xeon processor. To show the results much more directly, we protracted the histograms of the experimental results with CLASS A, B and C.

For all three problem sizes and for all the NPBs, the difference is the absolute execution time for each benchmark, but the speedup trend is accordant. Here we just take CLASS C as an example to analyze the impact of vector length and SIMD instruction sets when we use different processors

Figure 3 shows the speedups of all the NPBs with different vector length and SIMD instructions. Since the Intel AVX instruction set is compatible with all the SSE instructions and has the strategy to identify whether to use SSE-128 vector register or

AVX-256 vector register according to the loop iterations and data structure, the speedup of AVX-256 for the same benchmark generally is not lower than that of SSE-128. This can be obviously seen from figure 1, 2 and 3, ignoring the measuring error.

As shown in figure 3, for MG, the vectorizable version is faster than the non-vectorizable version and speedup increases with the length of vector growing. The reason is that the memory operation is contiguous in the innermost loops of the kernel functions which can be well vectorized. For BT, LU and SP, since the innermost loops are iterated only 5 times, gather and scatter instructions are used to deal with the non-contiguous memory operation as the IMC vector length is 512 bits. There is no performance improvement for EP and IS because their kernel functions have no appropriate loops to be vectorized.

Table 2. System configuration

	KNC	Xeon(R) CPU E5-2407
Vector width	512 bits	256 bits
Clock frequency	1.1 GHz	2.20GHz
L1 cache (per core)	32 KB/32KB I/D	32 KB/32KB I/D
L2 cache (per core)	512 KB	256 KB
L3 cache (per core)	N/A	10 MB
Compiler version	Intel compiler version 14.0.1	

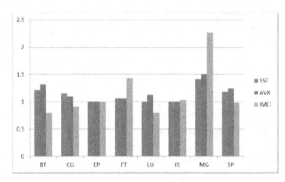

Fig. 1. Speedup of vectorization for NPBs CLASS=A with different SIMD extensions

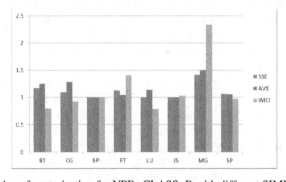

Fig. 2. Speedup of vectorization for NPBs CLASS=B with different SIMD extensions

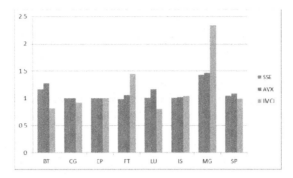

Fig. 3. Speedup of vectorization for NPBs CLASS=C with different SIMD extensions

3.3   Evaluation on SPEC CPU2006 Benchmarks

Evaluation on CFP2006

There are 17 benchmarks in CFP 2006 and 8 of them have no performance improvement after being vectorized. Indirect array access makes it difficult to vectorize the kernel loops of the kernel functions in 416.gamess. Other benchmarks that have no performance improvement are 433.milc, 444.namd, 447.dealII, 450.soplex, 453.povray, 459.GemsFDTD and 465.tonto.

mat_times_vec is a kernel function of 410.bwaves and it occupies 60% of the total runtime. It has a five-dimensional loop nest and the inner-most and inner loop iterations are both 5. SSE and AVX are appropriate to handle this case and have efficient 1.12 and 1.20 speedups respectively. The vector register length of IMCI is 512 bits and gather-scatter operation is used when dealing with the loop nests with 5 times iterations. Gather/Scatter memory access is slower than vector load/store access, so the speedup of IMCI version is low.

e_c3d is a kernel function of 410.bwaves and it takes 70% of the total runtime. The inner-most loop iterations is 3. SSE is appropriate to handle this case, and AVX only uses the lower 128 bit of vector register since AVX is compatible with SSE, so speedups on SSE and AVX are nearly the same. But IMCI generates gather/scatter instruction since the short trip count. Because of the vast instruction spending, the speedup of vectorization is 0.77.

mgau_eval and vector_gautbl_eval_logs3 are kernel function of 482.sphinx3, which take 70% of the total runtime. It contains reduction and control flow. The speedup is 1.38 on SSE and 1.66 on AVX. IMCI cannot deal with control dependence and reduction well, so it achieves a comparatively low speedup of 0.93.

Some programs are very suitable for longer vector register, such as 434.zeusmp, 435.gromacs, 436.cactusADM, 437.leslie3d, 470.lbm and 481.wrf. They all achieve very high speedups in a relatively long vector register. The speedup of 434.zeusmp is 1.34, 1.52 and 2.38 on the platform of SSE, AVX and IMCI respectively. The speedup of 435.gromacs is 1.84, 1.86 and 2.20 on the platform of SSE, AVX and IMCI respectively. The speedup of 436.cactusADM is 1.57, 1.92 and 2.36 on the platform of SSE, AVX and IMCI respectively. The speedup of 437.leslie3d is 1.52, 1.52 and 1.58 on the platform of SSE, AVX and IMCI respectively. The speedup of 481.wrf is 2.54, 2.65 and 2.81 on the platform of SSE, AVX and IMCI respectively.

Evaluation on CINT2006

There are 12 benchmarks in CINT 2000 and only 456.hmme and 462.libquantum have remarkable vectorizable performance improvement. Detailed analysis is shown as following.

P7Viterbi is kernel function of 456.hmmer and takes 92.52% of the total runtime together. The main operation is max, and SSE and AVX have the hardware instruction of max. While IMCI does not have hardware support for max operation, the speedup is lower than that of AVX.

Kernel functions of 462.libquantum are toffoli, sigma_x and cnot. They all contain reduction and xor operations. The speedup of SSE is 1.96 and AVX is 2.18, while the speedup of IMCI is 1.04. Though the data struct of the program is array of struct, the compiler will regroup array of struct to struct of array on platform of SSE and AVX with the option −ipa turned on to implement consecutive memory access. This optimization is not realized on IMCI and the speedup is low as a result.

400.perlbench is a cut-down version of Perl v5.8.7. It is part of a compiler and not a computation-intensive program, so the vectorizable performance is low. Similarly, 403.gcc is the same case.

Both 401.bzip2 and 464.h264ref contain lots of pointer operations, function calls as well as do-while loops which prevent exploiting SIMD parallelism。

429.mcf is derived from MCF which is a program used for single-depot vehicle scheduling in public mass transportation. The kernel functions contain a great deal of pointer operations and it is difficult to analyze the dependence during compiling time, so the kernel functions are not vectorized and there is no performance improvement.

The kernel function reading_cache_clear of 445.gobmk takes 83.26% of the total runtime and the nuclear statement is a function call. The vectorizable version does not have performance improvement.

458.sjeng is a program which plays chess and several chess variants based on Sjeng 11.2 (freeware). Due to the characteristics of chess game, there are many control flow and search operations, most of which are not vectorization candidates. In generally, the program is not suitable to be vectorized and has little speedup when running in vectorizable version.

471.omnetpp, using an Ethernet model which is publicly available, is the simulation of a large Ethernet network based on the OMNeT++ discrete event simulation system. There are many struct pointer operations in its kernel functions and the current Intel compiler version is unable to deal with such cases.

473.astar is derived from a portable 2D path-finding library used in game's AI. Two main kernel functions way2obj::fill and wayobj::fill take 37.46 and 33.23% of the total runtime. However, there is only one loop in the two bool functions and the single-layer loop consists of just one function call statement. For the loop, vectorization is applicable but seems inefficient.

483.xalancbmk is a modified version of Xalan-C++ which is an XSLT processor for transforming XML documents into HTML, text, or other XML document types. There is little parallelism through the program and the performance of vectorizable version is no better than the non-vectorizable version.

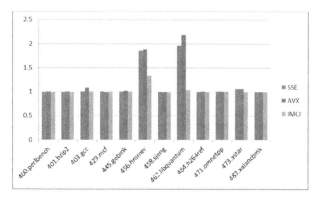

Fig. 4. Speedup of vectorization for SPEC CINT2006 size=ref

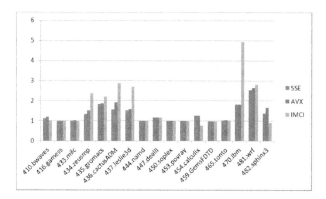

Fig. 5. Speedup of vectorization for SPEC CFP2006 size=ref

4   Analysis

In this section we analyze performance characteristics we observed and discuss approaches that can be taken to optimize the vectorizable code generation.

4.1   Suitable for Longer Vector Register

When the iterations of the innermost loop are larger than the vector factor and the memory access is contiguous, the vectorizable performance is commonly better than the non-vectorizable version. A typical benchmark is MG in NPS. Resid is a kernel function of MG and takes about 50% of the total runtime. The following is its code segment of the kernel loop.

```
do i3=2,n3-1
   do i2=2,n2-1
      do i1=1,n1
         u1(i1) = u(i1,i2-1,i3) + u(i1,i2+1,i3)
```

```
>                          + u(i1,i2,i3-1) + u(i1,i2,i3+1)
              u2(i1) = u(i1,i2-1,i3-1) + u(i1,i2+1,i3-1)
>                          + u(i1,i2-1,i3+1) + u(i1,i2+1,i3+1)
        enddo
      enddo
    enddo
```

The iterations is larger than the vector factor for the three instruction sets and the memory access is contiguous, and only additional operation is used in the innermost loop, so the speedup is high and larger with the growth of vector length. Similar benchmarks include FT of NPB and 462.libquantum, 436.cactusADM, 470.lbm, 481.wrf, 482.sphinx3 of SPEC CPU2006.

4.2 Indirect Memory Access

Indirect memory access makes the address reference non-contiguous and the dependence analysis difficult, so the vectorization exploition is difficult when indirect memory access exists. For IS in NPB, kernel function rank takes 61.76% of the total runtime and the code segment of its kernel loop is shown as follows.

```
for( i=0; i<NUM_KEYS; i++ ) {
    key = key_array[i];
    key_buff2[bucket_ptrs[key >> shift]++] = key;
}
```

A mass of indirect memory access make the kernel loops in the kernel functions fail to be vectorized and thus the performance of vectorizable version is no better than that of non-vectorizable version. CG in NPB and 416.gamess, 435.gromacs, 444.namd in SPEC CPU2006 have the similar problem in vectorization exploit. Since there is some intra-iteration parallelism in the kernel loops of 435.gromacs, a little performance improvement is got for the vectorizable version.

4.3 Short Trip Counts

The speedup for some benchmarks do not increase with the growth of vector register length since the iteration times of innermost loop are less than the vector factor. For example, the iterations of the innermost loops are in BT, LU and SP. While the vector factors for double type in SSE, AVX and IMCI are 2, 4 and 8 respectively, the speed up increases from SSE to AVX but decreases from AVX to IMCI. In addition, for 410.bwaves, the kernel function mat_times_vec takes 60% of the total runtime and its iteration times of the innermost loops are also 5. The experimental results show that the influence of the short trip counts obviously.

4.4   Reduction

Reduction operation is very common in the benchmarks but now there is no hardware support for this operation under the three instruction sets. The current solution to reduction is using shift operation at the software level. If hardware instruction is supported, the performance may get greater improvement. For example, mgau_eval and vector_gautbl_eval_logs3 are the top 2 kernel functions of 482.sphinx and take 70% of the total runtime. The code segment of the kernel loop is as follows.

```
for (i = 0; i < veclen; i++) {
        diff = x[i] - m[i];
        dval -= diff * diff * v[i];
}
```

4.5   Impact of Specific Operations

Unlike the Xeon processor, the Xeon Phi does not implement a hardware divide instruction in software. In the Gaussian elimination process in BT, a call to function binvcrhs takes 25% of the total runtime. This function computes 5 pivot elements for a 5x5 block matrix, which involves a division operation. This hurts the performance of the benchmark. 437.leslie3d and 434.zeusmp are the same cases. This is also true for the square root operation used in SP and 435.gromacs.

4.6   Non-unit Stride Access

Non-contiguous memory access has always been an important problem for SIMD vectorization. Well resolving non-adjacent memory access is crucial for improving SIMD performance. One method for resolving non-adjacent memory access is making use of data regrouping in the compile optimization to convert non-adjacent memory access to adjacent memory access. For example, ICC can convert array of struct to struct of array in 462.libquantum. The second method is to provide hardware support. For instance, ICC provides odd number bit operation and even number bit operation on SSE and AVX and also supports gather/scatter operation on IMCI. Current object of gather/scatter operation is cache-line, so the programs requiring regrouping obtain poor performance on IMCI. The third method is to utilize shuffle instruction of vector register. Load twice or more continuously and then regroup data to another vector register using shuffle instruction. The last method is to utilize mask registers by way of incomplete use of vector register slot to realize SIMD vectorization.

4.7   Control Flow

Control dependence is one of the crucial aspects of restriction on data parallelism exploited by SIMD extensions. Select instruction provides possibilities for control dependence vectorization. Current method converts control dependence to data dependence by way of if conversion and then makes use of select instruction. But select instruction is no longer required in presence of mask register. We can make the control conditions of the statements into predications directly.

5   Advice

5.1   SIMD Vector Length Design

The Longer vector length is a trend of SIMD development. Longer vector length provides possibilities for programs to achieve larger SIMD parallelism speedup. However, with the increase of vector length, the following three problems must be solved. The first problem is the ability of regrouping between vector registers, second is the mask memory operation and computation ability and third is the ability of irregular access.

The speedup of the program increases with the augment of vector length for the program with memory access continuous and iteration number larger than vector length. As for the program whose iteration number is smaller than the vector length, its speedup does not increase with the augment of vector length. For example, BT, LU and SP iteration number is five. Speedup increase from SSE to AVX, that is because vector length of AVX is 256-bit and it can process four statements at one time. But when it comes to IMIC which can process eight statements at one time, iteration number is smaller than vector length so the performance cannot keep on increasing. If mask write/read memory and mask computation are supported, the programs SIMD parallelism still can be exploited even the vector length is larger than iteration number of the programs.

At present, some programs cannot be vectorized owing to indirect memory access. Gather/scatter instruction provided by IMCI can vectorize programs with indirect memory access, such as CG, 416.gamess, 435.gromacs and 444.namd. Although CG and 416.gamess can be vecotized on KNC, but no performances is achieved. Maybe benefit analysis is needed when processing indirect memory access.

The data loaded by vector register increases with vector length. If fixed mode regrouping is supported rather than flexible mode regrouping, the number needed to regrouping grows exponentially. Hence for long vector register, supporting for flexible mode regrouping is important for improving program performance and decreasing exploiting difficulty.

When it comes to vector length design of SIMD extensions, it does not have to be the pursuit of pure length because the hardware design cost increase with the vector length. How to make full use of the certain length of vector register is important for promoting program performance and processor cost performance. For example, Intel AVX uses 256-bit length vector, but most of the time, it uses compatible 128-bit xmm vector registers of SSE rather than the 256-bit ymm vector registers. Currently, IMCI and AVX-512 presented by Intel both support 512-bit length vector and even 1024-bit length vector is proposed now. These provide theoretically possibilities for promoting vectorization performance on hardware, but how to make full use of the certain length of vector register is a concern.

5.2   SIMD Instruction Set Design

The alignment of memory access is an important factor influencing the program performance. The time of aligned memory access is greatly shorter than that of non-aligned memory access. Some advanced compiling techniques such as loop peeling can transform non-aligned reference into aligned ones. Many processors' SIMD extensions are integrated with shuffle or perm instruction to reorganize the two data obtained by twice aligned load operations. However, hardware support for non-aligned memory access instructions can not only shorten the time difference with aligned operations but also offer much convenience for vectorization exploitation. Nowadays the research of static aligned analysis is quite embedded, while dynamic and Inter-procedural analysis is still a challenge to be solved urgently. An available method is to improve the SIMD vectorizable performance by dealing with aligned memory access appropriately.

Reduction instruction ought to be supported at the hardware level. Currently there is no hardware instruction support for reduction operation and the reduction operation is realized through the operation using the entire vector register shift operand. The shift times of the entire vector register rise with the growth of vector register length. Reduction operation is very common in the Scientific Computing programs and the hardware support for reduction may contribute greatly to the performance improvement for such computing tasks.

Most SIMD extensions support all the data types, that is, int, long, float and double. For integer type, addition and subtraction as well as logical operations are supported. For floating type, addition, subtraction, multiplication and division operations are supported, so are the comparison, option and other special vector operations.

In the Gaussian elimination process in BT, a call to function binvcrhs takes 25% of the total runtime. This function computes 5 pivot elements for a 5x5 block matrix, which involves a division operation. This hurts the Xeon Phi, which unlike the Xeon processor, does not have a hardware divide instruction and implements it in software. This is also true for the square root operation used in SP.

SIMD extensions are static homogeneous. Dynamic configuration makes the hardware resources available to be sufficiently used across the slots. Isomerism makes the different slot have particular function. Dynamic isomerism components maybe like traditional SIMD extensions or a VLIW component as well. The Libra accelerator increases SIMD utility by blurring the divide between vector and instruction parallelism to support efficient execution of a wider range of loops, and it increases hardware utilization through the use of heterogeneous hardware across the SIMD lanes [6].

Since AVX is absolutely compatible with SSE, the low 128 bits vector register and SSE vector register are functionally equivalent. When dealing with shorter times iterations or data regrouping mode, the vectorizable exploitation capacity of SSE is higher than that of AVX. The current Intel compiler has the mechanism to choose SSE or AVX vector register according to the program characteristic. IMCI is not compatible with AVX or SSE. As a result, if the program is not appropriate to be vectorized with IMCI, the efficiency of vectorizable version is commonly no better than the regular version since SSE and AVX are not available to be used during the compiling.

5.3 Automatic Vectorization Compiling

The technology for realizing SIMD vectorization is classified as manual vectorization and automatic vectorization at present. Manual vectorization includes SIMD Directives supported by OpenMP 4.0, Cilk Plus, Array Notations and MKL and so on. Automatic SIMD vectorization is transparent to the programmers and thus becomes the important means of using SIMD vectorization. In the past, a lot of work is done to realize automatic vectorization in the compilers.

Loop-based vectorization is very mature. Commercial compilers like ICC as well as open-source compilers like GCC make use of innermost loop vectorization. Some study aims at outer loop [7] or nested loop vectorization [8, 9]. The vectorization within basic blocks, also known as straight-line vectorization, calls for enough parallelism in the basic blocks. Amarasinghe proposed superword Level Parallelism (SLP) algorithm which packs isomorphic instructions within a basic block starting from adjacent memory references [10]. Jun Liu [11] and Rajkishore Barik [12] give some improvement for SLP. Vectorization methods within basic block exploit   intra-iterations parallelism, while loop-aware vectorization methods based on loop-unrolling exploit both intra-iterations and inter-iterations parallelism[13]. Present ICC automatic vectorization module is loop-based without vectorzation methods within basic blocks. ICC supports a kind of rerolling technology, which can rewrite the similar multiply unrolling statements into loop. But the capacity of rerolling is far weaker than that of SLP. So at present, ICC fails to vectorize inl1311 of 435 and binvcrhs of BT which needs SLP vectorization.

The non-aligned and non-contiguous data reference makes it difficult to exploit vectorization within SIMD extensions. There are three applicable methods to handle the problem of non-aligned memory access. The first is shift or data regrouping after multiple aligned accesses [16]. The second are some compiling optimization methods such as loop peeling and array padding. The last is hardware instruction support for non-aligned memory access. As for non-contiguous memory access, two methods seem available. One is data permutation [17] after loading the data into the memory. The other is hardware instruction support for non-contiguous memory load and store operations [18]. In addition, control dependence is also a significant searching direction to exploit the data parallelism of SIMD [19].

Many regular problems such as loop transformation, control dependence, alignment and consistency are well solved by the current automatic vectorization technique oriented SIMD. These researches offer a firm foundation for the SIMD vectorization exploitation. However, the methods of automatic vectorization exploitation and optimization need to be developed with the development of SIMD extensions and instruction sets such as mask register and gather/scatter operation. Currently, many source codes are written with structs containing linked list, tree and graph or with function call and pointer operations. These programing characteristics make it much more difficult for the automatic vectorization exploitation. Function vectorization is proposed to vectorize the loops with function calls [15].Speculative parallelism is put forward to handle the parallelism exploitation for irregular memory access [14]. Now there are already many researches for the speculative parallelism oriented distributed memory structure and shared memory structure at the level of instruction parallelism, while little research is done for the vectorization exploitation of SIMD extensions.

6   Conclusion

We have examined the performance of the SIMD vectorizable versions of the NAS Parallel and SPEC CPU2006 Benchmarks on Xeon Phi and Xeon, and identified some common issues that may influence performance, or in other words, what is required to reach high performance.

Enough concurrency needs to be present in the levels of the loop nests to use the vector register adequately. If the concurrency is not sufficient, some optimization may be carried out to vectorize the loop nests, such as loop collapse. Gather/scatter operations are introduced when using IMCI to handle the loop nests whose iterations are less than the vector factor.

In general, higher vector length and more complex instruction set bring higher performance improvement when using the vectorization function of SIMD extensions. This is shown in the experimental results shown above. For workloads relying heavily on instructions not implemented in hardware, such as divide or square root , the performance is lower than those having hardware instruction support.

References

1. Ramachandran, A., Vienne, J., Van Der Wijngaart, R.: Performance evaluation of NAS parallel benchmarks on Intel Xeon Phi. In: 42nd International Conference on Parallel Processing (2013)
2. Pennycook, S., Hughes, C., Smelyanskiy, M., Jarvis, S.: Exploring simd for molecular dynamics, using intel xeon processors and intel xeon phi coprocessors. In: IPDPS (2013)
3. Liu, X., Smelyanskiy, M., Chow, E., Dubey, P.: Efficient sparse matrix-vector multiplication on x86-based many-core processors. In: Proceedings of the 27th International ACM Conference on Supercomputing, pp. 273–282. ACM (2013)
4. Huo, X., Ren, B., Agrawal, G.: A Programming system for Xeon Phis with runtime SIMD parallelization. In: ICS (2014)
5. Mytkowicz, T., Marron, M.: Single-Core Performance is Still Relevant in the Multi-Core Era
6. Park, Y., Park, J.J.K., Park, H.: Tailoring SIMD execution using heterogeneous hardware and dynamic configurability. In: Proceedings of the 45th Annual IEEE/ACM International Symposium on Microarchitecture (MICRO) (2012)
7. Nuzman, D., Zaks, A.: Outer-loop vectorization-revisited for short SIMD architectures. In: Proceedings of the International Conference on Parallel Architectures and Compilation Techniques (PACT) (2008)
8. Trifunovic, K., Nuzman, D., Cohen, A., et al.: Polyhedral-model guided loop-nest auto-vectorization. In: Proceedings of the International Conference on Parallel Architectures and Compilation Techniques (PACT) (2009)
9. Kong, M., Veras, R., Stock, K.: When polyhedral transformations meet SIMD code generation. In: Proceedings of the Conference on Programming Language Design and Implementation (PLDI) (2013)
10. Larsen, S., Amarasinghe, S.: Exploiting superword level parallelism with multimedia instruction sets. In: Proceedings of the Conference on Programming Language Design and Implementation (PLDI), pp. 145–156 (2000)

11. Liu, J., Zhang, Y., Kandemir, M.: A compiler framework for extracting superword level parallelism. In: Proceedings of the Conference on Programming Language Design and Implementation (PLDI) (2012)
12. Barik, R., Zhao, J., Sarkar, V.: Efficient selection of vector instructions using dynamic programming. In: Proceedings of the 43rd Annual IEEE/ACM International Symposium on Microarchitecture (MICRO) (2010)
13. Rosen, I., Nuzman, D., Zaks, A.: Loop-aware SLP in GCC. In: Proceedings of GCC Developers' Summit, pp. 131–142 (2007)
14. Kumar, R., Martínez, A.: Speculative dynamic vectorization for HW/SW codesigned processors. In: Proceedings of the International Conference on Parallel Architectures and Compilation Techniques (PACT) (2012)
15. Karrenberg, R., Hack, S.: Whole-function vectorization. In: Proceedings of the 9th Annual IEEE/ACM International Symposium on Code Generation and Optimization (CGO) (2011)
16. Eichenberger, A.E., Peng, W., O'Brien, K.: Vectorization for SIMD architectures with alignment constraints. SIGPLAN **39**(6), 82–93 (2004)
17. Kudriavtsev, A., Kogge, P.: Generation of permutations for SIMD processors. In: LCTES 2005, pp. 147–156. ACM, New York (2005)
18. Nuzman, D., Rosen, I., Zaks, A.: Auto-vectorization of interleaved data for simd. In: Proceedings of the ACM SIGPLAN Conference on Programming Language Design and Implementation, PLDI 2006, pp. 132–143. ACM, New York (2006)
19. Shin, J., Hall, M., Chame, J.: Superword-level parallelism in the presence of control flow. In: CGO (2005)

Prediction of High Resolution Spatial-Temporal Air Pollutant Map from Big Data Sources

Yingyu Li[1], Yifang Zhu[2], Wotao Yin[3], Yang Liu[4], Guangming Shi[1](✉), and Zhu Han[5]

[1] School of Electronic Engineering, Xidian University, Xi'an, China
yyli.xidian@gmail.com, gmshi@xidian.edu.cn
[2] Department of Environmental Health Sciences,
University of California, Los Angeles, USA
yifang@ucla.edu
[3] Department of Mathematics,
University of California, Los Angeles, USA
wotao.yin@gmail.com
[4] Rollins School of Public Health, Emory University, Atlanta, USA
yang.liu@emory.edu
[5] Department of Electrical and Computer Engineering,
University of Houston, Houston, USA
zhan2@uh.edu

Abstract. In order to better understand the formation of air pollution and assess its influence on human beings, the acquisition of high resolution spatial-temporal air pollutant concentration map has always been an important research topic. Existing air-quality monitoring networks require potential improvement due to their limitations on data sources. In this paper, we take advantage of heterogeneous big data sources, including both direct measurements and various indirect data, to reconstruct a high resolution spatial-temporal air pollutant concentration map. Firstly, we predict a preliminary 3D high resolution air pollutant concentration map from measurements of both ground monitor stations and mobile stations equipped with sensors, as well as various meteorology and geography covariates. Our model is based on the Stochastic Partial Differential Equations (SPDE) approach and we use the Integrated Nested Laplace Approximation (INLA) algorithm as an alternative to the Markov Chain Monte Carlo (MCMC) methods to improve the computational efficiency. Next, in order to further improve the accuracy of the predicted concentration map, we model the issue as a convex and sparse optimization problem. In particular, we minimize the Total Variant along with constraints involving satellite observed low resolution air pollutant data and the aforementioned measurements from ground monitor stations and mobile platforms. We transform this optimization problem to a Second-Order Cone Program (SOCP) and solve it via the log-barrier method. Numerical simulations on real data show significant improvements of the reconstructed air pollutant concentration map.

© Springer International Publishing Switzerland 2015
Y. Wang et al. (Eds.): BigCom 2015, LNCS 9196, pp. 273–282, 2015.
DOI: 10.1007/978-3-319-22047-5_22

Keywords: Air pollution · High resolution spatial-temporal concentration map · Big data · Heterogeneous data sources · SPDE · INLA · Sparsity · SOCP

1 Introduction

Air pollution is a major environmental risk to our health nowadays. This problem is happening all around the world. Many death-causing diseases, such as lung cancer, stroke, heart disease, and both chronic and acute respiratory diseases, including asthma, are closely related to air pollution . It will become even severe for children and the aged. In 2012, air pollution contributes to around 7,000,000 deaths worldwide [1]. Thus in order to reduce the damage caused by air pollution, it is of vital importance to investigate on its formation and transition process. But the acquisition of high resolution and accuracy air pollutant concentration map has always been a big barrier. Traditional ground monitor stations are one of the main ways to acquire air pollutant concentration information. They can provide accurate air pollutant concentration at specific locations, but the quantity of available information is quite limited, due to the high costs both in human resources and financial resources to run such monitor stations. In recent years, mobile air pollutant sensors [3][4] are becoming more and more popular. They are usually installed on mobile platforms or other public transport vehicles, causing a great reduction in costs. But the available data coverages are still restricted to their trajectories only. Satellites remote sensing techniques can provide the air pollutant concentration information all over their scanned areas, but the spatial and temporal resolutions are low. To summary, it is difficult to acquire high resolution spatial-temporal air pollutant concentration map from a single type of data source.

In order to remedy the limitations in sampling techniques, lots of works have been done to acquire high quality air pollutant concentration maps. Land-Use Regression (LUR) model [4][5][6] is a typical one, which simply assumes that air pollutant concentration is decided by meteorology, climate, geography, human activity and other relative covariates, and then uses regression models to fit the data. In 2013, Zheng et al. [7] proposed a semi-supervised learning approach based on a co-training framework that consists of a spatial classifier and a temporal classifier to predict the air pollution level in a fine defined grid. This method highly improved the prediction accuracy compared to LUR models, but it can only gives the index of air pollution instead of the exact concentration. Cameletti et al. [8][9] further improved the traditional LUR model by adding a spatial-temporal Gaussian field to it, and representing it as a Gaussian Markov Random Field (GMRF) through the Stochastic Partial Differential Equations (SPDE) approach. In order to make the computation more efficient, they also adopted the Integrated Nested Laplace Approximation (INLA) algorithms as an alternative to Markov Chain Monte Carlo (MCMC) sampling. However, the data source used in the above methods are all onefold, which limit their ability to improve the resolution and accuracy of the predicted spatial-temporal air pollutant concentration map.

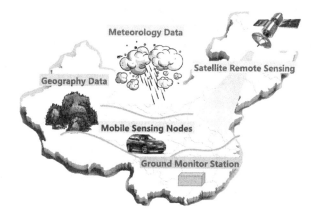

Fig. 1. Illustration of the big data sources used

With the arrival of an era of big data, new opportunities and challenges have been brought to this problem. As we mentioned above, we can have direct measurements from ground monitor stations, mobile platforms with sensing nodes and satellites, as well as many indirect measurements, including various meteorology and geography covariates. These data sources are "big" in both **Variety**, **Volume** and **Velocity**, and construct a typical example of the 3V model in big data techniques [2]. These properties of our data sources can provide us more information about the air pollutant concentration, but also presents new requirements on efficient algorithms and computation methods to handle them.

In this paper, we propose an air pollutant concentration prediction approach that takes advantage of big data sources, including ground monitor stations observed data, mobile platforms with mobile sensing nodes collected data and satellite observed data as shown in Fig.1. Firstly, we use the ground monitor stations observed data as fixed samples and mobile platforms collected data as random samples to predict a preliminary 3D high resolution air pollutant concentration map. The prediction result contains a large scale regression component, where various meteorology and geography covariates are used, and a small scale spatial-temporal process. It is fulfilled via the SPDE approach and solved by the INLA algorithms. In order to further improve the accuracy of the predicted concentration map, we then take satellite observed data into consideration and model the issue as a convex and sparse optimization problem, in which the objective function is built upon Total Variation. We recast the problem as a Second-Order Cone Program (SOCP) [18] and solve it by the log-barrier method. Simulation results show that our proposed methodologies can recover the air pollutant concentration map with much higher accuracy compared to traditional approaches.

The rest of this paper is organized as follows. The system model of the prediction of 3D high resolution spatial-temporal air pollutant concentration map is given in Section 2. Section 3 details the proposed algorithms in this paper. Section 4 illustrates the simulation results on real air pollutant data. Finally the conclusions and future works are given in Section 5.

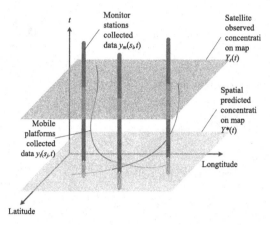

Fig. 2. Illustration of the system model. $y_{\mathrm{mon}}(s_i, t), i = 1, 2, ..., m$ are the observations from m monitor stations; $y_{\mathrm{taxi}}(s_j, t), j = 1, 2, ..., n$ are the mobile platforms collected measurements, and the blue curves show their spatial-temporal trajectories; $Y^{*(3)}$ is the preliminarily predicted 3D air pollutant map; $Y_s^{(3)}$ is the satellite observed low-resolution air pollutant concentration map

2   System Model

Consider a system model illustrated in Fig. 2. Since the distribution of ambient air pollutants closely follows a log-normal distribution [16], we use $y(s_l, t)$ to denote the logarithm of the air pollutant concentration measured at station s_l and day t, where $l = 1, ..., d, t = 1, ..., T$. Among these we have m fixed observations from ground monitor stations and n random observations from mobile platforms(i.e. taxis) with mobile sensing nodes, where $m + n = d$. In other words,

$$\{y(s_l, t) | l = 1, ..., d\} = \{y_{\mathrm{mon}}(s_i, t), y_{\mathrm{taxi}}(s_j, t) | i = 1, ..., m; j = 1, ..., n\}. \quad (1)$$

Then, we model the measurement equation as [8]:

$$y(s_l, t) = \mathbf{x}(s_l, t)\boldsymbol{\beta} + \xi(s_l, t) + \varepsilon(s_l, t), \quad (2)$$

where $\mathbf{x}(s_l, t) = (x_1(s_l, t), ..., x_p(s_l, t))$ denotes the vector consists of p covariates for station s_l at time t, and $\boldsymbol{\beta} = (\beta_1, ..., \beta_p)'$ is the corresponding coefficient vector. Thus, $\mathbf{x}(s_l, t)\boldsymbol{\beta}$ is regression part in the measurement equation (2), which is also called the large scale component [10][11] that models the dependence of air pollutant concentration on various meteorology and geography covariates.

The residual from the regression is partitioned into a spatial-temporal process, $\xi(s_l, t)$, and a measurement error $\varepsilon(s_l, t) \stackrel{i.i.d.}{\sim} N(0, \sigma_\varepsilon^2)$ which is defined by a both temporally and spatially uncorrelated Gaussian white-noise process. $\xi(s_l, t)$ is a Gaussian field that consists of a first order auto-regressive dynamics $a\xi(s_l, t - 1)$ that models the temporal correlation, and a temporally independent zero-mean Gaussian distribution $\omega(s_l, t)$ that models the spatial correlation.

Thus, $\xi(s_l, t)$ can be formulated as:

$$\xi(s_l, t) = a\xi(s_l, t-1) + \omega(s_l, t), t = 2, ..., T, \tag{3}$$

where $|a| < 1$ is the coefficient of the first order auto-regressive dynamics with $\xi(s_l, 1) \sim N(0, \sigma_\omega^2/(1-a^2))$. Here we set a to a constant under the consideration of the lack of data and computational efficiency.

The spatial correlation item $\omega(s_l, t)$ is characterized by a covariance function:

$$\text{Cov}(\omega(s_l, t), \omega(s_k, t')) = \begin{cases} 0, & \text{if } t \neq t', \\ \sigma_\omega^2 \mathcal{C}(h), & \text{if } t = t'. \end{cases} \quad \text{for } l \neq k. \tag{4}$$

where $h = \| s_l - s_k \| \in \Re$ is the Euclidean spatial distance between station s_l and s_k, and $\mathcal{C}(h)$ is a function of h defined by a Matérn covariance function [8][17]:

$$\mathcal{C}(h) = \frac{1}{\Gamma(\nu)2^{\nu-1}}(\kappa h)^\nu K_\nu(\kappa h), \tag{5}$$

where K_ν is the modified Bessel function of the second kind with a fixed order $\nu > 0$, which is also the smoothness parameter. And $\kappa > 0$ is the scaling parameter to be estimated [12][14].

Although the above spatial-temporal model has been widely used in the prediction of air pollutant concentrations [9], it has inherent limitations due to the amount direct measurements used. Thus, we propose a brand new model that can combine heterogeneous data sources and provide more reliable prediction results.

3   Proposed Algorithms

In this section, we will first predict a preliminary air pollutant concentration map from the above system model, then model it as well as the low resolution satellite observed concentration data cube in a convex and sparse optimization problem to further improve the accuracy of predicted concentration map.

3.1   Spatial Prediction

Let $\mathbf{y_t} = (y(s_1, t), ..., y(s_d, t))'$ denote all observations measured at time t, then we can rewrite the measurement equations (2)(3) as:

$$\mathbf{y_t} = \mathbf{x_t}\beta + \boldsymbol{\xi_t} + \boldsymbol{\varepsilon_t}, \quad \boldsymbol{\varepsilon_t} \sim N(\mathbf{0}, \sigma_\varepsilon^2 I_d),$$

$$\boldsymbol{\xi_t} = a\boldsymbol{\xi_{t-1}} + \boldsymbol{\omega_t}, \quad \boldsymbol{\omega_t} \sim N(\mathbf{0}, \Sigma), \tag{6}$$

where $\mathbf{x_t} = (\mathbf{x}(s_1, t)', ..., \mathbf{x}(s_d, t)')'$ is the covariance matrix at time t, $\boldsymbol{\xi_t} = (\xi(s_1, t), ..., \xi(s_d, t))'$, and I_d is the d-dimension identity matrix, Σ is the $d \times d$ dense correlation matrix with elements $\Sigma(l, k) = \mathcal{C}(\| s_l - s_k \|), l, k = 1, ..., d$.

In a Bayesian framework, we will need to estimate the parameter vector $\boldsymbol{\theta} = (\beta, \sigma_\varepsilon^2, a, \sigma_\omega^2, \kappa)$ given $\mathbf{y} = \{\mathbf{y}_t | t = 1, ..., T\}$, and then make spatial predictions based on $\boldsymbol{\theta}$. According to the Bayes's Rule, the posterior distribution of $\boldsymbol{\theta}$ given $\mathbf{y}$ is:

$$\pi(\boldsymbol{\theta}|\mathbf{y}) \propto \pi(\mathbf{y}|\boldsymbol{\theta})\pi(\boldsymbol{\theta}), \tag{7}$$

where $\pi(\cdot)$ denotes the probability density function.

The common approach for this estimation and prediction is MCMC sampling [9]. However, the computation cost of MCMC is quite heavy, making it almost impossible when it comes to large scale big data sources. In this paper, we will first represent the Gaussian field $\boldsymbol{\xi}_t$ as a GMRF [14], which can be characterized by a sparse precision matrix via the SPDE approach [12]. This allows us to further adopt the INLA algorithm [13] as an alternative of MCMC, which has significantly computational efficiency for the parameter estimation and spatial prediction.

Let $Y^{*(3)}(t)$ denote the spatially predicted air pollutant concentration maps for all $t \in \{1, 2, ..., T\}$, we can build a 3D air pollutant concentration map $Y^{*(3)} = (Y^{*(3)}(1); ...; Y^{*(3)}(T))$. Note that the prediction results are usually based on limited number of samples, so the accuracy of $Y^{*(3)}$ can not be guaranteed. Thus, we will further take advantage from our third data source, the satellite observed low resolution air pollutant concentration map $Y_s^{(3)}$.

3.2   Restoration via Sparse Optimization

Assume that the spatial resolution of each temporal slice $Y^{(3)}(t)$ in the 3D air pollutant concentration map $Y^{(3)}$ is $M \times N$, then we have $Y^{(3)} \in \Re^{M \times N \times T}$. For the ease of notation and computation, we represent the 3D air pollutant data tensor as a matrix $Y \in \Re^{D \times T}$, where $D = M \times N$. Let $\mathbf{e}_t := (0, ..., 1, ..., 0)$ be the column selector with the t-th element being 1 for $t = 1, 2, ..., T$. Thus we can denote the column vectors as $\mathbf{y}_t = Y\mathbf{e}_t$.

We find that for each temporal slice $Y_t^{(3)}$, it is sparse in the sense of small Total Variation [15][18]. Thus, we can formulate the problem of restoring each column y_t in the actual air pollutant concentration matrix Y as:

$$\min_{\mathbf{y}_t} \ TV(\mathbf{y}_t), \quad \text{s.t.} \quad \| L(\mathbf{y}_t^{(2)}) - Y_s\mathbf{e}_t \|_2 \le \epsilon_1,$$
$$\| K\mathbf{e}_t \cdot \mathbf{y}_t - Y_r\mathbf{e}_t \|_2 \le \epsilon_2, \tag{8}$$
$$\| \mathbf{y}_t - Y^*\mathbf{e}_t \|_2 \le \epsilon_3.$$

The physical meaning of each term in the above objective function is listed as follows:

- Total Variation: $TV(\mathbf{y}_t)$ is the Total Variation regularization of $\mathbf{y}_t$, where $TV(\cdot)$ denotes the Total Variation semi-norm;
- Satellite data: $\| L(\mathbf{y}_t^{(2)}) - Y_s\mathbf{e}_t \|_2 \le \epsilon_1$ is the constraint to keep the greatest similarity possible between $\mathbf{y}_t$ and satellite observed data $Y_s\mathbf{e}_t$, where ϵ_1 is

the error bound. In order to lower resolve each temporal slice $Y^{(3)}(t)$, we use $\mathbf{y}_t^{(2)}$ to denote the matricization of $\mathbf{y}_t$, i.e.,

$$\mathbf{y}_t^{(2)} = Y^{(3)}(t). \tag{9}$$

The operator $L(\cdot)$ is to lower-resolve $\mathbf{y}_t^{(2)}$ to the same size with $Y_s\mathbf{e}_t$, which is the t-th column in the 2D representation of the satellite data cube $Y_s^{(3)}$.

- Monitor stations and mobile platforms data: $\| K\mathbf{e}_t \cdot \mathbf{y}_t - Y_r\mathbf{e}_t \|_2 \leq \epsilon_2$ is the constraint to grantee the actual observed data can be preserved in the restoration result Y, where ϵ_2 is the error bound, and K is the mask matrix has the same size to Y:

$$K(q,t) = \begin{cases} 1, & \text{if there is a measurent,} \\ 0, & \text{otherwise.} \end{cases} \tag{10}$$

Y_r is the matrix marking the real observed data, i.e.,

$$Y_r(q,t) = \begin{cases} y(s_l,t), & \text{if there is a measurent,} \\ 0, & \text{otherwise,} \end{cases} \tag{11}$$

where is $y(s_l,t)$ is the ground monitor station or mobile platform collected data at the corresponding location, and $q \in \{1,2,...,D\}, t \in \{1,2,...,T\}$.

- Preliminary predicted data: $\| \mathbf{y}_t - Y^*\mathbf{e}_t \|_2 \leq \epsilon_3$ is the constraint to keep the similarity between $\mathbf{y}_t$ and the t-th column of preliminarily predicted data $Y^*\mathbf{e}_t$, and ϵ_3 is the error bound.

For the ease of computation, we first rewrite the above optimization equation (8) as follows:

$$\min_{\mathbf{y}_t} \quad TV(\mathbf{y}_t), \quad \text{s.t.} \quad \| A(\mathbf{y}_t) - b \|_2 \leq \epsilon, \tag{12}$$

where

$$A(\mathbf{y}_t) = \begin{pmatrix} L(\mathbf{y}_t^{(2)}) \\ K\mathbf{e}_t \cdot \mathbf{y}_t \\ \mathbf{y}_t \end{pmatrix}, b = \begin{pmatrix} Y_s\mathbf{e}_t \\ Y_r\mathbf{e}_t \\ Y^*\mathbf{e}_t \end{pmatrix}, \tag{13}$$

and ϵ is the total error bound. We can further recast equation (12) as a standard SOCP [18] problem via introducing an intermediate variable z:

$$\min_z \quad \sum_{u,v=1}^{D} z, \quad \text{s.t.} \quad \| V_{uv}\mathbf{y}_t \|_2 \leq z, \quad u,v = 1,...,D, \tag{14}$$

$$\| A(\mathbf{y}_t) - b \|_2 \leq \epsilon,$$

where $V \in \Re^{D \times D}$ is the matrix form of finite difference, i.e. $V\mathbf{y}_t = TV(\mathbf{y}_t)$, and D is the length of $\mathbf{y}_t$. Then we can easily solve (14) via the log-barrier method [19].

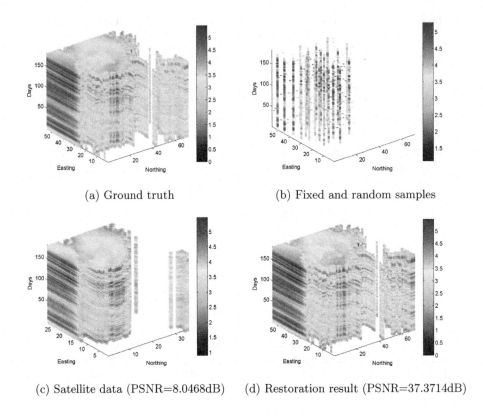

(a) Ground truth

(b) Fixed and random samples

(c) Satellite data (PSNR=8.0468dB)   (d) Restoration result (PSNR=37.3714dB)

Fig. 3. Simulation results

4   Simulation Results

We use the data set provided by [8] for our simulations, which is the daily PM10 concentration for the Piemonte area in Italy. The data set contains a total $T = 182$ days during the winter season from October 2005 to March 2006. The meteorology covariates taken in to consideration are: daily mean temperature (TEMP, K), daily maximum mixing height (HMIX, m), daily mean wind speed (WS, m/s), daily precipitation (P, mm) and daily emissions (EMI, g/s); the geography covariates taken in to cosidertaion are altitude (A, m) and spatial geographic coordinates (UTMX and UTMY, km).

In order to give a complete assessment of the performance of our proposed algorithm, we first perform a spatial prediction on the data set in [8] via the SPDE-INLA approach to get a 3D concentration map, which we treat as the ground truth as shown in Fig. 3a. We disregard the area with elevations above $1km$ in this concentration map, because there is a mountain in that region and has no monitor stations, thus the prediction result is not reliable. Then we sample from the ground truth map for 24 fixed measurements and 2 random measurements each day to serve as the labeled data, as shown in Fig. 3b. The parameter

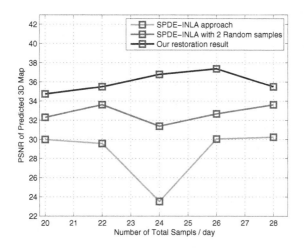

Fig. 4. PSNR curve for different number of samples

estimation and spatial prediction mentioned in Section 3.1 are based on these labeled data. We use the lower resolved ground truth data as satellite data shown in Fig. 3c, which has a very low Peak Signal-to-Noise Ratio (PSNR) of 8.0468dB. Finally we perform the algorithm in Section 3.2 to further improve the accuracy of the predicted concentration data. The restoration result is shown in Fig. 3d. We use PSNR of as the evaluation result instead of the traditional cross validation over several points. The PSNR of our restored concentration map is 37.3714dB, which shows a significant improvement.

The curves in Fig. 4 show the PSNR for different numbers of samples. The green curve is the PSNR of the prediction result using the SPDE-INLA approach in [8] with all measurements fixed. The significant decline at 24 samples is due to the inappropriately chosen samples. The red curve is the prediction result of SPDE-INLA approach with 2 fixed samples replaced by random ones, and it proves that mobile platform observations can help improve the prediction accuracy. Finally, the black curve is the prediction result of our proposed restoration algorithm. It shows that our proposed methodology, i.e. taking random samples and satellite data into consideration, can significantly improve the accuracy of the predicted air pollutant concentration map.

5   Conclusions

In this paper, we take advantage from heterogeneous big data sources to predict a high resolution and accuracy air pollutant concentration map. The final prediction result is the achieved by a spatial prediction via the SPDE-INLA approach and an optimization model based on Total Variation sparse assumptions. With 24 fixed samples and 2 random samples each day and some meteorology and geography covariates, we can improve the PSNR of low-resolution satellite data from 8.0468dB to 37.3714dB. Simulation results validate the efficiency of our algorithm.

References

1. WHO Air Pollution Estimates; WHO (2014)
2. Manyika, J., Chui, M., Brown, B., Bughin, J., Dobbs, R., Roxburgh, C., Angela, H.B., McKinsey Global Institute: Big data: The next frontier for innovation, competition, and productivity (2011)
3. Zhu, Y., Hinds, W.C., Kim, S., Sioutas, C.: Concentration and size distribution of ultrafine particles near a major highway. Journal of the air & waste management association **52**(9), 1032–1042 (2002)
4. Hasenfratz, D., Saukh, O., Walser, C., Hueglin, C., Fierz, M., Thiele, L.: Pushing the spatio-temporal resolution limit of urban air pollution maps. In: 2014 IEEE International Conference on Pervasive Computing and Communications (PerCom), pp. 69–77. IEEE, Budapest, March 2014
5. Hoek, G., Beelen, R., de Hoogh, K., Vienneau, D., Gulliver, J., Fischer, P., Briggs, D.: A review of land-use regression models to assess spatial variation of outdoor air pollution. Atmospheric Environment **42**(33), 7561–7578 (2008)
6. Ryan, P.H., LeMasters, G.K.: A review of land-use regression models for characterizing intraurban air pollution exposure. Inhalation Toxicology **19**(S1), 127–133 (2007)
7. Zheng, Y., Liu, F., Hsieh, H.P.: U-Air: when urban air quality inference meets big data. In: Proceedings of the 19th ACM SIGKDD International Conference on Knowledge Discovery and Data Mining, pp. 1436–1444. ACM, Chicago, August 2013
8. Cameletti, M., Lindgren, F., Simpson, D., Rue, H.: Spatio-temporal modeling of particulate matter concentration through the SPDE approach. AStA Advances in Statistical Analysis **97**(2), 109–131 (2013)
9. Cameletti, M., Ignaccolo, R., Bande, S.: Comparing spatio-temporal models for particulate matter in Piemonte. Environmetrics **22**(8), 985–996 (2011)
10. Cressie, N.A., Cassie, N.A.: Statistics for spatial data, vol. 900. Wiley, New York (1993)
11. Cressie, N., Wikle, C.K.: Statistics for spatio-temporal data. John Wiley & Sons (2011)
12. Lindgren, F., Rue, H., Lindstrm, J.: An explicit link between Gaussian fields and Gaussian Markov random fields: the stochastic partial differential equation approach. Journal of the Royal Statistical Society: Series B (Statistical Methodology) **73**(4), 423–498 (2011)
13. Rue, H., Martino, S., Chopin, N.: Approximate Bayesian inference for latent Gaussian models by using integrated nested Laplace approximations. Journal of the royal statistical society: Series b (statistical methodology) **71**(2), 319–392 (2009)
14. Rue, H., Held, L.: Gaussian Markov random fields: theory and applications. CRC Press (2005)
15. Rudin, L.I., Osher, S., Fatemi, E.: Nonlinear total variation based noise removal algorithms. Physica D: Nonlinear Phenomena **60**(1), 259–268 (1992)
16. Limpert, E., Stahel, W.A., Abbt, M.: Log-normal Distributions across the Sciences: Keys and Clues. BioScience **51**(5), 341–352 (2001)
17. Matérn covariance function - Wikipedia, the free encyclopedia. http://en.wikipedia.org/wiki/Mat%C3%A9rn_covariance_function
18. Goldfarb, D., Yin, W.: Second-order cone programming methods for total variation-based image restoration. SIAM Journal on Scientific Computing **27**(2), 622–645 (2005)
19. Boyd, S., Vandenberghe, L.: Convex optimization. Cambridge University Press (2004)

A Graph Community Approach for Constructing microRNA Networks

Benika Hall, Andrew Quitadamo, and Xinghua Shi[✉]

Department of Bioinformatics and Genomics,
University of North Carolina at Charlotte, Charlotte, NC, USA
{bjohn157,aquitada,x.shi}@uncc.edu

Abstract. Network integration methods are critical in understanding the underlying mechanisms of genetic perturbations and susceptibility to disease. Often, expression quantitative trait loci (eQTL) mapping is used to integrate two layers of genomic data. However, eQTL associations only represent the direct associations among eQTLs and affected genes. To understand the downstream effects of eQTLs on gene expression, we propose a network community approach to construct eQTL networks that integrates multiple data sources. By using this approach, we can view the genetic networks consisting of genes affected directly or indirectly by genetic variants. To extend the eQTL network, we use a protein-protein interaction network as a base network and a spin glass community detection algorithm to find hubs of genes that are indirectly affected by eQTLs. This method contributes a novel approach to identifying indirect targets that may be affected by variant perturbations. To demonstrate its application, we apply this approach to study how microRNAs affect the expression of target genes and their indirect downstream targets in ovarian cancer.

Keywords: Network integration · Graph community detection · Spin glass · microRNA networks

1 Introduction

With various types of data available at many layers in a biological system, it has become increasingly powerful to integrate data across multiple data sources towards understanding biological systems. Particularly, integrative methods provide useful insights into understanding how epigenetic factors, including microRNA (miRNA) expression, affects human health and disease. Recently, the functional influence of miRNAs on gene expression has become an important part of disease studies. Previous studies have shown that single nucleotide polymorphisms (SNPs) affect miRNA expression [1–3] through the technique of expression quantitative trait loci (eQTL) mapping [4]. These miRNA eQTL studies integrated SNP genotype data with miRNA expression by finding significant associations among SNPs and miRNA expression.

© Springer International Publishing Switzerland 2015
Y. Wang et al. (Eds.): BigCom 2015, LNCS 9196, pp. 283–293, 2015.
DOI: 10.1007/978-3-319-22047-5_23

Recent studies show that eQTL analyses provide invaluable findings in various diseases, such as discovering genetic variants that alter the susceptibility of lung cancer in non-smokers [17]. Although there has been much success with eQTL mapping studies, some factors hinder our understanding of how these loci contribute to diseases at a systems level. For instance, these studies integrate only two layers of data that are available, assuming that individual SNPs and miRNAs are independent on each other. Because it is known that genes do not act in isolation, we should investigate downstream effects caused by eQTLs. miRNAs have also shown before that they fine tune gene expression through post-transcriptional processes, albeit a lack of systematic analysis. With the availability of multiple layers of data, we should design methods to integrate these data together for a more systematic view of how miRNAs perturb gene expression.

In this paper, we integrate data from various sources such as miRNA expression profiles, gene expression quantifications, miRNA target interactions and protein-protein interactions. Using these data sources, we propose a network integration approach to expand eQTL associations using interactome networks, starting from miRNA eQTLs identified using standard eQTL mapping methods. We believe that by extending the eQTL network we can observe direct and indirect downstream targets of the eQTLs in our data. Such information is key in inferring relationships between miRNA expression and disease phenotypes. Particularly, we applied our approach to study how miRNAs affect gene expression and their downstream targets in their neighborhood in ovarian cancer. Our results show that our extended miRNA eQTL network cover miRNAs and genes related to diseases especially cancers, such as ovarian cancer.

2   Methods

2.1   Integrating miRNA eQTL Associations

Integrating various biological data sources provides a global outlook into the leading effects on gene expression. The Cancer Genome Atlas (TCGA)[8] data set provides rich data for network analysis across different data layers for many cancer types including ovarian cancer. The data used in this study consists of miRNA expression profiles, gene expression quantifications, miRNA interaction networks and protein-protein networks. In total we integrated 964 miRNAs and gene targets and 349,663 (genes) interactions from the protein-protein network. We integrate these various data types and construct a network which extends from miRNA eQTL perturbations on particular genes in the TCGA data set. Our goal is to integrate large-scale data in order to exploit the indirect effects of miRNA expression on gene expression in interaction networks in ovarian cancer.

Prior to our data integration, we performed an eQTL analysis between the miRNA expression and gene expression profiles. The eQTL analysis was performed using Matrix eQTL [11], an R package that uses matrix operations to perform eQTL mapping that assume miRNAs and genes are independent respectively. We performed *cis* eQTL analysis, whereby we only tested miRNA and gene pairs that were within 1MB of each other. To identify associations between the miRNA and genes, we used a linear regression model provided by Matrix eQTL. In some cases, we have to correct for multiple comparisons. To do so, we utilized a multi-test correction based on false discovery rate (FDR) and chose the significant eQTLs with a FDR cutoff at <0.01 for further analysis. begins with identifying the miRNA eQTL associations between miRNAs and target genes from the TCGA data set.

As a base network, we used a human protein interaction network, namely InWeb, produced in [16] that combines protein interaction networks from various resources and across diverse organisms. We first overlapped the matching genes from our miRNA eQTL analysis with the InWeb network. Then we removed interactions with an edge weight less than 0.10. To perform the eQTL analysis, we first used Matrix eQTL[11], We then merged the eQTLs identified from these two orthologous approaches and used the miRNA target genes to extend the eQTL network. The merged miRNA eQTL genes were considered as directly effected genes by miRNAs, and were used as seed genes to expand the network for an indirect effect analysis. Our overall workflow is shown in Figure 1.

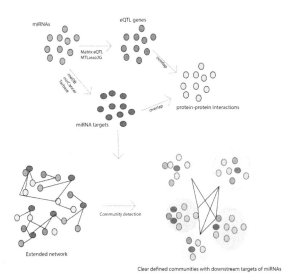

Fig. 1. The overall workflow of our study. The initial eQTL network construction starts with an eQTL analysis between miRNAs and gene expression with Matrix eQTL and MTLasso2G. We then performed database searches to identify immediate miRNA targets. Next, we overlapped those results with a protein-protein interaction network. This generated an extended network in which we performed community detection on to identify communities within an effective propogation distance from our miRNA eQTL genes. We visualized the extended network and the final network with communities in Cytoscape.

2.2   Network Integration

The initial network was created from the miRNAs and eQTL associations and then visualized in Cytoscape[7]. Using the miRNAs in this network, we collected known targets to add to our network. With the known targets and eQTL associations, we overlapped this network with protein-protein interactions collected from the InWeb network[16]. Using these two networks, we overlapped them by the genes they share in common and generated an edgelist to represent our integrated network. This allowed us to link the miRNAs to downstream targets by genes they share in common initially.

After generating this extended network, we performed community detection to identify downstream genes that interact directly with target genes affected by miRNAs and eQTLs. The community detection implementation can be found at https://github.com/shilab/community-detection. In particular, we used igraph[10] to employ the spin glass community detection algorithm, which is primarily based on thermodynamics[20,23,32]. The semi-supervised spin glass algorithm is coupled with simulated annealing and is less computationally intensive compared to other methods based on optimizing the modularity[13] of a network. Due to the large size of our network, the computational efficiency makes this algorithm attractive for large-scale networks. The communities provide a scope of interactions directly affected by genotypic perturbations on the miRNA level.

Using the spin model, the problem of community detection is mapped as identifying the spin states, which are the community indices, that lead to the ground state of the system. The model is based on a optimizing the qualifying energy function, known as the Hamiltonian, in Equation 2.2 such that it minimizes the energy and maximizes the modularity of the system. Compared with other methods that optimize modularity solely, the spin glass algorithm finds the spin configuration (the community indices) that minimizes the Hamiltonian. To minimize the Hamiltonian, this particular algorithm uses simulated annealing for optimization[18]. Simulated annealing is very efficient and easy to implement. Along with simulated annealing, the spin glass algorithm implements efficient update rules to improve computation when adding a node to its proper community. Another advantage of this algorithm is that it can detect overlapping communities without the network being affected by the degeneracy of Hamiltonian.

$$H(\sigma) = (A_{i,j}\,\gamma p_i, j)\delta(\sigma_i, \sigma_j), \tag{1}$$

where $A_{i,j}$ is the adjacency matrix, γ refers to the weights and σ_i and σ_j represent the spin states or community indices of the network. Once community detection process was completed, the detected communities were then used to expand the eQTL network with target genes from the InWeb network. The networks were then visualized in Cytoscape[7] for further analysis.

In this study, we illustrate the direct and indirect genes within an effective propagation distance[38,39] of the miRNAs in the ovarian cancer data set, generally at most six nodes. Nodes further from the source may not receive the perturbation signal, thus not being affected. With the primary focus on identifying indirect targets within close proximity, we chose to only visualize those genes that are within two neighbors away. The downstream effect can be further extended to include downstream genes further than two neighbors and thus creating a larger community network of downstream targets of interest.

3   Results

3.1   Integrated Network

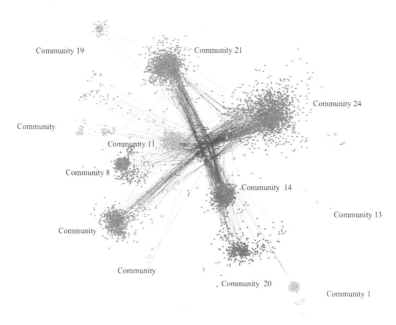

Fig. 2. The constructed miRNA eQTL network including the 12 communities containing our miRNAs retrieved from the spin glass community detection algorithm. Each community is color coded differently. Nodes are miRNAs and their affected genes, with edges representing the association or interaction among miRNAs and the affected genes.

To view indirect targets of our perturbed genes, we constructed an extended network consisting of miRNA eQTL genes as seed nodes, immediate target genes from the miRNA eQTL mapping results, and the indirect genes from the communities in the protein-protein interaction network. The edges of the constructed network include the miRNA eQTL associations between miRNAs and their affected genes and the interactions among miRNA directly affected genes and their downstream genes in a neighborhood. Our approach resulted in an integrated network consisting of miRNAs, their directly associated genes, protein-protein interactions and indirect targets from eQTL genes. Once completed, the spin glass algorithm detected a total of 25 communities. Of these 25 communities, we were only interested in communities that contained our miRNAs and eQTL genes, thus giving us 12 communities to further analyze. Communities obtained from the spin glass algorithm are shown in Figure 2.

The integrated network is essentially composed of genes that are directly or indirectly affected by miRNA expression in ovarian cancer using TCGA datasets. Using the eQTL genes as seed genes, we extended the miRNA eQTL network through the communities detected from the spin glass algorithm. For our network analysis, we viewed specific communities from the InWeb network that were linked to these miRNAs and their downstream targets within an effective propagation distance, within two neighbors of eQTL genes. By including these neighbor genes, we captured both direct and indirect effect of target genes affected by miRNA eQTLs. As an example of our extended network, please view Figure 3.

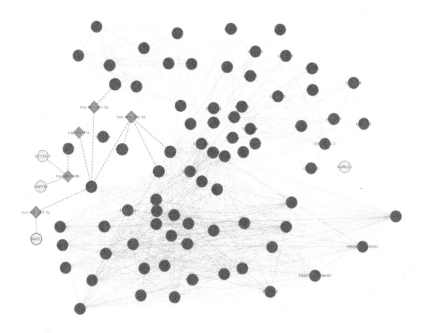

Fig. 3. An example of a community in the miRNA interaction network that contained miRNAs, their associated genes from eQTL mapping, their immediate targets, and downstream genes that are indirectly affected by the miRNAs. The miRNAs are denoted by green diamonds and their immediate targets are represented by dashed red lines while the indirect targets are shown as a solid gray line. The indirectly affected genes that overlapped with the OMIM database with disease associations are represented by red nodes.

3.2   Disease Associations

To further analyze our network for disease associations, we queried the eQTL genes in the Online Mendelian Inheritance in Man (OMIM) Disease database [6]. We found 59 eQTL genes with disease associations identified in the OMIM database. For example, $NUMA1$ is a nuclear mitotic asparatus protein found to be over-expressed in epithelial ovarian cancer, contributing to chromosal instability[25]. Our integrated network shows that $NUMA1$ is indirectly linked to $mir - 15b$. Therefore, further investigation is required to determine how $mir - 15b$ affects $NUMA1$. Please see Table 1 for other disease genes in our miRNA eQTL results.

We also queried the OMIM database with targets from our extended network. We found 2361 targets with disease associations in our extended network. An example of the community visualization, community 3, is shown in Figure 3. In this figure, the extended targets overlapped with the OMIM database are represented by red nodes. The miRNAs that affect gene expression are represented by green diamonds. The direct associations between miRNAs and their eQTL associated genes are denoted by dashed red lines.

Table 1. A summary table showing the genes affected by eQTLs that overlapped with disease genes according to the OMIM database

Gene	Description	Disease/Phenotype
ABCF1, ABC50	ATP-binding cassette 50, TNF-alpha stimulated	N/A
CD34	CD34 antigen	N/A
CDCA5, SORORIN	Cell division cycle-associated protein 5 (sororin)	N/A
CDKN1C, KIP2, BWS, IMAGE	Cyclin-dependent kinase inhibitor 1C (p57, Kip2)	IMAGE syndrome
CDKN1C, KIP2, BWS, IMAGE	Cyclin-dependent kinase inhibitor 1C (p57, Kip2)	Beckwith-Wiedemann syndrome
CENPO	Centromeric protein O	N/A
CLNS1A, CLNS1B	Chloride channel, nucleotide-sensitive, 1A	N/A
CMAS	Cytidine monophosphate N-acetylneuraminic acid synthetase	N/A
DGCR2, DGS2	DiGeorge syndrome chromosome region-2	DiGeorge syndrome/velocardiofacial syndrome complex-2
DHDDS, HDS, RP59	Dehydrodolichyl diphosphate synthase	N/A
DHX16, DDX16, DBP2, PRP8	DEAH (Asp-Glu-His) box polypeptide 16	N/A
ETNK1, EKI1	Ethanolamine kinase 1	N/A
FDPS, FPS	Farnesyl diphosphate synthase	N/A
FLOT1	Flotillin 1	N/A
GTF2H4	General transcription factor IIH, polypeptide 4	N/A
HIST1H1C, H1F2	Histone 1, H1c	N/A
HIST1H2BK, H2BK	Histone gene cluster 1, H2B histone family, member K	N/A
HNRNPA1, IBMPFD3, ALS20	Heterogeneous nuclear ribonucleoprotein A1	Amyotrophic lateral sclerosis 20
HNRNPA1, IBMPFD3, ALS20	Heterogeneous nuclear ribonucleoprotein A1	Inclusion body myopathy with early-onset Paget disease without frontotemporal dementia 3
HOXC6, HOX3C	Homeo box-C6	N/A
KIAA1279	KIAA1279 gene	Goldberg-Shprintzen megacolon syndrome
KIF1A, ATSV, UNC104, SPG30, HSN2C, MRD9	Kinesin family member 1A	Mental retardation, autosomal dominant 9
KIF1A, ATSV, UNC104, SPG30, HSN2C, MRD9	Kinesin family member 1A	Spastic paraplegia 30, autosomal recessive
KIF1A, ATSV, UNC104, SPG30, HSN2C, MRD9	Kinesin family member 1A	Neuropathy, hereditary sensory, type IIC
LDLRAP1, ARH, FHCB2, FHCB1	Low density lipoprotein receptor adaptor protein 1	Hypercholesterolemia, familial, autosomal recessive
MDC1, NFBD1, KIAA0170	Mediator of DNA damage checkpoint protein 1	N/A
MTERF4, MTERFD2	Mitochondrial trnascription termination factor 4	N/A
NACA	Nascent-polypeptide-associated complex alpha polypeptide	N/A
NDUFC2	NADH-ubiquinone oxidoreductase 1, subunit C2	N/A
NUMA1	Nuclear mitotic asparatus protein-1	Leukemia, acute promyelocytic, NUMA/RARA type
PCDHGA1	Protocadherin-gamma, subfamily A, member 1	N/A
PCDHGA12	Protocadherin-gamma, subfamily A, member 12	N/A
PCDHGA5	Protocadherin-gamma, subfamily A, member 5	N/A
PCDHGA9	Protocadherin-gamma, subfamily A, member 9	N/A
PCDHGB3	Protocadherin-gamma, subfamily B, member 3	N/A
PCDHGB4	Protocadherin-gamma, subfamily B, member 4	N/A
PCDHGB6	Protocadherin-gamma, subfamily B, member 6	N/A
PCDHGB7	Protocadherin-gamma, subfamily B, member 7	N/A
PDE2A	Phosphodiesterase 2A	N/A
PFKFB4	6-phosphofructo-2-kinase/fructose-2,6-bisphosphatase 4	N/A
POMC	Proopiomelanocortin (adrenocorticotropin/beta-lipotropin)	Obesity, early-onset, susceptibility to
POMC	Proopiomelanocortin (adrenocorticotropin/beta-lipotropin)	Obesity, adrenal insufficiency, and red hair due to POMC deficiency
RAB1B	Ras-associated protein RAB1B	N/A
RECQL4, RTS, RECQ4	DNA helicase, RecQ-like 4	RAPADILINO syndrome
RECQL4, RTS, RECQ4	DNA helicase, RecQ-like 4	Baller-Gerold syndrome
RECQL4, RTS, RECQ4	DNA helicase, RecQ-like 4	Rothmund-Thomson syndrome
RPL6	Ribosomal protein L6	N/A
RPL7A, SURF3	Ribosomal protein L7a (surfeit-3)	N/A
RPS6KA1, RSK1	Ribosomal protein S6 kinase, 90kD, 1	N/A
SCYL1, NTKL	SCYL1-like 1	N/A
SEPP1	Selenoprotein P, plasma, 1	N/A
SLC25A3, PHC	Solute carrier family 25 (mitochondrial carrier), member 3	Mitochondrial phosphate carrier deficiency
SQLE	Squalene epoxidase	N/A
STIP1, HOP	Stress-induced phosphoprotein 1	N/A
STMN1, LAP18, SMN	Stathmin	N/A
TRAF3IP3, T3JAM	TRAF3-interacting protein 3	N/A
TUBB, TUBB5, M40, CDCBM6	Tubulin, beta polypeptide	Cortical dysplasia, complex, with other brain malformations 6
ZNF79	Zinc finger protein-79 (pT7)	N/A

The mirCancer database was used to find cancer associated miRNAs[12]. We found eight miRNAs matching those in the mirCancer database that have significant roles in ovarian cancer. For example, $mir - 503$ plays a critical role in tumor progression and chemoresistance. In ovarian cancer, $mir - 503$ has been reported to be up-regulated in chemoresistant cancer cells [31]. Patients with increased expression of $mir - 200c$ have been shown to have decreased chance of relapse and increased survival[35, 36]. More examples of the miRNAs are shown in Table 2.

Table 2. A summary table showing identified miRNAs overlapped with miRNAs found in ovarian cancer from the mirCancer database

SNP/miRNA	Target	Disease	Profile	Role
hsa-miR-125b	PCDHGB3, PCDHGB6	ovarian cancer	down	Roles and Mechanism of miR-199a and miR-125b in Tumor Angiogenesis.
hsa-miR-192	STIP1	epithelial ovarian cancer	down	Regulates proliferation and apoptosis
hsa-miR-200c	SPSB2,PTPN6, RPL13P5, LRRC23, ACRBP	ovarian cancer	down	Prognostic marker
hsa-miR-200c	SPSB2,PTPN6, RPL13P5, LRRC23, ACRBP	ovarian cancer	up	Modulates metastasis in ovarian cancer
hsa-miR-23b	PCDHGA5, PCDHGB6	ovarian cancer	up	MicroRNAs overexpressed in ovarian ALDH1-positive cells are associated with chemoresistance.
hsa-miR-26a	PHF12,, ZNF830	ovarian cancer	up	Promotes prolifereation and tumorigenesis
hsa-miR-34a	RPL7A, RPS9	ovarian cancer	down	Downregulation in tumors and ovarian cancer
hsa-miR-503	MOSPD1	ovarian cancer	up	Implicated in tumor progression and chemoresistance
hsa-miR-98		ovarian cancer	down	Inhibits stem cell proliferation in ovarian cancer

4   Conclusion

We demonstrated in our study that by using a network extension approach to construct the miRNA eQTL network, we can view downstream targets of eQTLs. In our miRNA eQTL network, we gathered direct target genes for each eQTL and their indirect target genes. We thus constructed a network that captures the direct and indirect effect of eQTLs on gene expression. To investigate how the miRNA eQTLs and their affected genes are related to diseases, we overlapped our eQTL genes with the OMIM database. 59 of our eQTL genes and 2361 extended targets with disease associations were identified. For example, in Figure 3, gene $ASAH1$ is an extended target of $mir - 20a - 5p$ via gene $SGMS1$. We can investigate this community further in a different perspective such as a pathway analysis.

In addition, eight of our miRNA eQTLs were identified in the mirCancer database. Extended targets of these miRNAs were also identified in the OMIM database. For instance, $mir - 200c$, which modulates metastasis, targets gene $PTPN6$, which is a protein tyrosine phosphatase, nonreceptor-type, 6. Although the role of $PTPN6$ in cancer development is unclear, we know that it is targeted by $mir - 200c$, a common biomarker in epithelial ovarian cancer[35]. Using this knowledge and our extended network, we can investigate the downstream targets, such as $PTPN6$ and their neighbors. Such information can be vital in identifying new biomarkers for early prognosis in human diseases.

In this study, we used miRNA expression profiles, gene expression quantifications, miRNA target interactions and a protein-protein interaction network. We integrated data from all sources to create a community-based network based on eQTL perturbations. Hence, our resulted miRNA eQTL network includes the direct and indirect target genes that interact with miRNA eQTL affected genes.

In the future, we will include other types of biological networks such as regulatory networks in order to capture other types of relationships among genes. Additionally, we will include other types of interactions among miRNAs and genes. In addition to miRNA network construction, our method is applicable to build other networks by integrating other types of genomic and epigenomic data sets. Although we used genomic data in ovarian cancer, this can be used to identify downstream targets in any disease with similar genomic data available. An extensive network extracted from diverse data layers will be more informative when understanding the propagation of perturbation signals and the underlying mechanisms of complex diseases.

References

1. Gamazon, E.R., et al.: Genetic architecture of microRNA expression: implications for the transcriptome and complex traits. Am J. Hum Genet **90**(6), 1046–1063 (2012)
2. Lappalainen, T., et al.: Transcriptome and genome sequencing uncovers functional variation in humans. Nature **501**(7468), 506–511 (2013)
3. Huan, T., et al.: Genome-wide identification of microRNA expression quantitative trait loci. Nat Commun. **6**, 6601 (2015)
4. Tian, L., Quitadamo, A., Lin, F., Shi, X.: Methods for Population Based eQTL Analysis in Human Genetics. Tsinghua Science and Technology **19**(6), 624–634 (2014)
5. Chen, X., Shi, X., Xu, X., Wang, Z., Mills, R.E., Lee, C., Xu, J.: A two-graph guided multi-task lasso approach for eQTL mapping. Proceedings of the 15th International Conference of Artificial Intelligence and Statistics (AISTATS), Journal of Machine Learning Research (JMLR) W&CP **22**, 208–217 (2012)
6. Online Mendelian Inheritance in Man (OMIM). URL: http://omim.org/
7. Shannon, P., Markiel, A., Ozier, O., Baliga, N.S., Wang, J.T., Ramage, D., Amin, N., Schwikowski, B., Ideker, T.: Cytoscape: a software environment for integrated models of biomolecular interaction networks. Genome Research **13**(11), 2498–2504 (2003)
8. Cancer Genome Atlas Research Network: Integrated genomic analyses of ovarian carcinoma. Nature **474**, 609–615 (2011)
9. Ryan, B.M., Robles, A.I., Harris, C.C.: Genetic variation in microRNA networks: the implications for cancer research. Nat. Rev. Cancer **10**(6), 389–402 (2010)
10. Csardi, G., Nepusz, T.: The igraph software package for complex network research. InterJournal, Complex Systems **1695** (2006)
11. Shabalin, A.A.: Matrix eqtl: Ultra fast eqtl analysis via large matrix operations. Bioinformatics **28**(10), 1353–1358 (2012)
12. Xie, B., et al.: miRCancer: a microRNA cancer association database constructed by text mining on literature. Bioinformatics, btt014 (2013)
13. Ho, Y.-Y., Cope, L.M., Parmigiani, G.: Modular network construction using eqtl data: an analysis of computational costs and benefits. Frontiers in genetics **5**, 40–40 (2014)
14. Huang, Y., Wuchty, S., Przytycka, T.M.: Eqtl epistasis - challenges and computational approaches. Frontiers in Genetics **4**, 51–51 (2013)

15. Liu, C., Guo, J., Dung-Chul, K., Wang, J.: Inference of snp-gene regulatory networks by integrating gene expressions and genetic perturbations. BioMedical Research International

16. Lage, K., Karlberg, E.O., Størling, Z.M., Olason, P.I., Pedersen, A.G., Rigina, O., Hinsby, A.M., Tümer, Z.: A human phenome-interactome network of protein complexes implicated in genetic disorders. Nature biotechnology **25**(3), 309–316 (2007)

17. Li, Y., Sheu, C.-C., Ye, Y., de Andrade, M., Wang, L., Chang, S.-C., Aubry, M.C., Aakre, J.A., Allen, M.S., Chen, F., et al.: Genetic variants and risk of lung cancer in never smokers: a genome-wide association study. The lancet oncology **11**(4), 321–330 (2010)

18. Kirkpatrick, S.: Optimization by simulated annealing: Quantitative studies. Journal of statistical physics **34**(5–6), 975–986 (1984)

19. Liu, Y., Maxwell, S., Feng, T., Zhu, X., Elston, R.C., Koyutürk, M., Chance, M.R.: Gene, pathway and network frameworks to identify epistatic interactions of single nucleotide polymorphisms derived from gwas data. BMC systems biology **6**(Suppl 3), S15 (2012)

20. Eaton, E., Mansbach, R.: A Spin-Glass Model for Semi-Supervised Community Detection. In: AAAI (2012)

21. Quitadamo, A., Tian, L., Hall, B., Shi, X.: An Integrated Network of microRNA and Gene Expression in Ovarian Cancer. BMC Bioinformatics **16**(Suppl 5), S5 (2015)

22. Rachel Wang, Y.X., Huang, H.: Review on statistical methods for gene network reconstruction using expression data. Journal of theoretical biology **04**, 1–9 (2014)

23. Pan, L., Wang, C., Xie, J.: A spin-glass model based local community detection method in social networks. In: 2013 IEEE 25th International Conference on Tools with Artificial Intelligence (ICTAI). IEEE (2013)

24. Corney, D.C., Hwang, C.-I., Matoso, A., Vogt, M., Flesken-Nikitin, A., Godwin, A.K., Kamat, A.A., Sood, A.K., Ellenson, L.H., Hermeking, H., et al.: Frequent downregulation of mir-34 family in human ovarian cancers. Clinical Cancer Research **16**(4), 1119–1128 (2010)

25. Brüning-Richardson, A., Bond, J., Alsiary, R., Richardson, J., Cairns, D.A., McCormac, L., Hutson, R., Burns, P.A., Wilkinson, N., Hall, G.D., et al.: Numa overexpression in epithelial ovarian cancer. PloS one **7**(6), e38945 (2012)

26. Flutre, T., Wen, X., Pritchard, J., Stephens, M.: A Statistical Framework for Joint eQTL Analysis in Multiple Tissues. PLoS Genet **9**(5), e1003486 (2013)

27. He, J., Jing, Y., Wei Li, X., Qian, Q.X., Li, F.-S., Liu, L.-Z., Jiang, B.-H., Jiang, Y.: Roles and mechanism of mir-199a and mir-125b in tumor angiogenesis. PLoS One **8**(2), e56647 (2013)

28. Liu, T., Hou, L., Huang, Y.: Ezh2-specific microrna-98 inhibits human ovarian cancer stem cell proliferation via regulating the prb-e2f pathway. Tumor Biology **35**(7), 7239–7247 (2014)

29. Prokopi, M., Kousparou, C.A., Epenetos, A.A.: The Secret Role of microRNAs in Cancer Stem Cell Development and Potential Therapy: A Notch-Pathway Approach. Frontiers in Oncology **4**, 389 (2014)

30. Yan-ming, L., Shang, C., Yang-ling, O., Yin, D., Li, Y.-N., Li, X., Wang, N., Zhang, S.: mir-200c modulates ovarian cancer cell metastasis potential by targeting zinc finger e-box-binding homeobox 2 (zeb2) expression. Medical Oncology **31**(8), 1–11 (2014)

31. Park, Y.T., Jeong, J.Y., Lee, M.J., Kim, K.I., Kim, T.-H., Kwon, Y.D., Lee, C., Kim, O.J., An, H.-J.: Micrornas overexpressed in ovarian aldh1-positive cells are associated with chemoresistance. J. Ovarian. Res. **6**(1), 18 (2013)
32. Reichardt, J., Bornholdt, S.: Statistical mechanics of community detection. Physical Review E **74**(1), 016110 (2006)
33. Shen, W., Song, M., Liu, J., Qiu, G., Li, T., Yanjie, H., Liu, H.: Mir-26a promotes ovarian cancer proliferation and tumorigenesis. PloS one **9**(1), e86871 (2014)
34. Dernyi, I., Palla, G., Vicsek, T.: Clique percolation in random networks. Physical review letters **94**(16), 160202 (2005)
35. Prislei, S., Martinelli, E., Mariani, M., Raspaglio, G., Sieber, S., Ferrandina, G., Shahabi, S., Scambia, G., Ferlini, C.: MiR-200c and HuR in ovarian cancer. BMC Cancer **13**, 72 (2013)
36. Marchini, S., Cavalieri, D., Fruscio, R., Calura, E., Garavaglia, D., Nerini, I.F., Mangioni, C., Cattoretti, G., livio, L., Beltrame, L., Katsaros, D., Scarampi, L., Menato, G., Perego, P., Chiorino, G., Buda, A., Romualdi, C., D'Incalci, M.: Association between miR-200c and the survival of patients with stage I epithelial ovarian cancer: a retrospective study of two independent tumour tissue collections. The Lancet Oncology **12**(3), 273–285 (2011)
37. Lu, L.J., Xia, Y., Paccanaro, A., Yu, H., Gerstein, M.: Assessing the limits of genomic data integration for predicting protein networks. Genome Research **15**(7), 945953 (2005)
38. Nitzan, M., Steiman-Shimony, A., Altuvia, Y., Biham, O., Margalit, H.: Interactions between Distant ceRNAs in Regulatory Networks. Biophysical Journal **106**(10), 2254–2266
39. Huang, D., Zhou, X., Lyon, C.J., Hsueh, W.A., Wong, S.T.C.: MicroRNA-Integrated and Network-Embedded Gene Selection with Diffusion Distance. PLoS ONE **5**(10), e13748 (2010)

Sensor Networks and RFID

Distributed Multigrid Technique for Seismic Tomography in Sensor Networks

Goutham Kamath[1]([✉]), Lei Shi[1], Edmond Chow[2], and Wen-Zhan Song[1]

[1] Department of Computer Science, Georgia State University, Atlanta, USA
{gkamath1,lshi1}@student.gsu.edu, wsong@gsu.edu
[2] College of Computing, Georgia Institute of Technology, Atlanta, USA
echow@cc.gatech.edu

Abstract. The modern seismic sensor used to monitor volcanic activity can record raw seismic data for several years. This massive amounts of raw data recorded consists of information regarding earthquake's origin time, location, velocity the wave traveled etc. To extract this information from the raw samples, current state of the art volcano monitoring systems rely on gathering these high volume data back to a centralized base station. From these extracted information vulcanologist are able to compute tomography inversion and image reconstruction to understand the magma structure beneath. These high volume data are mostly gathered manually or sometimes relayed using powerful expensive broadband stations making vulcanologist unable to obtain real time information of the magma structure and also to predict the occurrence of eruption. Also, the sheer volume of raw seismic data restricts the deployment of large numbers of seismic sensors over the volcano making it difficult to obtain high resolution imagery. To overcome these challenges, a new in-network distributed method is required that can obtain a high resolution seismic tomography in real time. In this paper, we present a distributed multigrid solution to invert seismic tomography over large dense networks, performing in-network computation on huge seismic samples while avoiding centralized computation and expensive data collection. This new method accelerates convergence, thereby reducing the number of message exchanges required over the network while balancing the computation load (Our research is partially supported by NSF-CNS-1066391, NSF-CNS-0914371, NSF-CPS-1135814 and NSF-CDI-1125165.)

Keywords: Distributed multigrid · Cyber physical system · Big data · Seismic tomography · Sensor network · In-network computing

1 Introduction

Current volcano data collection and monitoring systems lack the capability of obtaining real time information and also recovering the physical dynamics of seismic activity with sufficient resolution. At present, the seismic tomography process involves aggregating raw data from seismic sensors into a centralized

© Springer International Publishing Switzerland 2015
Y. Wang et al. (Eds.): BigCom 2015, LNCS 9196, pp. 297–310, 2015.
DOI: 10.1007/978-3-319-22047-5_24

server for post-processing and analysis. To give some perspective on volume of data, the raw seismic samples are typically in the range of $16 - 24$ bit at $50 - 200$Hz. These high fidelity samples are generally primary (p) or secondary (s) wave, which consists of information such as earthquake origin time, location, speed etc embedded in it. This high fidelity sampling from each node makes it extremely difficult to collect raw, real-time data from a large-scale dense sensor network (e.g., hundreds to thousands of nodes) due to severe limitations on energy and bandwidth. Due to these restrictions many of the most threatening active volcanoes worldwide use fewer than 20 nodes [12], and require months to generate satisfactory tomography model. This limits our ability to understand volcano dynamics and physical processes in real-time. The centralized solution also introduces a bottleneck in computation. The risk of data loss also increases in case of node failures, especially at the base station. The centralized algorithm for these battery powered nodes, which have high risk of failures, are not suitable for volcano monitoring.

The high volume raw samples consists of sparse earthquake event information, however current technology requires station to transfer all the raw samples of p and s wave to centralized station for post processing. In [15] the data collected from 1980 - 2004 consists only of 19379 useful earthquake events and in addition 6916 events from october 2004 - december 2005. Fig. 1 shows the parse distribution of earthquake events obtained from 78 station placed on Mt St Helens (MSH). Few stations receive as few as 10 events while others receive more than 900. This sparse feature of raw samples have led researchers to adopt distributed techniques to perform in-network processing and avoid centralized computation. The advancement in current wireless sensor technology makes it possible to deploy and maintain a large-scale network for environmental monitoring and surveillance. However, seismic tomography algorithms commonly in use today cannot be easily implemented under this distributed scenario as it relies on centralized processing. Thus, real-time volcano tomography requires a practical approach which is distributed, scalable, and efficient with respect to tomography computation.

The advancement in current wireless sensor technology makes it possible to deploy and maintain a large-scale network for environmental monitoring and surveillance. However, seismic tomography algorithms commonly in use today cannot be easily implemented under this distributed scenario as it relies on centralized processing. Thus, real-time volcano tomography requires a practical approach which is distributed, scalable, and efficient with respect to tomography computation. Typically when solving large sparse linear systems, iterative methods tend to reduce high-frequency (oscillatory) components directly while not lower the errors caused due to low-frequency. Multigrid methods are often used to mitigate these low-frequency errors, as they reduce them by transferring the problem to lower grids. In this paper, we propose Distributed MultiGrid Tomography (DMGT) algorithm which accelerates the rate of tomographic calculation and thereby reducing the overall communication cost. This paper mainly focuses on the distributed tomography algorithm, while assuming the arrival time of

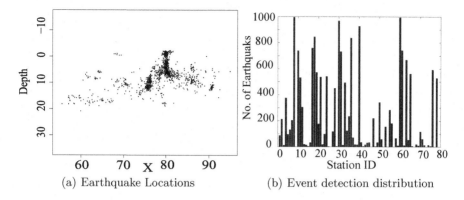

(a) Earthquake Locations   (b) Event detection distribution

Fig. 1. Non-uniform distribution of rays and events at Mt St Helens

events at each node has been extracted from the raw seismic data by each node itself [13]. The algorithm proposed here has application to fields far beyond the specifics of volcanology, e.g., oil field explorations have similar problems and needs.

The rest of the paper is organized as follows. In section 2 we provide background on seismic tomography inversion and present the problem formulation. Section 3 presents related work on distributed least squares, and distributed multigrid methods. In section 4 we first discuss mathematical developments that lead to the design of DMGT and then present the DMGT algorithm in detail. Simulation results are shown in section 5. Finally we conclude the paper in section 6.

2   Problem Formulation

Seismic Tomography: The methodology used in seismic tomography is borrowed from medical tomography where the travel time of elastic wave is used to probe internal structure. Although this idea is common in these two applications, there are significant differences, mainly pertaining to size of the structures and to event generation. The velocity model used in seismic tomography is nonlinear and the ray path of the waves traveling through the ground may be highly curved due to the size and complexity of the volcano. Typically, the ray source in volcano tomography is an earthquake event where the distribution of the ray path is highly non-uniform unlike uniform short distance rays generated in medical imaging. These differences indicate that special care must be taken when techniques borrowed from medical tomography are applied to seismic data.

The basic principle behind 2D or 3D seismic tomography is to use the arrival time of the P-wave to derive the internal velocity structure of the volcano. This approach is called *travel-time seismic tomography* and the model here is continuously evolving and refined as more earthquakes are recorded. Below we explain the three basic principles involved in travel-time seismic tomography.

i) *Event Location:* Once an earthquake occurs, seismic disturbances are detected by sensor nodes and arrival times are recorded. Using these estimated arrival times, Geiger [2] introduced a technique to estimate the earthquake location and origin time. This is a classic and widely used event localization scheme generally using Gauss-Newton optimization.

ii) *Ray Tracing:* This is the technique of finding the ray paths from the seismic source locations to the sensor nodes with minimum travel time. Given the source location of the seismic events and the current velocity mode of the volcano, ray tracing finds the ray paths from the event source location to the nodes as shown in Fig. 2(b).

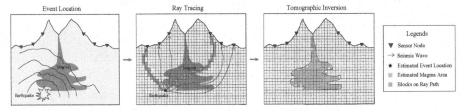

Fig. 2. Procedures of Seismic Tomography Inversion

iii) *Tomographic Inversion:* The ray paths traced in turn are used to estimate the velocity model of the volcano. The volcano is partitioned into small blocks as shown in Fig. 2(c). This allows us to formulate the tomography problem as a system of sparse linear equations. Suppose there are N sensor nodes and E earthquakes and x^* denotes the reference slowness (reciprocal of velocity) model of the volcano with resolution M blocks (eg. 32×32). Let x^* denote the sum of x^0, unperturbed model and x a small perturbation i.e., $x^* = x^0 + x$.

Let $b_i^* = [b_{i1}^*, b_{i2}^*, \cdots, b_{iE}^*]^T$, where b_{ie}^* be the travel time experienced by node i in the eth event. Based on the ray paths traced in step (2), the travel time of a ray is the sum of the slowness in each block times the length of the ray within that block, i.e., $b_{ie}^* = A_i[e, m] \cdot x^*[m]$ where $A_i[e, m]$ is the length of the ray from the eth event to node i in the mth block and x^* is the slowness of the mth block. Let $b_i^0 = [b_{i1}^0, b_{i2}^0, \cdots, b_{iE}^0]^T$ be the unperturbed travel times where $b_i^0 = A_i[e, m] \cdot x^0[m]$. In the matrix notation we have following equation,

$$A_i x^* - A_i x^0 = A_i x \tag{1}$$

where $A_i \in R^{E \times M}$. Let $b_i = [b_{i1}, b_{i2}, \cdots, b_{iE}]^T$ be the travel time residual such that $b_i = b_i^* - b_i^0$, equation (1) can be rewritten as,

$$A_i x = b_i \tag{2}$$

Since each ray path intersects the model at a small number of blocks, the design matrix, A_i, is sparse. For the system with N sensor nodes, the equation of the entire system would be,

$$Ax = B \tag{3}$$

where $B = [b_1, b_2, ...b_N]^T$, $b_i = [b_{i1}, b_{i2}, ..., b_{iE}]^T$ and $A = [A_1, A_2, ...A_N]^T$.

Now from the above equation, each seismic sensor $i \in (1, \cdots, N)$ contains at least E rows, i.e., earthquake events and travel time information. The column size of A denotes the resolution of the slowness model x being calculated. Our goal is to obtain the slowness model x without collecting the event information from each node in a centralized server, but only by exchanging partial slowness between the sensors.

3 Related Work

A popular iterative method for solving overdetermined systems was proposed by Kaczmarz (KACZ) [6] which is an alternating projection method. This method is also known under the name Algebraic Reconstruction Technique (ART) in computer tomography [5]. This algorithm does not require the full matrix to be in memory at one time and can incorporate new information (ray paths), on the fly. The vectors of unknowns are updated after processing each equation of the system and this cycle repeats until convergence. These iterative algorithms are distributed by averaging the boundary information, e.g., Component Averaging (CAV) [1] and Component-Averaged Row Projections (CARP) [3]. CA-DMET [7] involved modification of these algorithms for seismic tomography. The convergence of the iterative method used depended on spectral properties of the iteration matrix. Generally in iterative methods, convergence stalls once the error is smooth i.e. high-frequency errors are reduced. Multigrid methods provide great tool to prevent stagnation by transferring smooth errors from fine grids to coarse grids, resulting in overall acceleration of convergence [14], however, it cannot be applied to solve all the problems arising from systems of linear equations. In this paper, we analyze the tomography problem carefully and develop tools such as smoothers, intergrid operator etc satisfying the requirements of multigrid.

Multigrid has been parallelized on multicore computers and distributed memory clusters [16]. To perform multigrid in distributed networks, many new considerations arise, including high communication cost and the possibility of packet loss. For example, some existing parallel and distributed multigrid algorithms partition the multigrid levels among different cores/nodes and the intergrid operators communicate between each other to perform a multigrid cycle [11]. In case of seismic tomography, exchanging the rows of matrix A (ray information) between each nodes is expensive and defeats the whole purpose of the distributed approach. Thus, we cannot adopt all previous techniques for parallelizing multigrid and apply them to volcano tomography over sensor networks.

Iterative methods such as Jacobi, Gauss-Seidel, and SOR for many problems have the property of smoothing the error and are used as the "smoother" in multigrid methods [14]. However, for solving overdetermined systems, it is more natural to use Kaczmarz or ART as the smoother. This appears to be first considered in [10] for multigrid in medical image tomography in a centralized setup. In this paper we propose Distributed MultiGrid Tomography (DMGT) which accelerates the convergence of seismic tomography inversion over a network and balances the computation cost with reduced communication.

DMGT uses Bayesian ART (BART) as a smoother and we show that DMGT is applicable to seismic tomography. To the best of our knowledge, this work is the first attempt to distribute the multigrid computation of seismic tomography in sensor networks.

4 Distributed Algebraic Multigrid for Tomography

The tomography inverse problem involves finding a solution x which satisfies equation (4). Typically, the seismic tomography equation is quasi-overdetermined, inconsistent and contains measurement noise. Therefore, we need to use some form of regularization to avoid strong, undesired influence of small singular values dominating the solutions. This can be achieved by using a regularization parameter for the least-squares solution x_{LS}, i.e.,

$$x_{LS} = \arg\max_x \|B - Ax\|^2 + \lambda^2\|x\|^2 \qquad (4)$$

where λ is the trade-off parameter that regulates the relative importance we assign to models that predict the data versus models that have a characteristic, a priori variance.

A variant of ART called Bayesian ART (BART) [9] can be used for solving equation (4). Suppose the system $Ax = b$ is inconsistent, then we have $Ax+y = b$ where y is chosen from any given x. Then the system is transformed to a well-posed problem. Now x and y can be solved simultaneously using the following iterative algorithm [1], where e_i is a unit vector with the ith component equal to one, and λ is the regularization parameter.

Algorithm 1. Bayesian ART

1: **for** $k \leftarrow 0$ until convergence or maximum number of iteration **do**
2:    $k \leftarrow i \bmod m +1$
3:    $d^{(k)} = \rho^{(k)} \frac{\lambda b_i - (y_i^{(k)} + \lambda a_i^T \cdot x^{(k)})}{1+\lambda^2\|a_i\|^2}$
4:    $x^{(k+1)} = x^{(k)} + \lambda d^{(k)} a_i$
5:    $y^{(k+1)} = y^{(k)} + d^{(k)} e_i$
6: **end**

Multigrid methods are among the most efficient methods for solving a large sparse system of linear equations [14]. The core idea of multigrid is to reduce the error via transferring the problem between multiple levels and solving them over these levels. The residual equation is transferred to coarser grids and its solution is used to correct the finer resolution solution. This is performed recursively until convergence is met. The idea of multigrid aligns with multi-resolution techniques and we have shown in [7] that multi-resolution is essential in estimating volcano tomography.

The main components of multigrid are the smoother, prolongation and restriction operators, and wide variety of these are used in different scenarios. These components are chosen based on the type of the problem to optimize convergence. Prolongation and restriction operators mainly decide the construction of finer and coarser grids. In case of tomography the grids are constructed based on the principle of ray tracing and here we will show that ray tracing can be used for prolongation and restriction in multigrid. Prolongation and restriction are generally termed as intergrid operators as they define the transfer process between the grids. The tomography problem has a geometric structure and here we exploit this structure to define the intergrid operators. However, these intergrid operators must have certain properties and in this section we will show that our ray tracing satisfies these properties.

Let n be the number of columns in A and suppose that $n = 4p$ and let $P_1, \cdots, P_n$ be the pixels on the *fine grid*. The *coarse grid* is obtained by combining its 4 adjacent pixels of the fine grid. Let $S(j), j \in 1, \cdots, p$ be the set of indices of the fine grid that form the coarse grid P_j^H. i.e.,

$$S(j) = \{j_1, j_2, j_3, j_4\} \quad \forall j = 1, \cdots, p$$

where

$$j_1 < j_2 < j_3 < j_4$$

such that

$$P_j^H = \{P_{j1} \cup P_{j2} \cup P_{j3} \cup P_{j4}\}$$

From the above equation the coarse grid matrix A_p will be

$$A_p^{ij} = \sum_{k \in S(j)} A_{ik}, \forall i = \{1, \cdots, m\} \quad j = \{1, \cdots, p\} \tag{5}$$

Now the interpolation operator I_p^n is given by

$$I_p^n = \begin{cases} 1 \text{ if } i \in S(j) \\ 0 \text{ if } i \notin S(j) \end{cases} \tag{6}$$

We now see that $A = A_p \times I_p^n$ satisfying the interpolation property. We also observe that I_p^n has full column rank.

Now we describe the three-grid correction scheme used in our algorithm. If the finest resolution of our system to solve is of dimension 32×32, then resolution 16×16 is used as an intermediate grid and resolution 8×8 the coarsest grid. The coarsest grid is solved directly as the dimension is small, however we can also solve it by certain sweeps/iteration of BART. Later, the fine grid correction step is applied. The total number of iterations for one three-grid V-cycle will be equal to $4 \times l_1$. The three-grid V-cycle scheme is represented diagrammatically in Fig. 3.

Algorithm 2. $v^h \leftarrow Vcycle(v^h, b^h)$

1: $v^h = \text{BART}(A^h, b^h, v^h)$ *% Relax using l_1 sweeps of BART*
2: $r^h = b^h - A^h v^h$ *% Compute fine-grid residual*
3: $r^{2h} = I_h^{2h} r^h$ *% Restrict the residual to coarse grid*
4: $v^{2h} = \text{BART}(A^{2h}, r^{2h}, 0)$
5: $r^{4h} = r^{2h} - A^{2h} v^{2h}$
6: $r^{4h} = I_{2h}^{4h} r^{2h}$
7: $A^{4h} u^{4h} = r^{4h}$ *% Solve directly*
8: $e^{4h} = (A^{4h})^{-1} r^{4h}$
9: $e^{2h} = I_{4h}^{2h} e^{4h}$ *% Interpolate coarse grid error to fine grid*
10: $v^{2h} = v^{2h} + e^{2h}$ *% Correct the fine-grid approximation*
11: $e^{2h} = \text{BART}(A^{2h}, b^{2h}, v^{2h})$ *% Relax using l_1 sweeps of BART*
12: $e^h = I_{2h}^h e^{2h}$
13: $v^h = v^h + e^h$
14: $v^h = \text{BART}(A^h, b^h, v^h)$

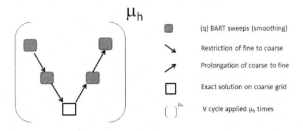

Fig. 3. V Cycle Scheme for three levels

Previously, we discussed the components of multigrid suitable for tomography. Here, we will put these ideas together to design a distributed multigrid scheme that can balance the computation load and compute the least-square solution for seismic tomography inversion over a sensor network. The seismic sensors are deployed on top of the volcano and each sensor gathers ray information after detecting earthquake events and forms a partial set of linear equations. Later, each sensor performs DMGT locally to obtain the partial slowness model ($\bar{x}^k$) which is then combined with the partial slowness model obtained from other nodes using component averaging to obtain the next iterate (x^{k+1}). This process is repeated until it converges to a threshold after which we obtain the global slowness model ($\mathbf{x}$). Here, we first show how component averaging can be used to combine the partial slowness from each node to form the next iterate. Later we discuss the working of distributed multigrid algorithm in detail.

Suppose there are N sensor nodes in the network and E ray paths are traced on each sensor node, following some earthquake events. From section 2 the seismic tomography model will be of the form

$$Ax = B \tag{7}$$

where $B = [b_1, b_2, ...b_N]^T, b_i = [b_{i1}, b_{i2}, ..., b_{iE}]^T$ and $A = [A_1, A_2, ...A_N]^T$.

Let the size of A be $m \times n$, where $\sqrt{n}$ denotes the resolution we are calculating in case of 2-D. Let $A_1, A_2 \cdots, A_N$ each contain $m_1, m_2, \cdots, m_E$ number of rows. Now in each node, we calculate the number of non-zero coefficients $\forall j$, where $1 \leq j \leq n$. Let I_j denote the index set of the blocks that contain an equation with a non-zero coefficient of x_j. Let $s_j = |I_j|$ (size of I_j). Partial slowness obtained from each node can be combined with others using component average operator given by

$$CA_A(\bar{x}^1, \cdots, \bar{x}^N) = \frac{1}{s_j} \sum_{t=1}^{N} \bar{x}_j^t$$

From the above discussion we understand that at each node we perform multigrid V-cycle to solve local linear system and then use component average operator to combine all the partial updates. This is later used as an initial guess for the next iteration and detailed algorithm is provided below.

Algorithm 3. Distributed Multigrid Tomography

Initialize

1: Node ID id,
2: Initialize the starting resolution dimension d
3: Initialize the number of seismic sensors N
4: Current resolution dimension $Q = d \times d$
5: Initial slowness model for ray tracing $\mathbf{x}^l$

Repeat

1: **Upon the detection of an event**
2:     Trace the ray path a_e for every node
3: **Upon the reception of a_e and b_e at each node start performing**
4: **calculation at each node**
5:     For each $1 \leq j \leq Q$, calculate s_j
6:     Where $s_j = |I_j| = \{1 \leq t \leq N | x_j$ has nonzero
7:     coefficient in some equation of node N
8:     $k \leftarrow 0, x^k \leftarrow 0$
9:         **while** not converged **do**
10:            In Every node t for $1 \leq t \leq N$ do in parallel
11:            $\bar{x}^t \leftarrow Vcycle(x^k, b^t)$
12:            Aggregate the partial slowness $\bar{x}^t$
13:            from all nodes and find the next iterate:
14:            $x_j^{(k+1)} = \begin{cases} \bar{x}_j^t & \text{if } s_j = 1 \\ \frac{1}{s_j} \sum_{t=1}^{N} \bar{x}_j^t & \text{if } s_j > 1 \end{cases}$
15:            Send $x_j^{(k+1)}$ to all the node N
16:            $k \leftarrow k + 1$
17:         **end while**
18:     $\mathbf{x}^l \leftarrow x^{(k-1)}$
19: **Upon the convergence obtaining final $\mathbf{x}^l$**
20:     Update slowness model: $\mathbf{x}^{(l+1)} = \mathbf{x}^l$
21:         TERMINATE

---

5   Evaluation and Validation

In this section, we evaluate the DMGT algorithm and present the simulation results. Typically, to test tomography inversion algorithm a synthetic model is used and we adopt a synthetic data of a fault model from [4] which has been widely used for cross bore-hole tomography. This fault model is created with velocities of $0.75V$ for the right fault and $1.0V$ for the left fault. We perform the simulation in a customized simulator where we have implemented event detection, ray tracing, etc, for the fault model. The test cases and convergence measure used in computerized tomography are adopted to measure the volcano tomography as these two processes are similar.

Our experiment setup has a network of 64 nodes which detects the earthquake event and traces the ray. A total of 512 earthquake events are generated at random and a data generator traces the ray to obtain the travel time at each node. In practice these processes are independent and can be performed at each node distributedly. After this, we obtain A and b on each sensor node and we add Gaussian noise to b to simulate the measurement noise. We perform experiments for the finest resolution of dimension 32×32 with the three-grid Vcycle scheme. For the iterative methods, the selection of relaxation and regularization parameters ρ and λ respectively are critical and in all of our experiments these parameters remain constant throughout the iterations, i.e., $\rho^k = \rho = 0.25$ and $\lambda^k = \lambda = 5$ for all $k \geq 0$.

In the implementation, 5 sweeps of BART are performed at each level except the coarsest where it is solved directly. This adds up to a total of 20 iterations for a single Vcycle. For fair comparison we run 20 iterations at each node for CA-DMET. We use the relative slowness updates of the estimation between the two sweeps (one sweep means that all partial slowness are averaged to calculate the next iterate) as the stopping criteria. Rate of convergence of different algorithms are compared using relative updates (ϕ), residuals (χ) and absolute error (ϵ) given by, $\phi = |x^{(k+1)} - x^{(k)}|/|x^{(k)}|$, $\chi = \|Ax^k - b\|$ and $\epsilon = \|x^* - x^k\|$, where x^* is the ground truth.

Firstly, we compare the relative performance of DMGT with two different algorithms: CA-DMET and MG-ART [8]. We use residuals and absolute error as the parameters for comparison and results are shown in Fig. 4. These plots demonstrate that there is a difference in the initial convergence behavior in these algorithms. Although the residuals of CA-DMET and MG-ART decrease at a similar rate, the absolute error of MG-ART tends to diverge from ground truth. This behavior is due to the lack of regularization parameter in this algorithm to handle inconsistent systems, whereas BART in DMGT takes care of this using appropriate λ. The iterations on x-axis denote the number of component averages required over a network i.e k as discussed earlier. We can see that DMGT converges faster (lesser k) compared to CA-DMET which means it requires lesser communication over the network.

A visual verification of these three algorithms is shown in Fig. 5. All the algorithms are run for the same number of iterations. The reconstructed images from different algorithms reveal that DMGT is able to obtain better reconstruction

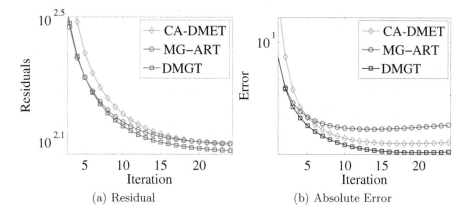

(a) Residual

(b) Absolute Error

Fig. 4. Comparing CA-DMET, MG-ART and DMGT

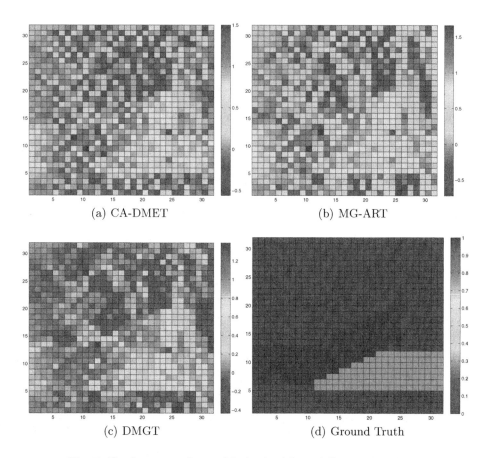

(a) CA-DMET

(b) MG-ART

(c) DMGT

(d) Ground Truth

Fig. 5. Final tomography model obtained from different algorithms

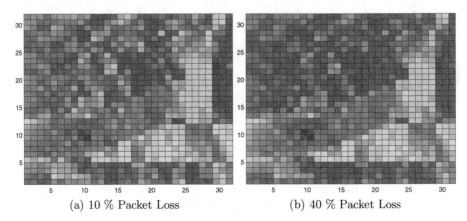

(a) 10 % Packet Loss   (b) 40 % Packet Loss

Fig. 6. DMGT under Packet Loss

Table 1. Robustness of DMGT

Cases	Relative Error (ϕ)	Absolute Error (ϵ)
No Packet Loss	0.0052	3.4606
10% Packet Loss	0.0386	3.5281
40% Packet Loss	0.0612	3.7411

compared to other algorithms. We also observed that CA-DMET and DMGT algorithms continued to improve its image reconstruction with further increase in iterations, however MG-ART's reconstruction deteriorated with increase in iterations. This is also because of the inconsistent system as mentioned earlier.

In the next set of experiments, loss tolerance and robustness of DMGT are evaluated. The algorithm runs with the same configuration for two different packet loss ratio of 10% and 40% in the simulator and the results are tabulated in Table 1. Fig. 6 gives the 2D tomography with packet loss and we can see that with 10% or even 40% packet loss, there is no significance difference in terms of the image reconstruction when compared to the results with no packet loss. Since the computation is distributed and all the nodes are involved in slowness calculation, the proposed algorithm is tolerant to a severe packet loss.

We also compare the efficiency of the algorithm by creating partitions and varying its size. Simulation results shown in Fig. 7 are run for total of 64 nodes, with partition number varying from 8 (each partition having 8 nodes) through 64 (each partition having one node). We can see that as partition number increases the convergence rate decreases. This is mainly due to the type of linear system each node has and the coefficient shared among the nodes. From this we can conclude that there is an optimal partition for a given set of nodes and given set of events. Also, in Fig. 7(b) we notice that for $P = 8$ case the solution diverges from ground truth. This phenomena is due to over smoothing/relaxation and to

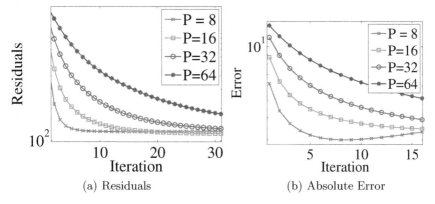

(a) Residuals   (b) Absolute Error

Fig. 7. DMGT with Different Partition Size

overcome this we need to dynamically select the parameters such as λ and ρ for a given partition size. We address these questions in our future work and it is beyond the scope of this paper.

6 Conclusion

In this paper, we presented a new algorithm to solve the seismic tomography problem over the sensor network. We also showed that BART satisfies the smoothing property and can also be used as a smoother in multigrid. We have also described the novel technique of performing multigrid in a distributed manner altogether forming the DMGT algorithm. This algorithm can distribute and balance the tomographic inversion computation load over the network, while computing real-time high-resolution tomography. The experimental evaluation also showed that our proposed method balances the computation load and is tolerant to data loss. Further enhancement of this algorithm can be done by applying the Full Approximation Scheme (FAS) and Full Multigrid (FMG).

References

1. Censor, Y., Gordon, D., Gordon, R.: Component averaging: An efficient iterative parallel algorithm for large and sparse unstructured problems. Parallel Computing **27**(6), 777–808 (2001)
2. Geiger, L.: Probability method for the determination of earthquake epicenters from the arrival time only. Bull.St.Louis.Univ. **8**, 60–71 (1912)
3. Gordon, D., Gordon, R.: Component-averaged row projections: a robust, block-parallel scheme for sparse linear systems. SIAM Journal on Scientific Computing **27**, 1092–1117 (2005)
4. Hansen, P.C., Saxild-Hansen, M.: AIR Tools A MATLAB package of algebraic iterative reconstruction methods. Journal of Computational and Applied Mathematics **236**(8), 2167–2178 (2012). http://dx.doi.org/10.1016/j.cam.2011.09.039

5. Herman, G.T.: Reconstruction from Projections: The Fundamentals of Computerized Tomography. Academic Press (1980)
6. Kaczmarz, S.: Angenäherte Auflösung von Systemen linearer Gleichungen. Bulletin International de l'Académie Polonaise des Sciences et des Lettres **35**, 355–357 (1937)
7. Kamath, G., Shi, L., Song, W.Z.: Component-average based distributed seismic tomography in sensor networks. In: IEEE DCOSS (2013)
8. Kostler, H., Popa, C., Rude, U.: Algebraic multigrid for general inconsistent linear systems: The correction step. Tech. rep., Lehrstuhl für Informatik 10 (Systemsimulation), FAU Erlangen-Nürnberg (2006)
9. Lees, J.M., Crosson, R.S.: Bayesian Art versus Conjugate Gradient Methods in Tomographic Seismic Imaging: An Application at Mount St. Helens, Washington. Institute of Mathematical Statistics **20**, 186–208 (1991)
10. Popa, C.: Algebraic Multigrid Smoothing Property of Kaczmarz's Relaxation for General Rectangular Linear Systems. Electronic Transactions on Numerical Analysis **29**, 150–162 (2008)
11. Smith, B., Bjorstad, P., Gropp, W.: Domain Decomposition: Parallel Multi-level Methods for Elliptic Partial Differential Equations. Cambridge University Press (1996)
12. Song, W.Z., Huang, R., Xu, M., Ma, A., Shirazi, B., Lahusen, R.: Air-dropped sensor network for real-time high-fidelity volcano monitoring. In: The 7th Annual International Conference on Mobile Systems, Applications and Services (MobiSys), June 2009
13. Tan, R., Xing, G., Chen, J., Song, W., Huang, R.: Quality-driven volcanic earthquake detection using wireless sensor networks. In: The 31st IEEE Real-Time Systems Symposium (RTSS), San Diego, CA, USA (2010)
14. Trottenberg, U., Oosterlee, C., Schuller, A.: Multigrid. Academic Press, San Diego (2001)
15. Waite, G.P., Moranb, S.C.: VP Structure of Mount St. Helens, Washington, USA, imaged with local earthquake tomography. Journal of Volcanology and Geothermal Research **182**(1–2), 113–122 (2009)
16. Yang, U.M.: Parallel Algebraic Multigrid Methods - High Performance Preconditioners, vol. 51, pp. 209–236. Springer-Verlag (2006)

Energy-Efficient and Smoothing-Sensitive Curve Recovery of Sensing Physical World

Qian Ma[✉], Yu Gu, Tiancheng Zhang, Fangfang Li, and Ge Yu

Institute of Computer Software and Theory,
Northeastern University, Shenyang, China
maqian_neu@163.com,
{guyu,lifangfang}@ise.neu.edu.cn,{tczhang,yuge}@mail.neu.edu.cn

Abstract. In recent years, sensing networks are widely used in the application of real-time monitoring. The change process of physical word is smoothing and continuous, but the sensing devices can only obtain the discrete data points. It is likely to lose the key points and distort the true curve if the discrete points are used simply to describe the physical world. Therefore, how to recover the approximate curve of physical world becomes a problem to be solved urgently. Based on this, an energy-efficient and smoothing-sensitive high-precision curve recovery algorithm for the sensing networks is proposed. Firstly, we recover the curve of physical world based on the existing physical-world-aware data acquisition algorithms preliminarily. And then a curve smoothing algorithm is proposed in order to acquire more key points (the inflexions are mainly considered in this paper) information which helps users better understand the change process of monitored physical world intuitively. Secondly, we propose an energy-efficient data source selection algorithm with residual energy of each data source and spatial correlation under consideration simultaneously. We select part of data sources to transmit data, maximize the lifetime of sensing network and minimize the error between the approximate curve and physical world. Finally, the effectiveness of our algorithms is verified by abundant experiments using both real and simulated data.

Keywords: Sensing network · Smoothing-sensitive curve recovery · Data source selection

1 Introduction

With the rapid development of IOT (the Internet of Things), a large number of sensing devices are deployed to acquire data continuously from monitored environments in many real applications[1–5] such as marine environment monitoring, air quality monitoring, crop growth monitoring and so on. Many related works have proposed data acquisition algorithms for sensing network based on *equi-frequency* sampling methods [6–8]. And the sampling frequency is usually low due to the constraint of energy consumption, which leads to the sparseness of sensed data. However, the change of physical world is continuous and

© Springer International Publishing Switzerland 2015
Y. Wang et al. (Eds.): BigCom 2015, LNCS 9196, pp. 311–324, 2015.
DOI: 10.1007/978-3-319-22047-5_25

smooth, so it is not enough to just output discrete sensed data to users. One of the reasons is that it may overlook critical points (such as extreme points and inflections) and distort the true curve only using discrete points to describe the continuous physical world. Another reason is that users can't observe the change process of physical world intuitively and it is detrimental to visual presentation. Therefore, how to precisely recover the physical world using the sparse discrete sensed data is a challenging and urgent problem to be solved. However, no sufficient effort has been made on this problem and the ground-breaking solution[9] mainly focuses on the accuracy of curve recovery. Based on this, we propose an energy-efficient and smoothing-sensitive curve recovery algorithm to reconstruct the high-precision approximate physical world for sensing networks.

Generally speaking, there are two stages during observing the physical world by sensor network. The first one is data acquisition in which sensed data was sampled from physical world through distributed sensor nodes. The second one is data transmission. In this stage, the sensed data is transmitted to sink nodes and then to upper applications after in-network processing such as integration, compression and so forth.

Data Acquisition. Several adaptive data sampling algorithms have been explored for energy saving. But they don't support the high-precision physical world recovery due to the worse accuracy. Cheng et al. [9] proposes two physical-world-aware data acquisition algorithms based on Hermit interpolation and Spline interpolation. But the results contain 2-order discontinuous points leading to the poor smoothness of approximate curves. However, the smoothness is one of the most important indicators for the approximate physical world recovery. There are two reasons: 1) the smoother of the approximate curve, the more intuitively for users to observe the change process of real physical world, and the more conductive to visual analysis for expert system. 2) More smoothness of approximate curve brings more accuracy of key points (extreme points and inflections) information contained in curve. The extreme points are easy to be found and many top-k query[10,11] techniques have been proposed for applications such as anomalous events detection[12]. But inflections which are the points from where the rate of curve starts to change have not been taken into consideration too much in existing works since they are difficult to obtain. And inflections can provide decision supports in many applications, especially when there is no law to be followed for curve changes. Maybe the irreparable damage has occurred when users are aware of the occurrence of a certain event. For example, in stock investment, inflections provide gist for the prediction of reverse trend occurrence. Investors usually buy or sell stocks at the inflection or its nearby to reduce risk. Product manufacturers often take control measures in advance to face the coming turbulence of market when product sales display inflection. And in sensing network, the effect of inflections is prominent too. We take red tide prediction as an example to illustrate.

Red tide endangers fisheries and aquaculture industries, which is one of the serious natural disasters. The heavy loss incurred by red tide can be avoided if we can predict it and take right measures in advance. The chlorophyll concentration is one of the indicators of red tide's occurrence. When it exceeds 12ug/L, red tide might happen. Figure 1 shows the chlorophyll concentration data of Xiamen offshore waters during June to September of 2007[13], where the points labeled red dot are part of inflections. From Figure 1, we can know that there is usually an explosive growth of chlorophyll concentration before the occurrence of red tide. If we give early warning and take effective measures at the beginning of explosive growth, probably the red tide can be remitted even be avoided because the measures need to take some time to work. And the beginning point of explosive growth is usually inflection which is the zero point of curve's second derivative in mathematics. So the continuity of second derivative of approximate physical world curve affects the coverage and accuracy of inflections. Based on the discuss above, a smoothing-sensitive approximate physical world recovery algorithm is proposed, which improves the smoothness of approximate curve to 2-order continuity from 1-order continuity.

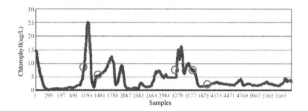

Fig. 1. Chlorophyll concentration data of Xiamen offshore waters

Data Transmission. Due to the high cost of deployment and replacement of sensors in special environments (such as ocean, forest), the sensor nodes are usually deployed densely to prevent the interference from surroundings and the invalidation of devices. Thus, the sensed data demonstrates the spatial correlations, and there is small deviation among sensed data of nearby sensors. Combing this feature of sensed data and the remaining energy of each sensor node, we propose an efficient-energy data source selection algorithm for lengthening the lifetime of sensing network. This algorithm selects part of data sources satisfying the constraints of remaining energy and data sharing to transmit data. The main contributions of this paper are stated as follows.

- We propose a smoothing-sensitive approximate physical world curve recovery algorithm. It concerns not only the accuracy but also the smoothness of the results during the physical world recovery. The approximate physical world curve generated by our algorithm is $O(\varepsilon)$-approximate and 2-order continuous.
- We present an energy-efficient data source selection scheme which takes the remaining energy of each sensor and data sharing into consideration simultaneously. The solution of this algorithm minimizes the loss of accuracy under the constraints and meanwhile maximizes the lifetime of network.

- The experimental results, using both real data and large simulated data, demonstrate the effectiveness of our proposed algorithms.

In the following section, the related works are discussed. Section 3 gives the problem statement and solution outline. In section 4, the curve smoothing algorithm for approximate physical world recovery is elaborated. And the method of energy-efficient approximate physical world curve transmission is derived in section 5. Section 6 shows the experimental results and finally the paper is concluded in section 7.

2   Related Work

Recently, a large number of researchers focus on the energy-efficient data acquisition methods for sensing networks. Papers [14–16] propose the adaptive sampling methods for energy saving based on *time-series forecasting*. The basic idea is to predict sensor readings in future prediction steps using *time-series forecasting* model, and then calculate the deviation between predicted values and observed readings. If the deviation is less than a given threshold, the sampling actions can be deleted for the reason that we think the predictions are accurate, otherwise sensors acquire data normally. Because the observations are unknown, the main problem to be solved is converted to judge whether the predictions are accurate enough. In [14], the predictions in forecasting steps are accurate if current prediction is precise. In [15], confidence interval is leveraged on the basis of work [14]. When the confidence interval of the lth-prediction is less than 2 times of the given threshold, the prediction could be regarded as accurate. Gupta et al. [16] provides an adaptive sampling algorithm based on exponential double smoothing techniques which ameliorate the prediction model in [14]. However, the assumption that sensed data can be denoted by a linear function of time t in above related works is too strong to perform quite well for practical applications. Alippi et al. [17] puts forward an adaptive sampling algorithm that adapts the sampling frequencies of sensors to the evolving dynamics of the process using Fourier transform technique. This method is efficient for abnormal event detecting rather than approximate physical world curve recovery. What's more, most of the existing adaptive sampling methods are based on some *equi-frequency* sampling method (EFS). The error in data acquisition is defined as difference between proposed method and EFS rather than difference between proposed method and real physical world.

In data transmission process, it's an efficient approach that part of but not all sensing devices are used to transmit data for energy saving because sensors are usually deployed densely. Aiming at aggregation queries, [18–21] propose approximate aggregation algorithms that reduce the amount of sensed data to be transmitted to save energy. Based on the similarity of sensed data and communication hops, the sensor nodes are clustered and a fringe of sensors in each cluster is selected to transmit sensed data in [22].

Literature [9] is the only one related work for approximate physical world curve recovery. It proposes two physical-world-aware data acquisition algorithms based on Hermit interpolation and Spline interpolation to construct the $O(\varepsilon)$-approximation physical world curve. Moreover, it proposes a distributed algorithm for reconstructing the physical world surface under the consideration of energy consumption. But the approximate physical world curve derived by [9], especially through the H-Based method contains 2-order discontinuous points. That is, the approximate physical world curve is not smoothing enough. In addition, it devotes to energy saving in stage of data transmission from the whole sensor network level instead of caring about each sensor's residual energy which concerns whether the sensed data can be transmitted successfully.

3   Problem Statement and Solution Outline

Suppose there are N sensors deployed in the monitored environment, and we use $\{l_1, l_2, \ldots, l_N\}$ to represent the location set of sensors. t_s and t_f are the start and end timestamps of monitored process respectively. Let $z_i(t), 1 \le i \le N$ be the real physical world curve changing with time t at location l_i. And then, the observation of sensor i at time t_c can be expressed as $z_i(t_c)$. As we know, data sampled by sensors is discrete, but the changing of physical world is continuous. It's insufficient to show the variation of physical world using the discrete data points simply. So we focus on the study of constructing the curve $\widetilde{z}_i(t)$ that approximates with the real physical world curve $z_i(t)$. We mainly evaluate the performance of recovery algorithm from two aspects: accuracy and smoothness. The accuracy is the deviation between approximate curve and real physical world curve. The smoothness is mutually influenced by the error of second deviation of approximate curve and the quality of inflection points contained in the approximate curve.

Definition 1: $O(\varepsilon)$-approximate curve. For time interval $[t_s, t_f]$, curve $\widetilde{z}_i(t)$ is called $O(\varepsilon)$-approximate curve of real physical world curve $z_i(t)$ if and only if there is a constant $C \ge 0$ satisfying $|z_i(t) - \widetilde{z}_i(t)| \le C\varepsilon, \varepsilon > 0$ for $\forall t_c \in [t_s, t_f]$. ε represents any error threshold specified by users.

Definition 2: n-order smoothing curve. $\widetilde{z}_i(t)$ is called n-order smoothing curve when equation $\widetilde{z}_i^{(n)}(t_c^-) = \widetilde{z}_i^{(n)}(t_c^+)$ is satisfied for any $t_c \in [t_s, t_f]$ where $\widetilde{z}_i^{(n)}(t_c^-)$ is the left n-th derivative of $\widetilde{z}_i(t)$ at t_c and $\widetilde{z}_i^{(n)}(t_c^+)$ is the right n-th derivative of $\widetilde{z}_i(t)$ at t_c .

There are three steps of our approximate physical world recovery. Firstly, we construct the continuous approximate curve through sensed data sampled by sensors based on the physical-world-aware data acquisition algorithms proposed by [9]. So far, the approximate curve only satisfies the accuracy but the smoothness is only 1-order continuous, which is not good enough. Secondly, we propose a curve smoothing algorithm to improve the smoothness of the approximate physical world to 2-order continuous. Now, the approximate curve is not only precise but also smooth, which is more suitable for describing the real physical world. Finally, due to the constraints of energy consumption, combining with spatial-correlation of sensed data, we propose a data source selection algorithm that select part of sensors satisfying both energy and spatial-sharing constraints to transmit their sensed data. We won't elaborate the detail of first step because the main techniques are proposed by [9]. The curve smoothing algorithm and data sources selection algorithm will be introduced in detail in the following sections.

4 Curve Smoothing Algorithm for Approximate Physical World Recovery

4.1 Approximate Curve Smoothing Algorithm

Suppose that $t_s < t_1 < \cdots < t_n < t_f$ are the segment points of approximate piecewise curve $\hat{z}_i(t)$ derived based on [9]. Assume that the second derivative $\hat{z}_i^{(2)}(t_c)$ of $\hat{z}_i(t)$ at segment point $t_c, 1 \leq c \leq n$ exists and let it be calculated according to equation (1).

$$\hat{z}_i^{(2)}(t_c) = \frac{\hat{z}_i^{(2)}(t_c^+) + \hat{z}_i^{(2)}(t_c^-)}{2} \tag{1}$$

Define

$$q_c(t) = h_c^2 d_c \psi_2(\alpha_c(t)) - h_{c+1}^2 d_c \varphi_2(\alpha_{c+1}(t)) \tag{2}$$

where

$$h_c = t_c - t_{c-1}$$

$$d_c = \frac{\hat{z}_i^{(2)}(t_c^+) - \hat{z}_i^{(2)}(t_c^-)}{2}$$

$$\alpha_c(t) = \begin{cases} \frac{t - t_{c-1}}{h_c} & , t \in [t_{c-1}, t_c] \\ 0 & , otherwise \end{cases}$$

$$\varphi_j(t) = \frac{1}{j!} t^j (1 - t)^{j+1}$$

$$\psi_j(t) = \varphi_j(1 - t)$$

Let $\tilde{z}_i(t) = \hat{z}_i(t) + q_c(t)$, then the smoothed approximate curve $\tilde{z}_i(t)$ is 2-order continuous in $[t_s, t_f]$, which is proved by Theorem 1.

Theorem 1. *For* $\forall t \in (t_{c-1}, t_{c+1})$, *let* $\tilde{z}_i(t) = \hat{z}_i(t) + q_c(t)$, *then* $\tilde{z}_i(t) \in C^2(t_{c-1}, t_{c+1})$.

Due to the space limitation, all the proofs of theorems and lemmas are omitted in this paper. From Theorem 1, the derivative at segment points of $\hat{z}_i(t)$ are improved to 2-order continuous through letting $\tilde{z}_i(t) = \hat{z}_i(t) + q_c(t)$. And then the smoothness of the whole approximate curve $\tilde{z}_i(t)$ achieves 2-order continuous. The pseudo code of the proposed curve smoothing method is shown in Algorithm 1.

Algorithm 1. Approximate Physical World Curve Smoothing Algorithm

---

Input : set of segment points $\{t_1, t_2, \ldots, t_k\}$ in $[t_s, t_f]$, and corresponding curve sections to be smoothed $\widehat{z}_i(t), 1 \leq i \leq k$ where the segment points are covered in

Output: set of smoothed curve sections $\widetilde{z}_i(t), 1 \leq i \leq k$

1 **while** $(t_1 < t_c < t_f)$ **do**

2   **for** $\forall t_c \in (t_{c-1}, t_{c+1}), 1 \leq c \leq k$ **do**

3   Set $q_c(t) = \frac{d_c}{2h_c^3}(t_c - t)^2(t - t_{c-1})^3 - \frac{d_c}{2h_{c+1}^3}(t - t_c)^2(t_{c+!} - t)^3$

4   where $d_c = \frac{\widehat{z_i^{(2)}}(t_c^+) - \widehat{z_i^{(2)}}(t_c^-)}{2}, h_c = t_c - t_{c-1}$ and then let $\widetilde{z}_i(t) = \widehat{z}_i(t) + q_c(t)$

5 **return** the approximate curves $\widetilde{z}_i(t)$

4.2 The Accuracy of the Approximate Curve

The error bounds of $\widetilde{z}_i(t)$, $\widetilde{z}_i^{(1)}(t)$ and $\widetilde{z}_i^{(2)}(t)$ will be discussed in this section.

Lemma 1. *For* $\forall t \in (t_{c-1}, t_{c+1})$, *the absolute value range of* d_c *is* $|d_c| \leq \frac{1}{16} \max\limits_{t_{c-1} \leq \xi \leq t_{c+1}} \left\{ z_i^{(4)}(\xi) \right\} \max\{h_c^2, h_{c+1}^2\}$

Theorem 2. *The error bound between 2-order continuous approximate curve* $\widetilde{z}_i(t)$ *and real physical world curve* $z_i(t)$ *satisfies* $|z_i(t) - \widetilde{z}_i(t)| \leq 1.75\varepsilon$ *for any* $t \in (t_{c-1}, t_{c+1})$, *where* $\varepsilon > 0$.

Theorem 3. *The error bound of first derivative between smoothed approximate curve* $\widetilde{z}_i(t)$ *and initial approximate curve* $\widehat{z}_i(t)$ *satisfies* $\left| q_c^{(1)}(t) \right| \leq \frac{9(8\sqrt{6}+3)\varepsilon}{125 \min\{h_c, h_{c+1}\}}$ *for any* $t \in (t_{c-1}, t_{c+1})$.

Theorem 4. *The error of second derivative between* $\widetilde{z}_i(t)$ *and* $z_i(t)$ *is less than that between* $\widehat{z}_i(t)$ *and* $z_i(t)$ *for any* $t \in (t_{c-1}, t_{c+1})$.

From Theorem 1, we can know that the final approximate physical world curve smoothen by our algorithm is 2-order continuous at any data point. This overcomes the shortages of approximate curve construction algorithms in [9]. And the max error is less than 1.75ε through the theoretical analysis in Theorem 2, which means that we lost 0.75ε accuracy at most to commutate smoothness. The max first derivative error between $\widetilde{z}_i(t)$ and $z_i(t)$ is more than one third of the first derivative error between $\widehat{z}_i(t)$ and $z_i(t)$ according to Theorem 3. At the same time, Theorem 4 evaluates that the smoothness of $\widetilde{z}_i(t)$ is better than $\widehat{z}_i(t)$. And the experiments show that the inflection points' qualities on precision and recall of $\widetilde{z}_i(t)$ is better than $\widehat{z}_i(t)$. What's more, in the smoothing algorithm, $q_c(t)$ is only needed to act on the adjacent intervals which 2-order discontinuous points fall in. If 2-order discontinuous points are fewer, the error arising by our smoothing algorithm is smaller for the whole curve which is verified by our experiments. The details are introduced in section 6.1. Additionally, the more smooth approximate curves help users observe the change process of physical world more conveniently. Meanwhile, it has a better assistant effect on decision making for experts.

5   Energy-Efficient Data Transmission Algorithm for Approximate Physical World Curve

The approximate physical world curve recovery algorithm has been introduced in last section in detail. The data transmission problem will be elaborated in this section.

In sensing network, the object being monitored is often a continuous area rather than a single point. For example, in the application of marine environment monitoring, the object being sensed is a sea area. And the deployment of sensors is usually dense, which incurs few different readings among proximity sensors due to the spatial-correlation of physical world. The energy consumption is quite large if all sensors transmit sensed data to upper applications. Furthermore, it is more likely to bring conflicts between adjacent data links in the process of data transmission. So we should avoid transmitting sensed data simultaneously for proximity sensors. Based on this, we define data sharing and propose a data sources selection algorithm.

Definition 3: Data Sharing. Sensor i and j are data sharing if $D(i,j) \le \delta_d$, where $D(i,j)$ is the distance between sensor i and j, δ_d is the distance threshold satisfying $\delta_d \ge 0$.

On the other hand, due to the complicated deployment environment, it is difficult to supply energy to sensor nodes. So the remaining energy of sensor is one of decisive factors to influence the lifetime of sensing network. If the remaining energy of sensor i is c_i, we specify that the sensor can transmit data only when $c_i \ge \delta_e$, where δ_e is the remaining energy threshold. In conclusion, in order to ensure the accuracy of sensed data to be transmitted and the lifetime of network both, we should select sensors those have higher reliability and satisfy data sharing and energy remaining constraints at the same time. The reliability r_i of sensor i is specified as the average error between approximate curve and real physical world. Intuitively, if the sensed data is from a reliable source, then it is more accurate. Based on the above discussion, we formulate the data source selection problem to an optimization problem P and solve it using Generic Algorithm (GA).

Let the reliability set of N sensors in $[t_s, t_f]$ be $\{r_1, r_2, \ldots, r_N\}$, and e_i is the error of sensor i between approximate physical world curve and real curve. Then, the reliability of sensor i specified as $r_i = \frac{1}{e_i}$. Combining with the data sharing and energy remaining constraints, the optimization problem P is formulized as

$$P: \quad \max \sum_{k=1}^{N} r_k x_k$$

$$s.t. \ c_k x_k \ge \delta_e, k = 1, 2, \ldots N$$

$$x_k x_s dis(k, \ s) \ge \ x_s \delta_d, k = 1, 2, \ldots N, s = 1, \ldots, N-1$$

$$x_k \in \{0 \ , \ 1\}, 1 \le k \le N$$

x_k is a binary variable. If sensor k is selected to transmit data, $x_k = 1$, otherwise $x_k = 0$. We can know that P is a nonlinear 0-1 knapsack problem which is admitted as NP-hard. We choose GA [23] to solve it. GA is one of the random global search and optimization methods that simulates natural biological evolution mechanism of the survival of the fittest. Compared with other approximate algorithms, GA has the advantages of easier calculation, and higher robustness. GA is more suitable for our problem because the encode and decode process during calculation can be omitted.

The sensed data is transmitted to upper applications after data source selection is finished. According to the introduction in section 4, the smoothed curve $\widetilde{z}_i(t)$ of sensor i is composed by a set of cubic or quintic polynomials, and $\widetilde{z}_i(t)$ in time interval $[t_{c-1}, t_c]$ can be expressed by eight-tuple $\left(t_{c-1}, t_c, z_i(t_{c-1}), z_i^{(1)}(t_{c-1}), z_i^{(2)}(t_{c-1}), z_i(t_c), z_i^{(1)}(t_c), z_i^{(2)}(t_c)\right)$. Since $\widetilde{z}_i(t)$, $z_i^{(1)}(t)$, $z_i^{(2)}(t)$ are continuous in $[t_s, t_f]$, and the segment points are shared by adjacent time interval, $\widetilde{z}_i(t)$ can be expressed by quadruple-tuple sets $\left\{\left(t_1, z_i(t_1), z_i^{(1)}(t_1), z_i^{(2)}(t_1)\right), \left(t_2, z_i(t_2), z_i^{(1)}(t_2), z_i^{(2)}(t_2)\right), \ldots, \left(t_n, z_i(t_n), z_i^{(1)}(t_n), z_i^{(2)}(t_n)\right)\right\}$. That is to say, in data transmission, we only need to transmit a series of quadruple-tuples to represent the approximate physical world curve.

6 Experiment

In this part, we evaluate our algorithms through a lot of experiments using the real-world data and abundant simulated data. The real-world data set is composed of sensed data of temperature varying with water depth sampled by Nansen station supplied by National Marine Information Center. For simulated data, we suppose sensors are placed into a 100m*100m rectangular region randomly. And then, temperature of some waters during 1000 continuous timestamps is generated independently for each sensor. Corresponding to the H-Based and S-Based approximate physical world reconstruction algorithms, we use HS-Based (Hermit Smoothing Based) and SS-Based (Spline Smoothing Based) to represent our Smoothing algorithms respectively. The evaluations of first three parts use the real-world data set and the simulated data is used in the last part.

6.1 Accuracy of Smoothing Algorithm versus ε

Figures 2, 3, 4 display the average error, 0.9-quantitile error and max error of approximate curve, first and second derivative of approximate curve varying with ε respectively. According to the theoretical analysis in [9], the 2-order discontinuous points are the segment points of time windows for S-Based method. And they are only a tiny portion of the whole approximate curve derived by S-Based method. They have scarcely any influence to the accuracy of the whole approximate curve. So we use a line written by S(SS)-Based to represent S-Based method and SS-Based method simultaneously.

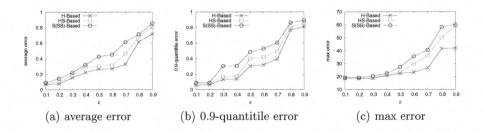

(a) average error   (b) 0.9-quantitile error   (c) max error

Fig. 2. Error of HS-Based and SS-Based vs. ε

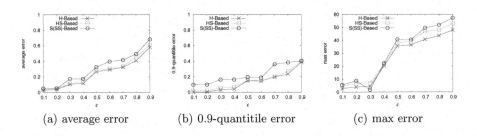

(a) average error   (b) 0.9-quantitile error   (c) max error

Fig. 3. Error of first derivative of HS-Based and SS-Based vs. ε

The accuracies of four methods are decreasing with the increase of ε shown as figure 2. Because the bigger of ε, the greater tolerance to error for all methods, and the fewer sampling times in the process of data acquisition, which lead to the increase of deviation. The accuracies of first derivative and second derivative act in the same way. And from Figures 2(a), 3(a), 4(a), we can find that the maximum average error of our approximate curve, first and second derivative generated by all algorithms are smaller than their theoretical upper bounds. The accuracy of HS-Based algorithm is between H-Based and S-Based, which indicates that HS-Based algorithm achieve 2-order continuity for whole approximate curve without too much accuracy loss. And Fig.2(b) shows that the 0.9-quantitile error of HS-Based method is small and at least 90% points in the approximate curve close to the real physical world.

6.2   The Accuracy of Inflections versus ε

To evaluate the effectiveness on inflections acquisition of our HS-Based and SS-Based algorithms, Fig.5 and Fig.6 are generated to show the accuracy of inflections from precision, recall, the average and max error between acquired inflections and true inflections varying with ε. Inflections are calculated through making the second derivative of approximate curve equals to zero. For each inflection calculated by the approximate curve, if its deviation between true value is less than the given threshold $maxD$, it is recognized as the accurate inflection.

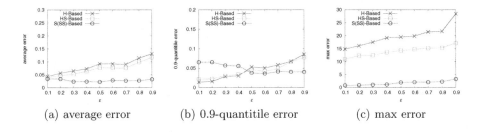

(a) average error   (b) 0.9-quantitile error   (c) max error

Fig. 4. Error of seconde derivative of HS-Based and SS-Based vs. ε

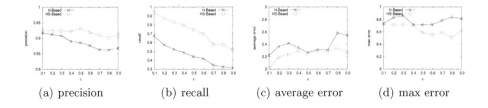

(a) precision   (b) recall   (c) average error   (d) max error

Fig. 5. Performance of HS-Based on inflection points vs. ε

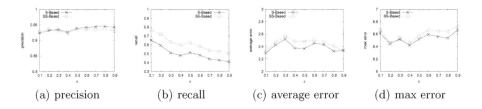

(a) precision   (b) recall   (c) average error   (d) max error

Fig. 6. Performance of SS-based on inflection points vs. ε

From the experimental figures, we can know that both the precision and recall of our two algorithms are improved compared with H-Based and S-Based. And the average and max error of HS-Base and SS-Based are closed to H-Based and S-Based. This means that we obtain more accurate inflection points. Besides, Figures 5(c), 5(d) and Figures 6(c), 6(d) show that the average and max error don't have too many fluctuations with the increase of ε. This illustrates that the inflection points obtained access high accuracy even though a small number of sensed data is sampled. And therefore, the stability of our smoothing algorithm can be verified.

6.3   The Accuracy of Inflections versus $maxD$

Fig.7 gives the accuracy of inflections versus the parameter $maxD$, which is the given threshold to judge whether the inflections obtained are true or not.

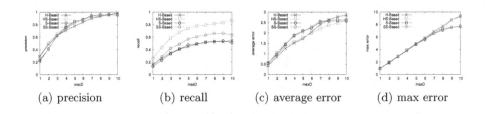

(a) precision   (b) recall   (c) average error   (d) max error

Fig. 7. Performance of HS-Based and SS-based on inflection points vs. $maxD$

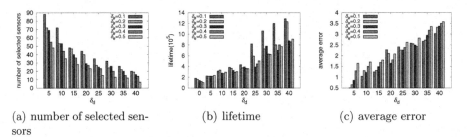

(a) number of selected sen-   (b) lifetime   (c) average error
sors

Fig. 8. Performance of data transmission affected by δ_d and δ_e

With the increase of $maxD$, both precision and recall increase gradually because the probability that obtained inflections points is true is proportional to the $maxD$. However, due to the increase of $maxD$, the deviations between obtained inflections and true inflections are also extended. So the average error and max error increase therewith. And in real applications, the appropriate $maxD$ should be chosen prudently.

6.4 The Performance of Data Source Selection Algorithm

The group of experiments in this section is to show the impact of δ_d and δ_e on the quantity of selected sensor nodes, the lifetime of network and the average error of data transmission respectively. The lifetime of network is calculated by $LifeTime = \frac{1(J) - \delta_e}{p * 5(\mu J)}$, where $1(J)$ is the initial energy of sensor node, $5(\mu J)$ is the energy consumption for each data transmission and p is the probability that sensor is selected to transmit data. And it can be estimated as the proportion of the selected nodes to the total. For upper applications, the sensed data of unselected sensors is obtained only through inference, and the resulting error is referred to the average error of this section. And in this paper, the sensed data of unselected sensor nodes is directly replaced by that of the shared sensor nodes. Fig.8 presents that the number of selected sensors decreases while both of the lifetime of network and average error of data transmission increase with δ_d growing from 5 to 40. This is because a lager δ_d will lead to more data sharing. Afterwards fewer sensor nodes need to be chosen to transmit data, and the selected probability is lower which lengthens the lifetime of the network.

Moreover due to the growth of quantity of unselected sensor nodes whose sensed data will be replaced by approximate values, the error will augment obviously.

On the other hand, it is observed that δ_e is proportional to average error of data transmission and inversely proportional to the number of selected sensors and lifetime of network. This is owing that with the increase of δ_e, more and more sensors are unsatisfied with the constraint on residual energy. And then fewer and fewer sensors are selected, which causes more error obviously. We can't distinguish the correlation between lifetime of network and δ_e because both of the numerator and denominator diminish with the increase of δ_e. But from Figure 8(b), we can know that δ_e is inversely proportional to lifetime of network. The lifetime of network are 180, 160, 140, 120, 100, 80 thousand rounds corresponding to the values of 0.1, 0.2, 0.3, 0.4, 0.5, 0.6 of δ_e when $\delta_d = 0$ which represents all of the sensors will transmit data. Notice that Fig.8 is generated based on HS-based algorithm and the results based on SS-Based algorithm are omitted due to the similarity between two algorithms.

7 Conclusions

An energy-efficient and smoothing-sensitive high-precision physical world recovery algorithm is proposed in this paper. The approximate physical world curve derived by our algorithm can achieve 2-order continuity at any point and $O(\varepsilon)$-approximation with real physical world. Moreover from the results of our experiments, there are more key points (inflections) contained in the derived curve, which proves the smoothing sensitivity of our algorithm further. In the stage of data transmission, we take the residual energy of each sensor node into consideration for the first time, and then define data sharing based on the spatial-correlation of sensed data. The data source selection problem is presented and formalized to an optimization problem solved through GA. Experiments are conducted on real-world data as well as simulated dataset. The results demonstrate the effectiveness and stability of our algorithms.

Acknowledgments. This work was supported in part by the National Key Basic Research Program of China (973) under Grant No. 2012CB316201, and in part by the National Nature Science Foundation of China under Grant Nos. 61202086 and 61272180.

References

1. Li, M., Liu, Y., Chen, L.: Non-threshold based event detection for 3d environment monitoring in sensor networks. In: ICDCS, p. 9 (2007)
2. Zhang, L., Wang, R., Cui, L.: Real-time traffic monitoring with magnetic sensor networks. J. Inf. Sci. Eng. **27**(4), 1473–1486 (2011)
3. Zhang, F., DiSanto, W., Ren, J., Dou, Z., Yang, Q., Huang, H.: A novel CPS system for evaluating a neural-machine interface for artificial legs. In: ICCPS, pp. 67–76 (2011)

4. Xu, G., Shen, W., Wang, X.: Marine environment monitoring using wireless sensor networks: a systematic review. In: SMC, pp. 13–18 (2014)
5. Kim, S., Pakzad, S., Culler, D.E., Demmel, J., Fenves, G., Glaser, S., Turon, M.: Health monitoring of civil infrastructures using wireless sensor networks. In: IPSN, pp. 254–263 (2007)
6. Cai, Z., Ji, S., He, J., Bourgeois, : Optimal distributed data collection for asynchronous cognitive radio networks. In: ICDCS, pp. 245–254 (2012)
7. Ji, S., Cai, Z.: Distributed data collection and its capacity in asynchronous wireless sensor networks. In: INFOCOM, pp. 2113–2121 (2012)
8. Dong, M., Ota, K., Li, X., Shen, X., Guo, S., Guo, M.: HARVEST: a task-objective efficient data collection scheme in wireless sensor and actor networks. In: CMC, pp. 485–488 (2011)
9. Cheng, S., Li, J., Cai, Z.: O(ϵ)-approximation to physical world by sensor networks. In: INFOCOM, pp. 3084–3092 (2013)
10. Akbarinia, R., Pacitti, E., Valduriez, P.: Best position algorithms for top-k queries. In: VLDB, pp. 495–506 (2007)
11. Zheng, J., Zhang, H., Song, B., Wang, H., Wang, Y.: Prediction-based filter updating policies for top-k monitoring queries in wireless sensor networks. In: IJDSN (2014)
12. Chen, M., Ge, Y., Yu, G., Jia, Z., Wang, Y.: An efficient method for cleaning dirty-events over uncertain data in wsns. J. Comput. Sci. Technol. **26**(6), 942–953 (2011)
13. Yufang, W.: Establishment of a chlorophyll forecast equation and its application in red tide forecasting in xiamen offshore area. Ocen Forecasting **29**(2), 39–44 (2012)
14. Chatterjea, S., Havinga, P.: An adaptive and autonomous sensor sampling frequency control scheme for energy-efficient data acquisition in wireless sensor networks. In: Nikoletseas, S.E., Chlebus, B.S., Johnson, D.B., Krishnamachari, B. (eds.) DCOSS 2008. LNCS, vol. 5067, pp. 60–78. Springer, Heidelberg (2008)
15. Law, Y.W., Chatterjea, S., Jin, J., Hanselmann, T., Palaniswami, M.: Energy-efficient data acquisition by adaptive sampling for wireless sensor networks. In: IWCMC, pp. 1146–1151 (2009)
16. Gupta, M., Shum, L.V., Bodanese, E.L., Hailes, S.: Design and evaluation of an adaptive sampling strategy for a wireless air pollution sensor network. In: LCN, pp. 1003–1010 (2011)
17. Alippi, C., Anastasi, G., Di Francesco, M., Roveri, M.: An adaptive sampling algorithm for effective energy management in wireless sensor networks with energy-hungry sensors. IEEE T. Instrumentation and Measurement **59**(2), 335–344 (2010)
18. Considine, J., Hadjieleftheriou, M., Li, F., Byers, J.W., Kollios, G.: Robust approximate aggregation in sensor data management systems. ACM Trans. Database Syst. **34**(1) (2009)
19. Deligiannakis, A., Kotidis, Y., Roussopoulos, N.: Processing approximate aggregate queries in wireless sensor networks. Inf. Syst. **31**(8), 770–792 (2006)
20. Cheng, S., Li, J.: Sampling based (epsilon, delta)-approximate aggregation algorithm in sensor networks. In: ICDCS, pp. 273–280 (2009)
21. Li, J., Cheng, S.: (ϵ, δ)-approximate aggregation algorithms in dynamic sensor networks. IEEE Trans. Parallel Distrib. Syst. **23**(3), 385–396 (2012)
22. Gedik, B., Liu, L., Yu, P.S.: ASAP: an adaptive sampling approach to data collection in sensor networks. IEEE Trans. Parallel Distrib. Syst. **18**(12), 1766–1783 (2007)
23. Mitchell, M.: Handbook of genetic algorithms (L. d. davis). Artif. Intell. **100**(1–2), 325–330 (1998)

Feedback-Based Reduplicate Complex Event Processing in IoT

Mingyue Cui$^{(\boxtimes)}$, Chunhong Zhang, Yuewen Su, and Yang Ji

Beijing University of Posts and Telecommunications, Beijing, China
{Cuimy,Zhangch,suyuewen,jiyang}@bupt.edu.cn

Abstract. Abundant sensors and smart devices deployed in the Internet of Things pose the potential for IoT applications to detect high-level meaningful events. Complex Event Processing technology offers solutions of event pattern(complex event) queries over streams in real time well timely. Yet when CEP is detecting complex events that are continuous for some time, it results in detecting out multiple reduplicate pattern matches, leading to the burden of high output throughput and unnecessary disturb to IoT applications. In this paper, we propose an efficient Event-Feedback Mechanism, to eliminate these reduplicate pattern matches via letting the first detected complex event feedback to the input stream and detecting each selected event based on "evenly spaced time window" and Poisson distribution. The Event-Feedback Mechanism is shown to achieve over three orders of magnitude performance in relieving output throughput, and a range of tested scenarios compared to a significant algorithm proves it practical and effective.

Keywords: CEP · The Internet of Things · Event-feedback mechanism · Reduplicate pattern matches · Continuous complex events

1 Introduction

With the advances of information and communication, many scenarios of the Internet of Things such as hospital application [19], supply chain [18], intrusion detection system(IDS) [4] and intelligent security and protection systems, etc., have set off a new wave of complex event processing(CEP) services by actively detecting meaningful information from high-throughput event streams in **real time** [3,4,5,6,7,8,10,12]. In IoT environment, data generated by billions of physical sensors, such as smoke sensors and RFID tags, form real-time and continuous simple-event streams. CEP technology provides a key paradigm to detect event pattern matches among simple-event stream generated by physical devices based on rules in real time. Typically, IoT application supports decision making to the CEP output through interface between CEP module and upper application [4,22], to make further action for complex events. Fig.1 shows an example of pattern query written in SASE+ language [1,2,4,14,15,16,17], to detect complex Fire events happened on certain floor of certain building within 90 seconds in intelligent security and protection environment.

© Springer International Publishing Switzerland 2015
Y. Wang et al. (Eds.): BigCom 2015, LNCS 9196, pp. 325–336, 2015.
DOI: 10.1007/978-3-319-22047-5_26

```
Query 1:
PATTERN SEQ(Fire+ a[], Fire b )
WHERE skip till next match
AND [alarmBuilding]
AND a[i].alarmFloor = a[i-1].alarmFloor
AND b.alarmFloor > a[1].alarmFloor
WITHIN 90
```

Fig. 1. An example of pattern query in SASE+

In this query, events in event stream will be **partitioned** by the **equivalence attributes**(presented in the equivalence tests [alarmBuilding] and $a[i].alarmFloor = a[i-1].alarmFloor$ in WHERE clause), and pattern matches with different equivalence attributes forms different complex events as output. For instance, a Fire event happened on the third floor of No.10 building and another Fire event on the sixth floor of No.2 building are supposed to be detected by this query immediately when events in event stream satisfy the rules defined in the query.

A true Fire event lasting for about 535 seconds happened on the third floor of NO. 10 building. During this period of time 12 relevant alarm events in NO. 10 building were produced whose alarmFloor values are (3, 3, 4, 4, 3, 3, 3, 4, 5, 3, 3, 5) and timestamp is (1, 2, 12, 39, 130, 131, 136, 143, 156, 447, 493, 535). For better explanation, we sign them as (3-1, 3-2, 4-3, 4-4, 3-5, 3-6, 3-7, 4-8, 5-9, 3-10, 3-11, 5-12), the latter number represents the timestamp order. When a pattern match (3-1,3-2,4-3) was detected according to the rules set in Query 1, the current complex Fire event is detected in real time.

However, current CEP has shortcomings for some scenarios in IOT. When some complex events are continuously lasting, i.e., complex Fire events to be detected by Query 1, multiple latter pattern matches of the same complex event will be detected, as more simple events are produced later for this complex event. We call this kind of complex events the ones with "**continuous temporal property**". Just as the true fire event mentioned above, after the first pattern match (3-1, 3-2, 4-3) is detected, four latter pattern matches (3-1, 3-2, 4-4), (3-5, 3-6, 3-7, 4-8), (3-5, 3-6, 3-7, 5-9) and (3-10, 3-11, 5-12) followed to be detected. Actually, these four pattern matches are regarded as reduplicate ones because their useless and even troublesome to IoT applications after the first pattern match is detected out.

More seriously, current CEP languages or syntax are incapable in realizing both real-time detection and avoiding reduplicate pattern matches in the following time no matter how the rules are organized or defined in queries. CEP only focuses on event pattern detection among event stream, it does not really care about whether pattern matches are reduplicate or not. In particular, the production of these reduplicate pattern matches enhances the burden of output throughput and even brings disturb to IoT upper applications when deciders have to affirm whether complex events really happen. Therefore, it is of great importance for CEP to develop a method to address the problem of these reduplicate pattern matches.

Therefore, it is challenging for CEP to implement the process of reduplicate pattern matches on the premise of real time detection requirement when processing complex events with continuous temporal property. In addition, the duration of such complex events are always with uncertainty, CEP did not support a method to judge when to stop detecting reduplicate pattern matches of such complex events, which remains another challenge in this paper.

To overcome the issues discussed above, we propose an **Event-Feedback Mechanism** around the notion of partitioning [1,9,11] of SASE+. Our solution allows to send the first affirmed complex event as **feedback** to the input stream, continuously caching events in latter reduplicate ones in the same partition(here we call the action of caching such satisfied events as "**absorbing**" to express the meaning of avoiding their output but saving the satisfied ones)and meanwhile based on "**evenly spaced time window**" and **Poisson distribution**, to provide an efficient basis of when to stop absorbing satisfied events of the current continuous complex events.

The remainder of the paper is organized as follows. We cover related work on CEP in section 2. We describe the problem formulation in section 3. In section 4 we illustrate the total of Event-Feedback Mechanism. Results of a detailed performance analysis are presented in section 5. We conclude the paper with remarks on future work in section 6.

2   Related Work

In this section, we survey related work on CEP, focusing in particular on systems and partitioning.

Numerous CEP systems have been proposed. NiagaraCQ [20] is one of the earliest CEP systems. It applies XML-QL as its query language and implements pattern matching via a graph of algebraic operators. However, NiagaraCQ did not mention partition. EventJava supports CEP as patterns guarding event methods [22], which detects equality conditions on keys as partitions and builds sophisticated index data structures based on them. Cayuga, is an algebraic CEP system that also supports partitions based on equality condition on keys [13]. The NEXT system also takes an algebraic approach to CEP, which is measured by automatically putting different operators of the same pattern on different hosts [21]. SASE+ is a CEP language implemented via NFA^b automaton (nondeterministic finite automaton with buffers) and a SASE+ 1.0 released system has been published. SASE+ performs partitioning based on equivalence test and directly supports partitioning by providing specialized syntax for it. We sum up that most of these CEP systems support partitioning in different forms, no of them support the processing of reduplicate pattern matches in partitions.

Our research is based on SASE+ language for it also supports skipping irrelevant events until selecting relevant ones(*skip till next match*, one of the four event selection strategy in SASE+). A pattern query processes a sequence of events that occur in order(not necessarily in contiguous positions) in the input stream and are correlated based on the values of their attributes. In last section, Query 1 shows such a query in SASE+. The PATTERN clause specifies a sequence pattern with two components: a Kleene plus on Fire events, whose results are in a[], and a separate single fire event, stored in b. An equality comparison across all events is referred to as an equivalence test(a shorthand for which is [alarmBuilding]). The predicate on a[bib1] addresses the initial alarmFloor. The predicate on a[i]($i > 1$) requires the alarmFloor number of the current event equal to the former one(a[i-1]). Finally, the query uses a WITHIN clause to specify a 90-second time window over the entire pattern, which compares the time difference between the first and last selected events against the specified time window [1]. Equivalence tests present in a query are used to define partitions. A query can customize the partition definition by underlining any predicated attributes in the WHERE clause. Events in specific partition own the same partition attributes.

3   Problem Formulation

3.1   The Property of Events in IoT Environment

Events in IoT have several properties in two aspects:
1. Complex events in IoT environment have **continuous temporal property**. It means that once a complex event happens, it lasts for a period of time. Fire event introduced in section 1 is a classic example of complex events with continuous temporal property. In contrast, some complex events happen in an instant, which will not last for a period of time. Namely, they do not own consistent temporal property. e.g., stolen events that happen in retail stores, they just happen and end at a sudden time.
2. For input of CEP, besides input of simple events, complex events are considered as another kind of input source. In this paper, complex events as input derive two situations: complex events generated by smart devices and ones as feedback from the output of CEP.

3.2   Why Feedback? and What is Event Feedback?

As mentioned in section 1, reduplicate pattern matches are produced when a complex event with continuous temporal property happens. From the new point of IoT applications, these reduplicate pattern matches are useless and even troublesome. And for output of CEP engine, these reduplicate ones are attributed to unnecessary output volume, which undoubtedly bring burden to CEP engine. Actually, when the first pattern match is output and affirmed by deciders of IoT application, latter reduplicate pattern matches will have been a trouble to upper application of IoT just as mentioned above.

Suppose that if we make full use of the first affirmed pattern match(the first output complex event), let it feedback to the input stream and select the same partition the complex event detected from, and then keep caching the latter reduplicate ones to buffer, will the output of these reduplicate ones be avoided and the burden of output volume be decreased?

4   Event-Feedback Mechanism Based on CEP

4.1   Optimization of Event Stream Models of SASE+

Input Event Stream. In IoT environment, atomic data are often generated by physical devices in high frequency. These atomic data, which are called simple events, compose a large part of input event stream of complex event processing system in IoT. Each event in a stream represents an atomic occurrence of interest at an instant time. A complex event is a composite event that is composited of a sequence of simple events. Each complex event represents a unique pattern match of the query, and contains the join of all the attributes of those simple events and attributes of the complex event itself. For complex events, there is a special parameter, *Tf* short for the time interval of feedback period detection, used to detect if a simple event is produced in the duration of the current continuous complex event.

In the paper, input event stream is composed of both simple events and complex events. When a affirmed complex event coming from the output of CEP is sent to the input of CEP as feedback, this kind of complex events are with people's decision and can be regarded as another source of input complex events besides those produced from smart devices. Our purpose of inputting complex events is to remind CEP engine of the occurrence of them and avoid outputting latter reduplicate complex events(reduplicate pattern matches).

Output Event Stream. For the output of CEP, event stream is a sequence of complex events. Each pattern match is output as a complex event, which is actually a composite event that contains the join of all attributes of the simple events and its event sequence.

4.2   Event-Feedback Mechanism

Our solution is based on the Basic Algorithm proposed in [1] that implements an optimized pattern matching algorithm over event stream. We assume that there is no delay of manual affirming phase, owing that the delay could be decreased or avoided by upper application layer of IoT.

Event-Feedback Mechanism

Input: event stream generated by both simple and complex events; query
Output: complex event stream

Step 1. Judge of simple/complex events. If it is a complex event, it will be proceeded in Feedback Process. If a simple event is input, it will be proceeded in Absorbing Process. The input complex events might come from two sources: feedback from the output of CEP engine and produced by smart devices in IoT environment. The only difference between processing of these two source is whether a feedback action to the complex events from the output of CEP engine is needed. Feedback Process performs uniformly for the input of these two sources. Thus in this paper, input complex events are called as feedback complex events uniformly for the convenience of explaining Event-Feedback Mechanism.

Step 2. Feedback Process. The key point of Feedback Process is creating **feedback partitions** on the basis of information of the input complex events. Firstly, the value of the **equivalence attribute** is extracted from the complex event; and so does the value of **Tf parameter**. Secondly, partition whose value equals to the extracted equivalence attribute value will be judged whether it exists, for a complex event that is feedback from the output of CEP engine, this partition is always existing. While for a complex event produced by smart devices, there might probably not exist such a partition for the reason that before its input, no same complex event belonging to the same partition is detected. Thus, the partition whose value equals to the extracted equivalence attribute value will be created. we call the partition as "**feedback partition**" in the paper, and other partitions as "general partition". If this feedback partition has been created for a previous same complex event, discard the current complex event, for that the input of it will not make any sense.

Each feedback partition is assigned a *buffer* for caching absorbed events and a field of *Tf* for usage in Absorbing Process. Assigning one buffer for each feedback partition benefits largely, because it focuses on absorbing satisfied events in certain feedback partition regardless of other irrelevant events.

Each query has a default value of *Tf*. Setting method of Tf might be various and the specific value of Tf could vary according to factors such as historical experience. Two basic principles of setting Tf can be as follows. Firstly, the value of Tf should not be smaller than the reporting frequency of physical sensors. Secondly, it should build relationship with the time window [1], namely, equal or close to the period of detecting a satisfied pattern match. we set the value of time window as the default value of Tf.

Notice that the default value of Tf is based on historical experience, it can not be guaranteed that it is always the best solution for the future complex events. Adding the dynamic adjusting result to historical experience is a good idea to obtain the best solution of Tf as well.

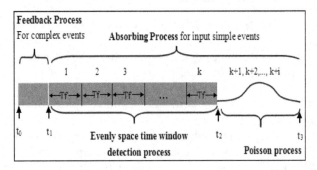

Fig. 2. The whole process of Event-Feedback Mechanism

Step 3. Absorbing Process. For any input of simple events, when it starts to be partitioned into a feedback partition and satisfies the proceeding condition of runs [1], it is identified to enter a period detection procedure to be detected whether it is within the time interval of detection, we call the procedure the "feedback period detection procedure". It aims to detect whether the current complex event is continuous or ended. We call this kind of simple events the "satisfied events". We implement it based on **"evenly spaced time window"** and **"Poissons Distribution"** to respectively detect when the current complex event is continuous and when no satisfied events are produced. We set t_1 as the start time of Absorbing time, which is shown in Fig.2. If a satisfied event is detected within $t_1 + Tf$, this satisfied event will be absorbed; or if a satisfied event is detected within $t_1 + 2 \times Tf$, then it will be absorbed and so on; if it is detected within $t_1 + k \times Tf$, it will be absorbed,but after this time point, no satisfied event is detected within $t_1 + (k+i) \times Tf$ in a continuous period detection based on Tf when i values from 1 to i in turn. Here i follows the random number produced by Poissons Distribution. λ in the formula is set to 3, to represent the average number value of the continuous period detection. Then we can judge concisely that the current complex event has been ended, which is corresponding to the ending point t_3 in Fig.2.

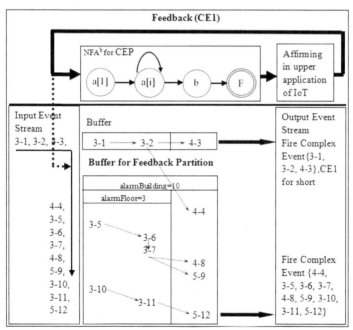

Fig. 3. Event-Feedback Mechanism of Query 1

Runs and matches in Event-Feedback Mechanism. A run of an NFA^b automaton is uniquely defined by i) the sequence of events that it has selected into the match buffer, ii) the naming of the corresponding units in the buffer, and iii) the current NFA^b state. Moreover, when a run has reached the final state [1,11], it is identified as an "accepting run", which will transform to one independent match in Basic Algorithm. Each match here represents a unique event sequence that satisfy the corresponding query. But in the Absorbing process, all accepting runs will be recycled instead of transforming to matches. So when the current complex event is judged ended, all the absorbed events in the buffer of the feedback partition will be output as a match. Notice that such output complex events are signed from feedback partition, they will not be feedback again to the input stream.

If the input simple event is partitioned into a general partition, it shows that the event is not produced for the current complex event, because any event can belong to only one partition. It will then be proceeded by the Basic Algorithm.

Therefore, we name these two sub-procedure as **evenly spaced time window detection process** and **poisson process**. The whole process of Event-Feedback Mechanism is illustrated in Fig.2. The time t_0 represents for the time when a complex event is sent into CEP, and t_1 as the start time of Absorbing Process. It is noted that t_1 is almost equal to t_0, which is to say, when a complex event is sent into CEP, CEP is ready to absorb any satisfied events in the following event stream. The duration of the poisson process is random for that the ending point t_3 of it is unfixed, which is caused by the randomness of the number of referenced Tf. We will also implement the poisson process based on exponential distribution or markov chain in the future.

Step 4. Recycle buffers. When a complex event is detected ended, the buffer of the relevant feedback partition will be recycled, the feedback partition that has accomplished absorbing work will also be destroyed, waiting for next time to be created. Fig.3 shows the Event-Feedback Mechanism in detail through the example of Query 1 in section 1.

Partition Query. In this paper, equivalence test is a key point that can make a difference in Event-Feedback Mechanism. As known, equivalence test makes events in event stream in different partitions. According to the research in equivalence attributes in [1,5,17,18,19], we classify queries into two kinds: **Partition Query** and **General Query**. Partition Query owns one or more equivalence tests in its WHERE clause. In contrast, General Query does not conclude any equivalence tests in its WHERE clause.

For Partition Query and General Query, the number of feedback partitions distinguishes, which follow such two rules logically. For Partition Query, $[feedback partition] = [complex event]$, where $[]$ represents the number of parameter in $[]$; for General Query, $[feedback\ partition] = 1$ when $[complex\ event] >= 1$. Logically a General Query is regarded as one that has only one partition. We will analyze the reasons of these two rules in section 5.

5   Performance Evaluation

5.1   Experimental Setup

We have implemented Event-Feedback Mechanism of CEP based on the SASE+. All measurements were obtained on a workstation with a Pentium 42.8 Ghz

CPU and 1.0 GB memory running Java Hotspot VM 1.5 on Linux 2.6.9, which is consistent with those in the workstation of Basic Algorithm [1].

5.2 Experimental Evaluation

In this experiment, we will illustrate the performance of Event-Feedback Mechanism based on Fire events. Queries for Fire events were generated from a template "$PATTERN(Fire + a[])WHERE\{Pi\ AND\ a[i].alarmFloor > a[i-1].alarmFloor\ AND\ a[i].alarmFloor < 10\}WITHIN\ 120$". The queries were designed to detect Fire events happened under tenth floor of certain building. The attributes alarmFloor and alarmBuilding represented the floor number and building number that a Fire event happened in. The iterator predicate, Pi used in Kleene closure was varied as listed in Table 1 to represent Partition Query and General Query respectively. Predicate constraints were defined in the WHERE clause, and time window was defined as 120 seconds in the WITHIN clause. Here we use a Kleene closure a[], to save more than one satisfied simple events whose event type is Fire.

Table 1. Values of Pi

Pi	P1:[alarmBuilding]
	P2:True

Table 2. Variables of CE1 and CE2

Variables	Tf(s)	Equivalence attributes	Duration(minutes)
CE1	90	alarmBuilding=6	10
CE2	90	alarmBuilding=10	7

We simulated Fire event streams in the following experiment. In each stream, all events have the same type, Fire, that contains two attributes, alarmBuilding and alarmFloor. Since effects of feedback of multiple complex events were alike to that of two complex events, we tested the process when there were two simulated complex events, CE1 and CE2 for short that will be feedback to the event streams. Main attributes and variables are shown in Table 2. The irrelevant attributes were omitted because they are irrelevant to performance of Event-Feedback Mechanism.

Fig.4 concludes three figures, showing the throughput of absorbing work only, the output throughput during CEP engine is doing absorbing work and the output throughput when finishing absorbing respectively, compared to that without feedback that tested in Basic Algorithm [1]. The X-axis of them shows the various values of Pi. The Y-axis shows the output throughput, whose unit is $events \times 100/second$.

In the absorbing process shown in the middle figure of Fig.4, we see that for both P1 and P2, the output throughput for both feedback CE1 and feedback CE2 is zero, which is corresponding to the Null bar. While the output throughput without feedback are 900 and 1200 events per second, which is to say, nearly three orders of magnitude of the throughput is relieved. As shown in Table 3, for P1, 89 runs were produced for the feedback CE1 and 63 runs were produced for the feedback CE2. All these runs when without feedback would be either output as matches(132) or discarded(152-132=20) owing to the extension of time window. And for P2, 237 runs were produced for the feedback CE1 in the absorbing process just as the number of without feedback, while no runs were produced for CE2. This is because for queries that own equivalence tests, different runs for different feedback complex events will be produced in each relevant feedback partition under Event-Feedback Mechanism, while for a query that does not own any equivalence tests, all runs are produced in the only one feedback partition created by the first feedback complex.

Table 3. Number of Runs and Matches

		Runs	Matches
P1	CE1(FB)	89	1
	CE2(FB)	63	1
	N/A	152	132
P2	CE1(FB)	237	1
	CE2(FB)	0	0
	N/A	237	206

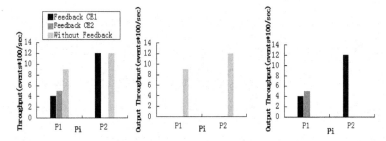

Fig. 4. Throughput for absorbing work only, Output throughput during absorbing and finishing absorbing

When the CEP engine is finishing absorbing shown in the right figure in Fig.4, we see that for P1, the output throughput increased to 400 and 500 events per second respectively for the feedback CE1 and CE2, while for P2, the throughput of only feedback CE1 increase to a high volume 1200, equal to the throughput of without feedback in the absorbing process. The reason for that is when finishing absorbing, all events in accepting runs will be output as one match, resulting in a high increase in the output throughput. The left figure in Fig.4 shows the throughput of absorbing work by itself, which proves the consistency of the output throughput when CEP is absorbing and the output throughput when CEP is finishing absorbing.

In summary, firstly for the query under P1 who belongs to Partition Query, more than one feedback partitions were created to absorb different runs for different feedback complex events, and only 2 matches for CE1 and CE2 respectively were output rather than 132 matches were output. While for the query under P2 who belongs to General Query, Event-Feedback Mechanism identified the feedback complex events as the same ones and runs were produced in just one feedback partition after processing the first feedback complex event, so only one match for CE1 was output, rather than 206 matches were output. Secondly, Event-Feedback Mechanism achieves over three orders of magnitude performance in relieving both output throughput during absorbing in the scenario of detecting Fire events.

6 Conclusion and Future Work

This work is the first to support complex event processing with the function of event feedback and under the consideration of complex events with continuous temporal property. We propose a Event-Feedback Mechanism solution based on partitioning of SASE+. It gracefully tackles the CEP-specific challenges including reduplicate pattern matches and input of complex events. Compared to the Basic Algorithm, Event-Feedback Mechanism is lightweight, thus achieving several orders of magnitude savings in both output throughout and output matches.

Currently, we assume that manual affirming process is of no delay, thus few reduplicate pattern matches will be produced during the manual affirming process. In the future we will also take the delay caused by manual affirming process into account.

Acknowledgments. The work is supported by the National Science and Technology Major Project (No.2015ZX03003012).

References

1. Agrawal, J., Diao, Y., et al.: Efficient pattern matching over event streams. In: SIGMOD, pp. 147–160 (2008)
2. Mozafari, B., et al: High-performance complex event processing over xml streams. In: SIGMOD, pp. 253–264 (2012)
3. Fengjuan, W., Xiaoming, Z., et al.: The research on complex event processing method of internet of Things. In: ICMTMA, pp. 1219–1222 (2013)
4. Yu , C., Jui, C., Fu, H., et al.: Complex event processing for the internet of things and its applications. In: CASE, pp. 1144–1149 (2014)
5. Zhang, H., Diao, Y., Immerman, N.: On complexity and optimization of expensive queries in complex event processing. In: SIGMOD, pp. 217–228 (2014)
6. Jun, C., Chi, C.: Design of complex event-processing IDS in internet of things. In: ICMTMA, pp. 226–229 (2014)
7. Govindarajan, N., Simmhan, Y., Jamadagni, N., et al.: Event processing across edge and the cloud for internet of things applications. In: Proceedings of the 20th International Conference on Management of Data. Computer Society of India, pp. 101–104 (2014)
8. Wang, Y., Cao, K.: A Proactive Complex Event Processing Method for Large-Scale Transportation Internet of Things. International Journal of Distributed Sensor Networks (2014)
9. Mayer, R., Koldehofe, B., Rothermel, K.: Predictable Low-Latency Event Detection with Parallel Complex Event Processing. IEEE Internet of Things Journal (2015)
10. Li, Y., Lee, J., et al.: A CEP-based smart residential service system. In: ICAST, pp. 233–237 (2014)

11. Hirzel, M.: Partition and compose: parallel complex event processing. In: DEBS, pp. 191–200 (2012)
12. Jing, X., Zhang, J., et al.: Oveview of complex event processing technology and its application in logistics Internet of Things. Journal of Computer Applications, 2026–2030 (2013)
13. Demers, J., Gehrke, J., et al.: Cayuga: a general purpose event monitoring system. In: CIDR, pp. 412–422 (2007)
14. Wu, E., Diao, Y., et al.: High-performance complex event processing over streams. In: SIGMOD, pp. 407–418 (2006)
15. Zhang, H., Diao, Y., et al.: Recognizing patterns in streams with imprecise timestamps. PVLDB **3**(1), 244–255 (2010)
16. Gyllstrom,D., Agrawal, J., et al.: On supporting kleene closure over event streams. In: ICDE, poster (2008)
17. Diao, Y., Immerman, N., et al.: SASE+: a agile language for Kleene Closure over event streams, UMass Technical Report 07–03 (2007)
18. Luckham, D.: Event Processing for Business: Organizing the Real-Time Enterprise. Wiley (2011)
19. Wang, D., et al.: Active complex event processing over event streams. PVLDB **4**(10), 634–645 (2011)
20. Chen, J., DeWitt, D.J., Tian, F., Wang, Y.: NiagaraCQ: a scalable continuous query system for internet databases. In: International Conference on Management of Data, SIGMOD, pp. 379–390(2000)
21. Schultz-Miller, N., Migliavacca, M., Pietzuch, P.: Distributed complex event processing with query rewriting. In: Conference on Distributed Event-Based Systems, DEBS (2009)
22. Jayaram, K.R., Eugster, P.: Scalable Efficient composite event detection. In: Clarke, D., Agha, G. (eds.) COORDINATION 2010. LNCS, vol. 6116, pp. 168–182. Springer, Heidelberg (2010)

An Identification Algorithm in Grouping and Paralleling for Data-Intensive RFID Systems

Duan Litian[1], Wang John Zizhong[2], and Duan Fu[1(✉)]

[1] Taiyuan University of Technology, Taiyuan, China
dylann@yeah.net, duanfu@tyut.edu.cn
[2] Virginia Wesleyan College, Norfolk, USA
zwang@vwc.edu

Abstract. In a radio frequency identification (RFID) system, the throughput is limited by tags collision, especially when the tags are distributed intensively. In this paper, we firstly analyze several representative Dynamic Frame Slotted ALOHA (DFSA) algorithms as a review. Then, we propose a novel DFSA algorithm based on grouping and parallel to identify large, unknown numbers of tags in RFID system with the data-intensive distribution. We would like to introduce our algorithm in estimated method for unread tags' population and grouping strategy for in-frame or out-frame identification processes. Additionally, parallel strategy is imported during the whole communication process to enhance the identification speed. In details, our estimated method is composed in some appropriate sub-methods to cover every possible scenario, and grouping strategy is based on twice-divided beginning with a roughly grouping firstly. As a result, the simulation running by MATLEB exhibits a satisfactory throughput in system efficiency and identification speed when comparing with the similar conventional algorithms.

Keywords: RFID · DFSA · Data-intensive · Grouping · Parallel

1   Introduction

Accompanied by the rapid development of the Internet of Things (IoT), radio frequency identification (RFID) is widely applied in object management with big data, such as warehouse and supply chain management, logistics tracking, animals' identification, and electronic toll collection, due to its long read range, low power and reduced cost. As a statistic, there are only 1.3 billion RFID tags produced in 2005 but this figure will rise to 120 billion in 2015[1]. What's more, the tag size of RFID system is sharply rising from hundreds to tens of thousands. In a data-intensive RFID system, an advisable anti-collision algorithm could assist the identification in a more efficient way. However, some of the existing algorithms could not satisfy the throughput of data-intensive RFID systems, with imprecise estimation methods for the population of unread tags and serial communications. In this paper, our algorithm is proposed to focus on the data-intensive RFID systems, designed in three parts to increase the system throughput and reduce the energy cost: 1) a series of precise

© Springer International Publishing Switzerland 2015
Y. Wang et al. (Eds.): BigCom 2015, LNCS 9196, pp. 337–346, 2015.
DOI: 10.1007/978-3-319-22047-5_27

estimation methods to satisfy each possible case; 2) a grouping-enhanced reading strategy; 3) two parallel models. By multi-aspect simulations, the proposed algorithm is superior to some conventional RFID algorithms.

The rest of our paper is organized as follows: Section 2 presents some conventional and existing DFSA algorithms. Section 3 proposes our grouping-enhanced algorithm detailedly. Simulations are showing in Section 4 involving the identification speed and system efficiency. At last, a conclusion and further work will be drawn in Section 5.

2   Overview of DFSA-Based Algorithms

In most RFID systems, the initial population of tags is indeterminate. On the other hand, the communication between the reader(s) and tags shares a same channel, which would easily cause the collision when more than one tag transmit the ID(s) to the same reader(s) simultaneously, and frequently collision enmeshes the system efficiency in a higher throughput. To deal with that, some protocols are proposed, such as TDMA (Time Division Multiple Access) [2], which is widely developed mainly in tree-based algorithms [3,4,5] and ALOHA-based algorithms [6,7]. From the development of pure ALOHA to dynamical frame slotted ALOHA (DFSA), the system efficiency is doubled to 36.8%. However, DFSA is not faultless; many DFSA-based algorithms are proposed in different perspectives to improve the throughput on different levels.

2.1   Estimated Method

Passion-Distribution-Based Method

In [7], Schoute claimed that every slot-selection obeys Poisson distribution with $\lambda = 1$, the average number of collision tags in every slot is about 2.39, written as formula (1).

$$n = 2.39 \times N_C \tag{1}$$

Maximum-Probability-Based Method

In [8], the author estimated the population of unread tags by partly observed results (M observed slots out of N slots), and got the optimal estimated population of tags by maximizing the probability

Collision-Ratio-Based Method

Based on repeating experiments, a collision-ratio-based algorithm was proposed in [9]. He offered a relationship between the collision factor β and the probability of collision C_m. For instance, β equals to 2 (minimum value) means that C_m is no more than 10%, and β equals to 7.2 (maximum value) means that C_m is between 99.24% and 100%.

2.2   Grouped Strategy for Large Population of Tags

As the number of unread tags is much larger than the frame size, it is testified that the probability of tags collision increases rapidly. To cope with this situation, the author in [10] proposed a grouped method for grouping the unrecognized tags. In this paper, the author assumed that grouped algorithm uses the power of two (2,4,8, ...) for grouping the tags, and the number of groups is decided by the modulo operation. For example, the unread tags would be grouped in 2 groups, the remainder of 0 after the division will be one group while the remainder of 1 will be the other group. In order to maximize the system efficiency, the author got an upper limit value by doing derivation, which equals to 354. At the beginning of an interrogation process, the reader would select one group of tags for identification, and select another group till all tags are recognized. EDFSA will cost a lot of time in the process of identification, because the reader could only select another group when all tags in selected group are identified successfully.

As an improved EDFSA [11], the author allowed the reader selects one group to identify like EDFSA algorithm first, and after the first cycle of identification, the reader will send a command to add the unread tags' number and tags' number of one group and make a comparison with 354. If it is larger than 354, the number of unrecognized tags will be identified according to DFSA separately, after one cycle of identification and repeat above steps; if it is smaller than 354, a new group will be combined with the unread tags and tags' number in the next group. GroupIEDFSA indeed achieved an improvement, but the author supposed that the number of tags is a certain quantity, which is only suitable for the known population of tags, since most of the practical applications are unknown for the tag number, we need to consider about the estimation method for tag number.

3   Proposed Algorithm

In this section, we would like to introduce our idea particularly. The skeleton of the proposed algorithm is shown in Figure 1. Since the number of tags is unknown in the most real applications, our algorithm would estimate the population of tags twice to enhance the estimation. First, an estimation (denoted by $Ntag_1$) for tags population is compelled to do by using the proposed estimation methods in the following subsection. And if the value of $Ntag_1$ is smaller than 256, there is no need to be grouped; otherwise, $Ntag_1$ tags will be grouped into $Ngroup_1$ groups. Then, m groups will be randomly selected from $Ngroup_1$ groups, for a second round tags population estimation (denoted by $Ntag_2$). Based on this more accurate estimation result, the tags will be re-grouped into $Ngroup_2$ groups. Lastly, the identification process will be finished with the assist of num_reader readers concurrently, in one of the two different kinds of parallel algorithms, which will be discussed in the following sub-section.

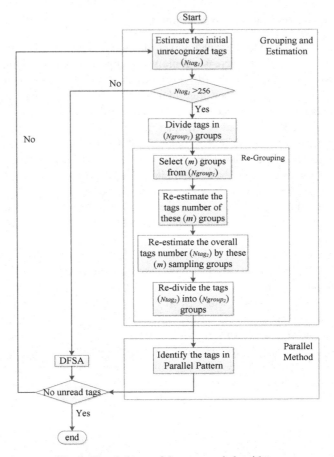

Fig. 1. The skeleton of the proposed algorithm

3.1   Estimation Methods

The core idea of our estimated method is based on using a sample to estimate the overall, by N_E, N_S, and N_C, where $N_E+N_S+N_C=M$, $M \leqq N$. The skeleton of the tags estimation method is shown in Figure 2. In order to enhance the accuracy of the estimation, we consider, as far as possible, any possible scenarios, each of them would have an appropriate method to achieve the optimal results.

1) $N_S+N_E=0$, means that all of the slots are collided. Therefore, we get the estimated M_{tag} by using formula (2):

$$M_{tag} = 7.2 \times N_C \qquad (2)$$

where the value of 7.2 is referenced from [9].

2) $N_C>N_S+N_E$, where the amount of collided slots are in the majority. Therefore, we get the estimated M_{tag} by using collision-ratio method.

3) $N_E>N_S+N_C$, where the number of idle slots are in the majority. Therefore, we get the estimated M_{tag} by using idle-ratio method.

4) Except for them above, we first get a roughly estimated M' by formula (3):

$$M' = 2.39 \times N_C + N_S \tag{3}$$

where the value of 2.39 is based on Schoute's in [7]. In this scenario, if idle slots are existed, we compute an adjusted-estimated M_{tag}' and then divide into two sub-scenarios:

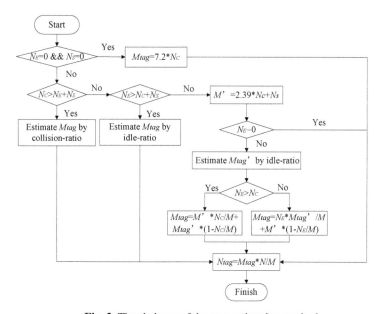

Fig. 2. The skeleton of the tags estimation method

a. $N_E>N_C$, we get the estimated M_{tag} by using formula (4):

$$M_{tag} = \frac{M' \times N_C}{M} + M_{tag}' \times (1 - \frac{N_C}{M}) \tag{4}$$

b. $N_C>N_E$, we get the estimated M_{tag} by formula (5):

$$M_{tag} = \frac{M_{tag} \times N_E}{M} + M' \times (1 - \frac{N_E}{M}) \tag{5}$$

After that, we input the formula (6) to estimate the tags population,

$$N_{tag} = \frac{M_{tag} \times N}{M} \tag{6}$$

In [12], the author proposed a relationship table of Q-value and n-value, helping us selects the appropriate Q-value after the tags population estimating.

3.2   Grouping Strategy

For a data-intensive distributed application in RFID system, classical DFSA algorithms are lamed since the frame size cannot be enlarged limitlessly. On the other hand, optimal system efficiency will be achieved only when the frame size is equal to the tags number. Therefore, the grouping strategy is proposed to ensure a satisfactory throughput in data-intensive distributed applications, the threshold of grouping number is calculated by an equation, which is built upon the same system efficiency in different frame size, showing in formula (7).

$$\frac{a_1^{L,n}}{L} = \frac{a_1^{L,2n}}{L}, \text{ expanding as: } n(\frac{1}{L})(1-\frac{1}{L})^{n-1} = n(\frac{1}{L})(1-\frac{1}{L})^{2n-1} \tag{7}$$

where a_1 is calculated by (8), and the frame size L is supposed as 256.

$$a_1 = \varepsilon\left(P_S\right) = L \times B\left(1\right) = \binom{n}{1}\left(\frac{1}{L}\right)^1\left(1-\frac{1}{L}\right)^{n-1} \tag{8}$$

After calculation, n equals to 354, means that the tags will be grouped only when the population exceeds 354. Based on this, the system efficiency is always beyond 35%, which is deemed as a high efficiency.

Since the tags are divided into some groups, Definition1 is written down as a set representation for grouping.

Definition 1: T→G is an onto mapping, where T is the set of tags, G is the set of groupings. Expressed as: T={$T_i | i \in [1, Ntag]$},G={$G_i | i \in [0, Ngroup]$}.

In our algorithm, we group the unread tags in two times, to get a more accurate estimation results. For the first time, the number of groups ($Ngroup_1$) is calculated by the estimation value of tags population ($Ntag_1$) module frame size (L), showing in formula (9).

$$Ngroup_1 = \left\lceil \frac{Ntag_1}{L} \right\rceil \tag{9}$$

Based on Definition 1, G = {G_0, G_1,..., $G_{Ngroup1}$}. The reader will select m groups from G randomly, and after re-estimating the number of tags in these m groups, a weighted average value is gained by formula (10), to re-calculate the overall value of tags number by running formula (11).

$$Atag = \frac{\sum_1^m NG_i}{m} \tag{10}$$

$$Ntag_2 = Atag \times Ngroup_1 \tag{11}$$

The second time grouping process is based on $Ntag_2$, similar with the first round, the number of groups ($Ngroup_2$) is calculated by the estimation value of tags population ($Ntag_2$) module frame size (L), showing in formula (12), expressing as G = {G_0, G_1,..., $G_{Ngroup2}$}.

$$Ngroup_2 = \left\lceil \frac{Ntag_2}{L} \right\rceil \tag{12}$$

In this sub-section, we only focus on the grouping strategy, the estimation methods for tags population is not mentioned, but we have described that detailedly in the previous sub-section.

3.3   Parallel Algorithms

In order to enhance the identification speed of data-intensive distributed applications, the parallel algorithm is considered after grouping the tags. Additionally, we propose 2 parallel models to cover the practical applications. Suppose there are $Nreader$ readers, expressing in a set as: $R = \{R_1, R_2,\ldots, R_{Nreader}\}$, and the set of grouping tags is $G = \{G_0, G_1,\ldots, G_{Ngroup2}\}$.

Classical Parallel
Since there are $Nreader$ readers, each reader will communicate with one group randomly, and only when the readers identify all the tags in the groups they selected, could they start the next identification round, till no unread tags left. Definition 2 describes this parallel process in set expression.

Definition 2. R is a nonempty set and G is also a nonempty set, existing an onto mapping G→R, where G=$\{G_0, G_1,\ldots, G_{Ngroup2}\}$ and $R = \{R_1, R_2,\ldots, R_{Nreader}\}$. Suppose G is composed by n disjoint nonempty subsets, written as subGi, and each of $n-1$ subset has $Nreader$ elements, and the last subset has l elements, where $Ngroup_2+1 = (n-1)\times Nreader + l$. Then, the onto mapping G→R could be divided as some onto mappings, denoted as: subGi→R, $i\epsilon[1,n]$.

Enhanced Classical Parallel
Different with the classical parallel model, the enhanced classical parallel model is proposed to ensure a higher identification throughput. Instead of reading all tags in the groups readers selected before, each reader will stop this round and communicate with other groups if the current identification throughput is lower than t, for instance, t equals to 32%. Definition 3 is the set expression of this model.

Definition 3. Based on Definition 2, each subGi is composed by two other sets, one is unGi, and the other is suGi, $i\epsilon[1,n]$. Suppose that unG is the set composed by unGi, unG could be divided into s disjoint nonempty subsets, written as subunGi, and each of $s-1$ subset has $Nreader$ elements, and the last subset has l elements, where $n +1 = (s-1)\times Nreader + l$.

4   Simulation

The comparison between our algorithm and others are tested by extensive simulations based on Monte Carlo method, running in MATLAB. To ensure the convergence and universality of the results, simulations are run repeatedly in 3000 experiments.

Our algorithm is mainly compared with the Group Improved Enhanced DFSA (GIED) [11], in a situation of data-intensive distribution.

In [13], the author provided a time cost of each operation in the communication between readers and tags: reader needs 0.4ms to detect a null time slot, 0.8ms to detect a collision time slot and 2.4ms to transmit a 96-bit ID. Based on this, we make a comparison in time cost between GIED and our proposed algorithms, listing in Table 1. Take 1000 tags as an example, GIED used 4,467ms to finish the identification, while CP could reduce the time cost to 2185.2ms and ECP could further reduce the time cost to 1523.6ms.

Table 1. Time Comparison of GIED, CP and ECP

Tags Population	Groups	GIED(ms)	CP(ms)	ECP(ms)
1000	4	4,476.0	2185.2	1523.6
3000	12	12,512.8	3928.8	3067.2
5000	20	20,491.2	6365.0	5833.0
7000	27	28,664.8	8084.5	7157.1
9000	35	36,508.4	12,312.7	9211.8
10000	39	40,642.8	15,471.0	10,572.1
11000	43	44,768.4	20,012.1	12,473.2
12000	47	48,542.0	27,153.5	14,367.1

Figure 3 displays the ratio of success, collision and idleness in different populations of tags, maximizing in 1.2K. Figure 3(a) is the result of GIED (since the number of tags is established, estimation is unnecessary), (b) is the result of classical parallel model in our proposed algorithm (abbreviated as CP), and (c) is the result of enhanced classical parallel model in our proposed algorithm (abbreviated as ECP), details of CP and ECP are introduced in section 3.4. GIED shows a satisfactory system efficiency, but it could not keep the same result if the number of tags is unknown. Therefore, we propose our algorithm for the tags with unknown quantity but intensive distribution situations. CP is one of our parallel algorithms, but the throughput is limited, because the reader has to read all of the tags in the group that it selected before. Moreover, the idleness ratio is bigger with number of tags increasing since the Q-value could not be adjusted dynamically. Therefore, we proposed ECP to ensure the throughput is always high. Comparing Figure 3 (a), (b) and (c), it is obvious that (c) has lower idleness ratio and estimation error than these in (b) with the number of tags increasing. In addition, the throughput of (c) is similar like (a), where the number of tags in former one is unknown.

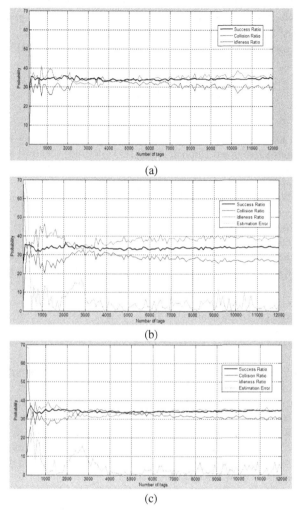

Fig. 3. Number of tags vs. Probability in: (a) GIED; (b) Classical Parallel; (c) Enhanced Classical Parallel

5   Conclusion

In this paper, a DFSA-based algorithm is proposed for data-intensive RFID system, involving in estimation method, grouping strategy and parallel algorithms. After the simulation, we testified that readers identify the large number of RF-tags in parallelization could shorten the time cost and enhance the system efficiency by using our grouping and parallel algorithm. However, we ignored the signal attenuation and influence in real applications of wireless network, which might lower the throughputs correspondingly. To strengthen the practicability of our proposed algorithm, we will do some improvements in the future researches, meanwhile, propose some new parallel models for RF-identification process.

Acknowledgements. We would like to thank all the reviewers for their helpful comments. This project is supported by National Natural Science Foundation of China (No. 61472271), Key Scientific and Technological Projects of Shanxi Province (Nos. 20130321001-09 and 2007031129).

References

1. Dheeraj, K., Kwan-Wu, R.: A survey and tutorial of RFID anti-collision protocols. Communications Surveys and Tutorials, IEEE **12**(3), 400–421 (2010)
2. Rom, R.: Multiple Access Protocols Performance and Analysis. Springer, London, Limited, p. 177 (2011)
3. Hush, D.R., Wood, C.: Analysis of tree algorithm for RFID arbitration. In: Information Theory, Proceedings of IEEE International Symposium, p. 107 (1998)   ╲
4. Myung, J., Wonjun Lee, S.: Adaptive binary splitting for efficient RFID tag anti-collision. IEEE Communications Letters **10**(3), 144–146 (2006)
5. Qian, Z., Chen, C., You, I., Lu, S.: ACSP: a novel security protocol against counting attack for UHF RFID systems. Computers & Mathematics with Applications **63**(2), 492–500 (2012)
6. Syed A., Ahson, M.I.: RFID Handbook: Applications, Technology, Security, and Privacy. USArCRC Press, pp. 18–40 (2008)
7. Schoute, F.: Dynamic frame length ALOHA. IEEE Transactions on Communications **31**, 565–568 (1983)
8. Chen, W.-T.: An accurate tag estimate method for improving the performance of an RFID anti-collision algorithm based on dynamic frame length ALOHA. IEEE Transactions on Automation Science and Engineering **6**, 9–15 (2008)
9. He, J.H.: Study on Anti-Collision Algorithm for Moving Tags in RFID System. Computer Application, 2048–2051 (2011)
10. Lee, S.-R., Joo, S.-D., Lee, C.-W.: An enhanced dynamic framed slotted ALOHA algorithm for RFID tag identification. In: Proceedings of 2nd Annual International Conference on Mobile and Ubiquitous Systems: Networking and Services, pp.166–174. IEEE Computer Society, Washington D.C. (2005)
11. Wang, H., Xiao, S., Lin, F., Yang, T., Yang, L.T.: Group Improved Enhanced Dynamic Frame Slotted ALOHA Anti-Collision Algorithm. The Journal of Supercomputer **69**(3), 1235–1253 (2014)
12. Bueno-Delgado, M.V., Vales-Alonso, J., Gonzalez-Castaño, F.J.: Analysis of DFSA anti-collision protocols in passive RFID environments. In: IECON 2009. 35th Annual Conference of IEEE, pp.2610–2617. Industrial Electronics Publishing (2009)
13. Li, T., Chen, S., Ling, Y.: Efficient protocols for identifying the missing tags in a large RFID system. IEEE/ACM Transactions on Networking **21**(6), 1974–1987 (2013)

I Know When to Do the Replenishment

Jumin Zhao$^{(\boxtimes)}$, Na Li, and Deng-ao Li

Taiyuan University of Technology, Taiyuan, China
{zhaojumin,lidengao}@tyut.edu.cn, clina0991@163.com
http://www.tyut.edu.cn

Abstract. RFID technology is an emerging one poised to extensive applications such as warehouse, retail, and object tracking. Replenishment is a thorny and efficiency-need problem, which requires to report the remaining information to the server seasonably to guarantee goods supply uninterruptedly. Traditional protocols are aimed at identification, detection and estimation. In this paper, we in turn propose two progressive protocols to prove the second one MMP is the best. We divide the MMP into two phase, firstly, we use the category ID to classify all the tags to serve preferably for laying the root of replenishment whole planning; secondly, we borrow the idea from micro view to estimate the macro, which only use the successful singleton slots, consequently avoiding the collision, and is a much easier design and novel enough. The simulation results show that our protocol significantly solves the replenishment problem and reduces the total scanning time, which is much more efficient.

Keywords: Replenishment · Classification · Efficiency · RFID

1 Introduction

Radio Frequency Identification (RFID) technology has attracted much more attentions nowadays with its small form and ultra-low power, and been widely employed in a wide range of applications such as warehouse supply chain management, item tracking, transportation and logistics [1]. In a RFID system, small tags each containing an ID in its memory can be attached to products and be read several meters away via the RFID readers, usually in the form of contactless, either a stationary gateway or a portable device.

Image a particular problem replenishment! In a large storage warehouse, thousands of goods attached tags are out of the warehouse for the supplying demand. Or in a large supermarket, kinds of goods are selected by customers from the shelves. Thus, the problem is whether the remaining inventories are sufficient or not, and when the staffs do the replenishment for uninterrupted supply chain. If just small portion of goods are away, do the supply, then it will be time-consuming, strength-consuming and unnecessary; if most items have been carried away, but don't do the replenishment in time, it will lead to the supply break or the demand exceed the supply. So when to supply is a thorny

© Springer International Publishing Switzerland 2015
Y. Wang et al. (Eds.): BigCom 2015, LNCS 9196, pp. 347–359, 2015.
DOI: 10.1007/978-3-319-22047-5_28

problem. Further, the replenishment threshold can be also different, for some best-sellings needing to keep efficient almost 80% remaining; while some flat-sellings are not necessary to supply continually, just keep the 40% left. Aiming at the particular and practical replenishment problem, and different requirements, adopting a one-to-one protocol is meaning.

However, current protocols are mainly aimed at the identification, the missing tags or the estimation. Firstly, they are all not the one-to-one protocols directing at the supply problem; secondly, even if can barely used, there are also some shortcomings existing such as efficiency, energy and complexity. Therefore, to address the above challenges, we propose an efficient and practical protocol micro-macro (MMP), which is firstly tailor-made for our problem, and then efficient with the category ID information.

The major contributions of this paper are briefly summarized as follows.

(1) We consider a practical replenishment problem in RFID systems and aim at a solution without collecting tag IDs;

(2) We propose two progressive protocols that use a compact structure to illustrate our MMP protocol is the efficient and tailor-made one.

(3) We analyze the performance of the proposed protocols theoretically and optimize the parameter settings to achieve the best time efficiency.

(4) The simulation results show that our protocol significantly solves the replenishment problem and reduces the total scanning time, which is much more efficient.

The rest of this paper is organized as follows. The related work is reviewed in Section 2. Section 3 presents the system model and problem formulation. The detailed progressive protocols are presented in Section 4. And section 5 evaluates the performance of the protocols. Finally, Section 6 concludes this paper.

2   Related Work

One closely related problem is RFID missing tag identification, in [2], Tan, Sheng and Li designed novel protocols to detect the missing-tag event with probability when the number of missing tags exceeds with m'. However, the protocols or the missing-tag protocols aim at the missing ones in the whole system rather than each categorys existence situations.

Another closely related problem is tag identification which identifies the tags through collision arbitration [3]. The anti-collision tag identification protocols are subdivided into two types, TDMA-based and CDMA-based protocols. The TDMA-based protocols can be further classified into two categories, that is, aloha-based protocols [4] and tree-based [5]. Aloha-based protocols avoid tag collisions by arranging tags to transmit their information in different time slots. But when multiple tags choose the same slot to respond, the methods are disabled. Tree-based methods resolve the collision problem by splitting collided tags into subsets, and then scheduling the subsets to transmit information separately. The CDMA-based protocols [6] use the code division multiplexing techniques to decode tag collisions. But, spectrum resource is limited. At the same time, the complex encoding and decoding techniques also make it not suitable for our storage system.

Instead of identifying the tags, some cardinality estimation protocols [7][8] are proposed. They mainly aim at estimating the number of tags, but can't be directly borrowed to our replenishment problem as they can't distinguish the categories, respectively.

3   Preliminaries

3.1   Problem Formulation

Based on the specification of the Philips I-Code system [5], it can be shown that $t_s = 0.4ms$ (for one bit) and $t_{tag} = 2.4ms$ (for a 96-bit tag ID). And we assume that one of the fields that actually 20-bit called category ID specifies the category the product belongs to, it can be as generic as a brand or a model number.

We consider a large RFID storage system with $T = t_1, t_2, ..., t_n$ representing all the RFID tags in the whole interrogation zone covered by all readers, where n is the total number of tags. $C = c_1, c_2, ..., c_m$ represents different categories of all tags, where m is the total number of all the categories. $C_{ij'} = c_{i1'}, c_{i2'}, ..., c_{ij'}$, represents the number of the i-th category in a reader's scanning area latter, where $i = 1, 2, ..., m$ and $j' = 1, 2, ...n$. More, we set $C_{ij} = c_{i1}, c_{i2}, ..., c_{ij}$, representing the initial number of category $i, j = 1, 2, ...n$. So, $Y(c_i) = t_1, t_2, ...t'_j$, represents the set of tags in a reader's scanning field, and it's clear the number of $Y(c_i)$ is j' or $c_{ij'}$. Clearly, $Y(c_i) \subseteq T$. The reader knows the T, C and C_{ij}, but in its scanning zone, it doesn't know which category is present and how many tags are still in, so it's hard to decide whether it's necessary to supply goods. For example, given the all information, the reader should find every set $Y(c_i)$ in its area and get the set of c_i and the number of $c_{ij'}$. Here, we define a parameter λ to represent the ratio between the remaining tags number and the primary. And we can also set different values such as 50%, 40% for different categories as lower thresholds to remind us the replenishment. If the ratio α is lower than λ (here, we set all categories' threshold are the same), we should make the replenishment. We can describe the problem in Fig.1 and show the key notations in table 1.

4   Protocol Design and Analysis

4.1   Categorized Protocol (CP)

First Phase. The first phase verifies the presence for a majority of categories with transmitting the category ID (20-bit) rather the whole ID (96-bit). The reader broadcasts all the categories ID one by one; if there are some responses aiming at the categories, then these categories exist in the scanning area. Repeating this process, until all the categories ID are broadcasted, then it is clear that the the backend server will known the specific categories in the reader's area.

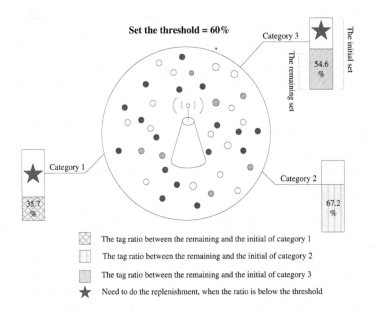

The tag ratio between the remaining and the initial of category 1

The tag ratio between the remaining and the initial of category 2

The tag ratio between the remaining and the initial of category 3

★ Need to do the replenishment, when the ratio is below the threshold

Fig. 1. When should do the replenishment

Second Phase. We identify the category information in the first phase, but how many tags are still in the confirmation of each category? The reader broadcasts the category ID in the first phase with a request $\langle\, r, f\,\rangle$, where r is a random number and f is the frame size. Each tag will be pseudo-randomly mapped to a time slot at the index $H(id, r)$ to affirm their presence, where H is a hash function whose range is $[0, f-1]$. Since the reader is acknowledge to all the hash data, it can verify the existence of the singleton slot tags. Aiming at the collision slot tags, the reader will broadcast their IDs one by one.

Analysis. In the first phase, the execution time is $T_{cat.1} = m \cdot t_c$, where t_c is the time for a category ID transmission. In the second phase, the execution time is $T_{cat.2} = m' \cdot T_{cat.2'}$, where m' is the present categories number and $T_{cat.2'}$ is the verification time for one category's present tags. And $T_{cat.2'} = c_{ij'_c} \times (t_{tag} + t_s) + f \times t_s$, where $c_{ij'_c}$ is the number of collision tags, t_{tag}, t_s is the time for an ID, a short response respectively, and f is the frame size.

So the total execution time of the categorized protocol is:

$$T_{cat.} = T_{cat.1} + T_{cat.2} = m \cdot t_c + m' \cdot T_{cat.2'} = m \cdot t_c + m' \cdot [c_{ij'_c} \times (t_{tag} + t_s) + f \times t_s] \quad (1)$$

It is noting that $c_{ij'_c}$ is the number of tags mapped to the collision time slots, which is a random variable value whose distribution is actually dependent on the frame size f. So is the $T_{cat.}$. Therefore, we get $E(T_{cat.})$ as follows:

$$E(T_{cat.}) = E(T_{cat.1} + T_{cat.2}) = m \cdot t_c + m' \cdot [E(c_{ij'_c}) \times (t_{tag} + t_s) + f \times t_s] \quad (2)$$

Table 1. Key notations

Symbols	Descriptions
T	The set of tags in the storage system
C	The set of different categories in the storage system
C_{ij}	The set of i-th category tag number in the initial phase
$Y(c_i)$	The set of tags in the i-th category in a reader's scanning area
t_i	Tag i
C_i	The i-th category
n	The total number of tags in the storage system
Q	The total number of tags in the reader's scanning area
m	The total categories' number of tags in the storage system
m'	The total categories' number of tags in the reader's scanning area
$c_{ij'_c}$	The collision tag number of i-th category in a reader's scanning area
$c_{ij'}$	The tag number of i-th category in a reader's scanning area
c_{ij}	The tag number of i-th category in the initial phase
k	The times of hash function
λ	The lower threshold for replenishment
α	The ratio between $c_{ij'}$ and c_{ij}

Aiming at an arbitrary time slot in the frame, the probability for exactly k tags mapped into a same slot is $p_k = \binom{c_{ij}}{k}(\frac{1}{f})^k(1 - \frac{1}{f})^{c_{ij}-k}$. When $k \geq 2$, the corresponding slot is a collision slot, and the expected number of tags mapped into this collision slot is $\sum_{k=2}^{c_{ij}} kp_k$; more, in the f slots, if all the tags are mapped into collision slots, the expected number is $f \cdot \sum_{k=2}^{c_{ij}} kp_k$, that is, $E(c_{ij}) = f \cdot \sum_{k=2}^{c_{ij}} kp_k$.

We consider the situation c_{ij}, f approach infinity, then:

$$E(c_{ij_c'}) = f \cdot \sum_{k=2}^{c_{ij}} kp_k = f \cdot \sum_{k=2}^{c_{ij}} k\binom{c_{ij}}{k}(\frac{1}{f})^k(1 - \frac{1}{f})^{c_{ij}-k} \approx c_{ij} - c_{ij} \cdot exp(-\frac{c_{ij} - 1}{f}) \tag{3}$$

It is clear $E(T_{cat.})$ is a function of f, if we want to get the minimum value $T_{cat.}$, do the derivation operation of $E(T_{cat.})$ and get the appropriate f, as follows:

$$\frac{dE(T_{cat.})}{df} = m' \cdot [\frac{dE(c_{ij_c'})}{df} \times (t_{tag} + t_s)] = 0 \tag{4}$$

Where $\frac{dE(c_{ij_c'})}{df} \approx \frac{-c_{ij}(c_{ij}-1)}{f^2} \cdot exp(-\frac{c_{ij}-1}{f})$, which can be derived from (3).

Therefore, we can find the optimal frame size f by solving the equation (4), and here we ignore the situation $m' = 0$.

4.2 Micro-Macro Protocol (MMP)

First Phase. At the beginning of this phase, the reader will do k-th different hash functions for each category in the system with seeds $r_1, r_2, ..., r_k$ and the frame f_1. So, the reader has known all the categories hash values in advance. For example, we can apply two hash functions: $H_1(c_1 \oplus r_1) = 5$, $H_2(c_1 \oplus r_2) = 7$, that is, category 1 c_1 will use its category ID to do hash functions H_1 and H_2, then store the two different values 5 and 7 in the reader's memory. In the frame, when the 5-th and 7-th time slots are '1' at the same time, namely, in the two slots, there are tags' responses simultaneously, then we can say the c_1 category is present in the scanning area.

Then, the reader issues enquiry with seeds r_1 and r_2, waiting for responses and putting corresponding information in order. Once the tags receiving the stimulating signal, they will only use their category ID to finish the hash functions with seeds r_1 and r_2, and respond in these hash slots. In the reader's scanning area, aiming at one category, there are many tags rather than one, and their hash values are the same. Here, we only verify the categories instead of each tag.

However, it is possible that some nonexistent categories are verified in. We denote such situation as false positive. To illustrate how the false positive happens, we can analyze it with the concept slot collision. If one absent category's two hash values are the same with some exist category's values, or maybe the combination of two different existent categories. As shown in Fig.2, it shows the situation in the first phase.

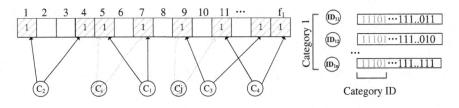

Fig. 2. The situation in the first phase of MMP

Second Phase. Known the existent categories in the scanning area, we will further get the information of the number of tags in each category, and compute the ratio α comparing with the given threshold λ to do the targeted replenishment in time. Traditional methods will identify each tag one by one to know which tag is in or out, or treat it as a tag missing problem, which is time consuming and collision tendency. We fully use the singleton slots among a frame, that is, the successful slots, and in which tags can be verified precisely. At the same time, we introduce the concept micro-macro—evaluating the macro situation from the micro information. Namely, we use the singleton slots of each category to get the whole present message. Therefore, the knowledge of ratio and probability are needed.

When the reader issues enquiry with a hash seed r_{k+1} and frame f_2, the tags are ready to do the hash function with their ID rather than category ID, and respond in the corresponding slot. The reader knows the whole slot values in advance, so it can successfully capture each category's tag number in singleton slots. Then, the ratio of singleton slots between one category and all categories in the scanning area, approximately equals to the ratio of whole tags number between the corresponding categories. We can perform several rounds to get the higher accuracy.

Analysis

(1) The False Positive Rate Analysis

As shown in table 1, k is the value of hash times, and m' is the number of categories in the scanning area. We can calculate P_{false} as follows:

$$P_{false} = (1 - (1 - (1 - \frac{1}{f_1})^k)^{m'})^k \tag{5}$$

From Taylor expansion, we have

$$1 - (1 - \frac{1}{f_1})^k = 1 - (1 - e^{-k/f_1}) = \frac{k}{f_1} + \frac{1}{2!}\frac{k^2}{f_1} - \frac{1}{3!}\frac{k^3}{f_1} + \cdots \tag{6}$$

Since $k \ll f_1$, we only take the first term and get

$$P_{false} \approx (1 - (1 - \frac{k}{f_1})^{m'})^k \approx (1 - (e^{-km'/f_1}))^k \tag{7}$$

We first decide the optimal number of hash functions k given m' and f_1, and we take the derivative of P_{false} with respect to k in Eq.(7). And derive the optimal value $k_{opt.}$ as follows:

$$k_{opt.} = \frac{f_1}{m'} \ln 2 \tag{8}$$

Plugging in Eq.(8) to Eq.(7), we can obtain the frame length f_1

$$f_1 = -\frac{m' \ln P_{false}}{(\ln 2)^2} \tag{9}$$

Therefore, we get two key observations, first, the frame size f_1 is linear with respect to the number of tags' categories m', given the false positive P_{false}. Second, f_1 is monotonically increasing with the false positive rate P_{false}. Hence, if we want to achieve smaller false positive rate, we can increase the frame size, and if we want to achieve time efficiency, we can decrease the frame size, for the total time in the first phase is $T_{MM_1} = f_1 \cdot t_s$.

(2)Time Analysis

In the first phase, when the reader issues enquiry with frame size f_1, tags will choose the corresponding slots to respond with k bits information, so the total time is as follows:

$$T_{MM_1} = f_1 \cdot t_s \tag{10}$$

Here, we further do the probability analysis about the successful verification as follows:

$$P_{veri.} = \prod_{l=0}^{k} \binom{m'}{1} (\frac{1}{f_1 - l})^1 (1 - \frac{1}{f_1 - l})^{m'-1} \tag{11}$$

Therefore, we can get the average time for verifying one category:

$$T_{MM_{1ave.}} = \frac{T_{MM1.}}{m' \cdot P_{veri.}} = \frac{f_1 \cdot t_s}{m' \cdot \prod_{l=0}^{k} \binom{m'}{1}(\frac{1}{f_1-l})^1(1 - \frac{1}{f_1-l})^{m'-1}} \tag{12}$$

In the second phase, all the Q tags in the scanning area will participate in the frame f_2, and we only care the singleton slots, so:

$$T_{MM_2} = f_2 \cdot t_s \tag{13}$$

As the probability of successful singleton slot is $P_{sig.} = \binom{Q}{1}(\frac{1}{f_2})^1(1 - \frac{1}{f_2})^{Q-1}$, we can get the proper f_2 through it.

Therefore, the total time in the phase two is as follows:

$$T_{MM} = T_{MM_1} + T_{MM_2} = f_1 \cdot t_s + f_2 \cdot t_s \tag{14}$$

(3)Micro-Macro Ratio Analysis

When the reader obtains the singleton slots information, how can it evaluate the whole situation? As shown in the following, which give the ratio relationship:

$$\frac{c_{is}}{c_s} = \frac{c_{ij'}}{\sum_{i=1}^{m'} c_{ij'}} \tag{15}$$

where c_{is} denotes the singleton slots number of category c_i, c_s is the total singleton slots number of the frame, and they all can be known by the reader. $\sum_{i=1}^{m'} c_{ij'}$ is the number of all present tags and we approximately treat it as Q. So:

$$c_{ij'} = \frac{c_{is}}{c_s} \cdot \sum_{i=1}^{m'} c_{ij'} = \frac{c_{is}}{c_s} \cdot Q \tag{16}$$

We can easily get the ratio $\alpha = \frac{c_{ij'}}{c_{ij}}$ and compare it with the value λ, then decide whether it is requirement to do the replenishment.

4.3   Discussion

The Scheduling of Readers. As we all known, in a storage system, it is not possible only one reader is existent. When multiple readers work at the same time, there must be the reader-collision problem which reduces the identification throughput greatly. Aiming at the efficiency of identification, we first construct a topological graph to present the physical position of all readers. Adopting the color method, each adjacent point in the topological graph is colored differently. Here, colors denote the different executive time, then all the readers will work based on their colors and this method ensures the adjacent readers will not work at the same time to produce reader-collision.

The Threshold Value λ**.** In this paper, we set λ denote the replenishment threshold, if the value α of each category is below the uniform value λ, then do the replenishment. Here, according to the practical situation, such as some categories are short of just a little, should be replenished in time; and some categories are tolerant, can be supplied at the 20% left. Based on the different conditions, we can set different threshold values λ, which is more practical and meaning.

Impact of Noise. The presence of channel noise can cause the error, which may be the false positives or the false negatives. For example, we image the expected singleton slot which actually is empty since the tag has been moved out, but it is the singleton slot. Due to the channel noise, a false positive will occur when the reader receives the high noise to mis-regard it as a tag response. Thankfully, the occasional false positive does not cause too much influence as our protocol can be executed multi-rounds periodically. And the false negative occurs when a tag replies, but the transmission signal is corrupted by the channel error or channel noise. In this case, the reader cant receive the response and the tag is mistaken for missing or moved out. The false negatives can be easily handled by doing the extra verification procedure, namely, we let the reader broadcast the uncertain tag ID, and wait for the corresponding response.

When tags are moved out of the system during the protocol execution time, both false probabilities may happen. On the one hand, they can be handled by the approaches mentioned above, and on the other hand, we should minimize the execution time greatly. This is the reason why our protocol is practical and meaning, and significantly improved. In the next section, we will demonstrate by the simulations.

5   Evaluation

In this section, we evaluate our protocols by some simulations. We compare the efficiency of our micro-macro protocol with the protocol CP mentioned in our paper, and the BP, ECMTI protocols which are the basic algorithm and one of the most effective aloha based identification protocols respectively.

5.1   Simulation Setup

In the simulation, we adopt the Philips I-Code specification and EPC C2G2 standard as basic simulation settings. Each tag ID is 96-bit and the tag bit rate is 53Kb/s, and each tag should wait for a period of 302 us called the waiting time before the transmission of its information, such as ID, category ID or one-bit response. Therefore, the time for ID transmission is $t_{tag} = 2.4ms$. Similarly, as the category ID is 20-bit long, the time for category ID transmission is $t_c = 0.8ms$ and the time for a short response is $t_s = 0.4ms$. The reader has a different transmission bit rate of 26.5Kb/s, as the advertisement message sent by the reader is very short, here its corresponding time is ignored.

5.2   False Positive

According to the discussion in the above section, we assume the total number of tags is $n = 1000000$, $m' = 50$, and the false positive is 10^{-4}, then can get the value k. Similarly, we can get the relationship between the execution time and the false positive; and the execution time with the frame. Fig.3 illustrates the execution time with different false positive rates. In the following simulation, we set the false positive is 10^{-4}.

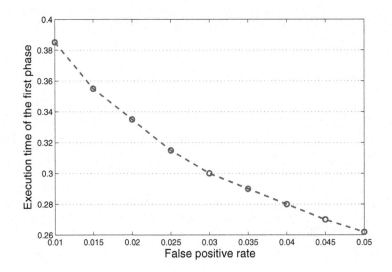

Fig. 3. The execution time with different false positive rates

5.3   Comparison of CP and MMP

Base on the above parameter settings, we will evaluate the average time for identifying a category's situation of category protocol (CP) and micro-macro protocol (MMP), respectively. According to different categories in the reader's scanning area, we all perform 200 simulations to get an average time for identifying one category's situation. Fig.4 shows the comparison of the average time of CP and MMP.

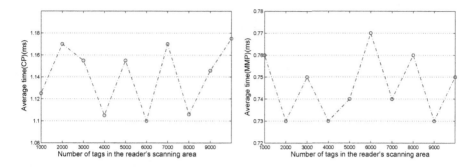

Fig. 4. The comparison of the average time of CP and MMP

5.4   Comparison of Protocols

Based on the setting value, we perform simulations of our two protocols and BP, ECMTI. Fig.5 shows the execution time of these protocols with the changing of Q, and the relationship with the value m'. It is clear that the MMP is about 34% of the time taken by the BP and around 70% of CP.

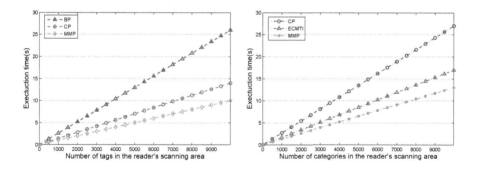

Fig. 5. The execution time with the changing of Q and m'

5.5   Identification Delay

With the number of tags fixed to 10000, we analyze the identification delay of our three protocols and get the relationship in Fig.6. Here, we denote the delay is the appendant of collisions, since the slots or tags collision will make more identification time. In BP, the reader will broadcast the tags' ID one by one, and once the information matching with some tag's ID, the tag will respond, so it is impossible two or more tags will collide, and its delay is constant. While in CP and MMP, they are ALOHA protocols and own probability to collide, therefore, the delay is changeable with the cumulative successful identification probability. Fig.7 presents the average delay of the protocols with the changeable tags number Q.

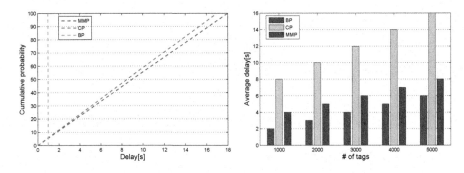

Fig. 6. The delay   **Fig. 7.** The average delay with Q

6   Conclusion

Motivated by the practical storage replenishment problem, we propose an efficient and reliable protocol aiming at the special need. In this paper, we propose two protocols with progressive time efficiencies. We use hash function of category to firstly ensure the corresponding categories' existence, for reducing the scanning range; then we use the micro-macro evaluation mind to quickly get each category's condition, which reduces the execution time greatly and is with practical value. We do extensive simulations to evaluate the performance of our micro-macro protocol. Both the theoretical analysis and the simulations show that our protocol significantly reduces the overall execution time and also improves the throughput.

References

1. Chen, S., Zhang, M., Xiao, B.: Efficient information collection protocols for sensor-augmented RFID networks. In: Proceedings of the IEEE International Conference on Computer Communications (INFOCOM 2011), pp. 3101–3109, April 2011
2. Tan, C.C., Sheng, B., Li, Q.: How to monitor for missing RFID tags. In: Proc. of IEEE ICDCS (2008)
3. Yang, L., Han, J., Qi, Y., Wang, C., Gu, T., Liu, Y.: Season: shelving interference and joint identification in large-scale RFID systems. In: Proceedings of the IEEE INFOCOM, pp. 3092–3100, Shanghai, China, April 2011
4. Roberts, L.G.: Aloha packet system with and without slots and capture. ACM SIGCOMM Computer Communication Review **5**, 28–42 (1975)

5. Myung, J., Lee, W.: Adaptive splitting protocols for rfid tag collision arbitration. In: Proc. of MobiHoc (2006)
6. Mutti, C., Floerkemeier, C.: Cdma-based rfid systems in dense scenarios: concepts and challenges. In: Proc. of RFID (2008)
7. Zheng, Y., Li, M., Qian, C.: PET: probabilistic estimating tree for large-scale RFID estimation. In: Proceedings of the 31st IEEE International Conference on Distributed Computing Systems (ICDCS 2011), pp. 37–46, July 2011
8. Han, H., Sheng, B., Tan, C.C., Li, Q., Mao, W., Lu, S.: Counting RFID tags efficiently and anonymously. In: Proceedings of the IEEE INFOCOM, San Diego, Calif, USA, March 2010

Social Networks and Recommendation

Tag-Based User Interest Discovery Though Keywords Extraction in Social Network

Ping Yang[(✉)], Yan Song, and Yang Ji

Mobile Life and New Media Laboratory, School of Information and Communication Engineering, Beijing University of Posts and Telecommunications, Beijing, China
{qianqiuyitong,songyan,jiyang}@bupt.edu.cn

Abstract. We consider the problem of exploiting to discover user interests from social network. User tags in social networks convey abundant implications of user interests,which great benefit various tasks ranging from user profile construction to user similarity calculation based recommendation. However,user interests extraction from social tags suffer from large diversity of word choices due to different user preference,especially the words that quite specific in minority knowledge domains. In addition,the deficiency of uniform concept hierarchy and lack of explicit semantic association between tags obscure the real interests of users. To obtain user interests from tags,we propose a tag normalization algorithm based on world knowledge to underpin the construction of common tags as well as the organization of user hierarchy interest. Experiments with Sina Micro-blog[1] show that our algorithm can infer user's interests better than traditional method based on contents.

Keywords: User interest · User profile · Keyword extraction · Social network

1 Introduction

Unsupervised clustering algorithms (e.g.,k-means, pLSI and LDA[1,7]) or weakly supervised models (e.g., supervised LDA[1],labeled LDA [14]) and some keywords extraction algorithms (e.g.,[10]) has been used to discovery the user's interests. A major limitation of these algorithms is that they ignored the fact that users always released posts about the fields which they are expert in, but not really interested in. What is more,a lot of person post very few contents which is impossible for us to discovery user's interests employing them. To infer user's interests accurately, in this paper, we focus on the tags instead of the contents posted by user. This task raises a set of unique challenges given tag's scale and the short, noisy and ambiguous natures of tags.

In social network, tags generated by himself hint what he likes or his interests, i.e. song, travel. Generally speaking, every user can possess 5 to 10 tags in social network which is enough to express one's interests sufficiently. So if we can

---

[1] http://weibo.com/

© Springer International Publishing Switzerland 2015
Y. Wang et al. (Eds.): BigCom 2015, LNCS 9196, pp. 363–372, 2015.
DOI: 10.1007/978-3-319-22047-5_29

employ the tags to discovery user's interests,we can infer the user's interests well and truly. Nonetheless we face three challenges as follows.

The first is,in the social network,there are hundreds of millions of tags while most of them appear only once or twice. We collect 30,012 user's tags on the Micro-blog and there are about 80,106 tags. The most frequently occurring tag is "movie" with 7,375 occurrences. Nearly 75.3% (43,615) of tags in the system only occur once and only about 10.1% of the tags occur more frequently than that average frequency(3.55). The detail will be extended in section 4. The second is that the tags are tousle and non-mainstream on account of the personalization. If someone has a tag "JonnyBuckland", do you know what is it?What is more,the tags is rough-and-tumble and have no hierarchies. If we can organized them by hierarchies(the higher-level tags are more general, while the lower-level ones are more specific),user interests can be expressed better.

In this paper, facing hundreds of millions of tags we create a hierarchical criterion tag set which can convey people's interests fair enough. As for the uncommon tags, we employ the world knowledge and keywords extraction algorithm to normalize them to the criterion tag set. At last, we construct a two-layer category of interests for the user which can be used to other systems which need structured data to provide better service.

The remainder of this paper is organized as follows. Section 2 discussed related work in user's interest discovery and key word extraction algorithm. Section 3 presents our method in detail. Section 4 gives the result of our experiments. Finally, a conclusion is drawn in section 5.

2   Related Work

There is an extensive body of work on discovering user's interests. However our work has a slightly different focus. One is that we do not aim to identify which users deal with same interest,but which interests are treated by a given user. The other is that we use the tags instead of the tweets and retweets and the world knowledge which is a knowledge repository. To discovery user's interests, the predecessors have done a lot of research as follows.

User interest hierarchy clustering algorithms[6] initially all objects start in one cluster,then each parent is forced to have two children in the hierarchy. However, it reserve all the words in the text so that it is difficult to understand for us. Our algorithm only retrain a few words which can convey the meaning of the tag and is universal. Liu [9,10] use the tweets and retweets mining the user's interests. The weakness is that they become invalid when user has few friends or post few tweets and retweets.

Kim [5] propose Explicit Semantic Analysis(ESA), a novel method that represents the meaning of texts in a high-dimensional space of concepts derived from Wikipedia[2] making use of the concept links. Nonetheless on account of the feature of the social network,the words on the Micro-blog is too orally so that

---

[2] http://www.wikipedia.org/

a part of words do not appear as a link in the Wikipedia. So we use the word instead of the links to discovery user's interests.

Starting with TextRank[5], graph-based ranking methods are becoming the state-of-the-art methods for keyword extraction. Keywords extraction of short text [15] propose a multi-feature,multi-step improved method after weighing both rate and efficiency of the extraction. Recently someone propose make use of the machine learning to extract keywords. The method of machine learning first establishes some rules and gets a prototype of training model, then trains the prototype with lots of annotated texts. If the training set is big and wide enough, good performance can be expected. But this method may consume great human resources on providing annotated text, and it only can deal with the text which is similar to training set. The performance will be dramatically degraded when facing with new field [18].

However, differing from these researches, our method focuses on normalizing tags and constructing hierarchical user interests and proposes a Tag Normalization Algorithm.

3 Tag Normalization Algorithm

The tags of user are represented by a multiset T and t_{ij} is the jth ($j = 1, 2, \ldots, n$) tag of user $i(i = 1, 2, \ldots, m)$. The Wikipedia – one form of world knowledge,which can relate the tags to the people's cognitive – is a huge database. So we employ Wikipedia to construct a document set D. Then we generate a keywords set k_{ij} for tag t_{ij}. Third,in order to express all people's interests by only common words we define a popular degree function to to generate a hierarchical criterion word set Θ. At the last,by means of the criterion tag set Θ,we can obtain the user's hierarchical interests M. Figure 1 illustrates the overview of our algorithm. In the following,we will describe our algorithm in detail.

3.1 Preprocessing

The purpose of this section is to construct the document d_{ij}. We opted to use Wikipedia because it is currently the largest knowledge repository on the Web. First according to [16,17],we construct index for contents of Wikipedia. Second we take the tag as the input making use of the search engine to return the top p pages. Then we compose the p pages to a document d_{ij}. A line is a sentence in document d_{ij}. Then we use the a Chinese word segmentation tool ICTCLAS to proceed word segment and filter them through a stop list on the document.

3.2 Word Similarity Net

If someone introduce his interests with a long article d_{ij}, would you like to listen to him? To describe one's interests briefly and straightforward,we select the keywords for each document which can convey the meaning of the tag. Fisrt

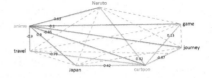

Fig. 1. Overview of Tag Normalization Algorithm

Fig. 2. RMSE Comparison

we calculate the words similarity and then they votes mutually to determine the keywords by score.

We employ a similarity function[3] to calculate how strongly two words are related. In general,the words are more related if they are in one sentence. So we set the window size as a sentence. There are many functions to calculate the similarity of two words. Mikolov and Collobert[4,12] employ neural networks which can get a good effect to represent every word as a vector. Whereas it need very large train corpus and polysemy arise in the large train corpus.

Among these similarity functions,We use *AEMI* (Augmented Expected Mutual Information) [5] as a similarity function. This method is still effective when the corpus is very small. What's more,the preprocessing guarantees the content in the document is related to the tag so that the similarity is more accurate and words ambiguity almost doesn't exist. *AEMI* is enhanced version of *MI* (Mutual Information) and *EMI* (Expected Mutual Information). Unlike *MI* which considers only one corner of the confusion matrix and *EMI* which sums the MI of all four corners of the confusion matrix,*AEMI* sums supporting evidence and subtracts counter-evidence. Chan [3] demonstrates that *AEMI* could find more meaningful multi-word phrases than *MI* or *EMI*. Concretely, $AEMI(a,b)$ is the similarity function between word a and word b. $P(a) = \frac{m}{n}$ is the probability of a document d_{ij} containing word a. In the expression,m is the number of sentences which contain word a and n is the total sentence number of d_{ij}.$P(\bar{a})$ is the probability of a document not having word a. Probabilities for b and joint probabilities a and b are defined similarly. $AEMI(a,b)$ is defined as:

$$AEMI(a,b) = P(a,b) \log \frac{P(a,b)}{P(a)P(b)} - P(a,\bar{b}) \log \frac{P(a,\bar{b})}{P(a)P(\bar{b})} - \log P(\bar{a},b) \log \frac{P(\bar{a},b)}{P(\bar{a})P(b)}$$

The first term computes the supporting evidence that a and b are related and the latter two term calculates the counter evidence. This expresses the strength of the relationship between the two words. As a result,each word represent a vertex and the weight of the edge is the *AEMI*. The similarity matrix is a all connection diagram. So the time complexity is $O(n^3)$.

For example,we select the tag "Naruto" which is a name of cartoon and calculate the *AEMI* of each two words in document d_{ij} responding to tag "Naruto". Then in order to sketch a simple we extract seven words which has different *AEMI* values as Fig 2.

As we can seen in Fig 2,the importance of the vertex is determined not only by itself but also based on the vertex which has relationship with it. So we propose the voting function. For document set D, we define a dictionary *Dic*.

We suppose that d_{ij} contains k words in Dic. We define the similarity matrix $S_{k \times k}$ in which s_{ij} represents the similarity degree between word w_i and w_j in Dic and the relative frequency set $\mathbf{V} = (v_1, \ldots, v_k)$ in which v_i is defined as follows: $v_i = n_i / \sum_{i=1}^{k} n_i$ in which n_i represents the times of w_i in Dic occurring in document d_{ij}. Each vertex is a word and the weight on the edge is $AEMI$. When one vertex links to another one, it is basically casting a vote for that other vertex. If the vertex has a higher similarity with another and the relative frequency responding to it is higher too, the vertex is more import to the document. At last, we define a vector $\overrightarrow{W} = \sum_{i=1}^{k} \sum_{j=1}^{k} s_{ij} \cdot v_j = S \cdot \overrightarrow{V}$ which represents the score of words.

We sort the $\overrightarrow{W}$ and select the top l words which can convey the meaning of the tag fully. For the tag "Naruto", through our keywords extraction algorithm we can get the keywords set {"cartoon","anime"}.

3.3 Tag Normalization

The purpose of this section is to construct the hierarchical criterion tag set Θ and user hierarchical interests set M. The criterion tag set is used to reduce the number of tags and normalize the uncommon tags. Let $\Theta = \{\{C_i; (f_{i1}, \ldots, f_{in})\}|_{i=1,\ldots,m}\}$, where f_{ij} in dictionary Dic represents the tag and C_i is the category which the tag belongs to.

First,we define a popular degree. Li [8] demonstrates the frequency of occurrence of a word is almost an inverse power law function of its rank and the exponent of this power law is very close to 1. Like the word distribution in the world,the tag distribution according with the long tail feature [13] also follows power laws. So we define the popular degree of tag as $g(r_{ij}) = c \cdot r_{ij}^{-\alpha}$,which is a power law function,where c is a constant and r_{ij} is the rank of the tag f_{ij} by frequency.

Second we construct the criterion tag set θ which is used to control the tag neither over-normalizating nor under-normalizating. For example,a TV program CCTV has keyword China and NARUTO has keyword Uchiha Sasuke which is a name of one person. They are over-normalizating or under-normalizating. To select the tags in Θ, we set a threshold β. K is the keywords set which has more words and it is a supplement of tag set T. So we define a big multiset $F = T + \{k_i | k_i \subset K, k_i \subset T\}$,which can express the tag's popular degree better than only by users's tag multiset T. If the popular degree of tag f_{ij} is greater than β,then $f_{ij} \in \Theta$. Hence Θ has only hundreds of tags so that we classifier the tags into two layers according to public website's partition manually. The first layer is highly abstract and the second layer is slightly specific. The first layer has 21 categories responding to 21 words and the second layer contains several tags for each category. That is to say we can confirm the category C_i for tag f_{ij} in the criterion tag set Θ. As a example,for tag f_{i1}="internet",f_{i2}="digital",f_{i3}="phone" we can set them to the category C_i="IT". As a result,the hierarchical criterion tag set Θ is formed.

For a user,we can get a two-layer interests category M. By means of keywords extraction,we can get a set of keywords k_{ij} for each tag t_{ij}. If the element in k_{ij} belong to the criterion tag set Θ,then element and its category can be put into M. That is to say each user can get a normalized two-layer tag set and the top layer is coarse-grained,the under layer is more detail. So the tag set M can convey user's interests greatly and it can be used to other systems which need structured or semi-structured data on account of be normalized.

In the end,the tag t_{ij} can be normalized. What is more,the tag is not only reflect user's interests but also is marked by themselves. So M can represents user's interests better.

4   Experiment

As the fundamental purpose of our work is to discovery user interest model on social network, our dataset is built by crawling data using Micro-blog API. In the same time,we let p – number of pages that d_{ij} contains – equal 30 for unification. In this section, first we will illustrate our keywords algorithm is effective by means of annotated news articles. Second threshold β of β will be given by experiments. Then we calculate our tag normalization algorithm making use of the HowNet to evaluate. Next we will do experiment on Sina Micro-blog to illustrate our algorithm effective. At last,we give a case on recommendation to demonstrate our Tag Normalization Algorithm further. HowNet is an on-line common-sense knowledge base unveiling inter-conceptual relations and inter-attribute relations of concepts as connoting in lexicons of the Chinese and their English equivalents. What is more,it has concept layer and can has a big word similarity dictionary.

4.1   Keywords Extraction Performance on Annotated News Articles

There is no gold standard answer for keyword extraction on Micro-blog. To perform quantitative experiments, we crawled a collection of 1000 Chinese news articles from news.sina.com.cn,one of the most popular news websites in China. The news articles are composed of various topics including science, technology, politics, sports, arts, society, and military. All news articles were manually annotated with labels by website editors,and all these labels come from the corresponding documents.

In this dataset, there are 54,600 unique words in the documents, and 805 unique words in the label set. The average lengths of the documents are 971.7 words. The average number of labels for each document is 2.4. To evaluate our keywords extraction method, we use the annotated labels from news.sina.com as the standard labels. Taking into the average number of the keywords consideration,we limit our keywords set to five words. If one of the keywords we suggested exactly matches the labels which is annotated by the editors,it is regarded as a correct suggestion. Each method can extract some keywords for each article so that it has not recall and we employ precision to evaluate the results. We use three representative unsupervised methods as baselines for comparison: TFIDF,

Table 1. A comparison of keywords extraction algorithms

algorithm	LDA	tfidf	TextRank	mine
count	782	801	809	836
Accuracy	78.2%	80.1%	80.9%	83.6%

Table 2. Accuracy evaluation of our algorithm from questionnaire

Accuracy	100%	70%-99%	50%-70%	< 50%
count	883	88	67	3
Percentage	84.8%	8.5%	6.5%	0.2%

TextRank,LDA[2]. Table 1 demonstrates that the our method is valid for keyword extraction. TFIDF which only considers the frequency and the length of the sentence does not regard the co-occurrence between the words. TextRank which consider the relationship ignore the particularity of scene our experiment applied to. The document d_{ij} is closely related to the tag so that the weight of frequency should be added. LDA based on probability but our article is too small. The probability of words is approximate and too little so that the topic model is invalid to some extend. In contrast to these methods,our method consider the relationship and frequency at the same time. What is more,the scale of the document has no influence on our method and our method conforms to the scene in which each article has only one topic. In conclusion,our keywords extraction method is valid and appropriate for our algorithm.

4.2 Determining the Threshold β

In this section,several experiments are executed to find the best threshold β. In order to let the popular degree range from 0 to 1 which is easy to evaluate for people,we make c equals 1 and the popular degree can match the tag distribution best when α equals 0.77. To perform quantitative experiments, we part the 80,106 tags in section 1 into training data $D_{train}(80\%)$ and testing data $D_{test}(20\%)$.

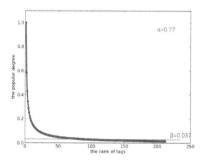

Fig. 3. Tag Popular Degree Trend

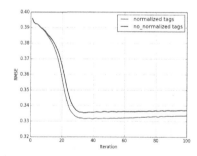

Fig. 4. RMSE Comparison

we construct the criterion tag set Θ on the training data. To get the best Θ,we define a measure function $R = \frac{|\Theta \cap H|}{|\Theta|}$ which can express the coverage rate of Hownet concept about the criterion tag set Θ among H represents the HowNet concept's words set. To get the best threshold,β has been adjusted from 1 to 0 with a step of 0.001. When β equals 0.037,the R can get the maximum value 0.92 and the number of the criterion tag set is 145. We plot the tags's popular degree of the training data in fig 3 where the x-axis is r and y-axis is $g(r)$.

To measure our criterion tag set Θ,we employ our algorithm on the testing data and define $P = \frac{q}{|D_{test}|}$ which can indicate the coverage rate of the criterion tag set Θ. We init $q = 0$,if $\{k_{ij} \cup \Theta\} \neq \emptyset$ then $q = q + 1$ for every element of K. On the testing data,the P can reach 0.94. This prove that our criterion tag set is universal.

4.3   Performance on Sina Micro-blog

How to measure the performance of interest modeling method is still a problem. A typical method to judge the performance of interest models is to list keywords and judge them by experience.

Table 3. Tags and keywords

User1		User2	
tags	normalized tags	tags	normalized tags
figure rapist	art, picture	DotA	play,game
Aragaki Yui	music,song, singer	2012EVA	cartoon, animation
design speciality	art	cats and dogs	pet,animal
bachata	dance,music	NBA	sports, athletic sports, basketball
linux	system, programmer	Adam Cheng	singer, film and television,movie

In our work, two representative users have been invited for experiments. We have show the tags which is annotated by the users. The detail is in Table 3. The left is the tag and the right is the keywords in criterion tag set Θ corresponding to the tag on the left. Fig 5 describes user's interests according to the level. The upper level is the abstract category and the lower level is the tags. To demonstrate the precision of our algorithm further,we issue 1,539 questionnaires to users of Sina Micro-blog and get 1,041 feedbacks. In the questionnaire,people can fill in their tags and pick their interests from our criterion tag set Θ. Then we let the users to evaluate the accuracy of our algorithm. 100% is exactly correct and 0% is totally wrong. The result illustrates in table 2.

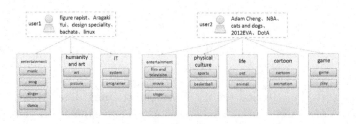

Fig. 5. User's Interests of Hierarchy Categories

4.4   Practical Application

In order to specify the task of the tag normalization algorithm, we take TV program recommendation as a case to study. Social TV creates valuable feedback about the targeted TV programs, which opened up new opportunities for TV recommendation. In our case study, we adopt the PMF recommendation algorithm[11] by adding a user regularizer, which is produced by user tag similarity. The user regularizer is presented as

$$\frac{\lambda_{sim}}{2} \sum_{i=1}^{m} \|U_i - \frac{\sum_{f \in S(i)} Sim(i,f)U_f}{\sum_{f \in S(i)} Sim(i,f)}\|_F^2$$

where $Sim(i,f) \in [0,1]$ is the similarity function to indicate the similarity between user u_i and user u_f where $\lambda_{sim} > 0$ is a parameter to control the strength of the item similarity regularization and $S(i)$ is the set of top N similar items of item j. Here,N is a parameter and a larger value for N may suggest that users are more likely to be interested in items similar to the ones they liked before.

In order to generate the regularizer, we calculate the users' tag similarity. However, the user-generated tags are personalized and abundant. Hence, we use our Tag Normalization Algorithm to reduce the number of tags. Based on the normalized tag set, we calculate the user similarity of the user regularizer. The experiment indicates that recommendation with the normalized tag has a better performance than the one with unnormalized tag, which is shown in Fig.4.

5   Conclusion

In this paper,we have proposed a tag normalization algorithm for discovering user's interests using world knowledge. Fisrt depending on bringing in the Wikipedia we make the tag not separate from the real world. By indexing Wikipedia and making use of the word similarity function and word relationship,we succeed in extracting the keywords which can express the meaning of the tag. Then by means of constructing criterion tag set Θ,we normalized the uncommon tag and construct hierarchical user interests. Our experiments on Sina Micro-blog show that our algorithm is effective and efficient for identifying user interests. The proposed algorithm is put into practice on the social TV scenario, and the implement of algorithm is used as a base of the recommendation which improves user satisfaction and experiences further.

Acknowledgments. This work is supported by the National Science and Technology Support Program that TV content cloud services technology integration and demonstration based on social network (No.2012BAH41F03).

References

1. Blei, D.M.: Probabilistic topic models. Communications of the ACM **55**(4), 77–84 (2012)
2. Blei, D.M., Ng, A.Y., Jordan, M.I.: Latent dirichlet allocation. The Journal of Machine Learning Research **3**, 993–1022 (2003)
3. Chan, P.K.: A non-invasive learning approach to building web user profiles (1999)
4. Collobert, R., Weston, J., Bottou, L., Karlen, M., Kavukcuoglu, K., Kuksa, P.: Natural language processing (almost) from scratch. The Journal of Machine Learning Research **12**, 2493–2537 (2011)
5. Gabrilovich, E., Markovitch, S.: Computing semantic relatedness using wikipedia-based explicit semantic analysis. In: IJCAI, vol. 7, pp. 1606–1611 (2007)
6. Kim, H.R., Chan, P.K.: Learning implicit user interest hierarchy for context in personalization. In: Proceedings of the 8th International Conference on Intelligent User Interfaces, pp. 101–108. ACM (2003)
7. Li, M., Zhang, C., Sun, L., Shao, X.: Topic extraction based on knowledge cluster in the field of micro-blog. In: Huang, D.-S., Jo, K.-H., Wang, L. (eds.) ICIC 2014. LNCS, vol. 8589, pp. 542–550. Springer, Heidelberg (2014)
8. Li, W.: Random texts exhibit zipf's-law-like word frequency distribution. IEEE Transactions on Information Theory **38**(6), 1842–1845 (1992)
9. Liu, Q., Niu, K., He, Z., He, X.: Microblog user interest modeling based on feature propagation. In: 2013 Sixth International Symposium on Computational Intelligence and Design (ISCID), vol. 1, pp. 383–386. IEEE (2013)
10. Liu, Z., Chen, X., Sun, M.: Mining the interests of chinese microbloggers via keyword extraction. Frontiers of Computer Science **6**(1), 76–87 (2012)
11. Ma, H., Zhou, D., Liu, C., Lyu, M.R., King, I.: Recommender systems with social regularization. In: Proceedings of the Fourth ACM International Conference on Web Search and Data Mining, pp. 287–296. ACM (2011)
12. Mikolov, T., Sutskever, I., Chen, K., Corrado, G.S., Dean, J.: Distributed representations of words and phrases and their compositionality. In: Advances in Neural Information Processing Systems, pp. 3111–3119 (2013)
13. Newman, M.E.: Power laws, pareto distributions and zipf's law. Contemporary Physics **46**(5), 323–351 (2005)
14. Ramage, D., Hall, D., Nallapati, R., Manning, C.D.: Labeled lda: a supervised topic model for credit attribution in multi-labeled corpora. In: Proceedings of the 2009 Conference on Empirical Methods in Natural Language Processing, vol. 1, pp. 248–256. Association for Computational Linguistics (2009)
15. Wang, J., Li, L., Ren, F.: An improved method of keywords extraction based on short technology text. In: 2010 International Conference on Natural Language Processing and Knowledge Engineering (NLP-KE), pp. 1–6. IEEE (2010)
16. Wang, Q., Wu, W., Gu, Y.: The application of lucene in information leakage monitoring and querying system. In: 2010 2nd International Conference on Information Engineering and Computer Science (ICIECS), pp. 1–4. IEEE (2010)
17. Xu, H., Li, J.M.: Design and implementation of web search engine based on lucene. Journal of Hebei Software Institute **1**, 024 (2009)
18. Xu, Z., Lu, R., Xiang, L., Yang, Q.: Discovering user interest on twitter with a modified author-topic model. In: 2011 IEEE/WIC/ACM International Conference on Web Intelligence and Intelligent Agent Technology (WI-IAT), vol. 1, pp. 422–429. IEEE (2011)

Implicit Feedback Mining for Recommendation

Yan Song$^{(\boxtimes)}$, Ping Yang, Chunhong Zhang, and Yang Ji

Mobile Life and New Media Laboratory, School of Information and Communication
Engineering, Beijing University of Posts and Telecommunications, Beijing, China
{songyan,qianqiuyitong,Zhangch,jiyang}@bupt.edu.cn

Abstract. Social media creates valuable feedback, either explicitly or implicitly, which can be used to develop an effective recommendation. Explicit feedback, like rating, allows users to explicitly express their preference on items. However, the reluctance of users to provide explicit feedback makes it difficult to get sufficient and representative explicit feedback. In contrast, implicit feedback has the advantage of being collected at much lower cost, in much larger quantities, and without burden on users. Thus, we mine the implicit feedback, including tweets and tags, to provide virtual rating and user similarity for recommendation. Taking the factor that tweets reflect users' sentiment on some item into consideration, we use sentiment analysis score as the virtual rating, and propose a Weighted Semantic Tag Similarity Method (WSTSM) to get user similarity. Experimental on a real SINA microblog dataset demonstrates that our method outperforms the traditional PMF in terms of RMSE by 8.55% due to the informative implicit feedback embedded in tweets and tags.

Keywords: Recommend · Implicit feedback · Sentiment analysis · Tag

1 Introduction

With the rapid increasing of information, recommender systems emerge and aim to help users filter the valuable information through a large variety of available information. Among the technologies behind these recommender systems, the most widely used and successful technology is Collaborative Filtering (CF), whose essential elements are users, items and ratings. "Item" is the general term used to denote what the system recommends to users, such as books, films, blogs, music, and so on. "Rating" is numerical rating that user rates on some item. However, in most social networks, explicit numerical ratings are rare and difficult to obtain for users have no initiative to make explicit feedback. Motivated by the phenomenon, we focus on the recommendation based on the implicit feedback in social networks.

In the social networks, there are plenty of implicit feedback used to obtain the ratings, such as user tweets, social relationship, users liking history, etc. Social relationship is commonly used for user-based recommender. This recommender has an assumption that user and user's social friends have similar preferences, which is not always true in a real scenario. While tweets about some item indicate

© Springer International Publishing Switzerland 2015
Y. Wang et al. (Eds.): BigCom 2015, LNCS 9196, pp. 373–385, 2015.
DOI: 10.1007/978-3-319-22047-5_30

user's sentiment to the item, which may be positive, negative or neutral. And tweets data are huge amount, rich of information and easily available. Therefore, we select tweets, instead of social relationship, as our basic data and regard sentiment analysis value of tweets as the ratings. This recommender has great application potential not only for social TV to enhance user experiences, but also for e-commerce sites, film company, etc., to realize social networking marketing.

Moreover, to improve recommendation performance, we take user tags into consideration, which mark user's interests from the perspective of user himself. With the assumption that users with similar tags may have similar preference, we use user tags to measure user similarity which is utilized by introducing a regularizer into recommendation algorithm. Specifically, we weight the tags using Term Frequency Inverted Document Frequency (*TF-IDF* [13]) with preprocessing by *K-means* [5] algorithm, and propose a Weighted Semantic Tag Similarity Method (WSTSM) depending on word vector expansion. Our proposed approaches are quite general, which can be applied to other recommendations based on tweets or tags.

In summary, the main contributions of our paper are as follows:

– In the absence of explicit user rating information, we propose that using implicit user social information (i.e., sentiment analysis of user-generated tweets) can also generate recommendation under the matrix factorization framework and demonstrate the effectiveness by experimenting on real dataset.
– We propose a 5-scale mapping rule of emotional value to rating, which perform better than 2-scale mapping showed in the experiment.
– To solve the problem of tags having low literal similarity, we present a Weighted Semantic Tag Similarity Method (WSTSM) for tags based on word vector expansion.

The rest of this paper is organized as follows. Section 2 reviews related work of recommendation algorithms and sentiment analysis algorithms. Section 3 introduces the proposed method for mining implicit feedback in detail. The experimental results are presented and analyzed in Section 4 followed by the conclusion.

2   Related Work

Recommender systems depend on different types of source. What mostly used are explicit feedback [4], [18], [23], like rating and implicit feedback [2], [11], [19], like tweets, tags, following behavior. Nevertheless, explicit feedback is not always available in social scenario, which leads to high attention to implicit feedback for implicit and explicit feedback are to some degree substitutable [17]. Vast majority of research is focused on implicit feedback, we will review them below.

The increasingly growing amount of tweets allows us to capture valuable information to understand user interests and build better recommendation model [7], [15]. Along with the popularity of sentiment analysis, there are several studies that exploit various aspects of sentiment analysis of user-generated content to recommend, which can be classified into two groups: supervised methods [10], [22], unsupervised methods [21] and hybrid solution [20].

The Explicit Factor Model [22] is proposed to generate explainable recommendations by utilizing explicit product features and user opinions by unsupervised sentiment analysis on user reviews. Pappas et al. [10] presented a sentiment-aware nearest neighbor model (SANN) for multimedia recommendations over TED talks, which makes use of user tweets. Motivated by the lack of rating information, Zhang et al. [21] presented a new recommender algorithm by fusing a self-supervised emoticon-integrated sentiment classification approach. Zhang et al. [20] proposed a new entity-level sentiment analysis method for Twitter and trained a classifier to assign polarities to the entities in the newly identified tweets to improve the F-score.

The other implicit feedback, users' tags, has also attracted considerable attention to recommender systems and tag similarity calculation is the main concern. Most existing methods [18], [23] are based on tag vector, computing tag similarity via various classic metrics including cosine similarity, Jaccard coefficient, etc. These techniques work well in dense tags, but they fail to do so when tag usage exhibits a power law distribution, as it often happens in real-life tags [12].

To tackle this issue, more researchers focus on defining and analyzing semantic similarity. Quattrone et al. [12] described an innovative tag similarity metric that is based on the mutual reinforcement principle, which induces the creation of a dense tag. Markines et al. [6] extended some of the traditional similarity methods of finding semantic relatedness to folksonomy. Analogous to our study, Bai et al. [1] stated a new clustering model focusing on users' tags based on word2vec mining, achieving a high coverage rate and low overlap rate.

In conclusion, all the results of sentiment analysis in research above are a scale with $\{0, 1\}$ or $\{-1, 0, 1\}$, while our work identifies a sentiment value mapping principle with 5-scale, which none of the previous works has focused on, to the best of our knowledge. And, differing from these studies, we propose a Weighted Semantic Tag Similarity Measure based on Word2Vec and combine it with K-means algorithm to get semantic similarity of tags, which further improves similarity calculation.

3   Mining Implicit Feedback

As the essential elements of CF, users, items and ratings separately represent users who tweet tweets, items which are mentioned in the tweets, and sentiment expressed in the tweet in social networking scenarios. In this section we illustrate how to prepare the three inputs of CF. Among them, the extract of item object is based on simple keyword filtering, where keywords are defined as words related to the target item. The generation of sentiment score is based on an improved lexicon-based sentiment analysis. And to further improve the accuracy of recommendation, we introduce tagging information into the procedure of recommendation. Figure 1 shows the overview of proposed mining algorithm. We will detail the two aspects above in the following sub sections.

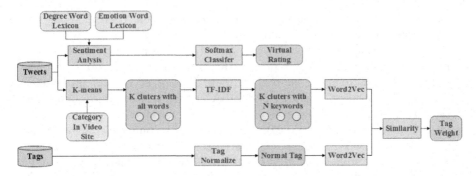

Fig. 1. Overview of Algorithm

3.1   Implicit Sentiment Analysis of Tweets

As a common online implicit resources, tweets provide many clues about what users like or dislike. And users tend to tweet more frequently on those items that they particularly care. Thus, we use sentiment analysis method to analyze tweets in order to construct virtual rating matrix for CF algorithm.

Among the large amount of sentiment analysis methods, the lexicon-based approach is not only simple, but also can achieve good performance. So we adopt the lexicon-based approach [16] as our base approach to calculate the sentiment. We first introduce some definitions of terms used in this algorithm. **Emotion words** are words that express users' sentiment, like "like", "hate". **Degree words** are words that modify emotion words, like "very much", "quite". And **negation words** are words that can reverse the sentiment polarity of a tweet, like "not", "hardly". **Emotion lexicon** $\{e_i, E(e_i)\}$ is composed of pairs of emotion word and corresponding emotion value. So is **degree lexicon** $\{d_j, D(d_j)\}$.

An example is shown to describe the basic idea of our algorithm. User u tweets a tweet tw "I like the Titanic very much" about film "Titanic". In the tweet tw, "I" and "Titanic" separately represents the current user and the item, and "like" and "very much" is separately emotion word and degree word. Given emotion value v_1 of word "like" and degree value v_2 of words "very much", the final emotion value of tweet tw is $v_1 \cdot v_2$.

Before detailing the improved lexicon-based algorithm, we make some assumptions, which are the basis of our algorithm. (1) Tweet represents user's sentiment to the item that the tweet describe. (2) User's sentiment is proportional to his rating. (3) The longer tweet, the stronger sentiment. Most of shorter tweets are neutral. (4) Sentiment of a single tweet cannot represent the sentiment of user to an item. (5) Most users express their sentiment using five positions: negative, somewhat negative, neutral, positive or somewhat positive and the extreme values were rarely used. Based on the basic idea and definition above, the final sentiment $ST_u(tw)$ of tweet tw for user u is as follows:

$$ST_u(tw) = (-1)^\alpha gram(tw) \sum_{j=1}^{l} E(e_j)D(d)^\sigma \tag{1}$$

where l is the number of emotion words (not all the words) contained in tweet tw after word segmentation, and σ is the indicator factor that is equal to 1 if emotion word e_j has a degree modifier d and equal to 0 otherwise. α also is an indicator factor that is equal to 1 if tweet tw has a negation word and equal to 0 otherwise. Function $gram$ is added to satisfy the assumption 3). The value of $gram$ is a variable, which grows exponentially with length of tweet.

Just as assumption 4) mentioned, the emotion value $S_u(i)$ of user u for item i is the average emotional value of all the tweets tweeted by user u about item i, which is formulized as $S_u(i) = \frac{1}{|T_{ui}|} \sum_{j=1}^{|T_{ui}|} ST_u(tw_j)$. T_{ui} is tweet set tweeted by user u to item i in historical dataset.

Taking assumption 5)[14] into consideration, we adopt a 5-class classification into these categories to capture the main variability of the sentiment. Moreover, recommendation should pay more attention to items user likes and avoid to recommend items user dislikes. To emphasize positive and negative preference, we adopt a $softmax$ classifier $S_{nor} = softmax(S)$ to normalize the eventual emotional score to a scale of 1 to 5, in which S and S_{nor} denotes separately the raw and normalized user-item sentiment matrix.

In addition, we know nothing about the users' preference to untweeted item. For item not tweeted by a user, it does not mean the item is not of the user's interest, and maybe the item is of the users' interest but she did not know about it before. In this scenario, how to initial the missing score is of great importance, named as One-Class Collaborative Filtering (OCCF) [9]. We initial the missing score using the weighted alternating least squares (wALS) for OCCF. More detailed information can be found in the original paper [9].

3.2 Implicit Tags Mining

In online communities, users usually choose words, phrases or any other informal vocabularies freely as their tags, which reflect interests of individual user. And users with similar tags are more likely to have similar interests. Thus, we introduce a Weighted Semantic Tag Similarity Measure (WSTSM) to characterize user tag similarities to improve the performance of recommendation.

Tags Normalization. The distribution of tag occurrence frequencies in our dataset is illustrated with log-log coordinate in Fig.2, which follows the Zipf's law [24], meaning that small number of tags frequently occurs and even 85% tags occur only once. Furthermore, tags are tousle and non-normative for personalization. Some tags express similar meaning despite occurring with different words, such as "travel" and "journey", "cartoon" and "animation".

Hence, we adopt a tag normalization algorithm to reduce the number of tags. The tags are represented by a multiset T and t_{ij} is the jth $(j = 1, 2, \ldots, n)$ tag of user $i(i = 1, 2, \ldots, m)$. First by indexing Wikipedia, we can get the document d_{ij} related to tag t_{ij}. We employ the function $\overrightarrow{W} = \sum_{i=1}^{k} \sum_{j=1}^{k} s_{ij} \cdot v_j = S \cdot \overrightarrow{V}$ to get the score of words in d_{ij} and select the top l words as keywords set k_{ij} of tag t_{ij}. In the equation, s_{ij} is the similarity between words w_i and w_j, and v_j is

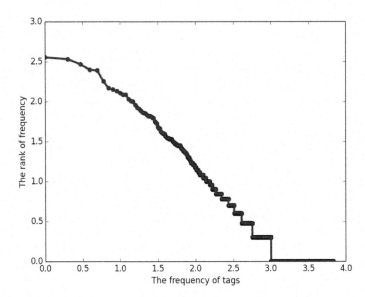

Fig. 2. The Distribution of Tags

defined as $v_j = n_j / \sum_{j=1}^{k} n_j$ in which n_j represents the times of w_j occurring in d_{ij}. Third, we get popular degree of tags by a power law function $g(r_{ij}) = c \cdot r_{ij}^{-\alpha}$, where c is a constant and r_{ij} is the rank of tag t_{ij} by frequency. Then we put tags with greater popular degree than predefined threshold β into a criterion tag set $\Theta = \{\{C_i; (t_{i1}, \ldots, t_{in})\} | i = 1, \ldots, m\}$, where C_i is the category which tag belongs to. According to k_{ij} and Θ, we can get user hierarchical interests, a coarse-grained layer and a detailed layer, which conveys user's interests greatly.

Weighted Semantic Similarity Measure. In actual scenario, not all the tags have equal importance and some tags more semantically relevant to items should be given higher weight. We regard semantic similarity between tag and tweets as the weight of tag. The bigger the similarity is, the higher the correlation is.

Traditional similarity measurements regard distance between text vector and tag vector as their similarity. While, *TF-IDF*, the mainstream text vector extraction algorithm, ignores the important term with a high occurrence in the same cluster. Thus, we implement *K-means* clustering before *TF-IDF* and confirm parameter K by introducing a predetermined category as prior knowledge. For example, for film recommendation, the predetermined category derives from partitioning strategy of film category in video websites. After filtering and integrating, we obtain a normal list with 70 categories as the final value of K.

After the pretreatment by *K-means*, we divide the tweets into K clusters, represented by G, and every genre is represented by some keywords w extracted by *TF-IDF*. The weight of tag t is the total similarity between t and all the genres showed in Equation (2).

$$W(t) = \frac{\sum_{g \in G} \sum_{w \in g} sim_{w2v}(t, w)}{\sum_{t' \in T} \sum_{g \in G} \sum_{w \in g} sim_{w2v}(t', w)} \tag{2}$$

where $sim_{w2v}(t, w)$ represents the word2vec similarity of tag t and keyword w included in genre g, and T is the set of all user tags. To simplified the calculation, we normalize the values of weight in Eq. (2) for each tag so that $\sum_{t \in T} W(t) = 1$.

In terms of normal tag set and tag weight, we represent the average similarity of tags of user u and user v as the users' tag similarity, the weights of tags added to stress the semantic relevance to the context, which is formulized below.

$$sim(u, v) = \frac{\sum_{t \in T_u} \sum_{\tau \in T_v} sim_{w2v}(t, \tau) \cdot W(t) \cdot W(\tau)}{\|T_u\| \cdot \|T_v\|} \tag{3}$$

4   Experiment

In this section, we implement our method using Probabilistic Matrix Factorization(PMF) with a user regularizer [4]. We regard the sentiment matrix of tweets of users on films as initial rating matrix of PMF, and treat similarity matrix of users' tags as similarity matrix used in the user regularizer. Our experiments are intended to address the following questions:

- Does sentiment mapping with 5-scale perform better than with 2-scale?
- Is implicit tagging information effective in improving traditional PMF method?
- How does the regularizer parameter influence the recommender performance?
- How does sentiment analysis influence the recommender accuracy?

We begin by introducing the Probabilistic Matrix Factorization with User Regularizer and experimental dataset, and report and analyze the experimental results to attempt to answer the above questions.

4.1   Probabilistic Matrix Factorization with User Regularizer

In the above section, we introduce our mining method in detail, which aims to generate effective recommendation when lack of explicit ratings. And we use Probabilistic Matrix Factorization [8] as the base model to evaluate the precision of our mining method. We introduce the PMF algorithm briefly in this section.

Suppose we have M programs, N users, and a rating matrix R in which R_{ij} indicates the rating of user i for item j. Let $U \in R^{D \times N}$ and $V \in R^{D \times M}$ be latent feature matrices for users and items, with column vectors U_i and V_j representing D-dimensional user-specific and item-specific latent feature vectors, respectively. The conditional probability distribution over the observed ratings $R \in R^{N*M}$ is given by Equation (4). Indicator function I_{ij} is equal to 1 if user i rated item j and equal to 0 otherwise. We place zero-mean spherical Gaussian priors on user and item feature vectors (Equations (5) and (6)).

$$p(R|U, V, \sigma^2) = \prod_{i=1}^{N} \prod_{j=1}^{M} [N(R_{ij}|U_i^T V_j, \sigma^2)]^{I_{ij}} \tag{4}$$

$$p(U|\sigma_U^2) = \prod_{i=1}^{N} N(U_i|0, \sigma_U^2 I) \tag{5}$$

$$p(V|\sigma_V^2) = \prod_{j=1}^{M} N(V_j|0, \sigma_V^2 I) \tag{6}$$

To estimate model parameters, we maximize the log-posterior over user and item features with fixed hyper-parameters. Maximizing the posterior with respect to U and V is equivalent to minimizing squared error with $L2$ regularization:

$$E = \frac{1}{2} \sum_{i=1}^{N} \sum_{j=1}^{M} I_{ij}(R_{ij} - U_i^T V_j)^2 + \frac{\lambda_U}{2} \sum_{i=1}^{N} \|U_i\|_{Fro}^2 + \frac{\lambda_V}{2} \sum_{j=1}^{M} \|V_j\|_{Fro}^2 \tag{7}$$

where $\lambda_U = \sigma^2/\lambda_U^2, \lambda_V = \sigma^2/\lambda_V^2$, and $\|\cdot\|_{Fro}^2$ denotes the Frobenius Norm. We use gradient descent to find a local minimum of the objective for U and V. Finally, we infer missing rating judgments in the user-item matrix R by taking the scalar product of U and V.

The users' interests can be implicitly reflected by the tags that they use. Based on this intuition, we use average-based social regularization term [4], making an assumption that every user's taste is close to the average taste of the users with similar tags.

$$\frac{\lambda_{sim}}{2} \sum_{i=1}^{m} \|U_i - \frac{\sum_{f \in S(i)} Sim(i,f)U_f}{\sum_{f \in S(i)} Sim(i,f)}\|_F^2 \tag{8}$$

where $Sim(i,f) \in [0,1]$ is the similarity function (3) to indicate the similarity between user u_i and user u_f. $\lambda_{sim} > 0$ is a parameter to control the strength of the item similarity regularization and $S(i)$ is the set of top N similar items of item j. Here, N is a parameter and a larger value for N may suggest that users are more likely to be interested in items similar to the ones they liked before.

4.2   Dataset and Evaluation Metrics

In order to evaluation our method, we perform experiments on online dataset obtained from a Twitter-like microblog platform, Sina Weibo. The dataset contains 1,451,139 unique users and 38 unique films with 4,694,891 tweets and 3,918,075 user tags. The statistics of dataset are summarized in Table 1 and Table 2.

We separately use *RMSE(Root Mean Square Error)* and *F1* [3] to measure rating prediction accuracy and evaluate system's quality. Given n pairs of actual rating y_i and predicted rating $\hat{y}_i$, $RMSE$ is defined as $RMSE = \sqrt{\frac{\sum_{i=1}^{n}(y_i - \hat{y}_i)^2}{n}}$. $F1 = \frac{2RP}{R+P}$, where R and P separately represent recall and precision. The definitions show that smaller $RMSE$ or bigger $F1$ indicates a better performance.

Table 1. Statistics of User-Item Matrix

Statistics	User	Item
Min. Num. of Tweets	1	1,823
Max. Num. of Tweets	734	1,446,561
Avg. Num. of Tweets	3.24	123,549.8

Table 2. Statistics of User Tags

Statistics	Tags per User
Min. Num.	0
Max. Num.	23
Avg. Num.	3

4.3   Sentiment Value Mapping

The first experiment attempt to answer the first question. Note that, sentiment value mapping is really difficult, which plays a core role in the quality of recommendation. As a primary step towards using implicit rating for *PMF*, we focus on the *F1* value using a 10-fold cross validation with stratified sampling, with final averaging on errors measured.

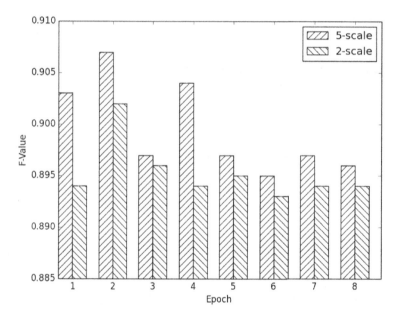

Fig. 3. Mapping's Impact

Figure 3 shows results of 8 times obtained by using 2-scale and 5-scale mapping principle. It indicates that tweets provide valuable information on user evaluation of a film, and implicit sentiment tendency reflecting users' implicit interesting degree is of great use when lack of explicit ratings. It can also be derived from the results that a scale of sentiment mapping between 1 and 5 proved to be more appropriate than 2-scale mapping. Based on the experimental results, experiments below put the 5-scale mapping to use.

4.4   Impact of User Tag Regularizer

To answer the second question, we compare the performance of *PMF* algorithm with *PMFReg* algorithm (*PMF* with a user regularizer). For both methods, we initialize the latent features to random numbers in $\{5, 10, 15\}$ and set the step size for gradient descent to 0.001. The parameters specific to our method are set as $\lambda = 0.06$. Actually, we find the performance is stable after about 30 rounds of gradient decent (see Figure 4). Hence, we set *Epoch* = 30 for all the following results, D=15 and make 80% of the dataset as training set size similarly.

Table 3. RMSE comparision between PMF and PMFReg

Training Set	$D = 5$		$D = 10$		$D = 15$	
	PMF	PMFReg	PMF	PMFReg	PMF	PMFReg
20%	0.4394	**0.3838**	0.4214	**0.3836**	0.4101	**0.3831**
40%	0.4131	**0.3706**	0.4117	**0.3696**	0.4094	**0.3668**
60%	0.3981	**0.3636**	0.3913	**0.3627**	0.3930	**0.3626**
80%	0.3948	**0.3554**	0.3898	**0.3513**	0.3830	**0.3502**

The results reported in Table 3 are the average *RMSE* values of *PMF* and *PMFReg*. The better results are shown in bold. It is clear that *PMFReg* achieves better performance than *PMF*. Because the main difference between *PMFReg* and *PMF* lies in the extra tagging information used by *PMFReg*, we can conclude tagging information is useful and *PMFReg* can utilize it effectively.

4.5   Sensitivity to Parameters λ_{sim}

As we saw in Section 3, the contribution of user tag similarity is controlled by the parameter λ_{sim}. If $\lambda_{sim} = 0$, we do not use user tagging information at all and hence our method degenerates to a special form of PMF; as λ_{sim} increases, we put larger weight on the tagging information. To examine how the regularization parameter affects the performance of our proposed method, we set $\lambda_U = \lambda_V = \lambda_{sim}$ for convenience, and λ_{sim} is set to begin in $[0.03, 0.10]$.

We observe from Fig.4 that the value of λ impacts the recommendation results significantly. When λ is less than 0.06, the curve is over-fitting, while when λ surpasses 0.10, the prediction accuracy decreases. Clearly in this figure, the best regularization parameter setting for dataset is $\lambda_U = \lambda_V = \lambda_{sim} = 0.06$.

4.6   Impact of Sentiment Error

To answer the last question, we experiment on two datasets, one without sentiment error and one with 16% sentiment error. As shown in Fig.5, sentiment error has impact on results, indicating that adopting sentiment analysis algorithm with better performance can improve recommender accuracy, which is one of our future focus. Based on the results, we calculate the average $F1$ value, getting a 0.055 decrease on the $F1$ value of recommender with a 16% sentiment error. The result demonstrates our method is resistant to sentiment error, and can put in practice effectively.

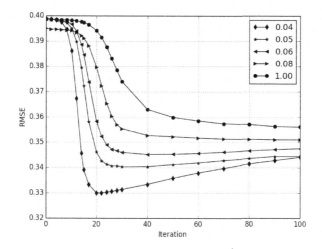

Fig. 4. Sensitivity to λ_{sim}

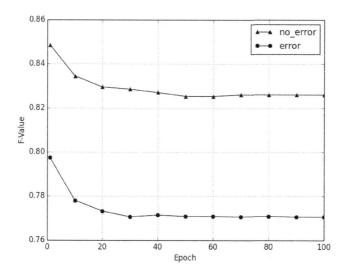

Fig. 5. Sentiment Error

5   Conclusion

In our paper, we attempt to analyze implicit feedback, specifically tweets and tags, for recommendation. To be concrete, we exploit the sentiment analysis value using a 5-scale mapping principle with good accuracy to obtain the virtual rating. Moreover, taking semantics of tags into consideration, we propose a Weighted Semantic Similarity Tag Method based on word2vec in order to build a better user similarity to improve the performance of recommender system. Experimental results on real dataset demonstrate social factors can greatly boost the performance of recommender systems on social network data. Furthermore, the proposed algorithm is put into practice on the intelligent remote control APP in the social TV scenario, and the implement of recommendation function improves user satisfaction and experiences further.

In the future, more studies will be performed to further optimize our algorithm and other implicit information, like demographics, follow history, should be considered to further improve performance of recommender.

Acknowledgments. The work is supported by the National Science and Technology Support Program that TV content cloud services technology integration and demonstration based on social network (No.2012BAH41F03).

References

1. Bai Xue, B.X., Chen Fu, C.F., Zhan Shaobin, Z.S.: A new clustering model based on word2vec mining on sina weibo users tags. International Journal of Grid and Distributed Computing **7**(3), 41–48 (2014)
2. Hu, Y., Koren, Y., Volinsky, C.: Collaborative filtering for implicit feedback datasets. In: Eighth IEEE International Conference on Data Mining, ICDM 2008, pp. 263–272. IEEE (2008)
3. Jianguo, L., Tao, Z., Qiang, G., Binghong, W.: Overview of the evaluated algorithms for the personal recommendation systems. Complex System and Complexity Science **6**(3), 1–10 (2009)
4. Ma, H., Zhou, D., Liu, C., Lyu, M.R., King, I.: Recommender systems with social regularization. In: Proceedings of the Fourth ACM International Conference on Web Search and Data Mining, pp. 287–296. ACM (2011)
5. MacQueen, J.: Some methods for classification and analysis of multivariate observations. In: Proceedings of the Fifth Berkeley Symposium on Mathematical Statistics and Probability, vol. 1, pp. 281–297. Oakland, CA, USA (1967)
6. Markines, B., Cattuto, C., Menczer, F., Benz, D., Hotho, A., Stumme, G.: Evaluating similarity measures for emergent semantics of social tagging. In: Proceedings of the 18th International Conference on World Wide Web, pp. 641–650. ACM (2009)
7. Messenger, A., Whittle, J.: Recommendations based on user-generated comments in social media. In: 2011 IEEE Third International Conference on Privacy, Security, Risk and Trust (PASSAT) and 2011 IEEE Third Inernational Conference on Social Computing (SocialCom), pp. 505–508. IEEE (2011)
8. Mnih, A., Salakhutdinov, R.: Probabilistic matrix factorization. In: Advances in Neural Information Processing Systems, pp. 1257–1264 (2007)
9. Pan, R., Zhou, Y., Cao, B., Liu, N.N., Lukose, R., Scholz, M., Yang, Q.: One-class collaborative filtering. In: Eighth IEEE International Conference on Data Mining, ICDM 2008, pp. 502–511. IEEE (2008)
10. Pappas, N., Popescu-Belis, A.: Sentiment analysis of user comments for one-class collaborative filtering over ted talks. In: Proceedings of the 36th International ACM SIGIR Conference on Research and Development in Information Retrieval, pp. 773–776. ACM (2013)

11. Parra, D., Karatzoglou, A., Amatriain, X., Yavuz, I.: Implicit feedback recommendation via implicit-to-explicit ordinal logistic regression mapping. In: Proceedings of the CARS-2011 (2011)
12. Quattrone, G., Capra, L., De Meo, P., Ferrara, E., Ursino, D.: Effective retrieval of resources in folksonomies using a new tag similarity measure. In: Proceedings of the 20th ACM International Conference on Information and Knowledge Management, pp. 545–550. ACM (2011)
13. Salton, G., Yu, C.T.: On the construction of effective vocabularies for information retrieval. ACM SIGPLAN Notices **10**, 48–60 (1973)
14. Socher, R., Perelygin, A., Wu, J.Y., Chuang, J., Manning, C.D., Ng, A.Y., Potts, C.: Recursive deep models for semantic compositionality over a sentiment treebank. In: Proceedings of the Conference on Empirical Methods in Natural Language Processing (EMNLP), pp. 1631–1642. Citeseer (2013)
15. Wang, J., Li, Q., Chen, Y.P.: User comments for news recommendation in social media. In: Proceedings of the 33rd International ACM SIGIR Conference on Research and Development in Information Retrieval, pp. 881–882. ACM (2010)
16. Wang, X., Zhang, C., Ji, Y., Sun, L., Wu, L., Bao, Z.: A depression detection model based on sentiment analysis in micro-blog social network. In: Li, J., Cao, L., Wang, C., Tan, K.C., Liu, B., Pei, J., Tseng, V.S. (eds.) PAKDD 2013 Workshops. LNCS, vol. 7867, pp. 201–213. Springer, Heidelberg (2013)
17. White, R.W., Jose, J.M., Ruthven, I.: Comparing explicit and implicit feedback techniques for web retrieval: Trec-10 interactive track report. In: Proceedings of the Tenth Text Retrieval Conference (TREC 2010), pp. 534–538. NIST (2002)
18. Wu, L., Chen, E., Liu, Q., Xu, L., Bao, T., Zhang, L.: Leveraging tagging for neighborhood-aware probabilistic matrix factorization. In: Proceedings of the 21st ACM International Conference on Information and Knowledge Management, pp. 1854–1858. ACM (2012)
19. Yu, X., Ren, X., Sun, Y., Sturt, B., Khandelwal, U., Gu, Q., Norick, B., Han, J.: Recommendation in heterogeneous information networks with implicit user feedback. In: Proceedings of the 7th ACM Conference on Recommender Systems, pp. 347–350. ACM (2013)
20. Zhang, L., Ghosh, R., Dekhil, M., Hsu, M., Liu, B.: Combining lexiconbased and learning-based methods for twitter sentiment analysis. HP Laboratories, Technical Report HPL-2011 89 (2011)
21. Zhang, W., Ding, G., Chen, L., Li, C.: Augmenting chinese online video recommendations by using virtual ratings predicted by review sentiment classification. In: 2010 IEEE International Conference on Data Mining Workshops (ICDMW), pp. 1143–1150. IEEE (2010)
22. Zhang, Y., Zhang, M., Zhang, Y., Liu, Y., Ma, S.: Explicit factor models for explainable recommendation based on phrase-level sentiment analysis. In: Proceedings of SIGIR, vol. 14 (2014)
23. Zhen, Y., Li, W.J., Yeung, D.Y.: Tagicofi: tag informed collaborative filtering. In: Proceedings of the Third ACM Conference on Recommender Systems, pp. 69–76. ACM (2009)
24. Zipf, G.K.: Human behavior and the principle of least effort (1949)

The Collaborative Filtering Algorithm with Time Weight Based on MapReduce

Hongyi Su[✉], Xianfei Lin[✉], Bo Yan, and Hong Zheng

Key Lab of Intelligent Information Technology, Beijing Institute of Technology, Beijing, China
{henrysu,xflin,yanbo,hongzheng}@bit.edu.cn

Abstract. As one of the most successful recommendation algorithm, collaborative filtering algorithm still faces many challenges, such as accuracy, extensibility, and sparsity. In the algorithm, ratings produced in different period are treated equally, so changes of users' interests have been ignored. This paper considers the influence of time factor on users' interests, and presents a new algorithm that involves time decay factor in the collaborative filtering algorithm, the new algorithm makes a more accurate recommendation by reducing the weight of old data. Deploying the collaborative filtering algorithm with time weight on parallel computing frame of MapReduce also achieves the extensibility of algorithm and improves the processing performance of large data sets.

Keywords: Collaborative Filtering · Time weight · Distributed application

1 Introduction

An outstanding recommendation system can provide the users information what appeals to them within a short time. Collaborative filtering algorithm (CF) is one of the most successful recommendation algorithms [1], it includes user-based CF [2] and item-based CF [3-4]. The traditional CF ignores the influence of time on user preferences. However, in fact, the users' interests may change. Therefore, for the old data, we should appropriately reduce their weight.

MapReduce framework is one of the currently popular distributed processing frameworks [5], it distributes the large scale data set on the slave nodes under the management of a master node, collects the intermediate processing results of the slave nodes, and integrates them to get the final result.

This paper proposes a method to include the time factor with weight in a distributed algorithm. With deep analysis on the influence of time factor for each user interest, it generates a time decay factor that adapts to the user's characteristics, implements the recommendation with time weight on MapReduce framework, and finally solves the problems of the traditional CF's can't reflect changes of the users' interests and extensibility. The rest of this paper is arranged as following: the second section introduces researches concerning CF; the third section describes CF with time weight based on MapReduce; the fourth section provides the experimental results and analysis; the last section makes a conclusion and suggests the directions for future work.

© Springer International Publishing Switzerland 2015
Y. Wang et al. (Eds.): BigCom 2015, LNCS 9196, pp. 386–395, 2015.
DOI: 10.1007/978-3-319-22047-5_31

2   Related Work

The basic idea of item-based CF is that a user will have similar ratings for similar items, according to the ratings that already exist, predict the ratings of the items haven't been rated, and recommend the items with higher predicted ratings to the users. Li, Y et al analyzed the limitations of the user-based CF and the item-based CF respectively and emphatically analyzed the deficiency of the user-based CF in user interest and content diversification, they put forward mixed CF based on the users and the items [6].

A simple method to adapt CF to changes of user interest is discarding the old data, but it will lead to more serious sparseness. Therefore, a more appropriate method is decaying the weight on old data. Yuchuan Zhang and Yuzhao Liu pointed out that the exponential function has better adaptability in measuring the changes of users' interests [7]. Liang Xiang and Qing Yang set up factorization model for four effects of time, that is the singular value decomposition (SVD) of time [8]. Yi Ding and Xue Li analyze different users' changing interests towards different item types, building CF with time weight [9]. These methods are realized in single machine environment.

The increasing amount of users and items is a huge challenge to the extensibility of traditional CF. Apache provides an open source of machine learning repository, namely Mahout, it successfully implements item-based CF on the MapReduce framework [10]. This calculation method distributes the computing of traditional CF to multiple nodes. Since traditional CF cannot transform into distributed CF completely, there are differences between them, the distributed CF also ignores the influence of time factor, still lacks in accuracy. Jing Jiang et al divided item-based CF into four MapReduce processes to realize parallel computing [11].Concerning the distributed implementation of the CF, Yang Shang et al tested in a distributed environment using simulated data and real data, the final result shows the performance improvement of the distributed system for large data processing [12].

3   Method

Our algorithm consists of three jobs: generating ratings with time weight by time decay factors, generating co-occurrence matrix, generating final recommendations. The choice of decay factors is extremely crucial to the final results' accuracy. In this section, we first introduce traditional item-based CF and distributed item-based CF, then present how to join the distributed algorithm with time decay factor and the choice of time decay factor. In addition, we describe the application of the new CF with time weight on MapReduce framework. Finally, summarize the whole process of the algorithm.

3.1   The Traditional Item-Based CF

The traditional item-based CF has two main steps: the calculation of similarity between the items and recommendation for the users.

The similarity's main measures are as follows:

Tanimoto coefficient, item A and item B have respectively been rated by user set a and user set b. The Tanimoto coefficient between item A and item B is

$$similarity_{AB} = \frac{a \cap b}{a \cup b} \tag{1}$$

Cosine similarity [13], the rating vectors of item A and item B are a and b, respectively, dimension of the vectors is the number of users n, user i's rating for item A is V_{iA}. The cosine similarity between item A and item B is

$$similarity_{AB} = cos(a, b) = \frac{a \cdot b}{|a||b|} = \frac{\sum_i^n V_{iA} \times V_{iB}}{\sqrt{\sum_i^n V_{iA}^2}\sqrt{\sum_i^n V_{iB}^2}} \tag{2}$$

Pearson correlation coefficient, the number of users is n, user i's rating for item A is V_{iA}, the average rating of user i is $\overline{V}_i$. The Pearson correlation coefficient between item A and item B is

$$similarity_{AB} = \frac{\sum_i^n (V_{iA} - \overline{V}_i)(V_{iB} - \overline{V}_i)}{\sqrt{\sum_i^n (V_{iA} - \overline{V}_i)^2}\sqrt{\sum_i^n (V_{iB} - \overline{V}_i)^2}} \tag{3}$$

In this paper, due to the need of distribution, we choose co-occurrence matrix to measure similarity between items [10]. The co-occurrence matrix use common occurrence time of item pairs to represent the similarity level of them, item A and item B have respectively been rated by user set a and user set b, the two items have been rated by m users at the same time, the co-occurrence value of item A and item B is m. The thought of this computing method is: if any two items are rated by more users at the same time, the two items are more similar. The co-occurrence matrix is symmetric matrix, the value of Ath line, Bth column is calculated as follows:

$$similarity_{AB} = a \cap b \tag{4}$$

When recommending for user u, check item i which user u haven't rated, respectively calculate the $similarity_{ij}$ of item i and item j, item j have been rated V_{uj} by user u, if user u has rated item set m, user u's predicted rating for item j is

$$V_{uj} = \frac{\sum_j^m V_{uj} * similarity_{ij}}{\sum_j^m similarity_{ij}} \tag{5}$$

Rank the items which haven't been rated by user u according to the predicted ratings, recommend the top n items to user u.

3.2   The Distributed Item-Based CF

The jobs of the distributed item-based CF mainly include the calculation of users' rating vectors, construction of co-occurrence matrix, generating the recommendation.

The basic unit of data set is quad $<U_i, I_j, V_{ij}, T_{ij}>$, U_i is the user whose id is i, I_j is the item whose id is j, V_{ij} is user i's rating for item j, T_{ij} is the distance between current time and the rating time when user i gave the rating V_{ij} for item j.

MapReduce1: The calculation of users' rating vectors.

Map1: Consider each quad as a unit of the input data set, change the quad into key/value pairs with U_i as the key, vector composed of I_i and V_{ij} vector as the value.

Reduce1: Collect the key/value pairs produced by map1 process according to the ids of users. If U_i has rated k items, then generate a key/value pair with U_i as key, k dimensional vectors composed of $< I_i: V_{ij}>$s as the value, each key/value pair is a user's rating vector.

MapReduce2: Construction of co-occurrence matrix. There are n items, the result is a matrix of $n * n$.

Map2: As the co-occurrence matrix is symmetrical matrix, we just calculate co-occurrence value once time for every items pair, rank each user's rating vector according to the items' ids, item I_a and I_b $(a < b)$ appear in the same user vector, change them into a key-value pair with I_a as the key, $< I_b: 1 >$ as the value.

Reduce2: Collect the key/value pairs produced by map2 process on the key, accumulate the co-occurrence results of item pairs. Co-occurrence vector whose key is I_a contains $m < I_b: 1>$, then its cumulative result will be $< I_b: m >$, calculate co-occurrence time between I_a and the k items which have co-occurrence relationship with it, generate I_a's k dimensional co-occurrence vector.

The co-occurrence matrix is composed of the co-occurrence vectors.

MapReduce3: Generating the recommendation. Through a MapReduce process, the users' rating vectors and the items' co-occurrence vector can be processed into key/value pairs that easier for multiplication, these key/value pairs' keys are the items' ids, and values are the key item's co-occurrence vector, an array composed of the users' ids who have rated the key item and an corresponding array composed of the key item's ratings.

Map3: We analyze the key/value pairs that easier for multiplication, divide them into new key/value pairs on the user ids, these new key/value pairs' keys are user ids, values are an item's rating and the item's co-occurrence vector. Each rating record is corresponding to a key/value pair.

Reduce3: For each user, multiply the item's rating with its co-occurrence vector, respectively, accumulate the result vectors as the molecular, accumulate the co-occurrence vectors of the rated items as the denominator, and recommend n items with the highest ratios to the user.

The above process can be summarized as formula (6). $I_i I_j$ presents the co-occurrence time of item i and item j, V_{ui} presents user u's real rating for item i, P_{ui} presents user u's predicted rating for item i.

$$\begin{bmatrix} I_1 I_1 & I_1 I_2 & \cdots & I_1 I_n \\ I_2 I_1 & I_2 I_2 & \cdots & I_2 I_n \\ \vdots & \vdots & \ddots & \vdots \\ I_n I_1 & I_n I_2 & \cdots & I_n I_1 \end{bmatrix} \times \begin{bmatrix} V_{u1} \\ V_{u2} \\ \vdots \\ V_{un} \end{bmatrix} = \begin{bmatrix} P_{u1} \\ P_{u2} \\ \vdots \\ P_{un} \end{bmatrix} \tag{6}$$

3.3   The CF with Time Weight Based on MapReduce

The Time Factor Decay Function of Ratings. Users' interests change over time, the more recent ratings could describe the users' current interests better. We construct a time factor decay function, the function's value reduces along the distance between the rating time and current time:

$$w_u(t) = e^{-d_u t} \tag{7}$$

$w_u(t)$ presents the weight of ratings that gave t days ago, d_u presents the user u's time decay factor, the ratings' weights are calculated by the factor, all users' time decay factor constitute a configuration file, with the format: userID, decay. Fig. 1 shows the changes of the function when the decay changes.

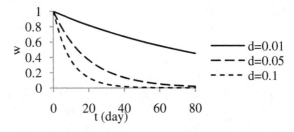

Fig. 1. The time factor decay function with different decay

This process needs two mappers, one mapper is used to process the configuration file of time decay factors, the other is used to process users' rating file. We built a new Writable class: DecayOrVectorWritable. This class can be both DoubleWritable class and another new built class: ItemIDValueTimeWritable class. DoubleWritable class can be used to store time decay factor, and ItemIDValueTimeWritable can be used to store user's rating vector. The new built class makes the different types results of two mappers can be received by the same combiner. During the reduce process, determine and classify the value, get the user's rating vector and time decay factor, we use formula (7) to calculate the ratings with weights. Fig. 2 shows the distributed implement of CF with time weight.

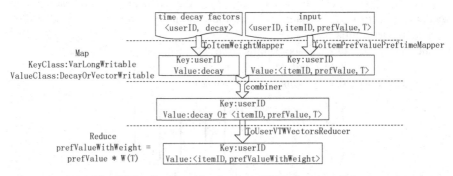

Fig. 2. The distributed implement of CF with time weight

The following pseudocode states the concrete implement of the algorithm.

```
class ToItemWeightMapper
  map(Long key, Text value):
    //key: a record id
    //value: a user's time decay factor record
    for each record<userID, decay> in value:
      EmitIntermediate(userID, decay);
class ToItemPrefvaluePreftimeMapper
  map(Long key, Text value):
    //key: a record id
    //value: a user's rating record
    for each record<userID,itemID,prefValue,T> in value:
      EmitIntermediate(userID, <itemID,prefValue,T>);
class ToUserVTWVectorsReducer
  reduce(Long key, iterator<DecayOrVector> values):
    //key: a user id
    //value: the user's personalized time decay factor or
    rating vector
    for each value in values:
      if (Double.isNaN(value))
        ratingList.add(value);
      else
        decay = value;
    for each ratingVector in ratingList:
      prefValueWithWeight
        = Math.exp((-1) * decay * T) * prefValue;
      Emit(userID, <itemID,prefValueWithWeight>);
```

The Choice of Users Interests' Time Decay Factors. Due to different users have different preferences habits, some users have a relatively stable interest, time doesn't impact them much, such kind of users have smaller time decay factors; while some users have diverse interests, they always keep exploring new items, the more recent ratings can represent the users' recent interests better, these users have bigger time decay factors. We generate personalized time weight for each user through analysis of the training set, these time weights produced in the preprocessing stage, and generate the users' configuration file: configuration file of time decay factors.

We partition the training set in three ways, generate 85%, 90% and 95% as training sets (TR_1, TR_2, TR_3) for training time decay factor, the rest is used to choose users' optimal personalized time decay factor as testing sets(TE_1, TE_2, TE_3). Decay factor starting from 0, 0.01 is chosen as the step length, calculate the results of three training sets, and test their accuracy by the corresponding testing sets, choose the time decay factor of the highest average accuracy as the user's personalized time decay factor.

We use precision and recall to calculate accuracy [14]. The recommendation list of N items is top - N, while the item list users really interested in is the test.

Precision is the ratio of the items users really like in top-N recommendation list.

$$precision = \frac{|test \cap top-N|}{N} \qquad (8)$$

Recall is the ratio of the recommendation items in the item list users really like.

$$recall = \frac{|test \cap top-N|}{|test|} \qquad (9)$$

The ultimate measure is F1, a combination of precision and recall.

$$F1 = \frac{2*recall*precision}{recall+precision} \qquad (10)$$

For the user u, every time decay factor could work out three F1s with the three training sets, these three F1s are called $F1_1, F1_2, F1_3$. We choose the time decay factor d_u which meets the condition $max\left(\overline{F1} = \frac{F1_1+F1_2+F1_3}{3}\right)$.

Implementation of the Algorithm. We present the CF with time weight based on MapReduce, gain the recommend speed by distributing computing of co-occurrence matrix and ratings that have decayed by the time factor decay function. The input of recommender is the users' rating file and the configuration file produced by time decay factors chooser. The output of recommender is users' ids and the users' corresponding recommendations. The final result is evaluated by F1. Fig. 3 shows the algorithm's whole process.

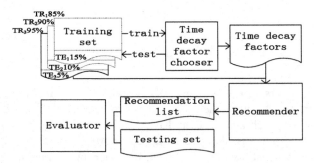

Fig. 3. Whole process of the CF with time weight based on MapReduce

4　Experiment

4.1　Data Sets

The three data sets of MovieLens (http://www.movielens.umn.edu) are used in our experiment. They respectively are data set of 100,000 ratings, data set of 1,000,209 ratings and data set of 10,000,054 ratings, the format of the data sets is: user id, item id, users' ratings for the items, UNIX timestamp of ratings time. Characteristics of these three data sets are shown in table 1.

Table 1. Characteristics of the data sets

Dataset	# of Users	# of Items	# of Ratings
100k	943	1,682	100,000
1m	6,040	3,900	1,000,209
10m	71,567	10,681	10,000,054

Rank each data set according to timestamp, each user's latest 10 ratings is divided into testing set, treat the remaining as training set. Regard the latest rating's time as current time, we update the training set's timestamps minus current time, and in calculation, we use day as the unit instead of second. The algorithm recommends for every user according to the training set, and the evaluator compares recommendation list with testing set. The evaluator uses precision, recall and F1.

Respectively classify each user's latest 5%, 10%, 15% ratings of the training set as the testing sets for the time decay factor chooser, classify the remaining as the training set. With different time decay factor, algorithm recommend for uses according to the three different training sets, compare recommendation list with the corresponding test sets, and choose each user's optimal time decay factor.

4.2   Experimental Design

Our experiment is depend on a Hadoop cluster with four computers. The master node's is equipped with Intel Core Xeon 2620 processor, 16 GB memory, CentOS 6.5 64bit operating system, the deploying Hadoop version is 2.4.0.

Compare the CF with time weight (CF-TW) with traditional item-based CF, their precisions, recalls and F1s are respectively shown in table 2, table 3 and table 4.

Table 2. The precisions of CF and CF-TW

Algorithm	Precision		
	100k	1m	10m
CF	0.0291	0.0251	0.0278
CF-TW	0.0842	0.0738	0.0928

Table 3. The recalls of CF and CF-TW

Algorithm	Recall		
	100k	1m	10m
CF	0.1534	0.1328	0.1451
CF-TW	0.2551	0.2398	0.2631

Table 4. The F1s of CF and CF-TW

Algorithm	F1		
	100k	1m	10m
CF	0.0489	0.0423	0.0467
CF-TW	0.1266	0.1128	0.1373

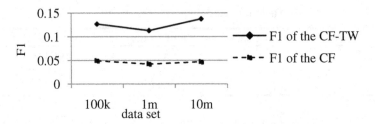

Fig. 4. The F1s' comparison of CF and CF-TW

Fig. 4 is the F1s' comparison of CF and CF-TW. The experiment shows that, compared with traditional item-based CF, precision, recall and F1 of the CF with time weight have improved greatly.

When dealing with three different data sets, deploying the CF with time weight on MapReduce can greatly shorten the run time, the comparison with run time of the CF with time weight running on the master is shown in table 5.

Table 5. The run time under two environments

Environment	Run time (s)		
	100k	1m	10m
Single node	334	8614	200256
MapReduce	107	971	12023

Relative to single node, speedup of running on MapReduce is shown as fig. 5.

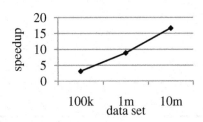

Fig. 5. The speedup of running on MapReduce

As we can see from fig. 5, deploying the CF with time weight on the MapReduce can significantly improve the speed, solve the problem of large data, and realize the scalability of the algorithm successfully. The larger the data sets are, the algorithm performance better.

5   Conclusion and Future Work

In this paper, we present a new CF with time. According to the variation of each user's interest, we choose appropriate time decay factor for each user, decay the weight of users' ratings in the remote past to meet the users' interests changes. In this article, we also deploy the algorithm on MapReduce, implementing distributed computing. This algorithm triples the F1 value of recommendation list, and solves the scalability problem of large data sets in the mean while.

In future work, we will focus on further accelerate computational speed by GPU. GPU is graphics processor that produced by Intel. Since it's a processor designed for a large number of parallel computing, it can performance excellent in these related fields.

References

1. Wang, G., Liu, H.: Survey of personalized recommendation system. Jisuanji Gongcheng yu Yingyong (Computer Engineering and Applications) **48**(7), 66–76 (2012)
2. Goldberg, D., Nichols, D., Oki, B.M., et al.: Using collaborative filtering to weave an information tapestry. Communications of the ACM **35**(12), 61–70 (1992)
3. Resnick, P., Iacovou, N., Suchak, M., et al.: GroupLens: an open architecture for collaborative filtering of netnews. In: Proceedings of the 1994 ACM Conference on Computer Supported Cooperative Work, pp. 175–186. ACM (1994)
4. Linden, G., Smith, B., York, J.: Amazon. com recommendations: Item-to-item collaborative filtering. Internet Computing, IEEE **7**(1), 76–80 (2003)
5. Dean, J., Ghemawat, S.: MapReduce: simplified data processing on large clusters. Communications of the ACM **51**(1), 107–113 (2008)
6. Li, Y., Lu, L., Xuefeng, L.: A hybrid collaborative filtering method for multiple-interests and multiple-content recommendation in E-Commerce. Expert Systems with Applications **28**(1), 67–77 (2005)
7. Zhang, Y., Liu, Y.: A collaborative filtering algorithm based on time period partition. In: 2010 Third International Symposium on Intelligent Information Technology and Security Informatics (IITSI), pp. 777–780. IEEE (2010)
8. Xiang, L., Yang, Q.: Time-dependent models in collaborative filtering based recommender system. In: IEEE/WIC/ACM International Joint Conferences on Web Intelligence and Intelligent Agent Technologies, 2009. WI-IAT 2009, vol. 1, pp. 450–457. IET (2009)
9. Ding, Y., Li, X.: Time weight collaborative filtering. In: Proceedings of the 14th ACM International Conference on Information and Knowledge Management, pp. 485–492. ACM (2005)
10. Anil, R., Dunning, T., Friedman, E.: Mahout in action. Manning (2011)
11. Jiang, J., Lu, J., Zhang, G., et al.: Scaling-up item-based collaborative filtering recommendation algorithm based on hadoop. In: 2011 IEEE World Congress on Services (SERVICES), pp. 490–497. IEEE (2011)
12. Shang, Y., Li, Z., Qu, W., et al.: Scalable collaborative filtering recommendation algorithm with MapReduce. In: 2014 IEEE 12th International Conference on Dependable, Autonomic and Secure Computing (DASC), pp. 103–108. IEEE (2014)
13. Sarwar, B., Karypis, G., Konstan, J., et al.: Item-based collaborative filtering recommendation algorithms. In: Proceedings of the 10th International Conference on World Wide Web, pp. 285–295. ACM (2001)
14. Herlocker, J.L., Konstan, J.A., Terveen, L.G., et al.: Evaluating collaborative filtering recommender systems. ACM Transactions on Information Systems (TOIS) **22**(1), 5–53 (2004)

Recommendation Specially for Fanatic Fans in SNS

Yuewen Su[✉], Mingyue Cui, Yang Ji, and Yunxu Yuan

Mobile Life and New Media Laboratory, School of Information and Communications
Engineering, Beijing University of Posts and Telecommunications, Beijing, China
{suyuewen,Cuimy,jiyang,yuanyunxu}@bupt.edu.cn

Abstract. In social network celebrities are always followed by millions
even billions of users, among whom some users are regarded as fanatic
fans in the sense that they spare no efforts to show their enthusiasm such
as retweeting and commenting every microblog the celebrities tweeted.
The fanatic fans are therefore targeted as powerful potential candidates
of follower recommendation for new emerging celebrities. To address the
problem of potential users filtering to benefit the recommendation accuracy improvement, the features including actions, keywords and tags of
users are explored to recognize the possible fanatic fans and a Latent
Factor Model (LFM) for recommendation is therefore proposed. The
experiment conducted on a data set of Tencent Microblog shows that
our model can achieve higher accuracy than traditional CF mechanism
by taking special user features into consideration.

Keywords: Fanatic fan · LFM · User feature

1 Introduction

Talking about social network which is well developed today, it is separated into
two kinds. One is bidirectional, users follow each other, like Face book, the other
one is unidirectional, users are mainly one way followed, like twitter or MicroBlog.
In our paper, we focus on the social relationship in the latter. MicroBlog is based
on users' pursuit for their followees, especially celebrities like Tencent and Sina
MicroBlog. Here exists one group of people, they retweet and comment everything
the celebrities tweet, they respond every topic associated with the celebrities, they
help their idols to be the top hot stars. We call them fanatic fans.

Study of fans culture has a long history. Years ago, the raise of Korean
Wave(K-wave) [22] is the contribution of fans. They watch TV dramas, listen to
pop music of Korean. Furthermore they followed the Korean stars' fashion and
make K-wave popular culture. That is the power of fans.T hey are an example of millions of fanatic fans. They not only behave crazy in the real world,
they also follow Korean popular stars in microblog to get closer interaction.
There is similarity and difference between fans in the real word and social network. The similarity is they spare no effort to show their enthusiasm to the star.

Y. Wang et al. (Eds.): BigCom 2015, LNCS 9196, pp. 396–406, 2015.
DOI: 10.1007/978-3-319-22047-5_32

The difference is in social network, the ways of expressing their love have more variety and convenience and lower price. Retweet, @ and comment , even attaching tags are among them. Our motivation of this paper is to provide appropriate recommendations for fanatic fans, fully inspire their potential of being crazy of celebrities. More accepted celebrities means more user content and more user interaction to make the social network more active and leads to increase users' loyalty to the service.

But there are two problems we need to face when we want to make accurate recommendation for the fanatic fans. The first is there is no precise definition of fanatic fans before. We can't pick out fanatic fans directly from the celebrity's followers. All we know is that they show their enthusiasm through different way: action, tag, keyword and so on. We have to explore the unique features of the fanatic fans. The second is how to make recommendation targeted. Collaborative filtering (CF) recommendation is widely used nowadays. There are two models in CF: the neighborhood approach model and matrix factorization model. Latent factor models(LFM) is derived from matrix factorization, which is well developed today. In this paper, We build extend LFM model including CF idea. We take advantage of features include actions, keywords and tags which is significant and efficient to define a fanatic fan.

In this paper, we choose a large-scale dataset from Tecent Weibo as our experiment data. To protect the privacy, the Chinese characters are turned into numbers, which increases the universal applicability. Our contribution lies on analyzing fanatic fans' specific features and adding these features into recommendation algorithm. We bring out different method to take advantage of the features and tried to get the best results.

The rest of the paper is organized as the following. Section 2 introduces related work of celebrity recommendation methods. Section 3 suggests the definition of fanatic fans, which is the base of this paper. Different latent factor models are created in Section 4. Section 5 mainly talks about the evaluation and summary. In the last section, we discuss the future work.

2   Related Work

Efforts have been made to study social network. Comparing MicroBlog and Facebook, the main difference is whether the follow is one-way or not. In Facebook, friends always know each other, but in MicroBlog, only the followers show interest to his followees. This suits the relationship between fans and their idols. In Silvia's paper [1], she pointed that a users influence is measured based on his/hers followers and retweets, which means fanatic fans' actions can expanded the celebrity's influence. LV Hai-xia et al. [2] studied recommendation algorithm for celebrity endorsement on social network. The followers group has their common attributes and features, but in this paper just by analyzing basic ones: the gender, location, age and other basic profiles, not including their behaviors. For the MicroBlog platform , more fanatic fans means more influence, more research value. As we can't deny the importance of fanatic fans, we have to excogitate ways to make specific recommendation for fanatic fans.

As to make followee recommendation more perfect, people's effort is mainly from two aspects. One is to add more features into the traditional rate matrix. Karen H.L eg [8] proposed a generic method that allows tags to be incorporated to standard CF algorithms. It proved to be effective. But the algorithm only included tags not taking users's behavior features into consideration which reflect the users' interest directly. The other one is to explore new ways to facilitate social network recommendation. Systems based upon collaborative filtering or content-based recommendation have been widely used and perfect [9]. But there is 'cold start' problem. Viewing collaborative filtering as a matrix completion process, researchers have proposed strategies based on matrix factorization (MF), which has proven to be one of the most successful solutions for collaborative filtering [6]. The key idea behind MF is to approximate the rating r_{ui} as the inner-product of a length-k vector p_u of user factors and another vector q_i of item factors, where k is a pre-specified parameter [19, 20]. Chung-Yi Li and Shou-De Lin conducted the regular matrix factorization for incomplete rating information [7]. Simon Funk proposed one matrix decomposition method which is called Latent Factor Model(LFM) by Koren. In its basic form, matrix factorization characterizes both items and users by vectors of factors inferred from item rating patterns. High correspondence between item and user factors leads to a recommendation [13]. Recently, latent factor models comprise an alternative approach to collaborative filtering with the more holistic goal to uncover latent features that explain the relation between users and items [10]. Those papers are focus on how to make predicted rate more accurate, without considering specific situations. In this paper, we combine the existing algorithm with our demand and features.

3   DataSet and Definition

When we decide to study fanatic fans in MicroBlog, we focus on their characteristics. If the user is obsessed with the star, he retweets the star's blog and comment it, so his action feature is explicit. Then we turn to the users' blog contents. Their retweet blogs nearly have the same resource- the celebrity, their own blogs mentioned or at(@) the celebrity at a very high frequency. That is the fanatic fan group may have the same keywords generated from their blogs content. Then we explore their profiles. In social network, people may not want to register with privacy information, such as birthday or gender. But they'd like to attach tags to show their interests. For example, the fanatic fans who are fond of opera star may have the tag "popular star".

In this paper, we take advantage of the data set released from Tencent Weibo. We got celebrity list and users' information consisting of user actions, tags and keywords. User action data contains the statistics about the at (@), comment and retweet frequency between the user to the others. For example, if user A has retweeted user B 5 times, has at B 3 times, and has commented user B 6 times, then there is one line A B 3 5 6 in the data set. User key word data contains the keywords extracted from the tweet/retweet/comment by each user. In order to protect the privacy of the users, user names, keywords and the tags are replaced by number string.

Why we Focus on Fanatic Fans: Intuitively, action data is one direct aspect to show the users' interest. We pick one celebrity's followers' corresponding action list, and rank them according to the total frequency. In MicroBlog, the amounts of @, comment and retweet suggests the current hot degree. We can see that fanatic fans make the star 'hot'. The statistics show in table 1 may suggest the result more intuitively. Pick one celebrity's fans ranked action list for example, we pick top 5%, 10% 20% users and count their behavior quantity proportion. The result show that a small number of people produce most of the actions. Considering the other functions of MicroBlog, there are other ways for the fanatic fans to show their passion, such as topic discussion and tweetting information about the star. They contributed large amounts of user information. Furthermore, we consider the fanatic fans' behavior as a habit, the habit of showing their affection in MIcroblog. Supposing that we recommend another proper star for him, he will also be fanatic about. For the celebrity, the fanatic fans are their strong support. For the development of MicroBlog, meetting the needs of fanatic fans means more active users and more user content. So it is necessary to improve the accuracy of recommendation for fanatic fans.

Table 1. The percentage of users' action

percentage	at@	comment	recomment
top5%	54.6%	42.6%	13%
top10%	65.2%	55.4%	23.6%
top20%	72.1%	68.8%	33.8%

What Is a Fanatic Fan: When we say someone is a fanatic fan, firstly, we mean he or she is fanatic about one or several particular celebrities. Intuitively speaking, the ratio of the users' retweet, comment and @ for this celebrity is quite high. This shows his loyalty and helps us to make recommendation more targeted. Secondly, a fanatic fan must be a positive user which means the quantity of their actions is large.

4   Extend Latent Factor Model

4.1   Latent Factor Model Summary

Since Netflix Prize, scholars pay more and more attention to matrix factorization(FM). Simon Funk published a algorithm based on SVD which is called Latent Factor Models(LFM) attracted everybody's attention. The main idea of LFM is to reduct decomposition of matrix R which represents user-item rates. r_{ui} is the rate user u scored item i. If there is no score, the rate is 0. LFM algorithm is to extract several themes from the data, as the connection between the user and item. The R matrix is expressed as P matrix and Q matrix multiplication, that is a user latent factor and an item latent factor. A standard method of MF is applied to learn the latent factors to complete P and Q and minimize the sum of squared errors to prevent overfitting [12]. r_{ui} is the original rate, $\hat{r_{ui}}$ is the predicted score.

$$R = P^T Q \tag{1}$$

The dimensionality of R is m × n, if we assume the number of latent factors is g, then P is g × m, R is g × n. The predicted rate is calculated as below.

$$\hat{r_{ui}} = \sum_g p_{ug} q_{ig} \tag{2}$$

However, in the actual circumstances, different people may have different habits of grading, maybe higher or lower than average. Considering this, we add bias items into equation 2. The predicted score is calculated by the equation below:

$$\hat{r_{ui}} = u + b_u + b_i + p_u^T q_i \tag{3}$$

Here u is the global average. b_u is the user bias, on behalf of the user's habit. b_i is the item bias, represents the item attribute. p_u represents the user latent factor vector, q_i is the item latent factor vector. We optimize the results by minimizing the following loss function [21]:

$$C(w) = \sum_{(u,i) \in Train} (r_{ui} - \hat{r_{ui}})^2 + \lambda(\|p_u\|^2 + \|q_i\|^2) \tag{4}$$

A modern optimization algorithm for solving the minimization problem is Stochastic Gradient Descent (SGD) [15]

$$\begin{aligned} p_{uf} &= p_{uf} + \alpha(q_{ik} - \lambda P_{uk}) \\ q_{if} &= p_{uf} + \alpha(p_{uk} - \lambda q_{ik}) \end{aligned} \tag{5}$$

α is learning rate. λ is regularization parameter. Both are adjustable. We build the matrix with rate generated from action feature. The result is shown in section 5.3.

4.2   Latent Factor Model with Keywords

Xing Zhao [12] builds LFM model for Keywords and Tags. We apply it into our situation.

$$\hat{r_{ui}} = u + b_u + b_i + q_i^T \left(\sum_{k \in K(u)} w_k p_k \right) \tag{6}$$

$$\begin{aligned} p_{uf} &= p_{uf} + \alpha(q_{ik} - \lambda P_{uk} w_{uk}) \\ q_{if} &= p_{uf} + \alpha(p_{uk} w_{uk} - \lambda q_{ik}) \end{aligned} \tag{7}$$

Here p_k is the user latent factor for keyword k. K_u is the keywords for user u. w_k represents the keyword vector weight, which is calculated by $\|K_{ui}\|^2$. Here the user latent factor is expressed through users keyword latent factors.

4.3   Latent Factor Model with Tags

Similarly, we have the tag based latent factor models.

$$\hat{r}_{ui} = u + b_u + b_i + q_i^T \left(\frac{1}{\sqrt{|Tag(u)|}} \sum_{t \in Tag(u)} p_t \right) \tag{8}$$

Here p_t is the user latent factor for tag t, Tag(u) represents the tags of user u, and $\frac{1}{\sqrt{|Tag(u)|}}$ is the result of normalization.

4.4   Latent Factor Model Combined with User CF

Among many recommendation algorithms, collaborative filtering (CF) [4] has been widely used in both social networks and online stores. But CF-based recommendation algorithms severely suffer from the cold start and data sparsity problems [4,5]. Matrix factorization tries to characterize the users and items on latent factor space, thus predicting users rating on items by calculating the similarity between the user vectors and the target item vectors in the latent factor space [11]. LFM tries to solve the problem from the point of maths while CF is based on the assumption that similar users have similar tastes. In our algorithm, we want to take advantage of merits from both methods. Besides, most traditional matrix factorization method mainly analyze the user item rating matrix thus they are context unaware. Contextual information has proved to be useful in recommender systems and already been widely studied [16]. In our paper, we add regular terms generated from keyword feature and tag feature as bias items to Equation 4.

$$C(w) = \sum_{(u,i) \in Train} (r_{ui} - \hat{r}_{ui})^2 + \lambda (\|p_u\|^2 + \|q_i\|^2)$$
$$+ w \sum_{(u,i) \in Train} (rk_{ui} - \hat{r}_{ui})^2 + \mu \sum_{(u,i) \in Train} (rt_{ui} - \hat{r}_{ui})^2 \tag{9}$$

In the equation, ω, μ and λ can be adjusted. We take matrix R generated from action features as the central rate matrix. Then, we take advantage of keyword feature and tag feature to get predicated score matrix RK and RT .

$$\hat{rk}_{ui} = \overline{r_{ui}} + \frac{\sum_{v \in S(u,K) \cap N(i)} wk_{uv}(r_{vi} - \overline{r_v})}{\sum_{v \in S(u,K) \cap N(i)} |wk_{uv}|} \tag{10}$$

rk_{ui} in RK represents the predicted rate of user u to item i calculated by keywords feature similarity. S(u,K)is the set of top K users which are similar with user u. N(i)is the set of users who have rated item i. r_{vi} is the rate user v gave to item i. $\overline{r_v}$ is the average value of user v's all rates. wk_{uv} is user similarity calculated by keyword features.

$$w_{uv} = \frac{\sum_{k \in K} (r_{uk} - \overline{r_u})(r_{vk} - \overline{r_v})}{\sqrt{\sum_{k \in K}(r_{uk} - \overline{r_u})^2 \sum_{k \in K}(r_{vk} - \overline{r_v})^2}} \tag{11}$$

K is the set of user keywords. users have different weights for different keywords. r_{uk} represents user u's weight on keyword k. rt_{ui} in RT represents the predicted rate of user u to item i calculated by tag feature. Equation 10 is the same, but wk_{uv} need to be changed. As for a certain tag, r_{ut} is 1 if user u has the tag, otherwise 0. We can get w_{uv} simply by the equation below:

$$w_{uv} = \frac{|N(u) \bigcap N(v)|}{\sqrt{|N(u)\|N(v)|}} \tag{12}$$

N(u) is the set of tags belong to user u, N(v) is the set of tags belong to user v. Accordingly, we need change the loss function. We combine Equation 2 with Equation 8, there are two sets of parameters in loss function: p_{uk} and q_{ik}. First, the steepest descent method need to solve the partial derivative respectively, we can get the equation below:

$$\begin{aligned} \frac{\partial C}{\partial p_{uk}} &= -6q_{ik} + 2\lambda P_{uk} + 2\omega P_{uk} + 2\mu P_{uk} \\ \frac{\partial C}{\partial p_{ik}} &= -6q_{uk} + 2\lambda P_{ik} + 2\omega P_{ik} + 2\mu P_{ik} \end{aligned} \tag{13}$$

According to the stochastic gradient descent method, we need to move forward in the direction of the steepest descent to get the perfect value. We calculated the rate of matrix P and Q by the equation below.

$$\begin{aligned} p_{uk} &= p_{uk} + \alpha(q_{ik} - \lambda P_{uk}) + \alpha(q_{ik} - \omega P_{uk}) + \alpha(q_{ik} - \mu P_{uk}) \\ q_{ik} &= q_{ik} + \alpha(p_{uk} - \lambda q_{ik}) + \alpha(p_{uk} - \omega q_{ik}) + \alpha(p_{uk} - \mu q_{ik}) \end{aligned} \tag{14}$$

The value of α, λ, ω and μ need to be obtained through trial and error. α is learning rate. If the learning rate is too small, it leads to a slow convergence, if the learning rate is too big, it will lead to oscillation cost function. We set it as 0.01, 0.05,0.1,0.25 and 0.5. When it increases from 0.01 to 0.1, the result is getting better, but then the result turns worse as the learning rate grows. Similarly, λ, ω, μ can be set as 0.01, 0.05 and 0.1. We need to undertake a variety of combinations to get optimal results.

5   Experiment

5.1   Features of Fanatic Fan

Selecting an informative subset of features has important applications in many data mining tasks especially for high-dimensional data [17]. Although we have the definition of a fanatic fan, our definition is subjective. We can see that there is no firm line between a fanatic fan or not. What we know that can distinguish them from normal followers is their action data. As we want to study their group characteristics, we don't have to mind the small portion of error judgment. We roughly pick top 10% of celebrity's fans ranked action list as fanatic fans.

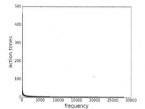 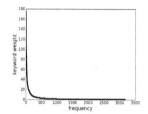 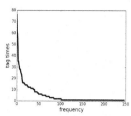

Fig. 1. Action frequency   **Fig. 2.** Keyword frequency   **Fig. 3.** Tag frequency

Feature 1: According to Figure 1, we can see that the fanatic fans have more interactions with the celebrity.

Feature 2: We calculate the total weights of each keyword belongs to the celebrity's fanatic fan. Figure 2 suggests that the keywords also have long tail effect and different celebrity's fanatic fan group have different top ones.

Feature 3: We count the number of occurrences of each tag belongs to the celebrity's fanatic fan. Figure 3 suggests that the tags also have long tail effect and different celebrities' fanatic fan group has different top ones.

5.2   Turn Features to Rate

In this paper, we must turn features into rate. Rate, in some way, is a more intuitive way to show love.

Action Feature: Accurate prediction of user behaviors is important for recommendation. The frequency that user i comments, retweets messages from the followee k and @ k highly correlates with the degree of trustiness that the user i has on user k [11]. We thus empirically define the degree of trustiness or rate T_{ik} to be:

$$RT_{ik} = \frac{RT_{ik}}{RT_i} \tag{15}$$

where RT_{ki} is the number of retweets, comments or @ from the user i to the celebrity k and RT_i is the total number of retweets comments or @ of the user i.

Key Word Feature: People often turn content to Vector Space Model (VSM) by individual words' frequency, and computes their similarity as the deviation of angles between them [18]. Analysis show that the total weights of keywords belong to different celebrities' fanatic fans presents long tail effect and the head ones are different. We pick top 10 keywords to compose the keyword vector represents the celebrity's fanatic fan group. Users have different weights to different keywords. K_{ui} represents the vector of user u belongs to celebrity i. K_{ui_n} is 1 if the user u has the tag n belong to celebrity i, otherwise 0. $\|K_{ui}\|^2$ represent the value.

Tag Feature: Similarly, we get the tag features in the same way as keyword feature. We pick top 10 tags to compose the vector represent the celebrity's fanatic fan group feature. T_{ui_n} is 1 if the user u has the tag n belong to celebrity

i, otherwise 0. $|T_{ui}|$ represents the vector that user's key words u belong to celebrity i. $\|T_{ui}\|^2$ represents the value.

5.3 Evaluation

Recommendation algorithm is used to predict what rate user u gives to item i. Generally, root mean square error(RMSE) is widely used to measure the accuracy of prediction.

$$RMSE = \frac{\sqrt{\sum_{(u,i)\in T}(r_{ui} - \hat{r_{ui}})^2}}{|Test|} \tag{16}$$

In our paper, we adopt different methods to predict the rate. We want to get the best method for fanatic fan recommendation.

Table 2. The results of different methods

standard	LFM(action)	LFM(keyword)	LFM(tags)	LFM with CF
RMSE	0.282712	0.282824	0.282812	0.272014

In the dataset, the words are replaced by specific number, so we didn't know the meaning of the keywords or tags vector for each celebrity. What we can tell is their frequency. We couldn't tell which feature is more important or why it is important, but by comparing the RMSE of different methods, we can confirm which feature is effective to our recommendation.

From table 2, we can see that action feature is more important than the other two features, but the difference is not so big. The tag feature and keyword feature is basicly the same. It's not hard to figure out that for a fanatic fan the action feature is the most obvious. Then we associate LFM with userCF algorithm, the result turned out to be the best. Our evaluation demonstrated that when we recommend a celebrity to fanatic fans, we take the action feature as main feature, the keyword feature and tag feature as complementary component to generate accurate recommendations. The three features can behave rather well separately, when we combine them together, we can get better result.

6 Conclusion and Future Work

In this paper, we explore features of fanatic fans and how to recommend celebrities to them. We build Laten Factor Models with their action, tag and keyword features which can describe the general characteristics of them from the results of their experiment. Accurate recommendations for fanatic fans bring new vitality to the social network. According to the statistical data, the positive users of registers in Tecent Weibo is only about 50%. Here still has room to improve. Fanatic fans' actions make their followees exposed to their followers, which may potentially motivate new following activity [3]. In the future, we want to focus on fans in different craziness level, we hope our recommendation can stimulate low level ones and keep high ones. In the social network, user's single action shows his different attitude, it we capture it accurately, we can make a success. Xin Wayne Zhao,et al., even attend to detect users' purchase intents from their MicroBlogs and makes product recommendation [16]. The tendency is that the analysis of SNS data is more and more important.

Acknowledgments. The work was supported by the National Science and Technology Support Program that TV content cloud services technology integration and demonstration based on social network (No.2012BAH41F03).

References

1. Ciotec, S., Dascalu, M., Trausan-Matu, S.: A comprehensive study of Twitter social networks. In: RoEduNet Conference 13th Edition: Networking in Education and Research Joint Event RENAM 8th Conference, pp. 1–7 (2014)
2. Hai-xia, L., Guang, Y., Xian-yun, T.: A matching recommendation algorithm for celebrity endorsement on social network. In: International Conference on Management Science and Engineering, vol. 20(4), pp. 72–77 (2013)
3. Sorathia, V., Prasanna, V.K.: Predict whom one will follow: followee recommendation in microblogs. In: 2012 International Conference on Social Informatics (Social-Informatics), vol. 8258(1), pp. 260–264 (2012)
4. Yang, B., Lei, Y., Liu, D., Liu, J.: Social collaborative filtering by trust. In: IJCAI 2013, pp. 2747–2753 (2013)
5. Luo, C., Pang, W., Wang, Z.: Hete-CF: Social-Based collaborative filtering recommendation using heterogeneous relations. In: 2014 IEEE International Conference on Data Mining (ICDM), pp. 917–922. IEEE (2014)
6. Dror, G., Koenigstein, N., Koren, Y., Weimer, M.: The yahoo! music dataset and kdd-cup 11. Journal of Machine Learning Research-Proceedings Track (2012)
7. Chung-YiLi, Shou-Delin.: Matching users and items across domains to improve the recommendation quality. In: KDD 2014 Proceedings of the 20th ACM SIGKDD International Conference on Knowledge Discovery and Data Mining, pp. 801–810 (2014)
8. Tso-Sutter, K.H.L., Marinho, R.B., Schmidt-Theme, L.: Tag-aware recommender systems by fusion of collaborative filtering algorithms. In: Proceedings of the 2nd ACM Symposium on Applied Computing (1995)
9. Bell, R.M., Koren, Y., Volinsky, C.: Modeling relationships at multiple scales to improve accuracy of large recommender systems. In: Proc. KDD 2007 (2007)
10. Salakhutdinov, R., Mnih, A.: Probabilistic Matrix Factorization, NIPS (2007)
11. Zhang, Y., Chen, W., Yin, Z.: Collaborative filtering with social regularization for TV program recommendation. Knowledge-based Systems **54**(4), 310–317 (2013)
12. Zhao, X.: Scorecard with Latent Factor Models for User Follow Prediction Problem (2012)
13. Koren, Y., Bell, R., Volinsky, C.: Matrix Factorization Techniques for Recommender Systems. Computer **42**(8), 30–37 (2009)
14. Guan, L., Lu, H.: Recommend items for user in social networking services with CF. In: 2012 International Conference on Computer Science and Service System, pp. 1346–1350 (2012)
15. Cheng, C.: Gradient boosting factorization machines. In: Proceedings of the 8th ACM Conference on Recommender Systems, pp. 265–272 (2014)

16. Zhao, X.W., et al.: We know what you want to buy: a demographic-based system for product recommendation on microblogs. In: KDD 2014 Proceedings of the 20th ACM SIGKDD International Conference on Knowledge Discovery and Data Mining, pp. 1935–1944 (2014)
17. Xiang, S., Yang, T., Ye, J.: Simultaneous feature and feature group selection through hard thresholding. In: KDD 2014 Proceedings of the 20th ACM SIGKDD International Conference on Knowledge Discovery and Data Mining, pp. 532–541 (2014)
18. Tsourougianni, E., Ampazis, N.: Recommending Who to Follow on Twitter Based on Tweet Contents and Social Connections. Social Networking (2013)
19. Guan, L., Lu, H.: Recommend items for user in social networking services with CF. In: 2012 International Conference on Computer Science and Service System (CSSS), pp. 1347–1350. IEEE (2012)
20. Koren, Y., Bell, R.M., Volinsky, C.: Matrixfactorization techniques for recommender systems. Computer **42**(8), 30–37(2009)
21. Xiang, L.: Recommendation system practice, pp. 187–191. Posts and TELECOM Press (2012)
22. Ahn, J., Oh, S., Kim, H.: Korean pop takes off! Social media strategy of Korean entertainment industry. In: 2013 10th International Conference on Service Systems and Service Management (ICSSSM), pp. 774–777. IEEE (2013)

Signal Processing
and Pattern Recognition

In-Line Monitoring of Belt Transport with Adaptive Bandwidth Mean-Shift Hazard

Tiezhu Qiao[1], Yanfei Duan[1(✉)], and Yusong Pang[2]

[1] Key Lab of Advanced Transducers and Intelligent Control System,
Ministry of Education and Shanxi Province, Taiyuan University of Technology,
Taiyuan 030024, China
qtz2007@126.com, 1304352153@qq.com
[2] Section of Transport Engineering and Logistics, Faculty of Mechanical,
Delft University of Technology, Delft, The Netherlands
Y.Pang@tudelft.nl

Abstract. For the disadvantage of long online monitoring processing time of hazards of coal transportation belt infrared image, the adaptive bandwidth Mean-Shift monitoring is proposed. To establish an infrared spectroscopic imaging hazard model, according to the difference of hazards and the coal reflective background radiation in the best bands, hazards image be extract. Using the automatic bandwidth selection method based on backward tracking and object centroid registration, gray histogram is established only for hazards within the kernel bandwidth and tracking it. Experiments show that this method can effectively identify and track hazards and reduce the processing time compared to the conventional image for the hazard line monitoring with strong Real-Time.

Keywords: Mean-shift · Kernel-bandwidth selection · Motion hazard identification · Image processing

1 Introduction

Infrared spectral imaging technology to detect the coal transport belt hazard identification is a new detection technology. As usual, the Infrared image was processed in order to correctly identify hazards from infrared image and tracking it. For the detection of the Infrared image, the optimal threshold was given according to the information of gray scale of each image pixel, the mean and mid-value [1]. The hazard histogram was achieved by the image smoothing filtering, edge enhancement and segmentation, and the hazard information using the hazard histogram was given [2]. Single threshold image segmentation was introduced by bacteria foraging algorithm based on maximum entropy histogram mode [3].

Mean-Shift as an efficient pattern matching algorithm does not require exhaustive search. The process of the following picture of per frame only faces to hazards within the kernel bandwidth after the hazard is identified. So the hazard recognition was

National Natural Science Foundation Director of the Fund (61450011).

achieved. The kernel bandwidth size determines the number of iterations to participate Mean-Shift samples, and reflects the size of the hazard. Typically, the kernel bandwidth is determined by the size of the initial tracking window, and the change in the whole tracking process. However, the size of the hazards into the camera field is showed a process of first increase then reduction, especially if the target is gradually increased to the over range of the kernel bandwidth, the fixed kernel bandwidth will loss the targets. Currently, there is no good way to solve the problem of the kernel bandwidth automatically selected. The calculated moments feature based on invariant moments have a serious impact on the Real-Time of Mean-Shift tracking algorithm [4].

In this paper, a kernel bandwidth automatic selection algorithm is proposed and used to identification tracking the increasing size of the hazard, and the method of after tracking can be used to registries the target centroid of two consecutive frames images because of hazard movement in consecutive frames to meet affine model. And extracting feature points on the basis of the registration and regression calculations. It can effectively eliminate false matches and ensure the accuracy of the return.

2   Bandwidth Adaptive Mean-Shift Tracking Algorithm

2.1   Mean-Shift Tracking Algorithm

Let A be a finite set embedded in the n dimensional Euclidean space X, Mean-Shift vector definition in $x \in X$ is [5]:

$$ms = \frac{\sum_a k(a-x)w(a)a}{\sum_a k(a-x)w(a)} - x, a \in A \tag{1}$$

Where k is the kernel function, w is the weight. In x calculated Mean-Shift vector point ms convolution surface gradient reverse direction:

$$J(x) = \sum_a g(a-x)w(a) \tag{2}$$

Where g is k`s shadow kernel. Constantly moving along ms direction of the kernel function center position until convergence is found near the location of pattern matching [6].

In the one-frame image, the image area existed the target is provided as a target area denoted F. The background region denoted B is outside the image area F. Including F, the center of the smallest circular area is the target centroid. The target image region F and region B are tracking round window T. If a given trace window T, let $\{X_i\}_{i=1,2,...,n}$ be the pixel coordinates of the center of its origin, Contains T definition of the image kernel histogram is:

$$p_\mu = C\sum_{i=1}^{n} k\left(\left\|X_i \Big/ r\right\|^2\right)\delta[q(X_i)-\mu]\qquad(3)$$

Where δ is Kronecker delay function. Mapping $q:R^2 \rightarrow \{1,...,m\}$ make the color of the pixel at the corresponding position be m stage quantization. C is the normalization constant. $C = 1\Big/\sum_{i=1}^{n} k(\|X_i/r\|^2)$ can be obtained by constraints $\sum_{\mu=1}^{m} p_\mu = 1$. r is called the kernel function k`s bandwidth, also is track window $T$`s radius.

If the similarity of the kernel histograms P_i and P_j that having m components can be represented by Bhattacharyya coefficien $\rho = \sum_{l=1}^{m}\sqrt{P_l^i P_l^j}, i \neq j$ t. where P_l^i and P_l^j are value corresponding to the component in the kernel histograms.

When the target image uses the kernel histogram to give the model and the candidate image, the similarity between the kernel histograms can be measured by Bhattacharyya coefficient. Comaniciu et based on the above ideas designed corresponding $J(x)$ and proved Mean-Shift iterative convergence, so that the tracking problem into a matching problem of Mean-Shift mode [7].

The two frame images i and $i+1$ have different scales and in the presence of the same target in the literature. The three radii same tracking window T_1, T_2 and T_3 is given, and the kernel histogram of them are P_1, P_2 and P_3. If it meets that: (1) T_1 is the tracking window of the image i, and T_2 and T_3 is the tracking window of the image $i+1$;(2) F_1`s centroid coincides with the $T_1$`s center, the distance from the F_2`s centroid to the $T_2$`s center of the circle is less with F_3`s to $T_3$`s; Then the Bhattacharyya coefficient of P_1 and P_2 is greater than P_3[8].

The theorem shows that the image as the center of the target centroid tracking window (set T_1') contains is most similar with the image in T_1 in the $i+1$ frame. According to Mean-Shift theory, the panning is within the scope of T_1 (the kernel bandwidth range) as long as the target in the frame $i+1$ stretching, then the best match window T_1' of frame $i+1$ can be obtained through the Mean-Shift iteration. That is, center tracking window will always lock the target centroid.

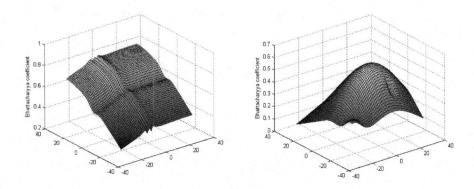

Fig. 1. Bhattacharyya coefficient of the windows around the object centroid

When the target is increasing and the size is greater than the kernel bandwidth, Mean-Shift tracking algorithm will produce spatial orientation bias. We can reverse the process after getting the error in the position. In this case, the reverse sequence of frames is equivalent to shrinking the size of the target. Thus, Mean-Shift tracking algorithm can accurately lock the frame corresponding to reverse bias point, and makes it possible to compensate for the deviation.

2.2   Automatically Selected of the Kernel Bandwidth

Consecutive frames in motion hazard meet affine model assumptions. First, the use of after tracking can make the registration for the centroid and Compensation spatial orientation bias. On the basis of the registration, feature point coordinates of the trace window inside the adjacent frames is normalized to the centroid of the target as the origin of the coordinate system. Thus, compared with the two matching features point directly from the no registration tracking window, the method can effectively reduce the mismatch, The telescopic amplitude estimated accurately laid a good foundation based on the model of the target affine. Finally, the kernel bandwidth based telescopic amplitude updated.

In the captured image sequence of hazards, suppose in the frame i, the centroid of the hazards target O_i is selected by the initial tracking window T_i the center of $c_i = O_i$. For the current hazards target centroid, there should be an offset $d = c_{i+1} - O_{i+1}$ increasing the size of the hazards target appears in the frame $i+1$, which c_{i+1} is the center of the frame $i+1$ inside the tracking window T_{i+1}. The deviation is the spatial orientation deviation from the fixed kernel bandwidth Mean-shift tracking algorithm. Because the current size of the hazards target is larger than the size of the T_{i+1}, therefore, only some part of the hazards target is contained within T_{i+1}. For the registration of two hazards target centroid, a new kernel function histogram is used to indicate the part of the hazards image contained by T_{i+1}. In fact, the centroid of this part is indicated. From frame $i+1$ to i, the size of the image area reduced. Its centroid c'_{i+1} is accurately found by Mean-shift tracking algorithm in frame i. In this way, there is an additional offset $d' = c_i - c'_{i+1}$ between c_i and c'_{i+1}. The adjacent inter frame motion is assumed small and approximation, and d' can be used to compensate d. Finally, the centroid position of the frame $i+1$ hazards target is estimated as follows [9]:

$$o_{i+1} \approx c_{i+1} - d' \qquad (4)$$

Therefore, when the image of hazards is captured, the equation (4) in the current frame is used to a registration for hazards target centroid. Figure 2 is a schematic diagram to a registration for the centroid of the hazards target by after tracking.

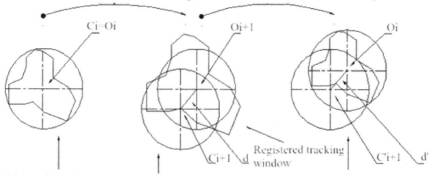

From frame i to i+1(forward tracking)From frame i+1 to i(backward tracking)

Ci–Oi   Oi+1   Oi

Registered tracking
Ci+1 d   window   C'i+1   d'

Initial tracking window Unregistered tracking windowTracking window in backward tracking

Fig. 2. Illustrations of backward tracking and object centroid registration

When the centroid registration of hazards target is completed, the corner matching is followed. For rigid objects such as coal gangue and anchor rod, corner point can characterize the spatial structure. So the use of Corner point matching two frames as a sample to estimate the parameters of the affine model.

Assuming tracking window T_i in the frame i has N corner points and tracking window T_{i+1} in the frame $i+1$ have N' corner points. T_i and T_{i+1} have the same radius, which coincides with the centers corresponding to the centroid of the target. The hazards target centroid in the adjacent two frames has been in the registration. Given a corner point $P_c|T_i$ in T_i, in T_{i+1}, it's corresponding Corner point $P_c|T_{i+1}$ should satisfy that [10]:

$$I(P|T_{i+1}) = \min\left\{\left|I(P|T_i) - I(P_j|T_{i+1})\right|\right\}_{j=1,2,\dots,n} \qquad (5)$$

Where I is the pixel brightness. n candidate points located in the center $P_c|T_i$ of the frame T_{i+1} to the position of a given widget G. Because after registration and normalization process, the small size G can be taken so that $n < N'$. Each corner point $P_c \mid T_i$ in T_{i+1}, its matching point in the frame T_{i+1} using equation (5) can be found.

Thereafter, the telescopic amplitude in hazards target affine model is accurately estimated, the kernel bandwidth based telescopic amplitude updated. The hazards exists only translation telescopic movement during the belt in operation. Thus, the affine model target is given by the following form [11]:

$$\begin{pmatrix} x' \\ y' \end{pmatrix} = \begin{pmatrix} s_x & 0 \\ 0 & s_y \end{pmatrix} \begin{pmatrix} x \\ y \end{pmatrix} + \begin{pmatrix} e_x \\ e_y \end{pmatrix} \tag{6}$$

Where $(x, y)'$ and $(x', y')'$ are the position of the same target feature point in the frame i and $i+1$. $e = \{e_x, e_y\}$ is the translation parameters, $s = \{s_x, s_y\}$ is the telescopic amplitude of target horizontal and vertical directions. The kernel bandwidth can be updated in accordance with the following methods [12]:

$$r = r \cdot \max(s_x, s_y) \tag{7}$$

As used herein, a rectangular area as a display window, then the scope of kernel function (Trace window) is circumscribed circle area of the rectangular. In this case, the algorithm should read:

$$r = r \cdot \sqrt{(s_x)^2 + (s_y)^2} \tag{8}$$

3 Infrared Spectral Imaging Detection

3.1 Infrared Band Selection

In order to effectively detect hazards, the active infrared spectral imaging detection system is used. The purpose of selecting the optimal band is to seek the differences from the reflect background radiation with the biggest band of the coal in many types hazards. Hazards of belt conveyor generally are mainly large pieces of metal, metal rod bolt and gangue. Since the single metal material mainly containing a chemical bond and the infrared measuring a covalent bond, the infrared absorption spectrum is not formed after infrared light irradiation for the elemental metal.

To describe the reflective background radiation difference between the coal and hazards, the coal and hazards of background radiation at a wavelength reflectance difference is defined:

$$\Delta\alpha(\lambda) = |\alpha_c(\lambda) - \alpha_i(\lambda)| \tag{9}$$

Where $\alpha_c(\lambda)$ is coal reflectance, $\alpha_i(\lambda)$ is hazards reflectivity $i = 1, 2, \cdots$.

The optimal band is to consider reflecting differences of the coal and multiple hazards, when evaluating the best bands of coal and multiple hazards, three principles should be considered. First of all, the mean sum of squares of $\alpha_i(\lambda)$ is higher in order to facilitate the overall make multiple hazards and coal separately. Then, the minimum absorption difference in $\alpha_i(\lambda)$ should not be too small, or not conducive to distinguish single hazards. At last, the data fluctuation of the absorption different should not be too much to avoid the large absorption differences between the coal with some hazards but small different with another hazards.

According to the first requirement, the definition of absorption mean sum of squared difference of the coal and hazards as follows:

$$\alpha_e(\lambda) = \{\sum_{i=1}^{n}[\Delta\alpha_i(\lambda)]^2\}^{\frac{1}{2}}\Big/ n \tag{10}$$

According to the above second requirement, the definition of minimum absorption difference is defined as:

$$\alpha_{\min}(\lambda) = \min(|\Delta\alpha_i(\lambda)|) \tag{11}$$

According to the third requirement, the standard deviation of the difference in absorption is defined as:

$$\Delta\alpha_\sigma(\lambda) = \{\frac{\sum_{i=1}^{n}[\Delta\alpha_i(\lambda) - \overline{\Delta\alpha_i(\lambda)^2}]}{n}\}^{\frac{1}{2}} \tag{12}$$

$$\overline{\Delta\alpha_i(\lambda)^2} = \sum_{i=1}^{n}\Delta\alpha_i(\lambda)/n \tag{13}$$

Considering the above three factors, the optimal band evaluation function as follows:

$$\Delta\alpha(\lambda) = \Delta\alpha_{\min}(\lambda)\bullet\frac{\Delta\alpha_e(\lambda)}{\Delta\alpha_\sigma(\lambda)} \tag{14}$$

$$= \frac{\min(|\Delta\alpha_i(\lambda)|)\bullet\{\sum_{i=1}^{n}[\Delta\alpha_i(\lambda)]^2\}^{\frac{1}{2}}}{\{n\times\sum_{i=1}^{n}[\Delta\alpha_i(\lambda) - \overline{\Delta\alpha_i(\lambda)}]^2\}^{\frac{1}{2}}} \tag{15}$$

Obtained at different wavelengths $\Delta\alpha(\lambda)$ is defined as the coefficient of the integrated absorption difference. Infrared absorption spectra of the coal are shown in Figure 3. The infrared absorption spectrum of gangue is shown in Figure 4.

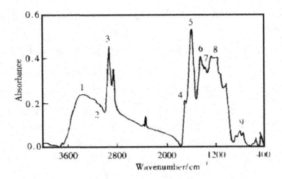

Fig. 3. The infrared absorption spectra of the coal

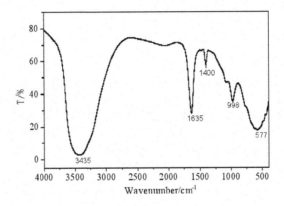

Fig. 4. The infrared absorption spectrum of the gangue

A higher absolute value $\Delta \alpha(\lambda)$ is required to make variety of hazards and separate from the coal in a whole. In 2000~2800 cm^{-1}, the coal and hazards reflection have a largest difference in figure 3 and 4. By Equation (15), the dominant wavelength 2400 cm^{-1} to irradiate the coal and hazards is selected.

3.2   Infrared Spectral Imaging Detection System Design

Imaging detection system consisted of the array light, infrared CCD camera, the image processing computer, coal, metals and coal gangue and other hazards. For light design, 10×10 array infrared diode with the main wavelength 2400 cm^{-1} are selected and light intensity can be adjusted. The two symmetrical plane array light source is fixed at detected upper sides. And each forms an angle $45°$ with the vertical direction. IR CCD camera lens vertically downward obtain images from the top to be detected. The experimental apparatus is shown in Figure 5.

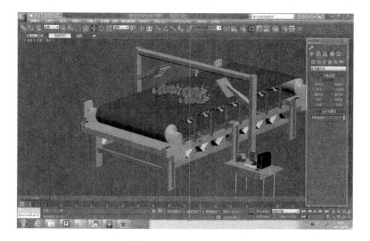

Fig. 5. The detected structural diagram of infrared spectral imaging

The gangue, metal bolt and other dangerous source into coal and started belt conveyor is consided. The main wavelength received by IR CCD is 2400 cm^{-1}.

4   Algorithm Experiment

In this paper, the algorithm is based on the Labview and compared with the traditional processing of image. The traditional processing of image is that pixel difference algorithm used to obfuscate the image which is captured and median filter be used to attenuate its noise. Then build histogram for the entire image, gray value of coal essentially unchanged after image enhancement, and hazards gray value is pulled high or down. Since the test substance in small occupied hazards, the maximum probability pixel of captured image is the set of pixel of the detected object. So a self adaptive image enhancement method for hazards is used, and hazards identification and tracking is achieved. Then in this paper adaptive bandwidth Mean-shift is used that gray histogram is established only for hazards within the kernel bandwidth.

4.1   Algorithm Description

The current kernel bandwidth can be update by the formula (8) after the affine model of the target obtained by regression calculation. The updated kernel bandwidth can correct the current tracking window size and reduce the scale positional deviation. The mean-shift iterative process samples are determined in the next frame tracking. Thus, the system can adapt well to changes in target size, and overcome the limitations of a fixed kernel bandwidth. Specific algorithm is as follows:

1.   The target is selected in the frame i to get the initial tracking window T_i and T_{i+1} is got in the frame $i+1$ using Mean-shift track;

2.   By the initial tracking window T_{i+1}, $T_i{}'$ is got in the frame i using Mean-shift track;

3. Expand its size: $r = r \cdot \varepsilon, \varepsilon > 1$ according to the difference of the center position moving T_{i+1} between T_i and T_i';

4. Extracted the corner point from T_i, T_{i+1} and matched it;

5. The horizontal and vertical coordinates of matching points were regressed to obtained s_x and s_y;

6. $r = r \cdot \sqrt{(s_x)^2 + (s_y)^2}$ is used to update the size T_{i+1}.

4.2   The Experimental Results

The infrared spectral imaging detection system is demonstrated. When hazards enter the field of vision, its image has obvious image features. As can be seen from Figure 6(a-i), the infrared imaging detection system built in this paper includes the size and the spatial accuracy of the hazard capturing. Compared with a conventional processing the whole image, the adaptive bandwidth Mean-Shift algorithm used in this paper shortens the processing time.

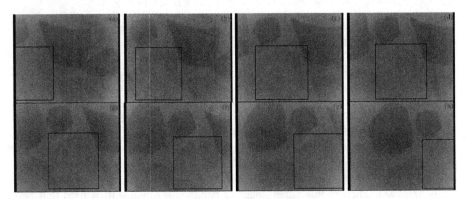

Fig. 6. Hazards capture tracking results

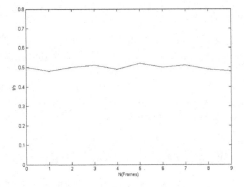

Fig. 7. Traditional algorithm for image processing time

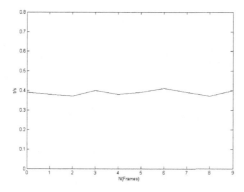

Fig. 8. Adaptive bandwidth Mean-Shift algorithm for image processing time

5   Conclusion

In this paper, Mean-shift with adaptive bandwidth monitoring methods of the coal transport belt hazards is proposed. To establish an infrared spectroscopic imaging hazard model, hazards image be extract according to the difference of hazards and the coal reflective background radiation in the best bands. Based on this hazards kernel histogram at different scales is analyzed by the similarity of Bhattacharyya coefficient. Using the automatic bandwidth selection method based on backward tracking and object centroid registration, the gray histogram is established only for hazards within the kernel bandwidth and tracking it. The experimental results show that the method compared with the conventional image processing can reduce the processing time for the hazard line monitoring with strong Real-Time.

Acknowledgements. YanfeiDuan thanks to the help and advising of TiezhuQiao, and thanks to the support of National Natural Science Foundation Director of the Fund, also thanks to Taiyuan University of Technology support of this research.

References

1. Comaniciu, D., Ramesh, V., Meer, P.: Real-time tracking of non-rigid objects using mean shift. In: Proc. IEEE. Conference on Computer Vision and Pattern Recognition, pp. 142–149 (2000)
2. Li, Y., Lu, Z., Gao, Q., et al.: Particle filter and mean shift tracking method based on multi-feature fusion. Journal of Electronics & Information Technology 32(2), 411–415 (2010)
3. Kang, Y., Xie, W., Hu, J., et al.: Target scale adaptive mean shift tracking algorithm. Acta Armamentarii 32(2), 210–216 (2011)
4. Ma, D.-H., Zhu, B., Fan, X., et al.: Method of tracking target images based on multi–feature fusion and particle filter. Journal of Detection & Control 31(4), 39–43 (2009)

5. Peng, N.-S., Yang, J., Liu, Z., et al.: Automatic selection of Kernel-bandwidth for Mean-Shift object tracking. Journal of software **16**(9), 1542–1550 (2005)
6. Zhang, H., Zhao, B., Tang, L., et al.: Infrared object tracking based on adaptive multi–features integration. Acta Optica Sinica **30**(5), 1291–1296 (2010)
7. Collins, R.T.: Mean-Shift blob tracking through scale space. In: Danielle, M. (ed.) IEEE Int'1 Cone on Computer Vision and PaRern Recognition, vol. 2, pp. 234–240. Victor Graphics, Baltimore (2003)
8. Hu, W., Wang, S., Lin, R.S., Levinson, S.: Tracking of object with SVM regression. In: Jacobs, A., Baldwin, T. (eds.) IEEE Int'1 Conf.on Computer Vision and Pattern Recognition, vol. Z, pp. 240–245. Victor Graphics, Baltimore (2001)
9. Mohammad, G.A.: A fast globally optimal algorithm for template matching using low-resolution pruning. IEEE Trans. on Image Processing **10**(4), 533–626 (2001)
10. Comaniciu, D., Ramesh, V., Meer, P.: ReabTime tracking of non-rigid objects using mean shift. In: Werner, B. (ed.) IEEE Int'1 Proc.of the Computer Vision and PaRern Recognition, vol. 2, pp. 142–149. Printing House, Stoughton (2000)
11. Zuo, J., Liang, Y., Zhao, C., et al.: Researches On scale adaptation strategy in Mean Shift tracking algorithm. Journal of Image and Graphics **13**(9), 1750–1757 (2008)
12. Zhang, H., Li, L., Yu, Q.: Scale and direction adaptive Mean Shift tracking algorithm. Optics and Precision Engineering **16**(6), 167–173 (2008)

A Method for Automated J Wave Detection and Characterisation Based on Feature Extraction

Dengao Li$^{(\boxtimes)}$, Yanfei Bai, and Jumin Zhao

College of Information Engineering, Taiyuan University of Technology,
Taiyuan 030024, China
{lidengao,zhaojumin}@tyut.edu.cn, baiyanfei0236@link.tyut.edu.cn

Abstract. J waves are low-amplitude, high-frequency waveforms which look like notches or slurs appearing in the descending slope of the terminal portion of the QRS complex in electrocardiogram (ECG). J wave is related to early repolarization syndrome (ERS), idiopathic ventricular fibrillation (IVF) or Brugada syndrome (BrS). Patients with the three syndromes are susceptible to cardiac arrhythmias and sudden cardiac death. Accordingly, J wave detection presents a non-invasive marker for some cardiac diseases clinically. In this report, 12-lead ECG record with higher signal-to-noise ratio (SNR) is formed using multi-beat averaging method. Then, we define five feature vectors including three time-domain feature vectors and two wavelet-based feature vectors. Those feature vectors are processed by principle component analysis (PCA) to reduce its dimensionality. Finally, a Hidden Markova model (HMM), trained by a proper set of these feature vectors, is employed as a classifier. Compared with other existing methods, the results show the proposed method reveals high evaluation criteria (accuracy, sensitivity, and specificity) and is qualified to detect J waves, suggesting possible utility of this approach for defining and detection of other complex ECG waveforms.

Keywords: Automated J wave detection · ECG · Feature extraction · Hidden Markova model (HMM)

1 Introduction

In the last few years, there has been renewed interest in the significance of J waves which are notches or slurs appearing in the descending slope of the terminal portion of the QRS complex in ECG [1]. At the junction between QRS complex and ST segment, J wave with specific amplitude, form and interval deviates from the baseline apparently. Meanwhile, J wave is called hump wave, osborn wave and late pre-shock wave. When the pattern, duration and scale of J-point and/or ST-segment have significant change, which is characterized by an elevation of the J-point and/or ST-segment from the baseline by at least 0.1 mV in two adjoining leads and duration of up to 20 ms forming a peak, hump, dome morphology

© Springer International Publishing Switzerland 2015
Y. Wang et al. (Eds.): BigCom 2015, LNCS 9196, pp. 421–433, 2015.
DOI: 10.1007/978-3-319-22047-5_34

known as J wave [2]. J wave is closely linked to the J wave syndrome which includes ERS, IVF and BrS. Three clinical conditions share many common ECG features [3]. Due to the presence of noise and minute morphological parameter values, it is very difficult to identify ECG classes accurately by the naked eye. There are still major unanswered questions relating to medicine limited ability to determine which individuals with common ECG variant are at risk for sudden death [4]. As the appearance of J waves is associated with ventricular tachycardia (VT) in many cases, there is a clinical interest in detection of J waves in order to anticipate the occurrence of this arrhythmia. In other words, J wave automated detection is a non-invasive diagnostic marker of VT for ERS, IVF or BrS patients.

In this paper, to approach the automated detection strategy and to assess the accuracy of such automated detection of J wave phenomenon by comparing manual and computer detection for a set of ECGs, we first reduce the noise to improve signal-to-noise ratio (SNR) and locate R point and endpoint of the QRS complex. Next, depending on the position of the two points, we define and extract five feature vectors. Those features respectively describe J wave from the different angle and it can reflect time-domain features and frequency characteristic of J wave accurately. Last, we use these extracted feature vectors to train a hidden markov classifier which is used to detect J wave automatically.

This paper is organized as follows. Section 2 presents the proposed method for automated J wave detection. Section 3 presents experiment results and comparison. Section 4 presents discussion and conclusion. Section 5 presents limitations of this paper. Section 6 presents acknowledge of this paper.

2   The Method for Automated J Wave Detection

The method proposed in this study for automated J wave detection finds a proper structure in order to enhance the J wave discrimination against other components of the ECG. This method, like many of the detection techniques in signal processing, consists of four stages:data preparation, preprocessing, feature extraction and classification.

2.1   Data Preparation

In this research, a ECG database consisting of 2000 patterns is employed in the evaluation procedure. The ECG records of the J wave-negative group including 100 healthy volunteers (29 women and 71 men) are acquired from MIT/BIH database, sampling frequency is 500 Hz. We select three minutes of each ECG recording which have no obvious variation by naked eye observation as the normal J wave-negative signals. However, the rarity of the J wave-positive variants and even greater rarity of adverse events call for an automated system of detection which can be applied to a large number of subjects. So, the ECG signals of the J wave-positive group are obtained by simulation. To simulate J wave-positive signals, a basic simulated J wave waveform is added to proper regions (QRS complex end part) of a basic ECG signal (a normal pattern) to obtain a J wave-positive signal. In this way, the basic J wave waveform should be simulated using different waveforms such as different amplitudes Gaussian white noise or random signals. We depend on the definition and characteristics of J wave to simulate the J wave signal. Based on the above facts, we choose a proper processes resembling the basic J wave signal. In order to avoid sharp transitions to the QRS complex end part, we make the starting point of J wave signal simulation starts from zero. As shown below, we simulate the two basic J wave.

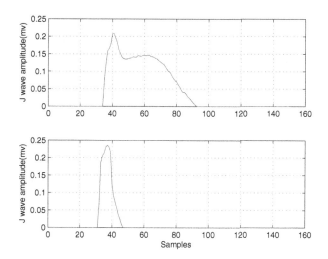

Fig. 1. The two typical simulated basic J wave waveforms

After generating the basic J waveform, we need to add it randomly to the corresponding position of QRS complex end part in each lead of the ECG recording. The way to determine the specific location is in accordance with the following three steps:

Step1. Locate the end part of QRS complex of each lead (the detailed methods are given in the preprocessing stage).

Step2. Determine the time interval between QRS complex endpoint and R wave peak.

Step3. Two simulated J wave signals is placed $\sigma + t$ ms after the R wave of each lead ECG recording, where t is a Gaussian random variable with mean value of -5ms and standard deviation of 1ms. Therefore, J wave position constantly change to the reference point of R wave peak.

According to these processes of simulation, we form a database including 200 ECG signals pattern. The ECG database is divided into two different sets. The first set is a training set comprised of 80 ECGs, which 40 ECGs had J wave and the other 40 did not. Another set is the test set, which contains 60 J wave-positive ECG recording and 60 J wave-negative ECG recording. These data are not enough for training a classifier. In order to improve the performance of classifier, we use the following method to expand the database. Ten beat groups

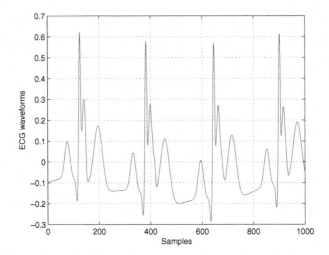

Fig. 2. The simulated II lead J wave-positive pattern of using the first basic J wave

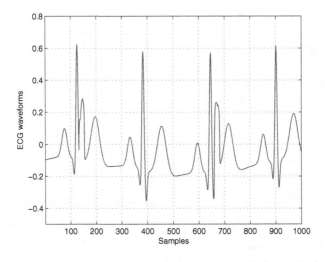

Fig. 3. The simulated II lead J wave-positive pattern of using the second basic J wave

are selected from every ECG record of the training set. We select time length of 3 minutes in each ECG recording and each ECG recording approximately contains 180 beats. 150 heartbeats are selected randomly from each ECG record without replacement to generate the first beat group. The second group is also generated by another random selection, and so on. Finally, ten beat groups are generated from each record. Therefore, an expanded training set consisting of 800 patterns is obtained. To evaluate the performance of the HMM classifier more completely, the test set is expanded in the same way separately. Thus, an expanded test set including 1200 patterns is used to investigate the HMM classifier performance.

Because the expansion process is applied to the primary sets after choosing their members, it did not affect the correlation between the training and test sets.

2.2 Pre-processing Stage

The purpose of preprocessing is locating the feature point of ECG signal and denoising. We locate the R wave peak and QRS complex endpoint which prepare for feature extraction in the next stage. The pre-processing stage is divided into three steps:

Step1. Locate the R wave peak. We detected R-peaks using standard methods [5].

Step2. Multibeat average is used to improve the signal-to-noise ratio (SNR).

A ECG signal can be expressed as a superposition of a pure ECG signal and noise. Therefore, the kth ECG beat can be expressed as:

$$\overrightarrow{b_k} = \overrightarrow{b'_k} + \overrightarrow{n_k} \tag{1}$$

where $\overrightarrow{b'_k}$ is the pure kth beat of the ECG signal and $\overrightarrow{n_k}$ is the noise. J wave is a part of $\overrightarrow{b'_k}$ and is buried in noise.

The most traditional and successful de-noising method is multibeat average technology for the ECG signal detection. Ideal conditions and assumptions of this method is that J wave of beat-to-beat ECG signal is invariant and uncorrelated with noise. If we compute the average of m beats, we have

$$\overrightarrow{b_{av}} = \frac{1}{m} \sum_{k=0}^{m-1} \overrightarrow{b} = \frac{1}{m} \sum_{k=0}^{m-1} (\overrightarrow{b} + \overrightarrow{n_k}) \tag{2}$$

where $\overrightarrow{b_{av}}$ is the average of m beats. Consequently, if the noise is stationary random process with the mean value of zero, we have

$$\lim_{m \to \infty} \overrightarrow{b_{av}} = \lim_{m \to \infty} \frac{1}{m} \sum_{k=0}^{m-1} (\overrightarrow{b} + \overrightarrow{n_k}) = \overrightarrow{b} \tag{3}$$

As can be seen from Eq.(3), if there is an infinite number of beats, the original signals $\overrightarrow{b}$ can be recovered with no distortion.

Step3. Locate the QRS complex endpoint.

Theoretically, the signal after the QRS complex in the ECG is "flat". However, due to the presence of noise, the QRS complex endpoint is not stable, which increases the difficulties of locating its position. We adopted the method that evaluates the mean value and variance of the noise voltage in ST segment (iso-electric segments) to locate the QRS complex endpoint. If there is a assumption that the noise is a random variable with mean value of a and standard deviation of b. We set up a 5ms moving time window. When it advances from left to right, in which the mean voltage surpasses a+3b, the midpoint of the time window is the QRS complex endpoint. In this way, we locate the QRS complex endpoint.

2.3   Feature Extraction Stage

We define five feature vectors which contain three time-domain feature vectors and two wavelet-based feature vectors. Those feature vectors defined describe the characteristics of J wave from the angle of time-domain and frequency-domain respectively and help to capture the variation of J wave comprehensively, improving the detection accuracy.

•Time-domain feature vector I

In the preprocessing stage, we locate the R wave peak. According to the characteristics of J wave, J wave appears in the descending slope of the terminal portion of the QRS complex in electrocardiogram. So we choose 100ms interval as the region which is used to extract time-domain feature vector following the R wave peak. To improve the robustness of detection method, we consider a 60ms interval left shift before the R wave peak. The time-domain feature vector reflects the dynamic changes in the 160ms interval of ECG signal. Due to the sampling frequency, we chose 80 sampling points including 29 points before the R wave peak, and 50 points immediately following the R wave peak. In this way, we form a 80 dimensions time-domain feature vector I which reflects the morphology state of the descending slope of the terminal portion of the QRS complex.

•Time-domain feature vector II

Based on the preprocessing stage, QRS complex endpoint has been determined. J point is a sudden turning point between the QRS complex and ST segment of the surface electrocardiogram. The J point's elevation can also cause J wave, so we select 40ms interval after the J point as "monitoring" object. Also, to improve the robustness of the detection method, a 8ms left shift is considered. In this way, we form a 24 dimensions time-domain feature vector II which includes 4 points before the point,20 points immediately following the J point.

•Time-domain feature vector III

Time-domain feature vector I and time-domain feature vector II can detect J wave variance from the morphology integrally. In order to detect morphological changes more precisely, we define the third time-domain feature vector which subdivide the 48ms interval of time-domain feature vector II into 4ms intervals and calculate the mean value and standard deviation of each interval. In this way, we form a 24 dimensions time-domain feature vector III.

•DWT-based feature vector

The selection of relevant wavelet is an important task before building the feature vector. The choice of wavelet depends upon the type of signal to be analyzed. Daubechies6 (Db6) Wavelet has been found to give details more accurately [6]. Most of the energy of the ST segment is related to frequency range 0.7-5 Hz. To complete the frequency information, we consider detail coefficients at levels 6 and 7 obtained with discrete wavelet decomposition. Therefore, we can use the corresponding coefficients to reconstruct the ST segment. After that, the power of each ST segment can be computed and used as a classification feature. In fact, signal variance represents the averaged power in that band as:

$$\sigma_x^2 = \frac{1}{N} \sum_{n=1}^{N} [x(n) - \bar{x}] \qquad (4)$$

where $x(n)$ is the discrete signal of the reconstructed ST segment, $\bar{x}$ is the sample mean of the signal and N is the number of samples in the considered segment.

In this way, we form a one-dimension DWT-based feature vector which reflects the state of ST segment. When the J wave appears, the energy of ST segment changes. From the perspective, the DWT-based feature vector can detect the presence or absence of J wave.

•CWT-based feature vector

These feature vectors above have been able to detect areas of J wave occurring from time domain and frequency domain. But it is detected from the time domain or frequency domain separately, easily affected by the other variance of ST segment elevation. Therefore, it needs to extract a feature vector consisted of scalogram, which can reflect the energy distribution and characteristic of time-frequency. Scalogram is defined as following:

$$|CWT_x(\tau, a)|^2 = \left| \frac{1}{\sqrt{|a|}} \int_{-\infty}^{\infty} x(t) \varphi^*(\frac{t - \tau}{a}) dt \right|^2 \qquad (5)$$

The CWT is applied to the 48ms interval including 40ms interval after the J point and a 8ms left shift. We select the scale range of 1-32 and the Morlet wavelet as the mother wavelet. In order to refine the characteristics of feature parameter, the time axis is divided into 9 sub-regions and the scale axis is divided into 12 sub-regions. Therefore, the time-scale plot is divided into 108 areas, and then we calculate the sum of the squared wavelet coefficients $CWT(\tau, a)$ of each subdivision. In this way, we form a 108 dimensions CWT-based feature vector which reflects the energy distribution of J wave generating area.

In Fig.4, the time-scale plots resulting from applying the CWT to the QRS complex end part in the typical J wave-positive and J wave-negative patterns are presented. As shown in Fig.4, the squared wavelet coefficients $CWT(\tau, a)$ for J wave-negative and J wave-positive patterns are obviously different, and in Fig.4(b) apparently greater than that in Fig.4(a). Accordingly, a proper feature extraction based on the differences between the time-scale plots shown in Fig.4 can be a reliable tool to recognize the J wave-negative and J wave-positive patterns.

2.4 Classification Stage

The five feature vectors are composited into a big feature vector which is a 237 dimensions feature vector synthesizing characteristics of time and frequency. The big feature vector is used as the input of HMM classifier [7]. However, the dimension of the input vector is too large, resulting in the data redundancy and

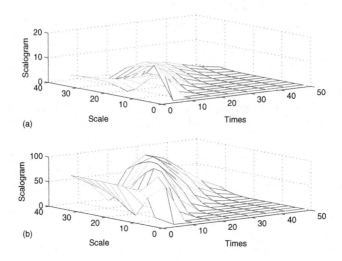

Fig. 4. The scalogram of J wave-positive and J wave-negative. (a) J wave-positive pattern. (b) J wave-negative pattern.

large amount of calculation. Therefore, we need to reduce the dimensionality of the input vector. In this paper, we choose PCA technology is to reduce the dimension of the input vector. PCA is a statistical techniques of reducing dimensionality by using the orthogonal linear transformation. The orthogonal linear transformation is defined by

$$y_{d \times 1} = T^T x_{n \times 1} \qquad (6)$$

where x is an original feature vector and y is the resultant vector.

The main task of PCA is to compute the transformation matrix T [8]. Before computing the transformation matrix, the original vector needs to be normalized in order to keep the most of the discriminatory information.

The normalization procedure used in this study is defined by Eq.(7)

$$\widehat{x}_j = \frac{x_j - \overline{x}_{j-train}}{\sigma_{j-train}} \qquad (7)$$

where $\overline{x}_{j-train}$ and $\sigma_{j-train}$ are, respectively, the mean value and standard deviation of the jth elements of the original training feature vectors.

Aiming to compute the transformation matrix T, the covariance matrix of the original feature vector is first calculated, and then calculate the eigenvalues of the covariance matrix. The characteristic value is from big to small order as $\lambda_1 \geq \lambda_2 \geq \cdots \lambda_n \geqslant 0$. Calculate the d orthogonal unit eigenvectors associate with the first d orthogonal eigenvalues $(\lambda_1, \lambda_2, \cdots, \lambda_d)$. Last, these units eigenvectors are formed into the transformation matrix T. In this research, the dimensionality of HMM classifiers input vector is reduced to 118.

After reducing the dimensionality of the input vector, the classification stage is divided into five steps:

Step1. The HMM model is initialized. In this research, the number of state of HMM is four and the initial state probability is $\pi = [1, 0, 0, 0]$ [9]. The state observation probability $\boldsymbol{B}$ is expressed by a set of mixture Gauss density. State transition probability $\boldsymbol{A}$ is set as following:

$$\boldsymbol{A} = \begin{bmatrix} 0.5 & 0.5 & 0 & 0 \\ 0 & 0.5 & 0.5 & 0 \\ 0 & 0 & 0.5 & 0.5 \\ 0 & 0 & 0 & 1 \end{bmatrix}$$

Step2. Extract feature parameter from the training set and reduce its dimensionality to 118 as observation sequence $\boldsymbol{O}$ of the training model. That is to say, $\boldsymbol{O} = (o_1, o_2, \cdots, o_{118})$. And then, the model is trained by Baum-Welch algorithm, to determine the HMM model $\lambda_i (i = 1, 2)$ associate with different observation sequence $\boldsymbol{O}$.

Step3. Extract feature parameter from the testing set and reduce its dimensionality to 118 as observation sequence $\boldsymbol{O}'$ of the training model. That is to say, $\boldsymbol{O}' = (o'_1, o'_2, \cdots, o'_{118})$. Using the forward backward algorithm, The conditional probability $P(\boldsymbol{O}'|\lambda_i)$ of the HMM model $\lambda_i (i = 1, 2)$ which have been trained is calculated.

Step4. The use of Viterbi algorithm to solve the $P(\boldsymbol{O}'|\lambda_i)$ maximum, calculate the model parameters λ_i corresponding $\boldsymbol{O}'$.

Step5. Gain the corresponding output of the recognition results by the model parameters $\lambda_i (i = 1, 2)$, which λ_1 respond to the J wave-positive mode and λ_2 respond to the J wave-negative mode.

The HMM classifier for J wave automated detection, as shown in Figure 5.

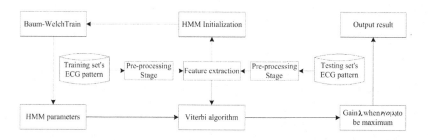

Fig. 5. The HMM classifier for J wave automated detection

Where the dotted line is the HMM training process, the solid line is the recognition process.

3   Experiment Results and Comparison

3.1   Experiment Results

To evaluate the performance of J wave detection techniques, three conventional criteria i.e. the accuracy ACC, sensitivity SE, and specificity SP defined by Eq.

(8) are used in this research.

$$ACC = 100 \times (TP + TN)/N$$
$$SE = 100 \times TP/(TP + FN) \qquad (8)$$
$$SP = 100 \times TN/(TN + FP)$$

Where N, TP, TN, FP, and FN are, respectively, the total number of patterns, the number of true positive, the number of true negative, the number of false positive, and the number of false negative. In short, $TP(TN)$ represents the number of J wave-positive (J wave-negative) modes which are detected truly. Similarly, $FP(FN)$ represents the number of J wave-positive (J wave-negative) modes which are detected wrongly.

In this research, we first utilize each single feature vector as the input of the HMM classifier to test the performance.

Table 1. The results of applying the single feature vector of J wave detection

J wave-positive test criteria	$ACC(\%)$	$SE(\%)$	$SP(\%)$
Time-domain future vector I	72	91	73
Time-domain future vector II	76	85	70
Time-domain future vector III	68	73	70
DWT-based feature vector	60	72.5	74
CWT-based feature vector	85	90	83

The results of applying the single feature vector of J wave detection to the test set are shown in Table 1. As indicated formerly, the highest ACC and SP are obtained by choosing CWT-based feature vector of the J wave-positive test, while time-domain feture vector I brings about the highest SE This is mainly decided by the properties of these feature vectors. Time-domain feature vector I can cover the largest detection region and that's the reason why it is very sensitive and low accuracy. Compared with the time-domain feature vector I, time-domain feature vector II detect the region of J wave mainly occurring and it has higher accuracy. Time-domain feature vector III and DWT-based feature vector are used for helping find details in J wave detection and making them as the index solely may cause bad performance. The essence of CWT-based feature vector is to detect the J wave signal in time domain and frequency domain synchronously, so it has a more harmonious accuracy, sensitivity and specificity.

In addition, in order to obtain a better sensitivity and specificity, we use the five feature vectors to compose a large feature vector and reduce its dimensionality as an input of the classifier.

3.2   Comparison

The arrhythmic events that are potentially related to J wave syndrome ECG events are rarely encountered, only three or four per 100,000 individuals per year.

Table 2. The result of the proposed J wave automated detection method

Method	ACC(%)	SE(%)	SP(%)
The proposed method	92.5	93	92

A single physician may practice a lifetime without encountering a single patient with this problem. Thus information gathering about such events is difficult. In the field, there is almost no research to realize the automatic detection of J wave. So far, only two groups of researchers have studied the automated J wave detection.

Clark et al. report a method for J wave detection, which is based on locating a break point in the descending limb of the terminal QRS and had a sensitivity of 90.5% and specificity of 96.5% based on its small scale of the dataset, which contains one hundred resting 12 lead ECGs from young adult men (mean age 24.8 ± 3.2 years) [10]. As the experimental results shown, the method has high sensitivity and specificity. However, the scale of the dataset is too small, and the results need to be further discussed.

Yi (Grace) Wang et al. develop a method for J wave detection which use signal processing and functional data analysis techniques and had a sensitivity of 89% and specificity of 86%. The dataset is also too small, and the training set is comprised of 100 ECGs and the test set contains 116 ECGs [11].

The two methods only set two parameters including sensitivity and specificity. The limitation of these two methods is the small range of the database, which results in the experiment results having no generality. In order to verify the validity of the three methods, we verify the three methods in this paper's database. The experimental results are shown in Table 3.

Table 3. The performance of three automated J wave detection method for the test set

Method	TP	TN	FP	FN	ACC(%)	SE(%)	SP(%)
Clarks method	531	549	51	69	90	88.5	91.5
Yi(Grace) Wangs method	507	525	75	93	86	84.5	87.5
The proposed method	558	552	48	42	92.5	93	92

As the results shown, the proposed method causes better classification of J wave-positive and J wave-negative patterns in comparison with the other methods.

In summary, as the results of the proposed method (last row in Table 3) shown, extracting the five feature vectors and using a HMM classifier may result in the favorable ACC, SE, and SP. Notably, our results of automated detection method could be improved by utilizing other appropriate machine learning tools [12] and by incorporating a spatial vectorcardiographic approach.

4   Discussion and Conclusion

The purpose of this research is to investigate the performance of a J wave detection method based on five feature vectors defined. This method firstly use multibeat averaging technique to improve the signal-to-noise ratio and locate R wave peak and QRS complex endpoint in the preprocessing stage. And then, we define five feature vectors reflecting J wave variation from the different details in the feature extraction procedure. Finally, we combine the five feature vectors into a large feature vector and reduce its dimensionality as an input of the classifier. The detection process is completed by applying an HMM structure, as a classifier, to the resultant feature vector.

In summary, this study has demonstrated that J wave can be detected by automated methods utilizing five feature vectors defined and HMM classifier with high degree of accuracy, sensitivity and specificity. The J wave automated detection method which we present in our report has the potential to overcome the current inconsistencies in studies investigating the prognostic significance of J wave. The papers automated J wave detection method has very important realistic significance for clinical diagnosis and treatment of patients. This could open the door for systematic evaluation of the prognostics significance of J wave using large databases and to help investigations into the prognostic value of J wave syndrome.

5   Limitations

First, all the ECGs used in the study are recorded at 500 samples/sec. Automated amplitude measurements and changes of slope in the recordings could vary slightly with different sample rates, but 500 samples/sec remains the accepted standard for automated ECG analysis. This sample set may not have included all possible morphologies of slurred or notched complexes. Seconddue to the lack of data in this research, the evaluation procedure is done by a ECG database including real and simulated signal; However, the proposed method should be applied to a larger ECG database consisting of real signals with and without J wave to complete the evaluation. These two aspects are the main limiting factors. Therefore, the performance of the method proposed in this paper in real J wave variation database remains to be further discussed.

Acknowledgments. This work was supported by National Natural Science Foundation of China (Grant No. 61371062); Scientific Research Project for Shanxi Scholarship Council of China (Grant No. 2013-032); International Cooperation Project of Shanxi Province (Grant No. 2012081031); International Cooperation Project of Shanxi Province (Grant No.2014081029-01).

References

1. Junttila, M.J., Sager, S.J., Tikkanen, J.T.: Clinical significance of variants of J-points and J-waves:early repolarization patterns and risk. European Heart Journal **33**(21), 2639–2645 (2012)
2. Juanhui, P., Jielin, P.: J wave syndrome. Advances in Cardiovascular Diseases **32**(4), 483–486 (2011)
3. Taboada Crispi, A.: Improving ventricular late potentials detection effectiveness, Ph.D. Thesis. The University of New Brunswick, Canada (2002)
4. Rezus, C., Floria, M., Dan Moga, V., Sirbu, O., Dima, N., Ionescu, S.D., Ambarus, V.: Early Repolarization Syndrome: Electrocardiographic Signs and Clinical Implications. Annals of Noninvasive Electrocardiology **19**(1), 15–22 (2014)
5. Clifford, G.D., Azuaje, F., McSharry, P.E.: Advanced methods and tools for ECG data analysis. Artech House, Norwood (2006)
6. Mahmoodabadi, S.Z., Ahmadian, A., Abolhasani, M.D.: Feature extraction using daubechies wavelets. In: Proceedings of the Fifth IASTED International Conference on Visualization, Imaging and Image Processing, pp. 343–348 (2005)
7. Eddy, S.R.: Hidden markov models. Current Opinion in Structural Biology **6**(3), 361–365 (1996)
8. Abdi, H., Williams, L.J.: Principal component analysis. Wiley Interdisciplinary Reviews: Computational Statistics **2**(4), 433–459 (2010)
9. Rabiner, L., Juang, B.H.: An introduction to hidden Markov models. IEEE ASSP Magazine **3**(1), 4–16 (1986)
10. Clark, E.N., Katibi, I., Macfarlane, P.W.: Automatic detection of end QRS notching or slurring. J. Electrocardiol. **47**(2), 151–154 (2014)
11. Wang, Y.G., Wu, H.T., Daubechies, I.: Automated J wave detection from digital 12-lead electrocardiogram. Journal of Electrocardiology **48**(1), 21–28 (2015)
12. Johnstone, I.M.: High dimensional statistical inference and random matrices. In: Proceedings of International Congress of Mathematicians (2006)

Metadata Organization and Retrieval with Attribute Tree for Large-Scale Traffic Surveillance Videos

Yi Tang$^{(\boxtimes)}$, Haitao Zhang, and Bin Xu

Beijing Key Lab of Intelligent Telecommunication Software and Multimedia,
Beijing University of Posts and Telecommunications, Beijing 100876, China
{piggy,zht}@bupt.edu.cn, zihuang1991@163.com

Abstract. Video metadata is an underlying basis for intelligent traffic applications such as detection, recognition and segmentation in traffic surveillance scene. However, it is still a challenge to organise efficiently the large-scale data and retrieve some important traffic surveillance metadata. In this paper, we introduce a novel metadata organization and retrieval approach for massive surveillance video. Firstly, we propose a tree-like structure called attribute tree, which is based on the characteristics of balance tree to organise and index video metadata. In the metedata organization phrase, we combine the spatio-temporal attributes (e.g.,camera location, time interval) and the visual semantic attributes (e.g.,vehicle type, vehicle color) to build up a hierarchical metadata organising structure with our attribute tree. Secondly, by taking use of the semantic attribute tree and visual feature vector, we propose an efficient retrieval method to query metadata. Thirdly, combining by Hadoop distributed processing and Hbase database, we implement a video metadata organization and retrieval system. The experimental results with the real traffic surveillance metadata demonstrate the retrieval efficiency with our approach.

Keywords: Video metadata · Metadata organization · Metadata query · Attribute tree

1 Introduction

In order to detect traffic emergency and cope with some abnormal events, traffic surveillance system have been widely installed in many urban cities. For example, in Chicago, the major declared the the city would be installed the cameras in every corner by 2016 [1]. The Global Eye system in China is another instance [2]. Now, the China Telecom Company have deployed thousands of digital cameras around many Chinese urban cities. For the reason that the traffic surveillance cameras have been widely deployed, massive video data have been produced in a very short period. In many traffic applications, we cannot keep looking at these surveillance videos, which is large-scale, unstructured and too complex to

© Springer International Publishing Switzerland 2015
Y. Wang et al. (Eds.): BigCom 2015, LNCS 9196, pp. 434–443, 2015.
DOI: 10.1007/978-3-319-22047-5_35

find something useful. Therefore, we need to extract video metadata from these large-scale videos. The metadata we extract can help us to solve some traffic application questions such as suspect escape car in traffic emergency, judging traffic flow rate and detecting other traffic events. Although the extraction and correction of the metadata is solved in these papers [12] [10], there are two challenges in organisation and retrieval of the traffic surveillance metadata:

1) Less semantic organization. In traditional database storage, the spatio-temporal attributes and visual attributes are stored in the database with feature vector, which is less semantic description. The query is the simple similarity matching with the feature vector. These problems can reduce the retrieval speed.

2) High volume. Taking the Globe Eye as an example, thousands of cameras have been deployed around the city. Each camera can produce 6M video data in one second. We collect the video data from all the cameras in the city during one day. The volume of the data can reach 30TB.

In order to address these challenges, we propose a semantic attribute tree to organise and retrieve the massive traffic surveillance video metadata. In the organization phrase, the attributes of metadata we extract are divided into two parts, the spatio-temporal attributes and visual attributes. The spatio-temporal attributes can be obtained from the digital smart camera and the visual attributes (e.g.,vehicle color, vehicle type) is obtained by some video analytic algorithms. We combine both of them to build a tree-like attribute hierarchical structure to organise the video metadata. In the retrieval phrase, traditional database query mainly depends on feature matching though all the database. However, with the increase of data, the efficiency of feature matching decreases obviously. Therefore, the attribute tree we proposed can filter some irrelevant metadata with their semantic attribute, and then the retrieval space can be pruned and the query efficiency can be improved.

The rest of this paper is organised as follows. Section 2 introduces some related work about attribute extraction, construction and retrieval. Section 3 describes the structure of our attribute tree and how to complete metadata retrieval with it. Section 4 presents our experimental results and the evaluation of our approach. Section 5 concludes this paper.

2   Related Work

In the recent years, Hadoop that can process data in distributed way has been utilized in many work. It has been used in text process at the beginning. But in order to cope with multimedia information, some new interfaces have been proposed. HIPI [7] is a distributed process interface. They bind the images into a bundle to enhance the efficiency of Hadoop distributed process. In the aspect of video process, because of the relationship between adjacent frames in the video like frame difference method, Zhao et al [3] proposed a new method to process videos by modifying Hadoop API. In source code of Hadoop API, the file should be split in map phrase, but considering the relationship between the frames in video, their method do not split the video into maps so that a video can be processed in an independent map function.

In the field of video analysis and video organization, the clustering have been introduced into many papers. IGroup [5] and Zeng's paper [4] have made use of text clustering to organise the image data. What's more, Wang et al [6] extracted the information from Wikipedia and combine video analysis to organise the video retrieval results. However, in some situation, we are short of the instruction and some relative text description about the videos. At this time, the authors in [8] and [9] extract the attribute by visual recognition algorithm, which use visual information to make up the shortage of the text description.

In the field of the metadata querying, many kinds of database have been used in systems. The work [10] [12] take use of traditional relational database to retrieve metadata with SQL query statement. With the development of distributed process and distributed database, Zhang et al [11] implements the data query by combining Hadoop and HBase framework. In Zhao's work [3], the authors also take use of the HBase to implement the query method by time interval and vehicle number plate. In this paper, we also use the HBase database linking our attribute tree to store our traffic surveillance metadata.

3   Metadate Organization and Retrieval

3.1   Overview

At first, we give the metadata model of the surveillance. As for a video file, we mainly care about the moving objects in video. So one object, containing its all kinds of attributes, is a record of metadata in this paper. The form of our video metadata we extract can be represented as a set:

$$M = \{L, T, A\}$$

where the L is the camera node location, which is the Geohash code encoded by the latitude and longitude coordinates of the location. T is the time interval set that we divide one day into several intervals according the traffic flow rate, and the A is the semantic visual attribute set:

$$A = \{a_1, a_2, ..., a_i\}$$

where a_i $(1 \leq i \leq n)$ represents a semantic attribute such as vehicle color.

The workflow of our metadata organization is displayed on Fig 1. The left of the figure is the video data acquisition and metadata extraction. First, we cope with the videos transmitting from the traffic surveillance camera and extract video metadata with the Hadoop video processing API [3]. The right of figure is metadata organization component. A record of metadata m_i appends the corresponding leaf node that links a HBase database through its own attribute. All of metadata can be stored in the HBase database with our insertion algorithm (described in Algorithm 1).

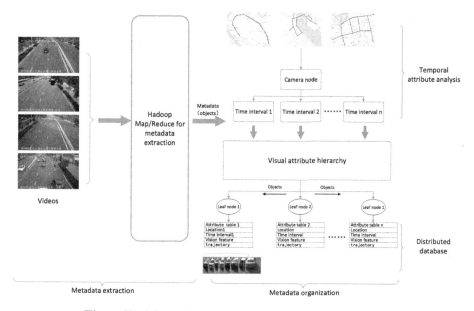

Fig. 1. Workflow of metadata extraction and organization

3.2 Metadata Organization

As shown in the Fig 1, we define a tree-like data structure to organise the metadata. The attribute tree combines the spatio-temporal attributes and visual attributes. Each leaf node in the attribute tree links one table in Hbase distributed database. First, we define the camera nodes as the set $G = (N, L)$. $n_i \in N$ $(1 \leq i \leq n)$ is a camera node. $l_i \in L$ $(1 \leq i \leq n)$ is Geohash code of the corresponding node location. As for each camera node n_i, we design an attribute balance tree data structure, and each tree node is below:

$$T = \{node, a_i, P\}$$

where node is current node, $a_i \in A$ $(1 \leq i \leq n)$ is the attribute expression of the node, and P is its child node set.

Therefore, the root of the tree is its spatial attribute, which represents camera location coordinate of the node in set G. The first layer nodes in the tree represent the time interval. The number of the time interval nodes is decided by the scale of metadata.

With the increase of metadata, the attribute tree has many leaf nodes. Because of the capacity of hardware is finite, we arrange a regular task to check the capacity. If the hard drive capacity is full, our task can delete the oldest data in database and remove correspond leaf node. Moreover, if there is no or less data in one table, the table is removed and the leaf node is pruned.

Secondly, in the middle layers of the attribute tree, we construct a hierarchical semantic structure. Every layer stands for a kind of visual semantic attribute.

Algorithm 1. Insert metadata m_i into HBase with attribute tree

---

Input: Metadata: $m_i = \{l_i, t_i, A\}$; Attribute tree T; Camera node set $G(N, L)$;
Output: Leaf node;
  1: $< l_i, t_i > \leftarrow m_i$;
  2: **for** $< l_i, t_i >$ in the $L \in G$ and $t_i \in T$ **do**
  3:     **if** $l_i = l_j$ **then**
  4:         $m_i \leftarrow m_i - l_i$;
  5:         $m_i \leftarrow m_i - t_i$;
  6:         Match the metadata m_i to corresponding tree in the node $n \in G$ and find the
          time interval $t_i \in T$ in the first layer;
  7:     **end if**
  8: **end for**
  9: **while** $attributeLength(m_i) \neq 0$ **do**
 10:     $a_i \leftarrow m_i$;
 11:     $(a_j, node) \leftarrow T$;
 12:     $compare(a_i, a_j)$;
 13:     $node_j \leftarrow T$;
 14: **end while**
 15: $leaf \leftarrow node_j$;
 16: Find the database table which links leaf node and insert m_i into it;
 17: **return** $node_j$;

In practice system, we mainly focus on two attributes, the vehicle type and the vehicle color, in traffic surveillance videos. The vehicle type and the vehicle color are extracted for metadata organization. The judgment for the vehicle type adopts the SVM [13] and HOG feature [14]. And the vehicle color is defined by the color name [16].

At last, after the analysis of the spatio-temporal attributes and visual semantic attributes, the metadata appends the corresponding leaf node. Every leaf node links a HBase database table. Before the metadata is inserted into the database, we also extract some feature vector for similarity matching in retrieval phrase. So the visual feature vector is also inserted into the database with the metadata information at the same time.

In conclusion, for a record of metadata m_i, we analyze its spatio-temporal and visual attributes, and then insert it into the database which is linked by the leaf node in the attribute tree through the inserting algorithm (described in algorithm 1).

3.3   Metadata Retrieval

For metadata retrieval, we implement a query method depending on the semantic attribute tree and the visual feature. We define a query like that: $Q = \{S, F\}$, where S is the semantic information including camera location, time, vehicle type and vehicle color. F is the visual feature vector including LBP [17] and SIFT [18].

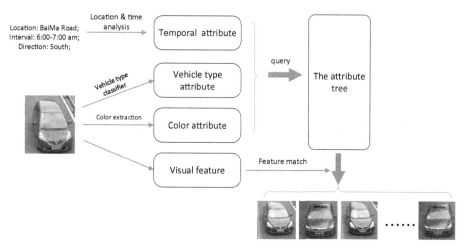

Fig. 2. The workflow of retrieval phrase

When we obtain the attribute of vehicle type, over 10 thousand vehicle samples (detected by Gaussian mixture model [15]) have been imported the SVM classifier. The classifier evaluation is displayed on the experiments section. The color attribute is obtained by the color name descriptor [16]. We define a set $C = \{c_i\}$, $i \in N$, c_i represent the a color name, which is a 3 dimensions vector containing RGB three channels color value. We use the Algorithm 2 complete

---

Algorithm 2. Extract domain color from an image I

Input: Image I; Mask M; Color vector C;
Output: The Color Name Index;
1: **for** each pixel $p_i \in$ the I and mask $m_i \in M$ **do**
2:     **if** $m_i == 255$ **then**
3:         **for** each color $c_k \in C$ **do**
4:            $\hat{k} = \arg_{c_k \in C} \min \|f(p_i) - f(c_k)\|$;
5:            $result[\hat{k}] \leftarrow result[\hat{k}] + 1$;
6:         **end for**
7:     **else**
8:         **continue**;
9:     **end if**
10: **end for**
11: $\hat{k} = \arg_{k \in strlen(result)} \max \|result[k]\|$;
12: $index \leftarrow \hat{k}$;
13: **return** $index$;

---

the extraction of the domain color. Where $f(\cdot)$ is a function to obtain RGB color value for each pixel in image.

After completing the extraction from the image metadata, the query $Q = \{S, F\}$ above can turn into $Q = \{l, t, a_1, a_2, f_s, f_l\}$, a_1, a_2 are vehicle type and color, f_s, f_l respectively stand for the SIFT and LBP feature vector. So we can take use of the method that is similar with the workflow of the the Algorithm 1 to find the HBase table. And then the query results are obtained by similarity matching of the visual feature vector. The method of similarity matching is represented by the equation 1.

$$Sim(q, m_j) = \sum_{f_i \in F} \alpha_i \cdot ||F(q) - F(m_j)|| \qquad (1)$$

where q is metadata for a query, m_j is metadata that storing in the database. $F(\cdot)$ is the feature vector of the metadata. α_i $(1 \leq i \leq n)$ is the normalization term. We utilize the Euclidean distance as the standard of similarity matching. The whole retrieval workflow is display on Fig 2.

4   Experiments

In this section, we use real traffic surveillance data to validate the efficiency and accuracy of our system in real environment. Our experiments mainly focus on the metadata organization with visual attributes and metadata retrieval with the attribute tree. The detail is as below:

Dataset: Our dataset contains traffic surveillance videos captured by 9 cameras in Fuzhou city China and 20 cameras in Yongtai town. In order to evaluate our system, we collect the videos during a day(from 6:00 a.m to 19:00 p.m), which includes 39 sequences(20 min for each) for one camera. The video's resolution is 1920×1080 and frame rate is 25fps. We change geographic coordinates of the camera into GeoHash in order to organise the videos.

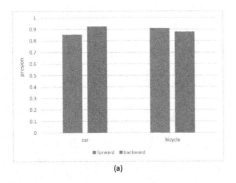
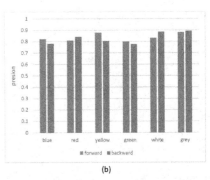

(a)   (b)

Fig. 3. (a) is the accuracy of the vehicle type classifier with different view: the forward and the backward and (b) is accuracy of the domain color extraction with different view: the forward and the backward

Cloud Platform: In metadata extraction, our previous work [3] provides an interface to process the video sequences in distributed computing platform, which depends on Hadoop. We build 20 Hadoop nodes cluster in our openstack cloud platform. Each nodes is an 8 cores of Intel CPU and 32 memory. In other to store the metadata, we use the HBase interface in Hadoop reduce phrase. The vision of the Hadoop and HBase is respectively 2.4.0 and 0.98.

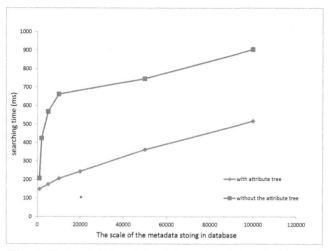

Fig. 4. The metadata retrieval time with different data scale

4.1 Visual Attribute Evaluation

In order to ensure the accuracy in the metadata query, we need to evaluate the accuracy of extraction of semantic attribute, which have significant effect on the retrieval accuracy. We evaluate the accuracy of visual semantic attribute focusing on the vehicle type classification and vehicle domain color extraction. As for the vehicle type classifier, we need to classify the car and bicycle, so we train four classifier in the light of the different view of the cameras. They are car forward view, bicycle forward view, car backward view and bicycle backward view. As for the domain color extraction, we pick up six kinds of color according to vehicle of real traffic surveillance videos. They are blue, red, yellow, green ,white and grey. And the normal metric accuracy is used to measure the vehicle type classifier and the domain color extraction. The result is shown on Fig 3.

4.2 Retrieval Efficiency with Attribute Tree

In the phrase of retrieval metadata, we implement our experience with the increase of the scale of the metadata. We test 6 group metadata, which is displayed on the Fig 4. As shown in the Fig 4, we can obtain the result that our attribute tree can reduce the retrieval time. Because our attribute tree can filter the the number of the metadata, the number of similarity matching with feature vector can lower much. In order to validate our approach in database query, we evaluate query efficiency with our attribute tree and mysql database. The experiment shows our work increase about 20 percent in query speed. The result shows in Fig 5.

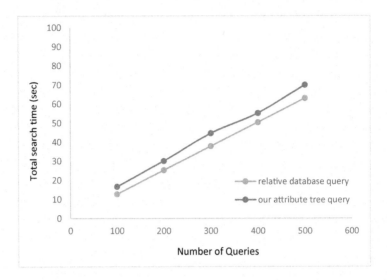

Fig. 5. The metadata retrieval time with different number of queries

5 Conclusion

In this paper, we introduce a attribute tree approach in metadata organization and retrieval, which guarantees that the metadata can be stored efficiently in HBase database and can be searched in high speed. We use spatio-temporal attributes and visual semantic attributes to organise the large-scale traffic surveillance metadata, which reduce retrieval space and improve the query efficiency. The Hadoop and Hbase are used for our attribute tree, which increase our processing speed.

Acknowledgments. This work is supported by the National High Technology Research and Development Program of China(No. 2014AA015101); National Natural Science Foundation of China(No. 61300013); Doctoral Program Foundation of Institutions of Higher Education of China (No. 20130005120011); Special Fund of Internet of Things Development of Ministry of Industry and Information Technology.

References

1. Schwartz, A.: Chicagos video surveillance cameras: A pervasive and poorly regulated threat to our privacy. Northwestern Journal of Technology and Intellectual Property **11**(2), 47 (2013)
2. Ding, F.: Real-time video surveillance across locations (2013). http://www.intel.com/content/www/us/en/software/intel-distribution-for-apache-hadoop-shanghai-ideal-study.html
3. Zhao, X.M., Ma, H.D., Zhang, H.T., et al.: Metadata extraction and correction for large-scale traffic surveillance videos. In: Proceedings of IEEE International Conference on Big Data, pp. 412–420. IEEE (2014)

4. Zeng, H.J., He, Q.C., Chen, Z., et al.: Learning to cluster web search results. In: Proceedings of the 27th Annual International ACM SIGIR Conference on Research and Development in Information Retrieval, pp. 210–217. ACM (2004)

5. Jing, F., Wang, C., Yao, Y., et al.: IGroup: web image search results clustering. In: Proceedings of the 14th Annual ACM International Conference on Multimedia, pp. 377–384. ACM (2006)

6. Wang, J., Jiang, Y G., Wang, Q., et al.: organizing video search results to adapted semantic hierarchies for topic-based browsing. In: Proceedings of the ACM International Conference on Multimedia, ppp. 845–848. ACM (2014)

7. Sweeney, C., Liu, L., Arietta, S., et al.: HIPI: A Hadoop image processing interface for image-based mapreduce tasks. University of Virginia, Chris (2011)

8. Liu, J., Yu, Q., Javed, O., et al.: Video event recognition using concept attributes. In: Proceedings of IEEE Workshop on Applications of Computer Vision, pp. 339–346. IEEE (2013)

9. Marszalek, M., Schmid, C.: Semantic hierarchies for visual object recognition. In: Proceedings of IEEE Conference on Computer Vision and Pattern Recognition, pp. 1–7. IEEE (2007)

10. Zhou, H., Pang, G.K.H.: Metadata extraction and organization for intelligent video surveillance system. In: Proceedings of IEEE International Conference on Mechatronics and Automation, pp. 489–494. IEEE (2010)

11. Zhang, J., Liu, X., Luo, J., et al.: Dirs: Distributed image retrieval system based on mapreduce. In: Proceedings of IEEE International Conference on Pervasive Computing and Applications, pp. 93–98. IEEE (2010)

12. Feris, R., Siddiquie, B., Zhai, Y., et al.: Attribute-based vehicle search in crowded surveillance videos. In: Proceedings of the 1st ACM International Conference on Multimedia Retrieval, p. 18. ACM (2011)

13. Chang, C.C., Lin, C.J.: LIBSVM: A library for support vector machines. ACM Transactions on Intelligent Systems and Technology **2**(3), 27 (2011)

14. Dalal, N., Triggs, B.: Histograms of oriented gradients for human detection. In: Proceedings of IEEE Conference on Computer Vision and Pattern Recognition, vol. 1, pp. 886–893. IEEE (2005)

15. Zivkovic, Z.: Improved adaptive Gaussian mixture model for background subtraction. In: Proceedings of the 17th International Conference on Pattern Recognition, vol. 2, pp. 28–31. IEEE (2004)

16. Shahbaz, K.F., Anwer, R.M., van de Weijer, J., et al.: Color attributes for object detection. In: Proceedings of IEEE Conference on Computer Vision and Pattern Recognition, pp. 3306–3313. IEEE (2012)

17. Ahonen, T., Hadid, A., Pietikainen, M.: Face description with local binary patterns: Application to face recognition. IEEE Transactions on Pattern Analysis and Machine Intelligence **28**(12), 2037–2041 (2006)

18. Lowe, D.G.: Distinctive image features from scale-invariant keypoints. International Journal of Computer Vision **60**(2), 91–110 (2004)

Development and Challenges of Crowdsourcing Quality of Experience Evaluation for Multimedia

Zhenji Wang[1], Dan Tao[1,2(✉)], and Pingping Liu[2,3]

[1] School of Electronic and Information Engineering, Beijing Jiaotong University,
Beijing 100044, China
dtao@bjtu.edu.cn
[2] Key Laboratory of Symbolic Computation and Knowledge Engineering of Ministry
of Education, Jilin University, Changchun 130012, Jilin, China
[3] College of Computer Science and Technology, Jilin University,
Changchun 130012, Jilin, China

Abstract. Crowdsourcing quality of experience (QoE) evaluation for multimedia are more cost effective and flexible than traditional in-lab evaluations, and it has gradually caused extensive concern. In this paper, we start from the concept, characteristics and challenges of crowdsourcing QoE evaluation for multimedia, and then summarize the current research progresses including some key technologies in a crowdsourceable QoE evaluation framework. Finally, we point out the open research problems to be solved and the future trends.

Keywords: Crowdsourcing · Quality of Experience · QoE · QoE evaluation · Multimedia

1 Introduction

In this era of information explosion, Internet and mobile end-users are more urgent to acquire the multimedia resources (mainly include text, image, voice and video) through the network. One of the most basic demands is to provide a satisfactory end-uses experience, which involves the technique that how to measure the quality of multimedia context efficiently and reliably [1]. Quality mentioned here refers to Quality of Experience (QoE), which is the measurement of overall acceptability of an application or service, as perceived subjectively by the end-users [2][1]. Obviously, QoE evaluation for multimedia requires a multi-disciplinary view that integrates some aspects including multimedia contents, end-users and implementation technologies.

Different approaches have been presented to measure and assess end-users' satisfaction of multimedia application. These approaches can employ either objective or subjective method [4].

---

[1] QoE is different from the concept of QoS (Quality of Service), which is an objective performance metric, such as bandwidth, delay, and loss rate of a communication network [1].

© Springer International Publishing Switzerland 2015
Y. Wang et al. (Eds.): BigCom 2015, LNCS 9196, pp. 444–452, 2015.
DOI: 10.1007/978-3-319-22047-5_36

- Objective methods: Peak Signal-to-Noise Ratio (PASN), Mean Square Error (MSE) and Perceptual Evaluation of Video Quality (PESQ) are convenient to use. However, these approaches are limited mainly because their quality measurements do not contain human vision perception [5].
- Subjective methods: Five-point Mean Opinion Score (MOS) scale [6], which may range from "Bad" (the worst) to "Excellent" (the best). MOS is tested in-lab environment. This method has been widely utilized since it is simple and intuitive. However, it can be time consuming and expensive because a large sample of participants are needed to achieve QoE evaluation results.

Until recently, QoE experiments have been conducted in controlled academic laboratories, which is both time-consuming and expensive [17]. Crowdsourcing, which is considered as a further development of the outsourcing principle, has emerged as an alternative to the traditional approaches to QoE evaluation [18], and has received increased attention recently. Ubiquitous Internet access, it is possible to select the group of end-users over the Internet instead of pre-designated employees or subcontractors to conduct specific evaluation experiments on their personal computers or mobile devices.

Crowdsourcing QoE evaluation for multimedia have gathered much attention recently due to their significant characteristics. In the multimedia domain, crowdsourcing became popular for the tasks, such as image/gesture annotation [19] [20], video summarization [21] [22], and so on. A series of front research results have been issued in IEEE/ACM Trans. (e.g. IEEE Trans. Multimedia), international conferences (e.g. IEEE INFOCOM, IEEE GLOBECOM, IEEE ICC; ACM SIGCOMM, ACM Multimedia). University of Wurzburg, Telecommunications Research Center Vienna, University of Bremen, Academia Sinica, National Taiwan University all are doing research on this interesting topic. However, QoE evaluation using crowdsourcing has not been widely explored yet. Since there are few mature theories and algorithms, a big gap exists between basic theory and practical application. In this paper, we will discuss the recent developments, challenging issues, open research areas and future trends of the crowdsourcing QoE evaluation for multimedia.

The reminder of this paper is organized as follows: Section 2 gives the characteristics and challenges of the crowdsourcing QoE evaluation. Section 3 discusses a crowdsourceable design framework and some key technologies involved. Finally, Section 4 points out the open research problems to be solved and the future trends.

2   Characteristics and Challenges

The reason of rapid development of QoE evaluation using crowdsourcing is due to many characteristics of crowdsourcable QoE evaluations.

- Firstly, it allows numerous anonymous end-users to participate QoE evaluation within short period from different backgrounds.

- Secondly, it provides an opening, realistic test environment including time, location, end-user, multimedia content (*e.g.* image, audio/video stream) and network access (wired, WI-FI, 3G/4G/5G, and so on).
- Thirdly, it reduces time and cost related with experimental facilities, in-lab personnel and traditional participant recruitment schemes.

Although crowdsourcing has several advantages and makes it possible for researchers to reach a wide audience for QoE evaluation, there still exist some challenges remain unresolved.

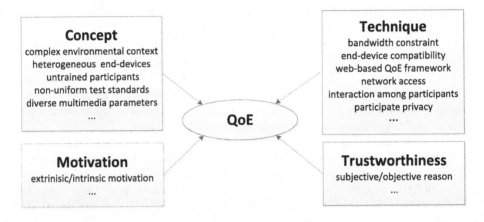

Fig. 1. The challenges of crowdsourceable QoE evaluation for multimedia

- From the conceptual perspective, almost all the factors of QoE evaluation using crowdsourcing are uncertain, such as complex environmental contexts, heterogeneous end-devices (*e.g.* hardware/software), untrained participates, non-uniform test standards, diverse multimedia parameters, and so on. How to balance among these factors to perform crowdsourcing QoE evaluation is one of the most fundamental issues to be solved.
- From the technical perspective, although in principle crowdsourcing could be used for any type of QoE evaluation, there are several practical limitations on the potential scope [8], such as bandwidth constraint, end-device compatibility, web-based evaluation framework, network access techniques, interaction among participants and participant privacy, these constraints will bring new technical challenges.
- From the motivational perspective, successful crowdsourcing QoE evaluation mainly depends on how many end-users can be motivated to participate. However, once end-users join in a test, the end-devices' resource has to be consumed and the privacy of end-users may be exposed. It is critical to investigate efficient incentive mechanisms to compensate participants for their costs in order to achieve enough high-quality participants.

– From the trustworthy perspective, the large number of end-users are in uncontrolled environments, it is hard to guarantee that each end-user is reliable or trustworthy. The end-users perform QoE evaluation tests without supervision, they may give erroneous feedback perfunctorily, carelessly or dishonestly [1]. Besides end-users' subjective reasons, some objective factors, for example, complex environmental contexts, technical errors, unclear test standards, will also lead to unreliable or untrustworthy QoE evaluation results.

3 Design Framework and Key Technologies Involved

In this section, we firstly discuss a crowdsourceable design framework about QoE evaluation for multimedia, which can be illustrated in Fig.1.

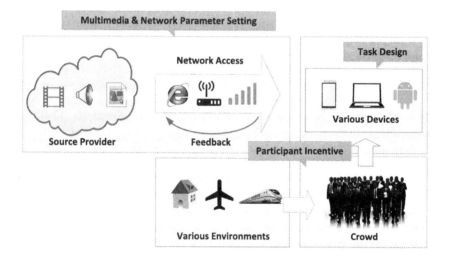

Fig. 2. A crowdsourceable design framework about QoE evaluation for multimedia

Firstly, multimedia resource providers release multimedia resources for crowd (*i.e.* participants with different end-devices) through their own cloud platforms, and crowd employ different Internet or mobile end-devices such as mobile computer, Ipad, mobile phone to access resources through diverse network access techniques from various environments. The crowd will response the QoE evaluation results when they finish the test experiment. All of these evaluation data will be feedback to the corresponding resource provider through the network.

From the design framework, we can conclude that the key technologies of crowdsourcing QoE evaluation for multimedia include multimedia & network parameter setting, task design and participant incentive.

3.1   Multimedia and Network Parameter Setting

Multimedia parameter setting may directly affect the QoE of evaluation results from participants [24].

- Instructions to subjects must be clear enough to guide participants to perform QoE evaluation. The instruction should include descriptions of the nature and purpose of the test what to evaluate, and how to rate the QoE evaluation results. For example, in a paired comparison experiment [16], a participant is simply asked to compare two stimuli simultaneously, and decide which one has the better quality based on his/her perception. In this way, the dichotomous decision of perceptual quality is clearly much simpler than the five-point rating in MOS method [13].
- Multimedia involves image, audio and video streaming media [23]. There exists a trade-off between keeping good quality and reducing the size of multimedia data [3]. The higher encoding bit rate is, the better the quality of output will become. However, the cost is a larger file which will increase the demand for data storage and network bandwidth. In paper [1], to assess the impact of codecs and compression levels on QoE of video, the authors compressed a raw video clip with three codecs, H.264, WMV3, and XVID, at two bit rates, 400Kbps and 800Kbps respectively. So, for each source clip, they obtained 6 test clips with different codec-and-bit-rate combinations.

As far as network parameters are concerned, they represent the Quality of service(QoS) [1] of the network, have great influence on users' satisfaction. The key factors are: (i) Packet loss: occurs due to the congestion in the networks and late arrival of packets at application buffers. (ii) Delay: the amount of time taken by the packet to travel from its source until its reception at the destination. (iii) Jitter: the variance of packet arrival times at the end-user buffer. There are some effective ways to decrease the effect of network parameters. For example, suitable video encoding techniques can be adopted, or QoE can be performed according to network parameters recorded on the client side. The improvement of network management (*e.g.* network protocol, network structure, SNMP, CMIP) can also achieve this objective.

Another main network parameter limiting the scope of QoE assessment is bandwidth constraint [7], which requires to consider the support of coding standards by the participants end-devices, as it is often not feasible to provide the uncompressed stimuli to the participants due to excessive bandwidth demands. In addition, it is necessary to introduce proper training and feedback mechanisms to increase the quality of QoE evaluation, since anonymous participants perform evaluation tasks lacking of supervision. It is obvious that direct feedback between test supervisors and participants will be the most effective. According to the real-time feature, feedback can be classified into two categories: non-real-time feedback may include comments, contact forms, or forums; while real-time feedback can include chat and social networking APPs [8].

3.2 Task Design

Generally, crowdsourcing task suffers from bad quality results [10]. For example, when candidate participants face the tasks which have little relationship with their own interests or have to spend lots of effort to participate, they will deal with it simply or ignore it directly. Recently, different task design schemes have been explored and proposed to ensure reliable evaluation results.

- Crowdsourcable QoE tests typically consist of multiple smaller sub-tasks, which can be called as basic test cells (BTCs for short) [7], since it is easy or fast to be selected to assess by participants. These separated BTCs could be grouped in one overall task, represent the complete test. The smaller and less time consuming a BCT is, the less time it takes to find enough participants to evaluate this BTC [11]. Similarly, the studies in [12] have suggested that the granularity of a task (*e.g.* video) can affect the degree of participation. In paper [1], to assess the QoE of video, the authors divided a complete video into several 12-second raw video clips. However, when the number of BTCs is large, it is impossible that all the BTCs can be evaluated by participants, which will result in than imbalance in the total number of evaluation per BTC.
- Even if the task is designed effectively, participants might still give incorrect results. As participants are typically remote anonymous users conduct the test with their own equipment at a place and time of their preference, therefore, the authors of paper [1] and [13] suggested that tasks should be equipped with paired comparison, consistency checking, verification question, treat detection, statistical analysis in order to perform trusted QoE assessment. Paired comparison of QoE evaluations, which takes advantage of simple comparative judgements to prioritize a set of stimili, has become a promising solution. After collecting the paired comparison results from a large number of tests performed by participants, the consistency of judgments across different tests should be assessed. Verification question is designed to detect untrustworthy inputs from participants and thus improve the evaluation quality.

3.3 Participant Incentive

In crowdsourcing QoE evaluation studies, it is important to provide necessary incentives so that enough participants are motivated to feedback high quality responses. Incentive mechanisms have been discussed extensively in works on peer production systems and reputation systems [14]. Generally, motivation can be divided into extrinsic and intrinsic method.

- Extrinsic motivation: to obtain better QoE evaluation, participants will normally be given certain incentives. Such as money, credit, or points. Considering that this situation may not guarantee the reliability of the final evaluation results, two kinds of inventive mechanisms [9], positive inventive mechanism and negative inventive mechanism, are proposed. For positive incentive mechanisms, if the evaluation results of participants match with those of experts, the participants will obtain pre-specified incentives. In contrast, in negative incentive mechanism, if the evaluation results of participants differs from those of experts, the incentives of participants will be reduced. Furthermore, differentiated rewards can also be proposed. Specifically, the reward for performing a task can be determined by the actual quality (*i.e.* consistency) of a participant's inputs [13] or the number of evaluated BTCs [4].

- Intrinsic motivation: participants can have intrinsic motives to participate in QoE evaluations. For example, gamification or games with a certain purpose (*e.g.* to help advance scientific research or support a particular community- or selfish) is an way to develop incentives for social factors, entertainment and altruism [15]. However, there are no general guidelines how to design a game, mainly because that it is strongly task related. It can be imaged that a creative and interesting task is more likely to be attracted or less likely to be cheated and also the time and cost is spent more efficiently.

4   Conclusion and Opportunities

This paper mainly discusses the development and challenges in QoE evaluation for multimedia by using crowdsourcing. In particular, we have investigated the crowdsourceable design framework and some key technologies.

However, there are many open issues in this emerging research area, including the following.

- QoE evaluation for multimedia using crowdsourcing is a kind of subjective method. However, objective methods perform QoE evaluation by analyzing multimedia content directly. So, how to achieve the actual fusion of the subjective assessment of people's perception with the objective measurement of packets and signals to provide effectual QoE evaluation for multimedia will become a trend.
- To achieve good evaluation results, many extrinsic incentive mechanisms have been studied till now. Lots of facts show that intrinsic incentive increases can result in higher data quality than extrinsic ones. Hence, designing efficient intrinsic incentive mechanisms (*e.g.* gamification) is particularly important. In addition, suitable punishment mechanisms are needed to pay attention for decreasing unreliable or low-quality results effectively.
- So far, web-based QoE models cannot support the mapping between typical network QoS parameters and QoE. As a consequence, the realization of web-based QoE models faces several practical challenges. In particular, QoE as perceived by end-users has the potential to become the guiding paradigm for managing quality in the cloud. One of the most important issues is to perform multimedia QoE management using cloud computing/service and big data technique.
- The quality of evaluation results is highly dependent on the environmental context and hidden influence factors. To ensure the objectivity and reliability

of results from participants, researchers should verify environmental condition. For example, temporal-spatial correlation or background illumination in visual QoE evaluation, the match of end-devices with test parameters, the subjective expectation, characteristics and possible impairments of participants [8].

– Various crowdsourceable QoE models, platforms or DEMO have been proposed or performed for different objectives or applications, and they have own advantages and disadvantages. However, there is no universal standard for these QoE theories and productions. In long term, it is necessary to work out a set of uniform standards.

Acknowledgments. This work was supported by the National Natural Science Foundation of China under Grant No.61271305, No.61202431, the Project-sponsored by SRF for ROCS, SEM. Beijing Higher Education Young Elite Teacher Project under Grant No.YETP0535, Jilin Provincial Natural Science Funds for Young Scholar under Grant No.20150520063JH, China Postdoctoral Science Foundation under Grant No.2015M571363, the open project program of Key Laboratory of Symbolic Computation and Knowledge Engineering of Ministry of Education, Jilin University.

References

1. Chen, K.-T., Wu, C.-C., Chang, Y.-C., Lei, C.-L.: A crowdsourceable QoE evaluation framework for multimedia content. In: Proceedings of the 17th ACM International Conference on MultimediaPages, pp. 491–500 (2009)
2. Jain, R.: Quality of experience. IEEE Multimedia **11**(1), 95–96 (2004)
3. Wang, H., Kwong, S., Kok, C.-W.: An efficient mode decision algorithm for H.264/AVC encoding optimization. IEEE Transactions on Multimedia **9**(4), 882–888 (2007)
4. Mushtaq, M.S., Augustin, B., Mellouk, A.: Crowd-sourcing framework to assess QoE. In: 2014 IEEE International Conference on Communications (ICC), pp. 1705–1710 (2014)
5. Anegekuh, L., Sun, L., Ifeachor, E.: A screening methodlogy for crowdsourcing video QoE evaluation. In: 2014 IEEE Global Communications Conference (GLOBECOM), pp. 1152–1157 (2014)
6. Methods for subjective determination of transmission quality. ITU-R Recommendation, p. 800 (1996)
7. Keimel, C., Habigt, J., Diepold, K.: Challenge in crowd-based ideo quality assessment. In: 2012 Fourth International Workshop on Quality of Multimedia Experience (QoMEX), pp. 13–18 (2012)
8. Hossfeld, T., Keimel, C., Timmerer, C.: IEEE Computer Society, Crowdsourcing quality-of-experience assessments (2014)
9. Hossfeld, T., Keimel, C., Hirth, M., Gardlo, B., Habigt, J., Diepold, K., Tran-Gia, P.: Best practices for QoE crowdtesting: QoE assessment with crowdsourcing. IEEE Transactions on Multimedia **16**(2), 541–555 (2014)
10. Hossfeld, T., et al.: Quantification of YouTube QoE via crowdsourcing. In: 2011 IEEE International Symposium on Multimedia, pp. 494–499 (2011)

11. Schulze, T., Seedorf, S., Geiger, D., Kaufmann, N., Schader, M.: Exploring task properties in crowdsourcing-an empirical study on mechanical turk. In: ECIS 2011 Proceedings (2011)
12. Faradani, S., Hartmann, B., Ipeirotis, P.G., Whats the right price? pricingtasks for finishing on time right price? pricing tasks for finishing on time. In: Workshops at the Twenty-Fifth AAAI Conference on Artificial Intelligence, August 2011
13. Wu, C.C., Chen, K.T., Chang, Y.C., Lei, C.L.: Crowdsourcing multimedia QoE evaluation: a trusted framework. IEEE Transaction on Multimedia **15**(5), 1121–1136 (2015)
14. Resnick, P., Kuwabara, K., Zeckhauser, R., Friedman, E.: Reputation systems. Commun. ACM **43**(12), 45–48 (2000)
15. Shaw, A.D., Horton, J.J., Chen, D.L.: Designing incentives for inexpert human raters. In: Proc. ACM 2011 Conf. Computer Supported Cooperative Work, CSCW 2011, New York, NY, USA, pp. 275–284 (2011)
16. David, H.A.: The Method of Paired Comparisons, 2nd edn. Hodder Arnold, London (1988). ISBN 0852642903
17. Yen, Y.-C., Chu, C.-Y., Yeh, S.-L., Chu, H.-H., Huang, P.: Lab experiment vs. crowdsourcing: a comparative user study on skype call quality. In: AINTEC 2013, Bangkok, Thailand (2013)
18. Hobfeld, T., Hirth, M., Korshunov, P., et al.: Survey of web-based crowdsourcing frameworks for subjective quality assessment. In: IEEE 16th International Workshop on Multimedia Signal Processing (MMSP) (2014)
19. Nowak, S., et al.: How reliable are annotations via crowdsourcing: a study about inter-annotator agreement for multi-label image annotation. In: Proceedings of the International Conference on Multimedia Information Retrieval, New York, USA, pp. 557–566 (2010)
20. Su, H., et al.: Crowdsourcing annotations for visual object detection. In: Workshops at the Twenty-Sixth AAAI Conference on Artificial Intelligence (2012)
21. Wu, S.Y., et al.: Video summarization via crowdsourcing. In: Workshops atthe Twenty-Sixth Extended Abstracts on Human Factors in Computing Systems, New York, USA, pp. 1531–1536 (2011)
22. Tang, A., Boring, S.: EpicPlay: crowdsourcing sports video highlights. In: Proceedings of the SIGCHI Conference on Human Factors in Computing Systems, New York, USA, pp. 1569–1572 (2012)
23. Ma, H., Liu, Y: Correlation based video processing in video sensor networks. In: IEEE International Conference on Wireless Networks, Communications and Mobile Computing, pp. 987–992 (2005)
24. Wang, H., Kwong, S.: Rate-Distortion optimization of rate control for H.264 with adaptive initial quantization parameter determination. IEEE Transactions on Circuits and Systems for Video Technology **18**(1), 140–144 (2008)

An Approach for J Wave Auto-Detection Based on Support Vector Machine

Dengao Li$^{(\boxtimes)}$, Xuebo Liu, and Jumin Zhao

College of Information Engineering, Taiyuan University of Technology,
Taiyuan 030024, China
{lidengao,zhaojumin}@tyut.edu.cn, liuxuebo0249@link.tyut.edu.cn

Abstract. Recent studies show that some types of J wave syndromes could indicate high-risk of malignant arrhythmias and sudden cardiac death. It has been a critical problem to identify J wave accurately and therefore effectively avoid misdiagnoses in clinical applications. Out of this purpose, an automatic J wave detection method is proposed in this paper. This technique explored for J wave will definitely stand out with a huge database. Firstly, a training set which contains J wave and normal ECG signals is needed. Method of feature point location could be used for picking up J-wave segment from abnormal ECG signals and non-J-wave segment from normal ECG signals as well. Secondly, feature extraction is accomplished by curve fitting and wavelet transform for vectors of extracted segments. The feature vectors are used to train an active learning support vector machine (SVM) classifier. Finally, a test set is used to assess the trained SVM classifier. The evaluation system shows that J wave segment could be detected quickly and accurately, which can help cardiologists make reasonable diagnosis in clinical cases.

Keywords: J wave · Support vector machine (SVM) classifier · Feature vectors

1 Introduction

J wave, also known as Osborn wave, refers to a dome or hump wave on the electrocardiogram [1,2,3]. It was occasionally found in a descending slope of the terminal positive wave of QRS complex. The amplitude of J wave is over 0.1mv and the period lasts over 20ms. The broadening and rising of J wave can indicate the occurrence of heart diseases, including malignant arrhythmia which may lead to sudden cardiac death. Therefore, enormous significance has been attached to J wave detection for the benefit of clinical diagnoses.

In recent years, medical community has found that many electrophysiological phenomena are closely related to the mutation of J wave. Benign variations including early repolarization syndrome and malignant variations such as Brugada syndrome, sudden death syndrome and idiopathic ventricular fibrillation are inclusively called J wave syndromes [1,2,4]. A recent research shows that

© Springer International Publishing Switzerland 2015
Y. Wang et al. (Eds.): BigCom 2015, LNCS 9196, pp. 453–461, 2015.
DOI: 10.1007/978-3-319-22047-5_37

the amplitude of J wave over 0.1mv accompanied by a variation of ST segment will lead to malignant early repolarization syndrome [5,6]. Patients with that syndrome are prone to suffer from malignant arrhythmia, fatal myocardial infarction, or even sudden death. It is not difficult to observe that J wave and J wave syndromes have become new high-risk indicators of malignant ventricular arrhythmia and sudden cardiac death. Therefore, its practically and operably of more significance to detect the J wave quickly in clinical researches.

Most of the previous researches on J wave are conducted from the perspectives of cellular electrophysiology, ion flow mechanism and genetics. Though J wave can be detected accurately, long time is needed for its analysis. In this paper, a method to detect J wave from a new perspective of signal processing is proposed. The block diagram of the algorithm presents as Fig.1. First, an ECG database containing normal and abnormal signals is needed. The method of feature point location is used to pick up J wave segment from abnormal signals after denoising. Mark the J wave segment as class 1. The same method is used to get non-J-wave (NJ) segment from normal signals and mark the segment as class 2. Then, curve fitting method and wavelet transform are used to extract feature vectors from the two classes. Finally, train active learning SVM with feature vectors and use the trained to detect J wave. Experimental results show that the algorithm can accurately detect J wave signal.

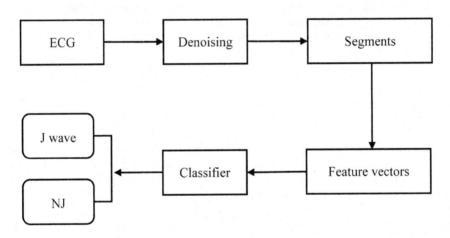

Fig. 1. The block diagram of the proposed algorithm

2   Theoretical Foundation

2.1   ECG Date

The annotated ECG records from the database of hospital are selected to this study. Each record lasts 30 min with 360Hz sampling frequency and contains a variety of waveforms. ECGs with J wave are randomly selected regardless of its location at the ECG leads. All ECGs were initially inspected by expert cardiologists.

2.2 SVM

The SVM, an excellent data mining method derived from statistical learning theory, can be very successful in handling regression and pattern recognition issues [7,8,9]. The ultimate goal of SVM is to find an optimal separating hyperplane that could meet the requirement of classification and whose marginal space on both sides must be maximized. We take binary classification as an example to illustrate the working mechanism [8]. The training sample set can be expressed as follows:

$$(x_i, y_i), \ i = 1, 2, \cdots, m, \quad x_i \subseteq \mathbf{R}^n, y_i \subseteq \{\pm 1\} \tag{1}$$

The expression of hyperplane can be expressed as follows:

$$f(x) = w^T x + b \tag{2}$$

w is the right of the vector. b is the threshold. The essential constraining condition is as follows:

$$y_i(w^T x_i + b) \geq 1 - \lambda_i \tag{3}$$

The original form of SVM classification algorithm can be summarized as the following:

$$\min_{w,\lambda} \frac{1}{2} \parallel w \parallel + C \sum_{i=1}^{m} \lambda_i \tag{4}$$
$$s.t \ y_i(w^T x_i + b) \geq 1 - \lambda_i$$

Where λ_i is non-negative slack variables, C is the penalty parameter controlling the tradeoff between the complexity of the decision function and the number of training examples. In order to solve the above optimal problem with constraint condition, the Lagrange function is introduced and it can be reformulated as follow:

$$\max \sum_{i=1}^{m} \alpha_i - \frac{1}{2} \sum_{i=1}^{m} \sum_{j=1}^{m} \alpha_i \alpha_j y_i y_j (x_i \cdot x_j) \tag{5}$$
$$s.t. \sum_{j=1}^{m} \alpha_j y_j \geq 0 \quad j = 1, 2, \cdots, m, \quad 0 \leq \alpha_j \leq C$$

$\alpha_j \geq 0$ is the Lagrange multiplier.

To overcome the limitations of nonlinear cases, SVM maps the original input vectors into a higher dimensional space through kernel functions. In this paper, the radial basis kernel function is used and the expression can be shown as in the following, where σ is a positive real number.

$$K(x \cdot x^{'}) = exp(- \parallel x - x^{'} \parallel^2)/\sigma^2 \tag{6}$$

3 The Proposed Algorithm

This paper presents a novel algorithm for detecting J wave, mainly including preprocessing, feature extraction, classification. The specific steps are shown in Fig.2:

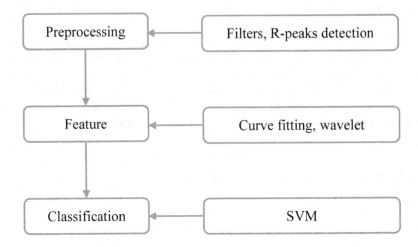

Fig. 2. The specific steps of the proposed algorithm

3.1   Preprocessing

ECG signals can be affected by many interfering signals such as power line noise, EMG signals and baseline wandering in the process of acquisition and transmission [10]. In order to ensure the integrity of the ECG signal, median filter presented in Literature [11] is utilized to filter out the noise signals. After the denoising process, the standard method proposed in Literature [12] is used to detect R-peak. For each segment, measurement for 80 points is obtained, including 29 points before R peak, R peak and 50 points immediately following the R peak.

3.2   Feature Extraction

In this paper, curve fitting and wavelets are adopted for segment feature extraction. Coefficient of curve fitting, instead of the original signal waveform, is taken into use as a feature vector. Its advantage lies in the reduced calculation in later classification by processing the smallest data instead of the original one without affecting the waveform. 8-order Fourier curve fitting is applied and the fitting precision can reach 99.72%. After an 8-order Fourier curve fitting, we can get a 18-dimensional feature vector. The fitting results are shown in Fig.3 and Fig.4.

Discrete wavelet transform is also used to extract the features because of its time-frequency localization properties. We use Daubenchies 6 (Db6) wavelet which can accurately give more details [13] to extract ECG features in this application. The power of the segments, mean energy, and coefficient of variation in the wavelet transform domain are chosen as the frequency domain parameters.

The quantities of the features may be quite different, so its of enormous necessity to normalize all features to the same level before classification. The

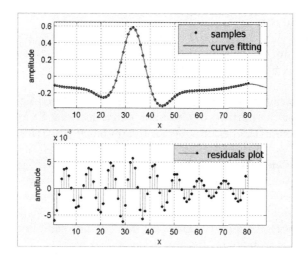

Fig. 3. The fitting waveform and residuals plot of NJ segment

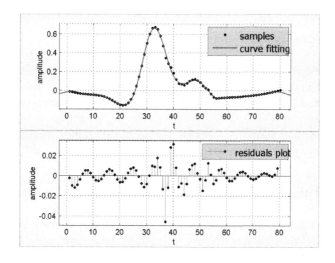

Fig. 4. The fitting waveform and residuals plot of J wave

formula of the normalization is defined as follows: [14]

$$x_{ij'} = tansig(\frac{x_{ij} - \overline{x_j}}{\sigma_{x_j}}) \tag{7}$$

x_{ij} is the jth component of the ith feature vector. $\overline{x_j}$ and σ_{x_j} are the mean and standard deviation of the jth component of the feature vectors respectively. $tansig(\cdot)$ is the hyperbolic tangent sigmoid transfer.

4   Classification

In order to improve the classification performance of SVM, this paper proposes an active learning SVM. The process is as follows:

Step 1

The feature vectors of the two classes were randomly divided into two sets: one is learning set for the purpose of obtaining active learning SVM and the other is test set to examine the performance of the obtained active learning SVM.

Step 2

Randomly subdivide learning set into training group and test group. The core of the proposed active learning SVM lies in similar support vectors which are actually the selected training samples. Its known that support vectors can be obtained after training. Similar support vector can be obtained by calculating the similarity between support vector and misclassified samples. The threshold in this paper is set to be 0.92, and the formula goes as:

$$sim(x,y) = cos(x,y) = \frac{(x,y)}{\| x \| \cdot \| y \|} = \frac{\sum_{i=1}^{n} x_i y_i}{\sqrt{\sum_{i=1}^{n} x_i^2 \sum_{i=1}^{n} y_i^2}} \tag{8}$$

x and y are n-dimensional vectors.

Step 3

Repeat Step 2 for two more times, and three groups of similar support vector are obtained. Exclude repeated samples by applying Formula 8, and add support vector, training samples for active learning support vector can be obtained.

Step 4

Train active learning support vector by the obtained training samples. Examine the classification performance with samples in testing group.

5   Results and Analysis

In order to evaluate the performance of the classifier, we use the following statistical measures. If the classifier correctly identifies the J wave, mark True Positive (TP). If a segment is NJ signal but classifier misinterprets the segment is J wave, mark False Positive (FP). If the classifier correctly identifies the NJ signal, mark True Negative (TN). If a segment is J wave but classifier misinterprets the segment is NJ signal, mark False Negative (FN).

Sensitivity is defined as the ability of the classifier correctly identifies the J wave

$$sensitivity = \frac{TP}{TP + FN} \times 100 \tag{9}$$

Accuracy is defined as the ability of the classifier correctly identifies the segments

$$accuracy = \frac{TN + TP}{TN + TP + FN + FP} \times 100 \tag{10}$$

The Matlab simulation is used to demonstrate the performance of the proposed algorithm in this paper. After detecting step, curve fitting and Db6 are adopted to obtain feature vectors. Finally, 365 segments are obtained in total, 285 of which are subsumed under learning group and 80 under test group. 65 segments from the learning group are drawn randomly to train SVM. Then, the remaining segments in learning group are used to select the similar support vectors. Repeat the above steps for three times and the performance of learning group shows in Table 1.

Table 1. Table 1 the performance of learning group

times	TP	FN	TN	FP	Sensitivity(%)	Accuracy(%)
1	104	16	86	14	86.7	86.3
2	99	21	85	15	82.5	83.6
3	106	14	90	10	88.3	89.1

32 feature vectors are finally selected as the similar support vectors after the above steps. After obtaining similar support vectors from learning group, testing group is used to examine the classification performance. To compare the performance of original SVM and proposed method, we randomly choose another 32 feature vectors to train the original SVM and examine it by testing group. It is not difficult to find that the performance of the proposed algorithm is better than the original SVM Classifier, as can be seen from Table 2.

Table 2. Comparison of the performance of original SVM and that of the proposed method

	TP	FN	TN	FP	Sensitivity(%)	Accuracy(%)
SVM	33	7	32	8	82.5	81.2
Proposed method	38	2	36	4	95	92.5

To achieve a better visual effect, a 3D map of the classification results after using the proposed method is designed. As shown in Figure 5, • represents the FP and red * shows the FN. ○ represents TP and blue + shows TN. With Fig.5, we can easily tell the details of the results.

Table 3 shows a comparison among classification accuracies and sensitivities of the proposed method, random tree (RT), and K-nearest neighbors (KNN). Performances of all methods were all evaluated on the same signals. It can be seen from the comparison table that the proposed method achieves a remarkable classification accuracy rate of 92.5% and it is superior to other methods.

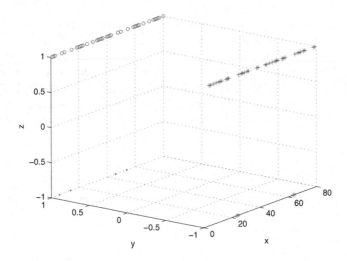

Fig. 5. 3D map of the classification results

Table 3. Comparison of proposed expert system with the studies in the literature

Method	Sensitivity(%)	Accuracy(%)
SVM	95	92.5
RT	87	84
KNN	85	82

6 Conclusion

To improve the diagnostic accuracy of J wave for clinical applications, we put forward an effective signal processing method based on SVM in this paper. After finding the segments required, wavelet transform and curve fitting are applied to extract the vectors. Finally an active learning SVM is proposed to detect the J wave. In the future work, we will study multiclass SVMs for a more detailed classification and explore other features that could probably help cardiologists apply better therapies to patients.

Acknowledgments. This work was supported in part by the Natural Science Foundation of China (Grant 61371062, Grant 61303207); Shanxi province science and technology development project, industrial parts (20120321024-01).

References

1. Wu, Q., Zhao, S., Wang, Y..: J wave syndrome: mechanisms and clinical significance. South China Journal of Cardiovascular Diseases 17(2), 94–94 (2011) (in Chinese)
2. Yan, G., Yao, Q., Wang, D..: Electrocardiographic J wave and J wave syndromes. Chinese Journal of Cardiac Arrhythmias 8(6), 360–365 (2005) (in Chinese)
3. Badri, M., Patel, A., Yan, G.X.: Cellular and ionic basis of J-wave syndromes. Trends in cardiovascular medicine 25(1), 12–21 (2015)
4. Mizusawa, Y., Bezzina, C.R.: Early repolarization pattern: its ECG characteristics, arrhythmogeneity and heritability. Journal of Interventional Cardiac Electrophysiology 39(3), 185–192 (2014)
5. Froelicher, V., Wagner, G.: Symposium on the J wave patterns and a J wave syndrome. J. Electrocardiol. 46(5), 381–382 (2013)
6. Wang, Y.G., Wu, H.T., Daubechies, I., et al.: Automated J wave detection from digital 12-lead electrocardiogram. Journal of Electrocardiology 48(1), 21–28 (2015)
7. Vapnik, V.: The nature of statistical learning theory. Springer (2000)
8. Shifei, D., Bingjuan, Q., Hongyan, T.: An Overview on Theory and Algorithm of Support Vector Machines. Journal of Electronic Science and Technology 40(1), 2–10 (2011) (in Chinese)
9. Joachims, T.: Text categorization with support vector machines: Learning with many relevant features. Springer, Heidelberg (1998)
10. Singh, B.N., Tiwari, A.K.: Optimal selection of wavelet basis function applied to ECG signal denoising. Digital Signal Processing 16(3), 275–287 (2006)
11. Lopez, A.D., Joseph, L.A.: Classification of arrhythmias using statistical features in the wavelet transform domain. In: 2013 International Conference on Advanced Computing and Communication Systems (ICACCS). IEEE, pp. 1–6 (2013)
12. Hamilton, P.S., Tompkins, W.J.: Quantitative investigation of QRS detection rules using the MIT/BIH arrhythmia database. IEEE Transactions on Biomedical Engineering 12, 1157–1165 (1986)
13. Mahmoodabadi, S.Z., Ahmadian, A., Abolhasani, M.D.: ECG feature extraction using Daubechies wavelets. In: Proceedings of the fifth IASTED International Conference on Visualization, Imaging and Image Processing, pp. 343–348 (2005)
14. Zidelmal, Z., Amirou, A., Ould-Abdeslam, D., et al.: ECG beat classification using a cost sensitive classifier. Computer methods and programs in biomedicine 111(3), 570–577 (2013)

Routing and Resource Management

Green and Fault-Tolerant Routing in Data Centers

Liang Shi[1], Xintong Guo[2], Lailong Luo[1]([✉]), and Yudong Qin[1]

[1] College of Information System and Management, National University of Defense
Technology, Changsha 410073, Hunan, China
{slranbo,luolailong09,didala.good}@163.com
[2] College of Information Science and Engineering, Northeastern University,
Shenyang 110819, Liaoning, China
gxttssl@163.com

Abstract. As large data centers become the dominant energy consumers, green has become increasingly significant. The mainstream
insight behind green routing designs is to aggregate diverse flows into
a subset of network devices such that idle devices can be shut off or
turned into sleep mode. However, this approach may not ensure the network performance primarily. In this paper, we present three dedicated
algorithms to provide each node pair k paths precisely, which not only
aim at realizing the green network eventually, but also try to ensure the
network performance originally. The results of comprehensive simulations
corroborate that three methods can accomplish the goal of energy-saving
from the perspective of insuring both fault-tolerance and appropriate
responsive-time.

Keywords: Green routing · Fault-tolerance · Responsive-time · k paths

1 Introduction

Generally, large-scale data centers are the basic infrastructure of applications,
leading to a great deal of energy consumption [1], [2], [3], [4]. It was reported
that 1.3% of the worldwide electricity was consumed by data centers and this
fraction would grow to 8% by the year 2020 [5], [6]. As a result, green becomes
an essential principle of designing and maintaining data centers.

The report illustrated that about 45% power of a data center is expended
by IT devices, such as servers and network devices [7]. To tackle the severe circumstance, a quintessential method is to aggregate the flows inside data center
networks (DCNs) into a subset of network devices (e.g., servers, switches and
links). According to this technique, the idle devices can be shut down or put

L. Luo—This work is partially supported by the National 973 Basic Research
Program under No.2014CB347800, the National Natural Science Foundation of
Outstanding Youth Fund under No. 61422214 and the National Natural Science
Foundation of China under No.61170284.

Y. Wang et al. (Eds.): BigCom 2015, LNCS 9196, pp. 465–478, 2015.
DOI: 10.1007/978-3-319-22047-5_38

into the sleep mode for energy-saving, while the active devices can be employed to transmit all flows with their best efforts [8], [9]. In this case, the method named workload consolidation strategies can result in 50% energy-saving at most [10]. Unfortunately, although these strategies have taken both the link bandwidth and capacity of network devices into consideration, the network performance cannot be guaranteed if any path breaks up beyond expectation. In other words, the active devices and links along the chosen paths may be encumbered or even overused. As a result, fault-tolerance cannot be pledged, let alone prompt responsive-time satisfy the demand, which can be a disaster for latency-sensitive applications, such as online business, voice over IP (VoIP) and content delivery networks (CDNs).

Indeed, the link abundance and device redundancy in DCNs, which are virtually oversubscribed, are designed for the purpose of fault-tolerance and prompt responsive-time. However, the extant strategies for green routing merely utilize a subset of network devices, which is contrary to the originally designed philosophy of DCNs more or less. Actually, even a few seconds latency may lead to a huge amount loss. As an example, Google reported that 20% revenue was lost with an experiment, which added an extra delay of only 500ms in displaying the searching results [7]. Similarly, Amazon experienced the phenomenon of 1% sales decreasing, as a result of 100ms additional delay [7].

In this paper, we argue that when aggregating the flows into a subset of network devices, the basic topology characteristics of DCNs should be remained for insuring fault-tolerance and appropriate responsive-time. In other words, if we only utilize part of topology, we may not attain the requirements mentioned above. Consequently, we put forward a novel model of green routing, depending on three interrelated algorithms.

The main contribution of this paper can be summarized as follows:

- We propose three beneficial algorithms to achieve the goal of energy-saving, along with assuring the basic network performance including both fault-tolerance and prompt responsive-time.
- Comprehensive evaluations have been conducted to evaluate the performance of our algorithms. The results demonstrate that the path-aware method outperforms other two algorithms in terms of time-consumption. Meanwhile, three algorithms present similar results in the aspect of active devices' quantity. Thus, it is advisable to put the third method into practice.

The remainder of this paper is organized as follows. Section 2 summarizes some related works concerning energy efficiency. Section 3 presents our various algorithms, which not only save energy to the greatest extent but also insure fault-tolerance and prompt responsive-time. Section 4 evaluates the performance of the proposals. Section 5 concludes this paper.

2   Related Work

Gupta et al. [11] advise that energy-saving is expressly imperative in their position paper. As to achieve the goal of providing efficient routing service, some network topologies are put forward, such as BCube [12], Dcell [13]. Unfortunately, these architectures are under 1:1 oversubscription ratio, where traffic workload cannot reach the peak value all the time and then waste large amount of energy. As a consequence, Shang et al. [8] provide an idea that network facilities should be utilized as few as feasible, in order to make little sacrifice during the transmission of flows and save energy.

In order to improve the efficiency of network, Li et al. [14] exploit a method called exclusive routing (EXR) for each network flow, which can inevitably allow a flow to occupy a specific routing path exclusively and save energy on the basis of the simulations. At the same time, Nedevschi et al. [15] design two categories of energy management scenarios to reduce the energy consumption, of which one idea relies on turning idle devices off during the absence of packets, while the other method is based on adapting the transforming rate when we process the packets. Moreover, Chiaraviglio et al. [16] present a new model to resolve the energy-saving problem according to minimizing the delay of users.

Besides, lowering data rate excessively and scheduling flows into a subset of network may trigger both low fault-tolerance and more latency. Faced with the unfavorable performances, Lombardo et al. [17] propose an autonomous measurement mechanism named G-Router, aiming at allowing an access node in the Internet to align its power consumption to either the maximum capacity or the minimum available capacity, of which the simulation results manifest G-Routing can minimize the energy consumption. In addition, Heller et al. [18] put forward ElasticTree, which lets operators adjust the set of active network factors more flexibly, in order to fulfil the goal of changing traffic loads and save power consumption. Wang et al. [10] put forward a time-aware model based on the framework, combining a slice of special characters on data centers with traffic project. And this method can reach 50% energy-saving.

In this paper, we mainly attempt to resolve the energy-consumption problem and assure the network performance simultaneously. Our methods may assist us to make sure the fault-tolerance in the first place. Next, we verify the prompt responsive-time in accordance with the detection of time-consumption in simulations. Eventually, we achieve the goal of energy-saving based on the examination of the least active devices' quantity.

3   Green Routing with k paths

3.1   Statement of Problem

In fact, the energy consumption captures increasingly individuals' attentions. Various techniques are presented to attempt to tackle this tough problem. In particular, quite a fraction of methods embrace that the idle devices can be shut off or put into the dormancy mode for energy-saving. However, the fault-tolerance and responsive-time cannot be insured once the fundamental topology is destroyed. Under this thought, we try to search for some approaches superior to the extant ideas.

With establishing the model of energy consumption in DCNs, we can draw a conclusion that a slice of variables can be adjusted to meet the requirement decreasing the energy consumption. For instance, in the expression (1), N stands for the number of total network devices. Then, in the expression (2), e_i and t_i represent the work and the time of work in energy consumption respectively. Meanwhile, in the expression (3), e_0 is the energy consumption when devices are idle, β is an index larger than 1 and x_ν is the transmitting speed of flows. According to these equations, we can briefly discover that if the number of devices under workload consolidation strategies (i.e., N), the adaptive link rate (i.e., x_ν) and another chance (i.e., t_i) are reduced, the energy consumption can effectively be lowered.

$$E_{Total} = \sum_{i=1}^{N} E_i \tag{1}$$

$$E_i = e_i \times t_i \tag{2}$$

$$e_i = e_0 + \alpha x_\nu^{\beta} \tag{3}$$

Nevertheless, certain new problems occur unpredictably. For instance, if N decreases, i.e., the idle network devices are turned off or put into sleep mode, the energy-saving may be guaranteed greatly while the basic topology may be broken up. As a result, the fault-tolerance cannot be assured with the blooey links. Also, if x_ν is lowered, it is likely to cause more latency and slows down the latency-sensitive applications further. So, we may concentrate on the finishing time, i.e., t_i, which can be shortened by scheduling flows properly.

Based on the discussions above, we may advocate that the idle network devices cannot be turned off or transformed into the sleep mode, i.e., we present our ideas under the integrated network topology. In accordance with this idea, we may realize green routing accompanied by assuring the fault-tolerance and appropriate responsive-time.

Compared with the existing approaches, we primarily major in finding out *amount* shortest paths from all paths and applying three diverse algorithms to select or aggregate k paths out of *amount* shortest paths in each node pair.

To be specific, the first algorithm aims at selecting k paths from *amount* paths of each node pair directly and then chooses the rational scheme with the least active network devices for all node pairs. However, although this method is accurate, it spends a great deal of time making a decision in real network. Consequently, we design another two algorithms to improve the characters. The second idea called link-aware method means that we firstly calculate the weight of links from *amount* paths in each node pair, which is decided by how many times that *amount* paths in each node pair have passed the link. Next, we aggregate k paths recursively in line with the link that occupies the largest weight in each node pair. The last algorithm is so-called path-aware method, which calculates the weight of paths from *amount* paths in each node pair via adding up the weight of each link along the path. Then, this algorithm aggregates k paths recursively according to the number of mutual links with the path that has the largest weight.

Table 1. Symbols and notations

Term	Definition
scale	Network scale
pair	Number of node pairs
adjacent[scale][scale]	Adjacent matrix of topology
node-pair[pair]	Each node pair
amount	Given number of shortest paths
k	Given number of selected shortest paths

In this section, we may describe three diverse algorithms in detail. To ease the description, we list the frequently utilized symbols and notations in Table 1.

3.2 First Algorithm: Accurate but Time-Consuming Method

Algorithm 1 depicts our first idea in detail. The main idea of first algorithm can be described as selecting k paths from *amount* paths of each node pair directly and choosing the ideal scenario with the least active network devices for all node pairs. In the very beginning, we require *adjacent[scale][scale]*, *node-pair[pair]*, *amount* and k. Meanwhile, some beneficial symbols are summarized as follows.

get_Amount() denotes the function of getting *amount-paths*; *amount-paths* storages *amount* shortest paths in each node pair; *get_Combinations()* denotes the function of getting C_{amount}^{k} diverse combinations in each node pair; *temp-c* records the temporary result of *get_Combinations()*; *combinations* records the final result of *get_Combinations()*; *get_Scenarios()* denotes the recursive function of getting $\prod_{i=1}^{pair} C_{amount}^{k}$ various scenarios in all node pairs that call for communication; *temp-s* records the temporary result of *get_Scenarios()*; *scenarios* records the final result of *get_Scenarios()*; *which* denotes each level of tree fabric.

Firstly, according to *get_Amount()*, we may acquire *amount-paths*, which denotes *amount* shortest paths in each node pair. (Line 1) Then, the 'for' circulation can obtain C_{amount}^{k} diverse combinations that include k paths in each node pair and storage them in *combinations*. (Lines 2-5) Secondly, we may get $\prod_{i=1}^{pair} C_{amount}^{k}$ various scenarios for all node pairs on the basis of recursive function *get_Scenarios()*. (Line 6) Also, the particular depiction is formulated in the last lines, which mainly utilize the recursive algorithm of tree fabric. (Lines 10-18) The tree fabric is merely utilized as an instrument, so we do not elaborate on it to much in this paper.

As an example, at first, the *pair*, *amount* and k can be set as 3, 5 and 2. Then, according to *get_Combinations()*, we can calculate $C_5^2 = 10$, which means that we have 10 combinations that cover 2 paths in each node pair. What is more, we can get $\prod_{i=1}^{3} C_5^2 = 1000$ scenarios. Faced with so many choices, we may calculate the number of active network devices in them and select the scheme that occupies the least devices. (Lines 7-8) At last, we return *method1-result* containing k paths in each node pair. (Line 9)

Algorithm 1. Accurate but time-consuming method

Require: $adjacent[scale][scale]$; $node\text{-}pair[pair]$; $amount$; k.

1: $amount\text{-}paths \leftarrow get\_Amount(adjacent, node\text{-}pair, amount)$;
2: **for** $i = 0$ to $pair$ **do**
3:    $temp\text{-}c \leftarrow get\_Combinations(amount, k)$;
4:    $combinations.push(temp\text{-}c)$;
5:    $temp\text{-}c.clear()$;
6: $get\_Scenarios(combinations, 0)$;
7: Calculate the number of active network devices in $scenarios$;
8: Select the scenario that occupies the least active devices and denote it as $method1\text{-}result$;
9: **return** $method1\text{-}result$;
10: **function** $get\_Scenarios(combinations, which)$
11:    **if** $which == combinations.size()$ **then**
12:        $scenario.push(temp\text{-}s)$;
13:    **else**
14:        **for** $count = 0$ to $combinations[which].size()$ **do**
15:            $temp\text{-}s.push(combinations[which][count])$;
16:            $get\_Scenarios(combinations, which + +)$;
17:            $temp\text{-}s.pop()$;
18:            $count + +$;

3.3 Second Algorithm: Link-Aware Method

However, although our first method is accurate, it takes a great deal of time to make a decision in real network. Consequently, we design another two algorithms to improve the capacity. The second algorithm called link-aware method means that we firstly calculate the weight of links from $amount$ paths in each node pair, which is decided by the times that $amount$ paths in each node pair have passed the link. Next, we aggregate k paths on the basis of the link that occupies the largest weight recursively in each node pair. Algorithm 2 describes our idea at length. In the very beginning, we also require $adjacent[scale][scale]$, $node\text{-}pair[pair]$, $amount$ and k. Meanwhile, some helpful symbols are summarized as follows.

$how\text{-}many$ records how many paths have been added in each node pair; $count$ records how many rounds the algorithm executes; $is\_Over()$ judges whether the number of paths has been k; $surplus$ records the number of paths that have not been added; $flag$ denotes whether a path has been added or not.

In the first place, according to $get\_Amount()$, we can acquire $amount\text{-}paths$, which denotes $amount$ shortest paths in each node pair. (Line 1) Next, in line with the times passing each link, we can calculate the weight of each link, which comes from $amount$ paths. In addition, we may rank the weight of those links and denote the result as $rank\text{-}links$. (Lines 2-3) Secondly, we utilize a 'for' circulation to initialize the enumerator $how\text{-}many$, which is in order to record the added paths in each node pair later. (Lines 4-5) In the end, the most imperative part aims at aggregating k paths in line with the 'while' circulation after initializing

Algorithm 2. Link-aware method

Require: $adjacent[scale][scale]$; $node\text{-}pair[pair]$; $amount$; k.
 1: $amount\text{-}paths \leftarrow get\_Amount(adjacent, node\text{-}pair, amount)$;
 2: Calculate the weight of each link based on $amount\text{-}paths$;
 3: Rank the weight and denote the result as $rank\text{-}links$;
 4: **for** $i = 0$ to $pair$ **do**
 5:     $how\text{-}many.push(0)$;
 6: $count = 0$;
 7: **while** $!is\_Over(how\text{-}many)$ && $count < rank\text{-}links.size()$ **do**
 8:     Aggregate all paths passing $rank\text{-}links[count]$ in each node pair and storage
them into $pass\text{-}links$;
 9:     **for** $i = 0$ to $pair$ **do**
10:         **if** $pass\text{-}links[i].size() > 0$ && $how\text{-}many[i] < k$ **then**
11:             $surplus \leftarrow min(k - how\text{-}many[i], pass\text{-}links[i].size())$;
12:             **for** $j = 0$ to $surplus$ **do**
13:                 Find out a path whose $flag = 0$ from $pass\text{-}links[i]$ and add this path
into $method2\text{-}result$;
14:                 Set the $flag$ of this path as 1;
15:     $count + +$;
16: **return** $method2\text{-}result$;

$count = 0$. (Lines 6-15) During every time the algorithm executes, we can push
new paths into the final result. At first, we should judge whether $is\_Over()$ is
false and $count$ is less than the size of $rank\text{-}links$, so as to decide if there are k
paths in each node pair. (Line 7) When the judgment is true, we may aggregate
all paths that pass $rank\text{-}links[count]$ in each node pair and storage them into
$pass\text{-}links$. (Line 8) However, the quantity of $pass\text{-}links$ may be not equal to
our goal k. So, we should utilize the symbol $surplus$ to determine the quantity
of remanent paths. (Lines 10-11) The latter 'for' circulation aims at searching
for a path whose $flag = 0$ from $pass\text{-}links[i]$ and add this path into $method2\text{-}$
$result$. Then, we should set the label $flag$ of this path as 1, which expresses that
this path has been added. (Lines 12-14) Eventually, we return $method2\text{-}result$.
(Line 16).

For instance, when $count = 0$, it means that we come to the first circulation. The $rank\text{-}links[0]$ denotes the link that has the max weight, which can
be regarded as a handle. According to this link, we may aggregate all paths
that pass it and storage them into $pass\text{-}links$. However, we should judge that
whether the quantity of these paths has reached the demand k. The following
part attends to aggregate paths until we have k paths for each node pair.

3.4 Third Algorithm: Path-Aware Method

Unlike the second algorithm, the last method is so-called path-aware method,
which calculates the weight of paths from $amount$ paths in each node pair via
adding up the weight of each link along the path. Then, this algorithm aggregates k paths recursively according to the number of mutual links with the path

---

Algorithm 3. Path-aware method

Require: $adjacent[scale][scale]$; $node\text{-}pair[pair]$; $amount$; k.
1: $amount\text{-}paths \leftarrow get\_Amount(adjacent, node\text{-}pair, amount)$;
2: Calculate the weight of each path based on $amount\text{-}paths$;
3: Rank the weight and denote the result as $rank\text{-}paths$;
4: **for** $i = 0$ to $pair$ **do**
5:     $how\text{-}many.push(0)$;
6: $count = 0$;
7: **while** $!is\_Over(how\text{-}many)$ && $count < amount * pair$ **do**
8:     Aggregate all paths that have the most mutual links with $rank\text{-}paths[count]$ in each node pair and storage them into $pass\text{-}paths$;
9:     **for** $i = 0$ to $pair$ **do**
10:        $surplus \leftarrow min(k - how\text{-}many[i], pass\text{-}paths[i].size())$;
11:        **for** $j = 0$ to $surplus$ **do**
12:            Find out a path whose $flag = 0$ from $pass\text{-}paths[i]$ and add this path into $method3\text{-}result$;
13:            Set the $flag$ of this path as 1;
14:     $count + +$;
15: **return** $method3\text{-}result$;

---

that has the largest weight. Algorithm 3 demonstrates this idea in detail. In the very beginning, we mainly require $adjacent[scale][scale]$, $node\text{-}pair[pair]$, $amount$ and k. Next, the main sequence of the idea can be described as follows.

Firstly, we can acquire $amount\text{-}paths$ on the basis of $get\_Amount()$. (Line 1) Then, we may calculate the weight of each path in accordance with the sum of links' weight, which means totaling up the weight of all links along the path. Meanwhile, we should rank the weight of paths in each node pair and denote the result as $rank\text{-}paths$. (Lines 2-3) Secondly, we utilize a 'for' circulation to initialize $how\text{-}many$, so as to record the added paths in each node pair later. (Lines 4-5) Lastly, we tend to rely on the 'while' circulation to add up to k paths for each node pair. (Lines 7-14) The judgement can be described as whether $is\_Over()$ is false and $count$ is less than all paths denoted by $amount * pair$. (Line 7) Then, the main train of thought is to aggregate all paths that have the most mutual links with $rank\text{-}paths[count]$ taken as a handle in each node pair. Then, we storage these paths into $pass\text{-}paths$. (Line 8) The symbol $surplus$ is also denoted as the quantity of remanent paths. (Line 10) During the last 'for' circulation, we can search for a path whose $flag = 0$ from $pass\text{-}paths[i]$ and add it into $method3\text{-}result$. Simultaneously, we set the $flag$ of this path as 1. (Lines 11-13) In the end, we return $method3\text{-}result$. (Line 15).

For example, when $count = 0$, $rank\text{-}paths[0]$ can be taken as a path having the max weight in each node pair. Next, we may aggregate all paths that have the most mutual links with $rank\text{-}paths[0]$ and storage them into $pass\text{-}paths$. However, the quantity of $pass\text{-}paths$ may not be equal to our goal k, so we should judge how many paths have to be added later and denote it as $surplus$. At last, we may append paths until we have k paths for each node pair.

4   Evaluation

4.1   The Sketch of Network Performance

Our simulations are under two typical topologies, 3D-Torus and Jellyfish [19]. Especially, we assure the integrality of these topologies, without any possible interference. Taking this design into account mainly aims at insuring the fault-tolerance. Although we do not conduct some necessary experiments, the insight behind this excogitation is feasible theoretically. For instance, when having a mission between two nodes, we can transmit the flows according to diverse k paths that generated from our methods. However, in the existing ideas that turn off the idle devices or put them into sleep mode, we cannot complete the tasks in case of appearing any faults in paths. Compared with this method, our ideas can provide a little more steady choices.

During this simulation part, we major in evaluating and comparing our three algorithms with each other on the basis of two performance metrics, time-consumption and number of active network devices. The indicator of time-consumption can measure our algorithms' time complexity, which may also reflect the responsive-time to a certain extant. Meanwhile, our second indicator of active network devices' quantity can be treated as the measurement of energy consumption, according to which we can multiply the quantity of active devices by energy per device, getting the total energy consumption. These performance metrics are evaluated under two topologies, 3D-Torus and Jellyfish, which are representative structured and random topology separately. In this section, we chiefly vary network *scale*, *amount* and k value respectively, in order to reveal their impacts on network performance.

4.2   The Impact of Network *scale*

As depicted in Section 4.1, the accurate but time-consuming method suffers high time-consumption due to its complicated computation. If there are dozens of node pairs call for data transmission, the first method spends hours or even days deciding the number of active network devices. Obviously, it is unwise to put our first idea into real applications. Thus, we just compare the link-aware method and the path-aware method further. Fig. 1 compares the performance of link-aware method and path-aware method in both 3D-Torus and Jellyfish. We vary the network *scale* of 3D-Torus and Jellyfish from 27 to 150, with *pair* = 10, *amount* = 20, $k = 3$.

Fig. 1(a) and Fig. 1(c) record the time-consumption, while Fig. 1(b) and Fig. 1(d) depict the number of active network devices. Concerning these two methods, both time-consumption and number of active network devices increase with the rising of network *scale*. Larger network *scale* means more involved links, which results in more computational time and more active devices reasonably. Note that, the path-aware method suffers less time than the link-aware method, as path-aware method needs less recursion. It is apparent that one chosen path in the path-aware method can aggregate paths more quickly than one selected

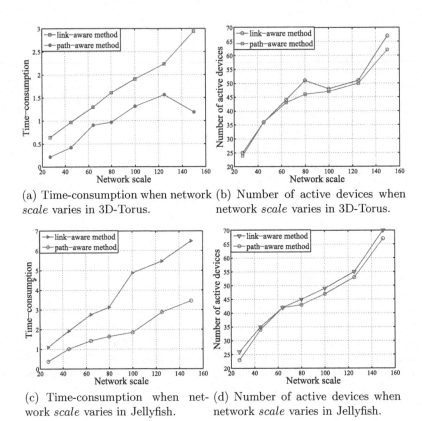

(a) Time-consumption when network scale varies in 3D-Torus.

(b) Number of active devices when network scale varies in 3D-Torus.

(c) Time-consumption when network scale varies in Jellyfish.

(d) Number of active devices when network scale varies in Jellyfish.

Fig. 1. The performance comparisons of two methods with $pair = 10, amount = 20, k = 3$ and network $scale$ varies under seven cases in both 3D-Torus and Jellyfish

link in the link-aware method. Also, according to the figures, the topology may not significantly affect the performance between the link-aware method and the path-aware method.

4.3 The Impact of *amount*

In this section, we evaluate the impact of the given number of shortest paths named *amount*. Fig. 2 illustrates the comparisons of all three methods, which are mainly about the performance under the circumstances that $scale = 27, pair = 3, k = 2$ and *amount* varies from 3 to 10 in both 3D-Torus and Jellyfish. Fig. 2(a) and Fig. 2(c) record the time-consumption of our three algorithms in 3D-Torus and Jellyfish respectively, while, Fig. 2(b) and Fig. 2(d) record the number of active network devices.

Note that the vertical axis of Fig. 2(a) and Fig. 2(c) is the logarithm value of actual time-consumption, i.e., log_{10}^{time}. It is obvious that the accurate but time-consuming method suffers more time-consumption than both the link-aware

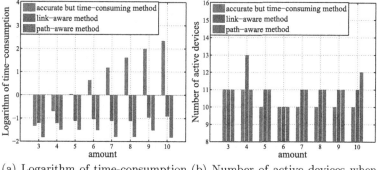

(a) Logarithm of time-consumption (b) Number of active devices when
when *amount* varies in 3D-Torus.   *amount* varies in 3D-Torus.

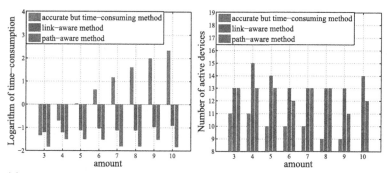

(c) Logarithm of time-consumption (d) Number of active devices when
when *amount* varies in Jellyfish.   *amount* varies in Jellyfish.

Fig. 2. The performance comparisons of three methods with $scale = 27, pair = 3, k = 2$ and *amount* varies from 3 to 10 in both 3D-Torus and Jellyfish

method and the path-aware method. Besides, the time-consumption of the first method increases dramatically with the increasing of *amount*, while in the other methods just fluctuates in a rather low level. In this experiment, with $pair = 3$ and $k = 2$, there exists $\prod_{i=1}^{3} C_{amount}^{2}$ feasible combinations, thus the computation of the results are of complexity. That is why the time-consumption of the first algorithm in Fig. 2(a) and Fig. 2(c) performs a linear growth. On the contrary, we cannot make a certain assertion when the link-aware and the path-aware methods are terminated. As a result, the time-consumption of the latter two methods fluctuates with no regulations but in a low level.

According to the number of active network devices shown in Fig. 2(b) and Fig. 2(d), our first method achieves the best solution, i.e., the quantity of active devices is no more than the other two algorithms. It is reasonable that the first method takes all possible solutions into its consideration and picks up the one that occupies the least network devices. What is more, the number of active network devices in the first method remains decreasing with the increasing of *amount*. However, the number of the active network devices in the link-aware

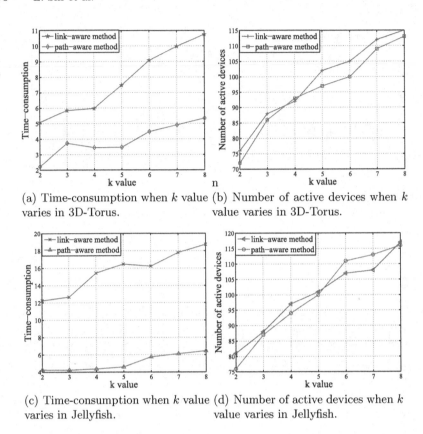

(a) Time-consumption when k value varies in 3D-Torus.

(b) Number of active devices when k value varies in 3D-Torus.

(c) Time-consumption when k value varies in Jellyfish.

(d) Number of active devices when k value varies in Jellyfish.

Fig. 3. The performance comparisons of two methods with $scale = 125, pair = 20, amount = 20$ and k value varies from 2 to 8 in both 3D-Torus and Jellyfish

method and the path-aware method still fluctuate without rules. This phenomenon is caused by the nature of the approximate algorithm. Note that, the number of active devices of path-aware method is no more than the link-aware method. Besides, Fig. 2(b) and Fig. 2(d) appear distinct from each other, due to the various topologies that 3D-Torus is inerratic and Jellyfish is generated randomly.

4.4   The Impact of k Value

There is no doubt that the value of k will affect the performance of each method, in this section, we conduct experiments to evaluate the impact of k value. Similarly, we just compare the latter two algorithms. Typically, with $scale = 125, pair = 20$ and $amount = 20$, we vary k from 2 to 8 and record the time-consumption and the number of active network devices in both 3D-Torus and Jellyfish.

Fig. 3(a) and Fig. 3(c) depict the time-consumption of these two methods in 3D-Torus and Jellyfish respectively. With the increasing of k value, the curves in Fig. 3(a) and Fig. 3(c) remain ascending. Since the larger k results in more paths and active network devices, the methods need more time to aggregate the paths. Besides, the path-aware method still costs much less time than the link-aware method. Note that, regarding the number of active network devices shown in Fig. 3(b) and Fig. 3(d), we can find out no apparent difference between these two methods. Thus, it is rational to conclude that the path-aware method outperforms the link-aware method when k value varies.

5   Conclusion

In this paper, we argue that the existing energy-saving strategies in DCNs cannot guarantee the fault-tolerance and prompt responsive-time. Thus, we advocate that it is essential to remain the topology characters of DCNs primarily. In the meantime, we present three diverse algorithms to achieve the goal of green routing, with acquiring k paths in each node pair that calls for data transmission. Also, we evaluate the performance of network's energy consumption in line with two metrics, time-consumption and the number of active devices. Comprehensive simulations demonstrate that the first method achieves the accurate solution, which needs the least active network devices. But it suffers the unacceptable time-consumption. The link-aware method and the path-aware method derive a little more active devices but spend much less time. Besides, the path-aware method costs the least time and achieves no more active network devices than the link-aware method. As a result, it is a wise to put the path-aware method into real DCNs.

References

1. Deke, G., Junjie, X., Xiaomin, Z., Wei, W.: Exploiting Efficient and Scalable Shuffle Transfers in Future Data Center Networks. IEEE Transactions on Parallel and Distributed Systems (TPDS) **26**(4), 997–1009 (2015)
2. Deke, G., Tao, C., Dan, L., Mo, L., Yunhao, L., Guihai, C.: Expansible and Cost-Effective Network Structures for Data Centers Using Dual-port Servers. IEEE Transactions on Computers (TC) **62**(7), 1303–1317 (2013)
3. Deke, G., Hanhua, C., Yuan, H., Hai, J., Chao, C., Honghui, C., Zhen, S., Guangqi, H.: KCube: A Novel Architecture for Interconnection Networks. Elsevier Information Processing Letter **110**(18–19), 821–825 (2010)
4. Deke, G., Chaoling, L., Jie, W., Xiaolei, Z.: DCube: A Family of High Performance Modular Data Centers Using Dual-Port Servers. Elsevier Journal of computer communication **53**, 13–25 (2014)
5. Koomey, J.: Growth in data center electricity use 2005 to 2010. A report by Analytical Press, completed at the request of The New York Times (2011)
6. Gao, P.X., Curtis, A.R., Wong, B., Keshav, S.: It's not easy being green. ACM SIGCOMM Computer Communication Review **42**(4), 211–222 (2012)

7. Greenberg, A., Hamilton, J., Maltz, D.A., Patel, P.: The cost of a cloud: research problems in data center networks. ACM SIGCOMM computer communication review **39**(1), 68–73 (2008)
8. Shang, Y., Li, D., Xu, M.: Energy-aware routing in data center network. In: Proceedings of the First ACM SIGCOMM Workshop on Green Networking, pp. 1–8. ACM (2010)
9. Beloglazov, A., Abawajy, J., Buyya, R.: Energy-aware resource allocation heuristics for efficient management of data centers for cloud computing. Future generation computer systems **28**(5), 755–768 (2012)
10. Wang, L., Zhang, F., Arjona Aroca, J., Vasilakos, A.V., Zheng, K., Hou, C., Li, D., Liu, Z.: GreenDCN: a general framework for achieving energy efficiency in data center networks. IEEE Journal on Selected Areas in Communications **32**(1), 4–15 (2014)
11. Gupta, M., Singh, S.: Greening of the internet. In: Proceedings of the 2003 Conference on Applications, Technologies, Architectures, and Protocols for Cmputer Communications, pp. 19–26. ACM (2003)
12. Guo, C., Lu, G., Li, D., Wu, H., Zhang, X., Shi, Y., Tian, C., Zhang, Y., Lu, S.: BCube: a high performance, server-centric network architecture for modular data centers. ACM SIGCOMM Computer Communication Review **39**(4), 63–74 (2009)
13. Guo, C., Wu, H., Tan, K., Shi, L., Zhang, Y., Lu, S.: Dcell: a scalable and fault-tolerant network structure for data centers. ACM SIGCOMM Computer Communication Review **38**(4), 75–86 (2008)
14. Li, D., Shang, Y., Chen, C.: Software defined green data center network with exclusive routing. In: INFOCOM, 2014 Proceedings IEEE, pp. 1743–1751. IEEE (2014)
15. Nedevschi, S., Popa, L., Iannaccone, G., Ratnasamy, S., Wetherall, D.: Reducing network energy consumption via sleeping and rate-adaptation. In: NSDI, vol. 8, pp. 323–336 (2008)
16. Chiaraviglio, L., Matta, I.: Greencoop: cooperative green routing with energy-efficient servers. In: Proceedings of the 1st International Conference on Energy-Efficient Computing and Networking, pp. 191–194. ACM (2010)
17. Lombardo, A., Panarello, C., Schembra, G.: Achieving energy savings and QoS in internet access routers. ACM SIGMETRICS Performance Evaluation Review **38**(3), 76–80 (2011)
18. Heller, B., Seetharaman, S., Mahadevan, P., Yiakoumis, Y., Sharma, P., Banerjee, S., McKeown, N.: ElasticTree: saving energy in data center networks. In: NSDI, vol. 10, pp. 249–264 (2010)
19. Singla, A., Hong, C.Y., Popa, L., Godfrey, P.B.: Jellyfish: networking data centers randomly. In: NSDI, vol. 12, pp. 17–17 (2012)

A Markov Chain Prediction Model for Routing in Delay Tolerant Networks

Shuai Liu, Fan Li$^{(\boxtimes)}$, Qian Zhang, and Meng Shen

Beijing Engineering Research Center of High Volume Language Information
Processing and Cloud Computing Applications, School of Computer Science,
Beijing Institute of Technology, Beijing 100081, China
fli@bit.edu.cn

Abstract. In Delay Tolerant Networks (DTNs), nodes seldom keep in contact with each other because of the disorder movements of nodes. Thus many traditional routing protocols may not be applicable in this emerging network. In recent years, many routing protocols have been proposed to improve the routing performance in DTNs, however, most of them do not fully mine the inherent regular pattern of the movements of nodes. In this paper, we propose a *Markov Chain Prediction (MCP)* model for routing in DTNs to mine the movements of nodes property with historical contact information and use this model to improve the delivery ratio of message transferring within the network.

1 Introduction

Delay Tolerant Networks (DTNs) are kinds of networks that lack continuous links between nodes due to limited wireless radio coverage, widely scattered mobile nodes, constrained energy resources, and high levels of interference. Examples of such networks are land mobile network, exotic media network, and pocket switched network.

DTNs have the following characteristics: (1) intermittent connection of nodes; (2) large transmission delay; (3) continual changes of network; and (4) limit of source. Because of no stable end-to-end transmission paths, researchers face more challenges in the routing design compared with traditional networks [1]. The main purpose of the DTN routing is to maximize the possibility of packet transmission. So many routing strategies increase encounter chances by generating multiple copies of the same messages. At present, routing algorithms in DTNs can be divided into two categories: routing based on multi copies and a single copy of the message.

Using multiple copies increases the possibility of a transfer, but it also increases the network traffic. Meanwhile, a node needs large storage space which

The work of F. Li is partially supported by the National Natural Science Foundation of China under Grant No. 61370192 and 61432015. The work of M. Shen is partially supported by Beijing Institute of Technology Research Fund Program for Young Scholars.

© Springer International Publishing Switzerland 2015
Y. Wang et al. (Eds.): BigCom 2015, LNCS 9196, pp. 479–490, 2015.
DOI: 10.1007/978-3-319-22047-5_39

could lead to network congestion and in turn affect the timely transmission of messages. Thus, it is important to decide a suitable number of replications when transferring messages. To improve the transmission probability, how to effectively control the number of copies and meanwhile maximize the efficiency have become one research direction in DTN routing [2,3].

The protocols based on single copy of a message follow the strategy that if there is no other nodes around the contact range, the current node stores and keeps the message itself. When the current node meets the next node, it makes a decision whether to forward the message or not. PRoPHET [5] and Fresh [6] are two examples.

Because the movements of nodes in the DTNs have certain regularities, similar to the social network in the form of human society, some properties of social networks can be used to characterize the properties between nodes. Many routing protocols use social attributes to achieve routing control in message transmission [15–17]. In this type of routing protocols, SimBet [9] and SEBAR [10] are two examples.

This paper proposes a new routing method based on Markov chain prediction model in accordance with the relationship of nodes and nodes' own attributes. Combing these characteristics, the proposed routing algorithm can improve message delivery ratio without introducing too much traffic in the network.

The rest of paper is organized as follows: Section 2 reviews existing routing methods for DTNs. Section 3 describes the network models we use. Section 4 presents the detailed design of the proposed *Markov Chain Prediction (MCP)* model for routing in DTNs. Section 5 describes simulation results and Section 6 concludes the paper.

2   Related Work

In the previous studies, many routing protocols have been proposed for DTNs. Epidemic Routing [4] is a classic and simplest routing protocol which allows multiple copies of the messages in the network. The main idea of Epidemic routing is that in the routing diffusion, the intermediate nodes send the copy of received messages to all their neighbor nodes. This method has the advantages of small transmission delay, and highest message delivery success rate, but the large number of message copies in the network needs a lot of cache and bandwidth resources.

In the aspects of routing strategy using single copy of a message, PRoPHET [5] predicts the delivery probability in the future network based on historical contacts, and then forwards the message according to the encounter probability. Fresh algorithm [6] forwards packets to the encountered node if it meets the destination node more recently than the current node does.

Recently, social-based routing has attracted lots of attention since most mobile devices (such as smart phones) are used and carried by people. The network behaviors could be better characterized by their social attributes. SimBet (similarity and betweenness) protocol [9] proposes two kinds of social attributes:

similarity and centrality. When current node encounters another node, the message is more likely to be forwarded to the node with higher social centrality and more similarity with the destination node. In SEBAR (Social Energy BAsed Routing) [10], a new metric *social energy* is introduced to quantify the ability of a node to forward packets to others. Social energy is generated via node encounters and shared by the communities of encountering nodes. SEBAR is in favor of the node with a higher social energy in its or the destination's social community.

Some researches explore that the contacts between nodes in the DTN environment has Markov property [13,14]. By using the Markov model, the meeting time span between nodes can be predicted. Therefore, in order to increase the delivery ratio, the relay node holding the shortest meeting time span with the destination node is the most efficient node to hold the message [13]. Also with resource constraints in DTNs, the message delivery process can be modeled as a Markov chain. Finding an optimal value for the message-replication limit of each message can reduce the delivery delay [14].

In this paper, we are committed to use Markov model to fully mine the regular pattern of movements of nodes in the network to improve the delivery ratio of the DTN routing. We propose a *Markov Chain Prediction (MCP)* model for routing which not only predicts the relationship between current node and a certain node, but also predicts the node activity degree. By combining these two aspects of nodes' attributes, we can predict a node's ability to encounter the target node.

3   Network Model

In DTNs, the movements of nodes have a certain regular pattern. The aim of our proposed MCP routing is to deeply mine this pattern of movements for higher message delivery ratio and more balanced traffic load. Notice that this pattern is not only reflected in the relationship between nodes, but also in the activity degree of the node itself.

We need to analyze the historical contacts of a node to discover its contact regularity with other nodes and its own attribute as well. Thus, the current node should store one sequence that saves activity degree value, and one matrix that records the historical relationship between the current node with all other nodes in the network. Then, by using the Markov prediction model, the node's corresponding state in the next time period can be predicted, consisting of both activity degree and relationship with the destination node. In the routing process, the node can decide whether to deliver the message to the encountered node or not by considering both these two attributes.

In the following parts, we explain how a certain node updates its storage of node activity degree sequence and this nodes' relationship matrix.

3.1   Node Activity Degree Sequence Update

Assume that $V = \{v_1, v_2, \ldots, v_n\}$ is a set of nodes in the network. Each node can send and receive a message when it encounters another node. There are a

---

Algorithm 1. UpdateActivityDegree(v_i)

Node v_i keeps a queue Q_i with length m, and an activity degree a_i. Assume current time is at t. t_0 is the timestamp of last Δt.

```
 1: if node v_i encounters node v_j then
 2:     Update a_i based on the chosen activity degree;
 3: end if
 4: if t - t_0 == Δt  then
 5:     Map a_i to s_i;
 6:     if Q_i.length¡m then
 7:         Q_i.add(s_i);
 8:     else
 9:         Q_i.remove();
10:         Q_i.add(s_i);
11:     end if
12:     Clear a_i;
13:     t_0 = t;
14: end if
```

---

lot of metrics can represent a node activity degree, such as the betweenness [9] and degree centrality [8]. Betweenness in SimbBet [9] measures the extent that a node lies on the paths linking other nodes. Degree centrality [8] is the number of one node's neighbors in social graph[8]. MCP uses betweenness to describe node activity degree as the example, the same as in SimBet.

We assume that the time is divided into discrete and equal time slots. Let Δt denote the time slot. In every Δt, for a certain node v_i, it calculates the value of betweenness during this time and adds the value to the sequence of the node activity degree sequence. Assume there are t states in total to describe the range of node activity degree. We partition the range of activity degree values equally and map it to t states. Then the sequence is transformed to a state chain $\{s_1, s_2, \ldots, s_t\}$. Considering of the load of every node, here we use the sliding window method to control the length of the sequence. Assume each node stores the sequence list with the length of m, for every period time of Δt, if the list is not full, node adds the new state, otherwise, it throws away the earliest state and adds the new information. Algorithm 1 shows the details of updating the activity degree of node v_i.

3.2   Node Relationship Matrix Update

For a certain node v_i, it also keeps a matrix which saves the relationship with other nodes. Here, we use encounter times to describe the relationship between other nodes as an example. The relation matrix of node v_i is presented in Figure 1. In the first matrix, each row represents every time interval Δt, and each column describes the encounter times with a certain node at different time interval. For instance, in the relation matrix of node v_i, $c_{k,l}$ represents encounter times of node v_i with node v_l in the time slot $k\Delta t$. Whenever v_i encounters another node v_l, the newest element $c_{k,l}$ will be updated. With this encounter matrix, we can

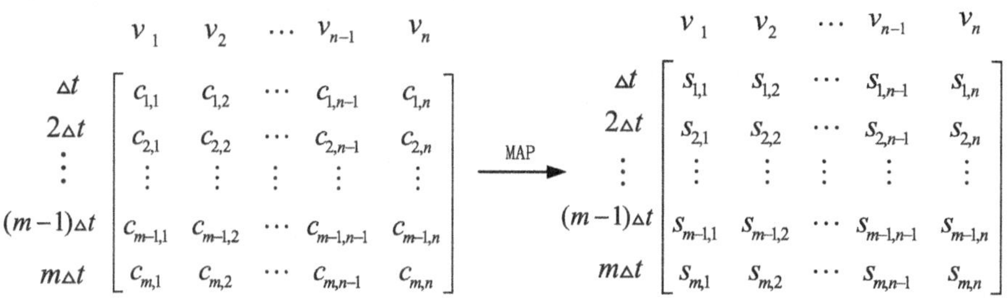

Fig. 1. Node v_i's relation matrix

Algorithm 2. UpdateNodeRelation(v_i)

Node v_i in network keeps a $m \times (n-1)$ matrix S and encounter times sequence $\{c_1, c_2, \ldots, c_m\}$. Assume current time is at t. t_0 is the timestamp of last Δt.

1: **if** v_i encounters node v_j **then**
2:     C_j ++;
3: **end if**
4: **if** $t - t_0 == \Delta t$ **then**
5:     map encounter times sequence $\{c_1, c_2, \ldots, c_m\}$ to state sequence $\{s_1, s_2, \ldots, s_m\}$;

6:     **if** S.length ¡ m **then**
7:       S.add($\{s_1, s_2, \ldots, s_m\}$);
8:     **else**
9:       remove the oldest state sequence row of matrix S;
10:      S.add($\{s_1, s_2, \ldots, s_m\}$);
11:     **end if**
12:     Clear encounter times sequence $\{c_1, c_2, \ldots, c_m\}$;
13:     $t_0 = t$;
14: **end if**

---

map the encounter times to t states to generate the state matrix like the second matrix in Figure 1, in which every state represents a certain range of encounter times. Algorithm 2 shows the details of updating nodes relationship matrix of v_i when it meets another node v_j .

4 MCP Routing

In this section, we will introduce our MCP routing process in details. When a node v_i with message M destined to v_d meets v_j, first v_i and v_j update their activity degree sequence and their node relationship matrix separately according to the method we discuss in Section 3.1 and Section 3.2. Then v_i and v_j predict their own activity degree states $S_{Act,i}$, $S_{Act,j}$ and the encounter times states $S_{Rel,i}$, $S_{Rel,j}$ with the destination v_d in the next time period.

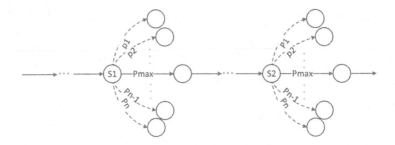

Fig. 2. Makov chain prediction process

Next, We will show how we use the sequence of activity degree stored in node v_i to predict its activity degree state in the next period of time. Assume the activity degree state sequence is $\{s_1, s_2, \ldots, s_m\}$. According to the property of the Markov chain, we have:

$$P(S_{m+1} = s_{m+1}|S_m = s_m, S_{m-1} = s_{m-1} \ldots S_1 = s_1) = P(S_{m+1} = s_{m+1}|S_m = s_m).$$

As shown in Figure 2, we focus on the current state to predict the future state of next period of time. Assume the current state of v_i is s_1. When v_i predicts the next state of activity degree, it scans the sequence and counts the number of S_1 transforming to all other states, then it derives the probability $\{p_1, p_2, \ldots, p_t\}$ which represents the chance of state S_1 transforming to all other states. It selects the state S_{max} which has the largest probability p_{max} as the next state of activity degree. We use a function named MCP-ActivityDegree(v_i) to denote this prediction process which returns the next state we predict using Markov chain. Similarly, we use function MCP-NodeRelaiton(v_i, v_d) to denote the prediction process of node relation state which returns encounter times state with the destination v_d.

Regarding the MCP routing process, the forwarding happens only when both activity degree of v_j and the relation between v_j and destination node v_d are greater than those of v_i. We believe that the neighboring node which is more activity in the network and meets the destination more times in the past has better chance to deliver the message to the destination successfully. Algorithms 3 gives the detailed routing method.

5   Simulations

We have conducted extensive simulation experiments over real-life wireless DTN traces to evaluate our proposed MCP routing method. In our simulations, we use the betweenness utility value [9] as our node activity degree metric. The encounter times is used to describe the relationship between nodes. We compare our proposed MCP with the following existing routing methods:

- **Epidemic[4]**: the node replicates a copy of the packet and forwards it to any encountered nodes.

Algorithm 3. Markov Chain Prediction (MCP) Model Based Routing

Node v_i with message M destined to v_d meets v_j which does not hold message.

1: **if** v_j is the destination **then**
2:     v_i forwards M to v_j;
3: **else**
4:     $S_{Act,i}$ = MCP-ActivityDegree(v_i);
5:     $S_{Act,j}$ = MCP-ActivityDegree(v_j);
6:     $S_{Rel,i}$ = MCP-NodeRelaiton (v_i,v_d);
7:     $S_{Rel,j}$ = MCP-NodeRelaiton(v_j,v_d);
8:     **if** $S_{Act,i} \leq S_{Act,j}$ and $S_{Rel,i} \leq S_{Rel,j}$ **then**
9:         v_i forwards M to v_j;
10:    **else**
11:        v_i holds the M and waits for the next encounter;
12:    **end if**
13: **end if**

- **SimBet[9]**: if the SimBet utility of node v_j is larger than that of node v_i, v_i sends the packet to v_j.
- **FRESH[6]**: the packet is forwarded from node v_i to node v_j if v_j has met the destination more recently than v_i does.
- **Greedy-Total[7]**: the packet is forwarded from v_i to v_j if v_j has a higher contact frequency to all other nodes than v_i does.

In all experiments, we compare the performance of each routing method using the following routing metrics:

- *Delivery Ratio:* the average percentage of packets that successfully delivered from the sources to the destinations.
- *Maximum Load:* the largest load among all nodes within a certain period of time.
- *Average Load:* the average load of all nodes within a certain period of time.
- *Average Hops:* the average number of hops during each successful delivery from the sources to the destinations.
- *Average Forwards:* the average number of forwarding times during each delivery (not only limit to successful delivery) from the sources to the destinations.
- *Average Delay:* the average duration of successfully delivered packets from the sources to the destinations.

5.1   Simulation Results with InfoCom 2006 Trace Data

First, we use InfoCom 2006 [11] to simulate a DTN environment. This trace data includes 78 mobile devices' connections for four days recorded in InfoCom 2006 conference in Barcelona, Spain. Each record in the dataset contains information about the ID of the device who recorded sightings and the device who was seen.

Also it contains the occurrence time of a certain encounter. Each node tries to send a packet to all other nodes. Here, we consider single-copy version of all routing methods. We use the contact information from the first 40 hours as historical data to generate node activity degree sequence and nodes' relationship matrix, then the performance of routing tasks are evaluated over the remaining time.

Δt is time interval of updating both activity degree and node relationships. If the time interval is too long, the state status may not reflect the time-variance of node social attributes accurately; on the other hand, if the time interval is too short, frequently exchanged messages will generate lots of network traffic. Thus, we choose an appropriate time interval by empirical.

We set the range of Δt changing from 1800 seconds to 20000 seconds for MCP. The delivery ratio with the change of Δt is shown in Figure 3. As Figure 3 shows, beginning with the increase of Δt, the delivery ratio of MCP increases significantly, indicating that Δt is too short to fully describe the movements of nodes. Later with the increase of Δt, the delivery ratio tends to be gentle, and began to decline, indicating the interval is too long and will loss the regularity. As can be seen from the figure, for the current data set, for nearly 6000 seconds as the time interval can better depict the current environment. Table 1 summarizes parameters we used in MCP Routing.

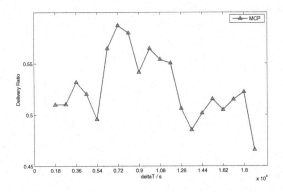

Fig. 3. Impact of Δt

Figure 4 demonstrates the performance comparison among MCP and other four existing routing methods. As Figure 4(a) shows, Epidemic has the highest delivery ratio because it offers the upper bound of the delivery ratio that any routing protocol can achieve. Comparing to the rest routing methods, MCP has better performance delivery ratio, since MCP mines more information of node movement regularity to make better decision. Meanwhile, MCP has good performance in terms of the maximum and average load as shown in Figure 4(b) and 4(c). Note that these two sub-figures do not include the results of Epidemic because its maximum load is usually higher than 14,000 and its average load

Table 1. Parameters used for MCP Routing

Parameter	Value or Type
Node activity degree	SimBet Betweeness
Δt for InfoCom 2006 trace data	6000s
Δt for Sigcomm 2009 trace data	7000s
Number of states of Node Activity Degree	5
Number of states of Node's Relationship	5

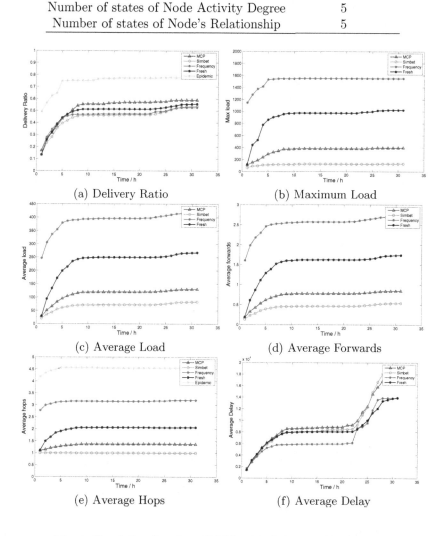

(a) Delivery Ratio

(b) Maximum Load

(c) Average Load

(d) Average Forwards

(e) Average Hops

(f) Average Delay

Fig. 4. Simulation Results of MCP using Infocom 2006 data set

is usually higher than 4,000. Because when MCP makes the routing decision, it considers both the relationship towards the destination node and also the node's own activity degree. It greatly reduces the number of forwardings, which

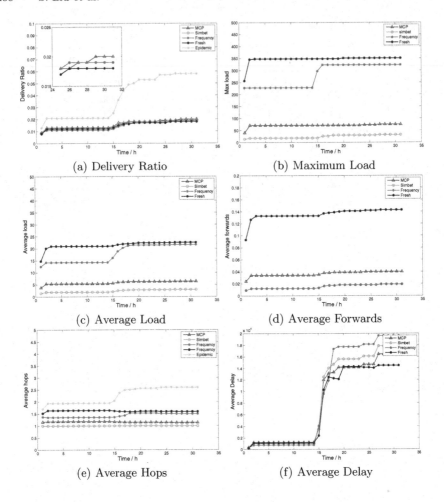

Fig. 5. Simulation Results of MCP using Sigcomm 2009 data set

can effectively control the network energy consumption. Figure 4(d) and 4(e) indicate that average forwards and average hops of MCP are competitive to SimBet.

5.2 Simulation Results with Sigcomm 2009 Trace Data

We also implement the routing methods in Sigcomm 2009 trace data[12]. This trace data contains data collected by an opportunistic mobile social application MobiClique, which was used by 76 persons during Sigcomm 2009 conference in Barcelona, Spain. The dataset contains data from 76 devices that shows significant activity during the four days experiment. We use the encounter information from the data which contains the ID of the devices and the occurrence time.

Similar to the above, we use the first 40 hours to initiate node activity degree sequence and nodes' relationship matrix, then the performance of routing tasks are evaluated over the remaining time.

Figure 5 demonstrates the performance comparison among MCP and other four existing routing methods in this dataset. Similar to the performance with InfoCom trace data, the delivery ratio of MCP routing has a better performance compared to other methods except for Epidemic according to Figure 5(a). The maximum and average load of MCP routing are also competitive to SimBet as shown in Figure 5(b) and 5(c).

6   Conclusion

Delay tolerant networks (DTNs) are partitioned wireless ad hoc networks, which usually cannot guarantee end-to-end paths between any pair of nodes. Thus, routing in DTNs is a challenging problem. A lot of routing protocols have been proposed to improve the performance in DTNs, however, most of them do not fully mine the inner regular pattern of movements of nodes. In this paper, we propose a Markov chain prediction model for routing in DTNs to mine the movements of nodes property with historical contact information. MCP routing not only predicts the relationship between nodes, but also the node's own activity degree. Considering both attributes, the routing selection can more accurately determine the capability of one node to forward a message to the destination. The simulation results from two real-life DTN trace data show that MCP can improve the delivery ratio of message transferring without introducing much traffic throughout the network.

References

1. Schurgot, M.R., Comaniciu, C., Jaffres-Runser, K.: Beyond traditional DTN routing: social networks for opportunistic communication. IEEE Communications Magazine **50**, 155–162 (2012)
2. Liu, M., Yang, Y., Qin, Z.: A survey of routing protocols and simulations in delay-tolerant networks. In: Cheng, Y., Eun, D.Y., Qin, Z., Song, M., Xing, K. (eds.) WASA 2011. LNCS, vol. 6843, pp. 243–253. Springer, Heidelberg (2011)
3. Zhu, Y., Xu, B., Shi, X., Wang, Y.: A survey of social-based routing in delay tolerant networks: positive and negative social effects. Communications Surveys & Tutorials, IEEE **15**(1), 387–401 (2013)
4. Vahdat, A., Becker, D., et al.: Epidemic routing for partially connected ad hoc networks. Technical Report CS-200006, Duke University, Tech. Rep. (2000)
5. Lindgren, A., Doria, A., Schelén, O.: Probabilistic routing in intermittently connected networks. ACM SIGMOBILE mobile computing and communications review **7**(3), 19–20 (2003)
6. Dubois-Ferriere, H., Grossglauser, M., Vetterli, M.: Age matters: efficient route discovery in mobile ad hoc networks using encounter ages. In: Proceedings of the 4th ACM International Symposium on Mobile Ad Hoc Networking & Computing, pp. 257–266. ACM (2003)

7. Erramilli, V., Chaintreau, A., Crovella, M., Diot, C.: Diversity of forwarding paths in pocket switched networks. In: Proceedings of the 7th ACM SIGCOMM Conference on Internet Measurement, pp. 161–174. ACM (2007)
8. Erramilli, V., Crovella, M., Chaintreau, A., Diot, C.: Delegation forwarding. In: Proceedings of the 9th ACM International Symposium on Mobile Ad Hoc Networking and Computing, pp. 251–260. ACM (2008)
9. Daly, E.M., Haahr, M.: Social network analysis for routing in disconnected delay-tolerant manets. In: Proceedings of the 8th ACM International Symposium on Mobile Ad Hoc Networking and Computing, pp. 32–40. ACM (2007)
10. Li, F., Jiang, H., Wang, Y., Li, X., Wang, M., Abdeldjalil, T.: SEBAR: Social energy based routing scheme for mobile social delay tolerant networks. In: Proceedings of the 2013 IEEE International Performance Computing and Communications Conference. IEEE (2013)
11. Scott, J., Gass, R., Crowcroft, J., Hui, P., Diot, C., Chaintreau, A.: Crawdad trace cambridge/haggle/imote/infocom2006 (v. May 29, 2009). http://crawdad.cs.dartmouth.edu/cambridge/haggle/imote/infocom2006
12. Kaisa, A.: CRAWDAD data set thlab/sigcomm2009 (v. July 15, 2012). http://crawdad.org/thlab/sigcomm2009
13. Wang, E., Yang, Y., Jia, B., Guo, T.: The DTN routing algorithm based on Markov meeting time span prediction model. International Journal of Distributed Sensor Networks, **2013**, Article ID 736796 (2013)
14. Ip, Y., Lau, W., Yue, O.: Forwarding and Replication Strategies for DTN with Resource Constraints. In: IEEE 65th Vehicular Technology Conference, 2007. VTC2007-Spring (2007)
15. Tian, C., Li, F., Jiang, L., Wang, Z., Wang, Y.: Energy efficient social-based routing for delay tolerant networks. In: Cai, Z., Wang, C., Cheng, S., Wang, H., Gao, H. (eds.) WASA 2014. LNCS, vol. 8491, pp. 290–301. Springer, Heidelberg (2014)
16. Li, F., Zhao, L., Zhang, C., Gao, Z., Wang, Y.: Routing with multi-level cross-community social groups in mobile opportunistic networks. Personal and Ubiquitous Computing **18**(2), 385–396 (2014)
17. Liu, G., Ji, S., Cai, Z.: Credit-based incentive data dissemination in mobile social networks. In: International Workshop on Identification, Information & Knowledge in The Internat of Things (IIKI 2013) (2013)

RAM: Resource Allocation in Mobility
for Device-to-Device Communications

Weiyang Lin, Cuibo Yu$^{(\boxtimes)}$, and Xu Zhang

School of Network Education, Beijing University of Posts and Telecommunications,
Beijing 100876, China
`linweiyang.1988@163.com, {cbyu,selina_zhangx}@bupt.edu.cn`

Abstract. As one of the key technologies of the Long Term Evolution (LTE)-Advanced networks, Device-to-Device (D2D) communication meets the requirement for higher data rate with the increase of local communication services. But it has to deal with new interference between cellular network and D2D devices. Several schemes have been proposed to cancel or mitigate such interference by proper power control or resource allocation. However, there is little discussion about the resource allocation considering user mobility. In this paper, RAM, a novel three-step resource allocation scheme, is proposed that makes D2D communications adapt well to the user mobility. In this solution, it first performs admission control based on the velocity judgment and then selects proper resource allocation mode for each admissible D2D pair, which includes the dedicated resource mode and the reused resources mode. The last step is to select the suitable cellular user (CU) partners for each admissible D2D pair in the reused resources mode. Simulation results show that RAM balances the velocity and user throughput and D2D mean throughput gain could improve 3Mbps compared with the dedicated resource mode.

Keywords: Long term Evolution-Advanced · Device-to-Device · Resource allocation · Mobility · System stability · Throughput

1 Introduction

To meet the requirements of IMT-Advanced system, 3GPP (Third Generation Partnership Project) mainly focuses on the development in the measures of spectral efficiency and throughput in recent years. LTE device to device (D2D) proximity services has been researched as a new theme for LTE-Advanced, which was approved in the RAN#58 plenary meeting. Generally, D2D communication is defined as a direct link establishment between devices in close proximity, which could transfers the data and necessary control signals directly. Hence, the direct communications between devices can save the power, increase the data rates and decrease the cell overload. Moreover, D2D communication described in [1] can improve the overall network spectral efficiency and throughput by reusing the resource of the cellular user.

However, D2D communications may introduce interference into the existing cellular network if less designed properly. Therefore, interference management becomes

© Springer International Publishing Switzerland 2015
Y. Wang et al. (Eds.): BigCom 2015, LNCS 9196, pp. 491–502, 2015.
DOI: 10.1007/978-3-319-22047-5_40

one of the most critical issues for D2D underlaying cellular networks. To limit interference to the existing cellular users (CUs), restricting the transmit power of D2D
links and the distance between the users of a D2D pair have been suggested in [2].
Furthermore, a distributed joint mode selection and resource allocation scheme have
been proposed in [3] to control D2D power and improve the overall system capacity.
Based on a predefined D2D resource pool and dedicated resources for D2D, a three-
alternative solution has been suggested in [4].

The above works have either aimed to increase the network throughput [3], or to
guarantee the reliability of D2D communications [4]. The works in [5]–[9] consider
both metrics simultaneously. In [5], they propose a resource allocation method that
can minimize the system interference, where there are N cellular users and N D2D
pairs, and one D2D pair can reuse the resource of only one cellular user in the system
model. In [6], throughput has been maximized for a network with a single D2D pair
and a single CU while the quality-of-service (QoS) of the CU is considered. In [7], a
novel resource allocation method that D2D can reuse the resources of more than one
cellular user is proposed. Although a D2D pair is still reusing one cellular user, it
proposes a resource allocation problem to maximize the overall network throughput
while guaranteeing the QoS requirements for both D2D users and regular CUs in [8].
And the scheme includes three steps: admission control for D2D pairs, allocating
powers for all users and resource allocation for each D2D pair.

The above-mentioned schemes require the perfect channel-state-information (CSI)
of all links at the eNodeB. They ignore the change of the channel quality, particularly
due to the movement of the users. Based on the assumptions of guaranteeing outage
probability for D2D pairs, a probabilistic access control for D2D pairs to satisfy all
the QoS requirements and power constraints is proposed in [9]. But it also does not
mention the resource allocation problem resulting from the user's mobility.

Since the mobility is inevitable in reality, we propose RAM, a novel resource allocation method considering the effect of user's velocity, for D2D communications
underlaying cellular networks. The scheme is aiming to maximize the overall network
throughput while guaranteeing the QoS requirements for both CUs and D2D pairs in
mobility. Simulation results show that RAM balances the velocity and user throughput and D2D mean throughput gain could improve 3Mbps compared with the dedicated resource mode.

The rest of the paper is organized as follows. In Section 2, we describe the system
model of D2D communication coexisting with cellular networks. The scheme of
resource allocation in mobility (RAM) is described in Section 3. Numerical results are
presented in Section 4 to demonstrate the performance of the proposed scheme.
Finally, conclusions are given in Section 5.

2   System Model

The system model includes a single cell environment as illustrated in Fig. 1, where M
D2D pairs $\langle R_x, T_x \rangle = \{D2D_1, D2D_2 ... D2D_M\}$ coexist with N cellular users CU $=$
$\{cu_1, cu_2 ... cu_N\}$. In particular, uplink (UL) resource sharing is considered since UL

spectrum is underutilized comparing to that of downlink (DL) in the frequency division duplexing (FDD) based cellular systems. Furthermore, UL resource sharing in D2D communications only affects the eNodeB and incurred interference can be mitigated by eNodeB coordination.

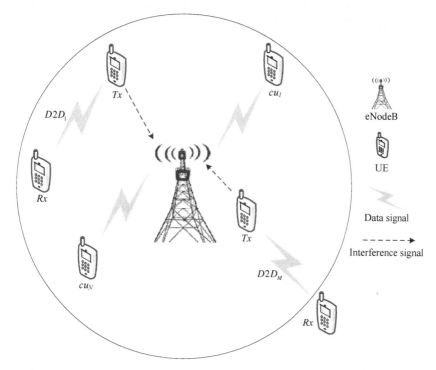

Fig. 1. System model for D2D communications reusing UL resources of cellular UEs

We assume a non-fully loaded cellular network scenario, i.e., N active CUs occupy the N orthogonal channels in the cell and there is still some spare spectrum. The D2D user distribution scenarios include two types: in coverage and partial coverage. In addition, we assume both CUs and D2D pairs have their minimum QoS requirements in terms of SINR, and the eNodeB has the necessary CSI information of the links and mobility state information of all UEs.

Besides, the resource allocation modes for D2D users can be divided into three classes: cellular mode, dedicated resources mode and reused resources mode. To take one D2D pair and N cellular UEs as an example, such resource allocation modes are defined as following:

- Cellular mode: The D2D users communicate with each other through the eNodeB like the cellular users. The resources which are shared by all D2D pairs and cellular users are divided into N+2 (N CUs and two D2D users) parts uniformly, and every user uses one part of them in this mode. It is similar to traditional communication way.

- Dedicated resources mode: The D2D users communicate with each other using dedicated resource. There is no interference between cellular communication and D2D communication since their resources are orthogonal. There are N+1 orthogonal transmission links at the same time. The resources which are shared by the D2D pair and cellular users are divided into N+1parts uniformly, and the D2D pair uses just one part of them in this mode.
- Reused resources mode: With guaranteeing the minimum QoS requirements of the cellular users, the D2D pair can reuse the resources of the suitable cellular users. The number of the cellular users can be one or more. The resources which are shared by the D2D pair and cellular users are divided into N parts uniformly, and every cellular user uses one part of them. The D2D pair reuses the resources of some of these CUs.

In the cellular mode and dedicated mode, both the CUs and the D2D pair do not interfere with each other with orthogonal resource. However, the eNodeB should control the interference between cellular communication and D2D communication by proper power control or resource allocation when D2D reuses the resources of one or more cellular users. When the transmitter and receiver of D2D pair is less than a certain distance, they can establish the D2D communication. Otherwise, the D2D UEs only can communicate with each other in the traditional cellular communication way, namely the cellular mode. According to the network load status, two UEs could choose the dedicated resources mode or the reused resources mode.

3   Resource Allocation in Mobility (RAM)

According to the system model and resource allocation modes for D2D users described in Section 2, we divide the procedure of RAM into three steps. The first is admission control for D2D pairs, where we determine whether a D2D pair in mobility can be admissible or not. The second step is resource selection for D2D pairs, where we determine whether the admissible D2D pair reuses the resource of the cellular UEs or not, according to the network load status. The last is the resource allocation for multiple D2D pairs, where we will choose the proper reuse partner for each D2D pair.

3.1   Admission Control for D2D Users

To solve the resource allocation problem, it should be determined whether a D2D pair can be admitted or not firstly. For this purpose, we consider the relative velocity between the transmitter and receiver of D2D pair before determining whether a D2D pair can meet the basic requirements of D2D communication. The basic requirements refer to D2D link quality and the distance between the transmitter and receiver.

By applying the distanced based pathloss model in [8], the channel gain between CU i and the eNodeB can be expressed as:

$$g_{i,B} = K\beta_{i,B}L_{i,B}^{-\alpha},$$ (1)

Where K is a constant determined by system parameters, α is the pathloss exponent, $\beta_{i,B}$ is the fading gain including the fast fading gain and the slow fading gain, and $L_{i,B}$ is the distance between CU i and the eNodeB. Similarly, we can express the channel gain of D2D pair j, g_j and the distance between the transmitter and the receiver of D2D pair, L_j.

$$g_j = K\beta_j L_j^{-\alpha},$$
(2)

If D2D pair can maintain the D2D link for a while to support the effective D2D communication, L_j in (2) must be satisfied a shorter distance, which means that the relative velocity between the transmitter and receiver of D2D pair in mobility must be less than a certain value. Otherwise, with the D2D users keep moving, the D2D pair will not satisfy the conditions for the D2D communication for a longer duration.

Let v_j denote the relative velocity between the transmitter and receiver of D2D pair.

v_j can be calculated by interacting between the transmitter and receiver of D2D pair. The transmitter sends the same signal to the receiver twice, and then receives the response from the receiver respectively. According to the two response time, we can calculate the relative velocity v_j between the two UEs. Then the transmitter informs it to the eNodeB, which can make the decision whether the two users could do the D2D communication. Furthermore, the D2D link quality and the distance between the two users should be considered. If the D2D link quality, the relative velocity v_j and distance between the two users all meet the requirement, the D2D pair could be admitted by the system.

3.2   Resource Selection for D2D Pairs

In the previous subsection, we have addressed admission control for D2D users. For the D2D users that are not admissible, cellular mode is the most suitable resource allocation mode. So they should choose the free orthogonal resources like the CUs. For the admissible D2D users, it needs further judgment to choose the reused resources or the dedicated resources.

According to the progress of the latest 3GPP meeting, a D2D capable UE can operate in two modes for resource allocation in [10]:

- Mode 1: eNodeB schedules the exact resources used by a user to transmit direct data and direct control information
- Mode 2: a user on its own selects resources from resource pools to transmit direct data and direct control information

Generally, Model is suitable to the users in coverage like the D2D pair 1, while Mode 2 is applying to the users of edge-of-coverage like D2D pair M in Fig. 1. The resource pool for Mode 2 should be pre-configured. So dedicated resources mode mentioned in the section 2 is more suitable for the D2D users of edge-of-coverage, compared to reused resources mode. What's more, when the network load is less than a predetermined threshold value, the users at the cell edge prefer to use the dedicated mode than the reused mode. Decisions on the used D2D resource allocation modes are taken at the eNodeB subject to existing network load status information. We define an indicator λ to express that.

When the admissible D2D pair is located in the edge of coverage or the indicator of network load is less than λ, the D2D users should choose the dedicated resources mode. Otherwise, the admissible D2D pair only can do the communication with the reused resources.

3.3   Resource Allocation for Multiple D2D Pairs

In this section, we study the resource allocation for multiple D2D pairs in the reused resources mode. According to the system model and resource allocation mode described in Section 2, by applying the Shannon capacity formula, the sum throughput of cellular communication and D2D communication can be expressed as in [8]:

$$\max_{\rho_{i,j},P_i^c,P_j^d}\left\{\sum_{i\in C}\sum_{j\in S}\left[\log\left(1+\xi_i^c\right)+\rho_{i,j}\log\left(1+\xi_j^d\right)\right]\right\} \tag{3}$$

Subject to
$$\xi_i^c=\frac{P_i^c g_{i,B}}{\sigma_N^2+\rho_{i,j}P_j^d h_{j,B}}\geq\xi_{i,\min}^c,\forall i\in C, \tag{3a}$$

$$\xi_j^d=\frac{P_j^d g_j}{\sigma_N^2+\rho_{i,j}P_i^c h_{i,j}}\geq\xi_{j,\min}^c,\forall j\in S, \tag{3b}$$

$$P_i^c\leq P_{\max}^c,\forall i\in C, \tag{3c}$$

$$P_j^d\leq P_{\max}^d,\forall j\in S, \tag{3d}$$

$$\sum_j\rho_{i,j}\leq 1,\rho_{i,j}\in\{0,1\},\forall i\in C, \tag{3e}$$

$$\sum_i\rho_{i,j}\leq N,\rho_{i,j}\in\{0,1\},\forall j\in C, \tag{3f}$$

where P_i^c and P_j^d denote the transmit power of CU i and that of D2D pair j, respectively, ξ_i^c and ξ_j^d denote the SINR of CU i and that of D2D pair j, respectively. S ($S \subseteq D$) denotes the set of admissible D2D pairs, and $\rho_{i,j}$ is the resource reuse indicator for CU i and D2D pair j, $\rho_{i,j} = 1$ when D2D pair j reuses the resource of cellular user i; otherwise, $\rho_{i,j} = 0$. $\xi_{i,\min}^c$ and $\xi_{j,\min}^c$ denote the minimum SINR requirements of CU i and D2D pair j, respectively, $P_{\max}^c$ and $P_{\max}^d$ denotes the maximum transit power of CU and D2D pair, respectively. We express the channel gains of the interference links, from the transmitter of D2D pair j to the eNodeB, $h_{j,B}$, and that from CU i to the receiver of D2D pair j, $h_{i,j}$. The power of additive white Gaussian noise on each channel is assumed to be σ_N^2.

In this formula, constraints (3a) and (3b) represent the QoS requirements of CUs and D2D pairs, respectively. Constraints (3c) and (3d) guarantee that the transmit powers of cellular users and D2D pairs are within the maximum limit. Constraint (3e) ensures that the resource of an existing CU can be shared at most by one D2D pair. Constraint (3f) indicates that a D2D pair shares at most N existing CU's resource.(the scheme in [8]refers to a D2D pair shares at most one existing CU's resource) Both constraints are used for reducing the complicated interference environment brought by the D2D communications.

Let $v_{i,j}$ denote the relative velocity between CU i and the receiver of D2D pair j. We select reuse candidates for D2D pair j based on $L_{i,j}$, the distance parameter between CU i and the receiver of D2D pair j. So the CU i will be a reuse candidate of D2D pair j, if $L_{i,j} \geq L_{i,j}^{\min} + \eta v_{i,j}$, where $L_{i,j}^{\min}$ is same as the $L_{i,jRx}^{\min}$ in [8], which means the shortest distance between the CU i and the D2D pair j. η represents the environmental parameter between CU and the receiver of D2D pair.

Accordingly, eNodeB can easily find suitable CU candidates for a D2D pair based on the $L_{i,j}$, which guarantees a targeted QoS requirement for both the CU and the D2D pair. We also can get the optimal power allocation for all users through[8].Then, we select several proper reuse partners for a D2D pair when more than one candidate could be obtained.

When CU i shares resource with D2D pair j, the maximum achievable D2D throughput gain, G can be expressed as in [8]:

$$G = \log(1 + \frac{P_i^c g_{i,B}}{\sigma_N^2 + P_j^d h_{j,B}}) + \log(1 + \frac{P_j^d g_j}{\sigma_N^2 + P_i^c h_{i,j}}) - \log(1 + \frac{P_i^c g_{i,B}}{\sigma_N^2}) \qquad (4)$$

The problem can be formulated as

$$\max \sum_{i\in C', j\in S} \rho_{i,j} G, \tag{5}$$

$$\text{Subject to} \quad \sum_{j} \rho_{i,j} \le N', \rho_{i,j} \in \{0,1\}, \forall i \in C', \tag{5a}$$

$$\sum_{i} \rho_{i,j} \le 1, \rho_{i,j} \in \{0,1\}, \forall j \in S, \tag{5b}$$

where C' is the union of all the reuse candidate sets of D2D pairs, and N' is the number of elements in C'.

It is obviously seen that the problem in (5) is difficult to obtain the solution directly. We will solve it by two steps. Firstly, we use the classic Kuhn-Munkres algorithm [11] to get one optimal reused CU for each D2D pair. We assume that the set of D2D pairs and the union of all the reuse candidate of D2D pairs are supposed as the two groups of vertices in the bipartite graph. Vertex i is joined with vertex j by an edge e_{ij}, when the user i is a reuse candidate of D2D pair j. D2D throughput gain G is considered as the weight of e_{ij}. Secondly, when one cellular user could not meet the service rate requirement of the D2D pair, we repeat the previous step until all D2D pairs are satisfied.

4   Numerical Results

In this section, we evaluate the performance of the RAM scheme by the numerical simulation. We consider a single cell network, where conventional cellular users are uniformly distributed in the cell while D2D users are distributed in a randomly located region with radius d. In the beginning of each loop, we assume that one of the D2D users in a pair is located in the center of the region, and the other is randomly located on the edge of the region. All users move randomly with the certain speed, and we update their locations in each cycle. In the simulation, we also assume all users include D2D users can share the system resources equally while the D2D users can also reuse the resources of the cellular users. The main simulation parameters are summarized in Table 1 and Table 2.

We use two metrics to evaluate the performance: user throughput and D2D throughput gain defined as the maximum increased throughput brought by the accessed D2D pairs. Moreover, we compare our scheme with the reused resource allocation scheme in [8]. For the reused scheme, it is assumed that all CUs share the total bandwidth equally, and the admissible D2D pairs only could reuse the resources of CUs. With the knowledge of the QoS requirements, it can adjust transmit power of all users, and determine the shortest distance between the D2D pair and its partner. In this way, all the D2D reuse candidates can be found and suitable powers for D2D pairs and the reuse partner CUs can be allocated. However, it does not consider the user mobility. With the change of the velocity, throughput gain brought by the D2D communication will be affected by more or less. Our scheme considers the impact of speed based on the method in [8].

Table 1. Main Simulation Parameters.

Parameter	Value
Cell radius	500m
System bandwidth	10MHz
Noise spectral density	-174dBm/Hz
Maximum transmit power P_{max}	23dBm
D2D radius (d)	20m
Number of active CUs (N)	20
Number of D2D pairs (M)	2
P_0	-60dBm
α	0.6

Table 2. Path-loss Models

Scenario	Path-loss [dB]	Shadow fading [dB]
D2D (LOS)	$18.7\log(d)+46.8$	3
D2D (NLOS)	$36.8\log(d)+43.8$	4
Cellular(LOS)	$103.4-24.2*3+24.2*\log(d)$	6
Cellular(NLOS)	$131.1-42.8*3+42.8*\log(d)$	8

Fig. 2 compares the performance for different resource allocation modes for D2D users with the velocity of 3km/h. For every D2D user, there are three resource allocation modes: cellular mode, dedicated resources mode and reused resources mode. The resource allocation scheme in [8] is typical reused mode, while the proposed scheme is hybrid mode of the reused mode and dedicated mode. From the figure, dedicated mode and reused mode perform better than cellular mode, while reused mode brings more throughput gain than dedicated mode. Since the characteristics of short direct communication, D2D communication can increase the system throughput effectively for two users in proximity. Moreover, the proposed scheme performs better than the dedicated resources mode, while it is a little bit worse than reused resources mode. It is somewhat surprising, but makes sense. When the network load is not high, our algorithm gives priority to use dedicated resources mode. Obviously, it will lead to less throughput gain than the pure reused mode. But we believe that it is worth to sacrifice some gain to maintain the communication of D2D users for a longer time.

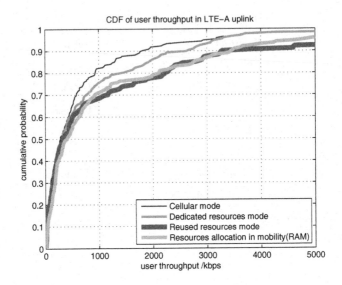

Fig. 2. User throughput for different resource allocation modes, where N = 20, M = 2, v = 3km/h and d = 20m

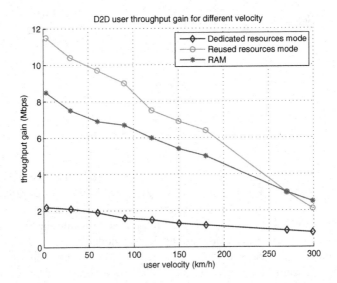

Fig. 3. D2D user throughput gain for different velocity and resource allocation modes, where N = 20, M = 2, d = 20m and v = 3km/h,30km/h,60km/h,90km/h,120km/h,150km/h, 180km/h, 270km/h, 300km/h.

Fig. 3 illustrates the D2D user throughput gain under different resource allocation modes and different user speed. In the figure, the D2D user throughput gain declines with the increasing of the user velocity. This is because that the increasing of the user velocity leads to the decline of the D2D channel gain and the increase of the interference. Therefore, the user throughput declines obviously. From the figure, it is also seen that the throughput gain of the RAM scheme declines a little more slowly than that of the reused resources mode, especially when the user speed is high. And the change of the dedicated resources mode is not obvious. Hence, we conclude that it is more stable to use the RAM scheme than the reused mode with the increase of the user velocity. For the D2D users, the proposed method can maintain the communication for a long time more effectively than the reused mode. Note that, when the user speed exceeds 270km/h, the proposed method performs better than the reused mode. That means the RAM scheme maybe more effective for the high-speed mobile users.

5   Conclusion and Future Work

In this paper, a resource allocation in mobility (RAM) scheme for D2D underlaying communications is presented. First, it is proposed a velocity-based D2D access metric to check whether a D2D pair can be access or not when satisfying its distance requirement. Second, the proper resources selection for the admissible D2D pairs is given. Finally, a joint power and channel allocation strategy to find the proper reuse partner for each admissible is derived D2D pair. Simulation results show the proposed scheme can adapt to the influence of the user speed change more effectively than the other resource allocation scheme, especially when the user speed is high. In future research, we will consider the more suitable scenes to use the proposed resource allocation scheme with various user velocities and network load status. Meanwhile, the scenarios when D2D channel information cannot obtain completely should be also concerned.

Acknowledgements. The study was partly supported by the National S&T Major Project no.2012ZX03001039-002 and the Disciplinary Construction and Postgraduate Education Project no. 081230. The authors highly appreciate all the comments and advice from anonymous reviewers.

References

1. Doppler, K., Rinne, M.P., Janis, P., Ribeiro, C., Hugl, K.: Device-to-device communications; functional prospects for LTE-advanced networks. In: The IEEE International Conference on Communications, June 14–18, 2009
2. Doppler, K., Rinne, M., Wijting, C., Ribeiro, C.B., Hugl, K.: Device-to-device communication as an underlay to LTE-advanced networks. IEEE Communications Magazine **47**(12), 42–49 (2009)
3. Belleschi, M., Fodor, G., Abrardo, A.: Performance analysis of a distributed resource allocation scheme for D2D communications. In: IEEE GLOBECOM Workshops, December 5–9, 2011

W. Lin et al.

4. Fodor, G., Dahlman, E., Mildh, G., Parkvall, S., Reider, N., Miklo, G.: Design aspects of network assisted device-to-device communications. IEEE Communications Magazine **50**(3), 170–177 (2012)
5. Janis, P., Koivunen, V., Ribeiro, C., Korhonen, J., Doppler, K., Hugl, K.: Interference-aware resource allocation for device-to-device radio underlaying cellular networks. In: Vehicular Technology Conference, pp. 26–29, April 2009
6. Chia-Hao, Y., Doppler, K., Ribeiro, C.B., Tirkkonen, O.: Resource Sharing Optimization for Device-to-Device Communication Underlaying Cellular Networks. IEEE Transactions Wireless Communications **10**(8), 2752–2763 (2011)
7. Bin, W., Li, C., Xiaohang, C., Xin, Z., Dacheng, Y.: Resource allocation optimization for device-to-device communication underlaying cellular networks. In: Vehicular Technology Conference, May 15–18, 2011
8. Daquan, F., Lu, L., Yi, Y.-W., Li, G.Y., Gang, F., Shaoqian, L.: Device-to-Device Communications Underlaying Cellular Networks. IEEE Transactions on Communications **61**(8), 3541–3551 (2013)
9. Daquan, F., Lu, L., Yi, Y.-W., Li, G.Y., Gang, F., Shaoqian, L.: Optimal resource allocation for device-to-device communications in fading channels. In: Global Communications Conference (GLOBECOM), pp. 3673–3678 (2013)
10. GPP TR36.843 (V0.1.0) Study on LTE Device to Device Proximity service, [2013-06-18]. DOI= http://www.3gpp.org/DynaReport/36843.htm
11. West, D., et al.: Introduction to Graph Theory. Prentice Hall (2001)

Group Signature Based Trace Hiding in Web Query

Jin Xu, Lan Yao$^{(\boxtimes)}$, and Fuxiang Gao

Northeastern University, Shenyang, Liaoning, China
422478972@qq.com, yaolan@ise.neu.edu.cn,
gaofuxiang@mail.neu.edu.cn

Abstract. To provide better service or push personalized advertisement, Internet companies collect users' browsing information intentionally to analyze their behaviors. However, users want to hide browsing traces sometimes because it involves personal privacy. In order to solve this problem, we propose trace hiding strategy based on group signature (THGS) in the paper. The principle of group signature is applied to create a group and HTTP proxy server uses the unified account which is shared between group members to login specific websites instead of a single group member. In addition, HTTP proxy server encrypts the response from web server with RSA and forwards it to client. Moreover, authentic membership, trustable proxy and the efficient forwarding are researched in this paper. In the experimental part, we achieved the establishment of groups based on group signature and HTTP proxy server. The experimental results show that web server can not get a single user's browsing trace by query history analysis and the response for each client is safe.

Keywords: THGS · Group signature · RSA · HTTP proxy server

1 Introduction

Internet service providers (ISPs) basically require users to register and many services are entitled only to login clients. This registration and login is part of hints for ISP to detect user's privacy. Most leading search engines log and analyse user queries [1] for business purpose. E-commerce service companies obtain the user's browsing and purchase histories in order to push ads more accurately. In addition, many IT companies have emerged account leak. In the 2011, CSDN leaked 600 million usernames, passwords and emails, thus users have to change the passwords and begin to focus on privacy protection. To protect Internet users from information leaking, these two issues have to be solved: (1) ISPs track individual browsing traces, and (2) hackers steal user account and data. Some existing solutions such as anonymizing network, P3P [2] and "TrackMeNot"(TMN) [3] tool has provided methods for this security issue, but they still need user account as initial data.

We propose a trace hiding algorithm based on group signature--THGS. Trust Center creates a group based on group signature. Each member in the group sends requests to HTTP proxy server and proxy server automatically logs on website with public account instead of members'. For a query collector, the queries are identified

© Springer International Publishing Switzerland 2015
Y. Wang et al. (Eds.): BigCom 2015, LNCS 9196, pp. 503–511, 2015.
DOI: 10.1007/978-3-319-22047-5_41

by a group id and a single member's trace is hidden. Additionally, thinking that query pattern from a group is a mixture of members, hackers can not get further information by data eavesdropping and analysis. In THGS, the response from the proxy to a member is encrypted to ensure the confidentiality.

In this paper, we present a solution for the problems above. In the next section, related work is discussed. Section 3 introduces the system architecture and database design. Section 4 presents communication process and algorithm design for Trust Center, HTTP proxy server and client. In Section 5, we conclude our work and the future work.

2   Related Work

Faced with the threat of privacy, many researchers have worked on a variety of privacy protection technologies and methods and we mainly summarize these works into the following three strategies.

The first solution is based on Private Information Retrieval (PIR) protocols [4]. It enables the user to retrieve the selected data from the database, while preventing database server to access to the identification information of user's retrieval data. PIR allows user to retrieve the data of di from server's n-bit data string $d = d_1 d_2 \ldots d_n$ without revealing any information of i[5]. Server sends all queries in database to user, but its communication complexity is $\Omega(n)$, which n is size of database [6]. Client and server must change infrastructure for the protocol to make sure privacy, however, it's meaningless for high communication and computation overload to deploy.

The second solution is based on Query Obfuscation [7]. Basically, the idea is that a client-side software injects noisy queries into the stream of queries transmitted to the search engine. Search engines track your query and establish the corresponding statistics based on query. TMN is a tool which implemented as a Mozilla Firefox plugin. TMN hides the user queries in a stream of programmatically generated search queries, which mimic or simulate the user's search behavior [1]. This extension will run in the background while you browse, and then randomly sends noisy queries to search engines.

The third solution is based on third-party platform. Third-party platform is a trusted service provider which independent of data owners and data users [8]. Proxy and anonymizing networks as third-party equipment play a different role. Microsoft Proxy Server is a agency between Internet service provider and personal networks which is responsible for forwarding legitimate network information, forwarding control and registration. Tor(The Onion Router) [9] is the second-generation onion routing. Tor users run a proxy server in computers, the agent periodically communicates with other Tor. The transmission has been encrypted by symmetric key between each router.

These three solutions are for different scenarios and have respective advantages and disadvantages. The first solution needs to deploy client and server infrastructures, and it is unfeasible for high communication because of low efficiency. The second solution requires constantly sending the queries to the search engine, increasing the pressure on the network transmission. The third solution usually sets up a large hard disk buffer in order to improve access speed and ensures safety of users.

In the paper, we propose a solution which is based on Trust Center and HTTP proxy server and is an inheritance of the third one. We establish a trust-centric group certification structure, which achieves authentication for group members, traces hiding and confidentiality of returned data.

3   Preliminary Design

ISPs track query behavior to provide personalized service, as well as the track data is provided to advertisement companies for customized advertisement push. Hackers attempt to get users' private accounts for illegal benefits. These actions are a violation of privacy protection.

To hide a single user's query behavior and pattern from web service, our solution try to avoid an individual to log in a website with his identified private account. Firstly, we hide an individual user in a group to confuse web server while it tries to collect and identify a user's query trace. To achieve this solution, three problems have to been involved:

(i)   status verification of each individual user by groups;
(ii) individual user identification in a group;
(iii) confidentiality of returned data.

3.1   Architecture Design

Fig.1 shows the system architecture, including Client, TC server, HTTP proxy server, Web server and database.

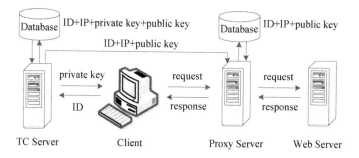

Fig. 1. System architecture

User (Client): application that builds a request connection to join a group. The user needs to register in a group and be certificated before it is hidden in this group.

TC server: Trust Center (TC) server is a third party authorization that can authenticate the users and issue keys.

HTTP proxy server/Group manager: a server that is a transit center between client and Web server and a transit station of network information.

Web server: a server to respond and handle HTTP requests exclusively, then transfer data to the browser.

Database: a DBMS to store and manage data. In this paper, the data is mainly keys.

We introduce the communication data stream between roles as following:

ID: a string less than 20 bytes and contains numbers and letters. It represents the identity of the user.

Private key and public key: an asymmetric key pair is generated by RSA algorithm. In this paper, we encrypt data with private key and decrypt data with public key.

Request: the request information including the method applied to the resource, the resource identifier and protocol version.

Response: the response information returned from the server including the HTTP protocol version, request status and MIME type of the document.

3.2   Database Structure Design

Database has three entities and $n+2$ tables, TC creates n groups and every group has m users. The *info* table records user information. The *client_info* table records all users and groups information. The *gro_mem_info* table records information of users belong to this group. And the *login_info* table records website' public account.

TC: *client_info* (*IP, ID, Pub_key, Pri_key*);

Client: *info* (*ID, IP, Pri_key*);

GM: *gro_mem_info* (*IP, ID, Pub_key*);

HTTP proxy server: *login_info* (*website, username, password, num*);

4   Algorithm Design

Fig.2 shows communication sequence diagram among TC, Client and GM/HTTP proxy server.

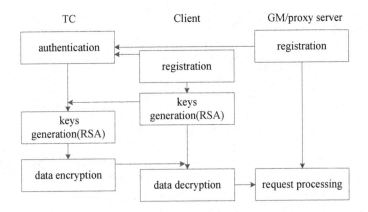

Fig. 2. Sequence diagram of THGS

4.1   RSA Cryptosystem in THGS

The generation procedure of keys as following:

(i)   Choose two large primes p and q randomly where $p \neq q$, then compute $n = p \cdot q$;

(ii) According to Euler function, calculate $\varphi(n) = (p-1)(q-1)$;

(iii) Choose an integer e where $1 < e < \varphi(n)$ and $\gcd(e, \varphi(n)) = 1$;

(iv) Compute $d = e^{-1}(\mathrm{mod}\, \varphi(n))$, where $0 \leq d \leq n$.

(n, e) is public key and (n, d) is private key. If Alice wants to receive Bob's message m, she sends (n, e) to Bob and saves (n, d) .

Encryption: Bob encrypts message m with e and $m < n$. Using formula (1) to encrypt:

$$c \equiv m^e \,(\mathrm{mod}\, n) \tag{1}$$

Bob computes c and sends it to Alice.

Decryption: Alice uses private d to decrypt c. Using formula (2) to decrypt:

$$m \equiv c^d \,(\mathrm{mod}\, n) \tag{2}$$

Alice computes n and recovers m.

4.2   Group Signature in THGS

Group signature [10] shown in Fig.3 is to meet the requirement: In a group signature scheme, any member can sign the message on behalf of the group anonymously. In addition, group signature can be verified with a single group public key by anyone.

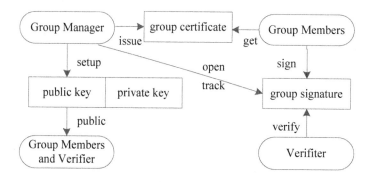

Fig. 3. The process diagram of group signature scheme

Group signature consists of six random polynomial time algorithms as follows:

(i) Setup: group manager (GM) generates group public key and private key, public key is open to all users.

(ii) Join: when users want to join the group, GM issues a group certificate to the member.

(iii) Sign: group members use certificate to generate group signature.

(iv) Verify: verifier only can use group public key to verify the correctness of the group signature, but can not determine the group signer.

(v) Open: GM use group private key to track the group signature and expose the identity of the signer.

4.3   Communication Based on Group Signature

In order to improve efficiency, we use multi-thread to receive clients' requests. There are three parties including TC server, client and HTTP proxy server (group manager) in this communication system.

TC server is responsible for authenticating the user's legitimacy and generating RSA keys for clients and proxy server. TC server as a third party ensures the security of data. Algorithm of TC server is as following:

Algorithm 1.

client generates keys (x, y), x is public key and y is private key.

TC server generates keys (u, v), u is public key and v is private key.

Require:

   user's ID;

   public key x.

Ensure:

   $x(v)$ to client ;

   (ID, u) to GM.

1. wait for users;
2. while (true)
   (a) ID→TC;
   (b) x→TC;
   (c) generate keys (u, v) ;
   (d) (ID, IP, u, v) →database;
   (e) return $encrypt(x(v))$ →client;
   (f) return (ID, u) →GM.

If a client wants to join the group, he connects to the TC server for registration. In order to ensure the transmission of keys is secure, we need to add a layer of encryption with keys. Only when the user is legitimate and obtains private key from TC server, it is considered that the user has joined the group.

Algorithm 2.

Require:
    client's *ID* and *x*;
    GM's *ID*.
Ensure:
    ID, *IP*, *u*, *v*;

1. connect to TC;
2. generate (x, y) ;
3. $(ID, x) \to$ TC;
4. TC $\to x(v)$;
5. decrypt $y(x(v))$;
6. return (ID, IP, u, v).

HTTP proxy server is also group manager that is a special group member. During the establishment of the group, group manager only achieves the function of receiving the ID, IP and public key.

Algorithm 3.

Require:
    GM's *ID*;
Ensure:
    Client' *IP*, *ID* and public key *v*;
1. connect to TC;
2. $ID \to$ TC;
3. $(ID, IP, u) \to$ proxy server;
4. $(ID, IP, u) \to$ database;
5.   return (ID, IP, v).

4.4   Communication Between HTTP Proxy Server and a Client

Some websites require users to login with their own username and password, but the registered account involves a lot of privacy. When a user uses account to login online, website companies and advertising agency can obtain accurate information about the user's query. HTTP proxy server in this experiment differs from conventional proxy server. To protect user's privacy, we use the way of public account login. Every website has many public accounts. When the proxy server picks a account to shield the individual account, the account which is used at least time will be chosen.

Algorithm 4.
___

Require:
   client's requests m;
Ensure:
   encrypted web server's response n;

1. create a connection;
2. connect to database;
3. verify ID;
4. client sends m;
5. If (m is login request)?
     (a) true→6;
     (b) false→8;
6. choose the *username* and *password* (min(num));
7. send *username* and *password*→Web server;
8. send m→Web server;
9. Web server→n;
10. return $encrypt(n(u))$ →client.

___

5   Experiment and Evaluations

In this study, the experiment includes two parts: building a group based on group signature and HTTP proxy server.

The first part of the experiment involves two tools: OpenSSL and Mysql database and three members as well: client, Trust Center and group manager. After TC starting, it connects to the Mysql database, then creates a thread for each user to handle related matters. TC identifies users and group manager by ID. OpenSSL is responsible for generating keys, encrypting and decrypting. When user gets private key and manager obtains public key, it represents that the group is built successfully.

In the second part of the experiment, proxy server is accomplished for two functions: (1) sending login request instead of client. (2) encrypting response from web server. Client adds a decryption plug in the browser. We adopt DebugView to view logs of proxy server. Log includes time, website, host, connected condition, post data and get data.

Many experimental results show users log in specific sites like Baidu, Weibo, Douban and CSDN without private account. ISPs track Web query through two ways: (1) private ID; (2) private ID and cookie ID. ISP records all behaviors with this user private ID and this ID as an identifier distinguishes from other users. We suspect that log of web server stores a query set of a group under a public account. Web server considers members as a user.

Due to the limited experimental conditions, we consulted many references to get comparative experimental results. At last, we selected results in the reference [1] to compare with. Comparative tests concentrated on TMN and a search engine. The result showed that user queries can be identified with an average true positive rate of 48.88%, while the average TMN query false positive rate was only 0.02% [1]. Above comparative experimental result of TMN shows user can be identified even without account logging in.

In the experiment, our algorithm effectively hides traces. If a client logins in a website with public account and group size is large enough, its probability of being identified is close to 0%. Websites analyze query set of a group because they do not know it is a group as a user.

6   Conclusions and Future Work

In this paper, propose a track hiding method for web query users--THGS. Users register to be a legal member of a group and are shielded by this group. HTTP proxy helps response the users and encrypt response. Decryption plug in client is responsible for decrypt data. Experiment shows that the web server does not track individual browsing traces and got a large query set of the group.

In the future, firstly we will simulate web server model. Because we can't get users' browsing logs by website, we must simulate a web server to obtain experimental data. Web server can collect and classify query information according to account. And it also has a simple function of analysis to identify client.

Secondly, we will improve operational efficiency. Encryption with RSA reduces the efficiency. We are going to increase the function of web caching and prefetching.

Acknowledgements. This research is supported by the National Natural Science Foundation of China under Grant No. 61173027 and the Fundamental Research Funds for the Central Universities (N140404006).

References

1. Peddinti, S.T., Saxena, N.: Web search query privacy: Evaluating query obfuscation and anonymizing networks. J. Journal of Computer Security **22**(1), 155–199 (2014)
2. P3P project. http://www.w3.org/P3P,2011/2014
3. TrackMeNot. http://www.mrl.nyu.edu/~dhowe/trackmenot/
4. Kushilevitz, E., Ostrovsky, R.: Replication is not needed: single database, computationally-private information retrieval. In: Symposium on Foundations of Computer Science, FOCS (1997)
5. Hua, C.: The Practical Private Information Retrieval and it's Application. D. AnHui University (2012)
6. Wang, J.: Research on Private Information Retrieval Algorithm. D. Huazhong University of Science and Technology (2012)
7. Peddinti, S.T., Saxena, N.: On the privacy of web search based on query obfuscation: a case study of TrackMeNot. In: Atallah, M.J., Hopper, N.J. (eds.) PETS 2010. LNCS, vol. 6205, pp. 19–37. Springer, Heidelberg (2010)
8. Jiang, W., Sun, Y.: Personelized Privacy Protection for Third-party Service Platform. J. Lanzhou University Journal (Natural Science) **4**, 85–90 (2012)
9. Dingledine, R., Mathewson, N., Syverson, P.: Tor: the Second-Generation Onion Router. J. Usenix Security Symposium. **40**, 191–212 (2004)
10. Huihui, S., Shaozhen, C.: An Efficient forward Secure Group Signature Scheme with Revocation. J. Journal of Electronics (China) **6**, 797–802 (2008)

Author Index

Bai, Yanfei 421

Chen, Dajiang 193
Chen, Jianwei 141
Chow, Edmond 297
Chu, Xiaowen 243
Cui, Mingyue 325, 396

Duan, Yanfei 409

Fu, Duan 337

Gao, Fuxiang 503
Gao, Wei 257
Gu, Yu 311
Guo, Lei 103
Guo, Pengxing 103
Guo, Xintong 465

Hall, Benika 283
Han, Junze 88
Han, Lin 257
Han, Zhu 273
Hawbani, Ammar 79
Hou, Weigang 103
Huang, Meng 55
Huang, Wenchao 126

Ji, Yang 325, 363, 373, 396
Jin, Naigao 14
Jing, Nan 28
Jung, Taeho 88

Kamath, Goutham 297

Lan, Jiewei 69
Leung, Yiu-Wing 243
Li, Dengao 347, 421, 453
Li, Di 14
Li, Fan 479
Li, Fangfang 311

Li, Na 347
Li, Peng 69
Li, Ping 43
Li, Qi 152
Li, Shufang 55, 217
Li, Shuyu 14
Li, Tianyuan 141
Li, Xiang-Yang 88, 193
Li, Yingying 257
Li, Yingyu 273
Lin, Weiyang 491
Lin, Xianfei 386
Litian, Duan 337
Liu, Chengjian 243
Liu, Daowei 55, 217
Liu, Hai 243
Liu, Pingping 444
Liu, Shuai 479
Liu, Xiyun 69
Liu, Xuebo 453
Liu, Yang 273
Long, Ying 217
Luo, Hong 69
Luo, Lailong 465

Ma, Huadong 141
Ma, Jun 115
Ma, Qian 311
Mao, Xufei 193
Mijumbi, Rashid 179

Ouyang, Kai 243

Pan, Haiwei 231
Pang, Yusong 409

Qiao, Tiezhu 409
Qin, Tong 141
Qin, Yudong 465
Qin, Zhen 115, 193
Qin, Zhiguang 193, 205
Qiu, Xiaofeng 55, 217
Quitadamo, Andrew 283

514   Author Index

Shen, Meng   179, 479
Shi, Guangming   273
Shi, Lei   43, 297
Shi, Liang   465
Shi, Xinghua   283
Song, Wen-Zhan   297
Song, Yan   363, 373
Su, Hongyi   386
Su, Jing   217
Su, Yuewen   325, 396
Sun, Huihui   257
Sun, Liang   14
Sun, Yue   231

Tang, Yi   434
Tang, Yuanyang   115
Tao, Dan   444
Tian, Feng   14
Tian, Xianzhong   3

Wang, Lei   14
Wang, Lin   28, 79
Wang, Weiyi   193
Wang, Xiaopu   126
Wang, Xingfu   79
Wang, Yilei   115
Wang, Zhenji   444
Wu, Chao   152
Wu, Shikun   205

Xie, Xiaoqin   231
Xiong, Hu   205
Xiong, Yan   126

Xu, Bin   152, 434
Xu, Jin   503
Xu, Ke   179

Yan, Bo   386
Yan, Yubo   43
Yang, Panlong   43
Yang, Ping   363, 373
Yao, Lan   503
Yin, Wotao   273
Yu, Cuibo   491
Yu, Ge   311
Yuan, Huaqiang   167
Yuan, Yunxu   396

Zhang, Chunhong   325, 373
Zhang, Haitao   434
Zhang, Qian   479
Zhang, Tiancheng   311
Zhang, Xu   491
Zhang, Zhiqiang   231
Zhang, Zusheng   167
Zhao, Bo   257
Zhao, Dong   141
Zhao, Jumin   347, 421, 453
Zhao, Rongcai   257
Zhao, Tiezhu   167
Zhao, Yudong   179
Zheng, Hong   386
Zhu, Chaoran   3
Zhu, Yifang   273
Zhu, Yi-Hua   3
Zizhong, Wang John   337

Printed in the United States
By Bookmasters